Phosphors for Radiation Detectors

Wiley Series in Materials for Electronic and Optoelectronic Applications

www.wiley.com/go/meoa

Series Editors

Professor Richard Curry, University of Manchester, Manchester, UK
Professor Harry Ruda, University of Toronto, Toronto, Canada
Professor Jun Luo, Chinese Academy of Sciences, Beijing, China

Honorary Series Editors

Professor Arthur Willoughby, *University of Southampton, Southampton, UK*
Dr Peter Capper, *Ex-Leonardo MWLtd, Southampton, UK*
Professor Sofa Kasap, *University of Saskatchewan, Saskatoon, Canada*

Published Titles

Bulk Crystal Growth of Electronic, Optical and Optoelectronic Materials, Edited by P. Capper
Properties of Group-IV, III—V and II—VI Semiconductors, S. Adachi
Charge Transport in Disordered Solids with Applications in Electronics, Edited by S. Baranovski
Optical Properties of Condensed Matter and Applications, Edited by J. Singh
Thin Film Solar Cells: Fabrication, Characterization, and Applications, Edited by J. Poortmans and V. Arkhipov
Dielectric Films for Advanced Microelectronics, Edited by M. R. Baklanov, M. Green, and K. Maex
Liquid Phase Epitaxy of Electronic, Optical and Optoelectronic Materials, Edited by P. Capper and M. Mauk
Molecular Electronics: From Principles to Practice, M. Petty
Luminescent Materials and Applications, A. Kitai
CVD Diamond for Electronic Devices and Sensors, Edited by R. S. Sussmann
Properties of Semiconductor Alloys: Group-IV, III—V and II—VI Semiconductors, S. AdachiMercury
Cadmium Telluride, Edited by P. Capper and J. Garland
Zinc Oxide Materials for Electronic and Optoelectronic Device Applications, Edited by C. Litton, D. C. Reynolds, and T. C. Collins
Lead-Free Solders: Materials Reliability for Electronics, Edited by K. N. Subramunian Silicon
Photonics: Fundamentals and Devices, M. Jamal Deen and P. K. Basu
Nanostructured and Subwavelength Waveguides: Fundamentals and Applications, M. Skorobogatiy
Photovoltaic Materials: From Crystalline Silicon to Third-Generation Approaches, Edited by G. Conibeer and A. Willoughby
Glancing Angle Deposition of Thin Films: Engineering the Nanoscale, Matthew M. Hawkeye, Michael T. Taschuk, and Michael J. Brett
Physical Properties of High-Temperature Superconductors, R.Wesche
Spintronics for Next Generation Innovative Devices, Edited by Katsuaki Sato and Eiji Saitoh Inorganic
Glasses for Photonics: Fundamentals, Engineering and Applications, Animesh Jha
Amorphous Semiconductors: Structural, Optical and Electronic Properties, Kazuo Morigaki, Sandor Kugler, and Koichi Shimakawa
Microwave Materials and Applications, Two volume set, Edited by Mailadil T. Sebastian, Rick Ubic, and Heli Jantunen
Molecular Beam Epitaxy: Materials and Applications for Electronics and Optoelectronics, Edited by Hajime Asahi and Yoshiji Korikoshi
Metalorganic Vapor Phase Epitaxy (MOVPE): Growth, Materials Properties, and Applications, Edited by Stuart Irvine and Peter Capper
Optical Properties of Materials and Their Applications, Second Edition, Edited by Jai Singh
Oxide Electronics, Edited by Asim Ray
Organic Semiconductors for Optoelectronics, Edited by Hiroyoshi Naito

Phosphors for Radiation Detectors

Edited by

TAKAYUKI YANAGIDA
Nara Institute of Science and Technology
Ikoma, Japan

MASANORI KOSHIMIZU
Tohoku University
Aoba, Japan

WILEY

This edition first published 2022

Registered Office(s)
John Wiley & Sons, Inc., 111 River Street, Hoboken, NJ 07030, USA
John Wiley & Sons Ltd, The Atrium, Southern Gate, Chichester, West Sussex, PO19 8SQ, UK

Editorial Office
The Atrium, Southern Gate, Chichester, West Sussex, PO19 8SQ, UK
For details of our global editorial offices, customer services, and more information about Wiley products visit us at www.wiley.com.

Wiley also publishes its books in a variety of electronic formats and by print-on-demand. Some content that appears in standard print versions of this book may not be available in other formats.

Library of Congress Cataloging-in-Publication Data Applied For
9781119583325 (hardback)

Cover Design: Wiley
Cover Image: Courtesy of Go Okada

Set in 10/12 pt TimesLTStd-Roman by Straive, Pondicherry, India

10 9 8 7 6 5 4 3 2 1

Contents

List of Contributors

Yutaka Fujimoto Graduate School of Engineering, Tohoku University, Sendai, Japan

Safa Kasap Department of Electrical and Computer Engineering, University of Saskatchewan, Saskatoon, Canada

Takumi Kato Nara Institute of Science and Technology, Ikoma, Japan

Noriaki Kawaguchi Nara Institute of Science and Technology, Ikoma, Japan

Naoki Kawano Graduate School of Engineering Science, Akita University, Japan

Hiromi Kimura Nara Institute of Science and Technology, Ikoma, Japan

Masanori Koshimizu Graduate School of Engineering, Tohoku University, Sendai, Japan

Hirokazu Masai National Institute of Advanced Industrial Science and Technology, Ikeda, Japan

Daisuke Nakauchi Nara Institute of Science and Technology, Ikoma, Japan

Hidehito Nanto Co-creative Research Center of Industrial Science and Technology, Kanazawa Institute of Technology, Hakusan, Japan

Go Okada Co-creative Research Center of Industrial Science and Technology, Kanazawa Institute of Technology, Hakusan, Japan

Kenji Shinozaki National Institute of Advanced Industrial Science and Technology, Ikeda, Japan

Kiyomitsu Shinsho Department of Radiological Sciences, Tokyo Metropolitan University, Tokyo, Japan

Kenichi Watanabe Department of Applied Quantum Physics and Nuclear Engineering, Kyushu University, Fukuoka, Japan

Takayuki Yanagida Nara Institute of Science and Technology, Ikoma, Japan

Preface

The idea of writing this book was first proposed by Professor Safa Kasap of the University of Saskatchewan in 2019. Up to now, some books have been published focusing separately on scintillators or dosimeters. Although the purpose is the same, detection of ionizing radiation by luminescent materials, these two topics have been developed in different communities by different conferences and research groups. Recently, the inverse proportional relationship between scintillation and ionizing radiation-induced storage luminescence for dosimeters has been introduced, and we feel that a text book treating these two topics comprehensively will be useful for people studying radiation detectors by using luminescent materials.

Luminescent materials for ionizing radiation detectors are mainly scintillators, thermally stimulated luminescent (TSL) materials, optically stimulated luminescence (OSL) materials which are sometimes called photo-stimulated luminescence (PSL) materials, and radiophotoluminesnce (RPL) materials. Collectively, we call the latter three (TSL, OSL, and RPL) materials as dosimeter materials. Actually, these luminescent mechanisms at around localized luminescence centers are different from each other, but generation and transportation of electrons and holes are common to all. Thus, the determination item of scintillators and dosimeter materials is a branching ratio of carriers generated by ionizing radiation toward luminescence centers or trapping centers, and we can understand ionizing radiation-induced luminescent phenomena and materials with clearer outlook by studying scintillators and dosimeter materials.

One of the features of the present book is not only the introduction of many scintillators and dosimeter materials but also description of a typical synthesis method. Such a synthesis method is categorized in chemistry, and such explanations will be useful for readers in related fields.

It should be noted that the dynamics of the excited states produced by ionizing radiation is different from those by optical excitation. The excited states dynamics should be considered in the wider scales of energy, space, and time. From the energy point of view, electrons having energies on the order of keV to MeV are generated by ionizing radiation. Subsequent generation of many secondary electrons results in the generation of lower-energy excited states, i.e., generation of conduction electrons and holes in valence bands or deeper core levels. Subsequent energy transfer and radiative recombination results in scintillation, whereas charge trapping results in a TSL, OSL, or RPL. From the viewpoint of spatial distribution, blocking of X-rays or gamma-rays needs bulk materials of mm or cm. The transport of initial high-energy electrons and secondary electrons should be considered. Energy transfer or charge trapping processes within the nanometer scale should also be considered. As for the dynamics in a wide time scale, generation of secondary electrons occurs within the ps order, energy transfer and charge trapping occur typically in the ns order, and radiative recombination occurs from ns to ms. The trapped electrons and holes survive from hour to day or month. To design novel materials with excellent properties, materials designed based on the understanding of the dynamical processes are desirable.

Chapter 1 describes some fundamental physics of inorganic scintillators and storage phosphors for ionizing radiation detectors, and following these fundamental topics constructed up to now, the idea and experimental confirmation of the complementary relationship of these materials will be explained.

Chapter 2 focuses on organic scintillators because an analytical model has been well established in chemical physics and photophysics fields, and is very useful for understanding luminescent mechanisms. The following two chapters treat the fundamental aspects, where we will move on to details of materials and properties.

Chapters 3, 4, and 5 focus on inorganic oxide, fluoride, and halide scintillators, respectively, since the number of inorganic scintillators is very large due to the large market size of practical applications. In addition to the introduction of common scintillators, typical synthesis methods are introduced.

Chapter 6 focuses on the semiconductor scintillator. In this chapter, a very hot topic, organic–inorganic layered perovskite materials for scintillation detectors are introduced.

From Chapter 7, the topic of the book moves to dosimeter (radiation-induced storage phosphor) materials. In this chapter, TSL materials and common analysis methodology for TSL phenomenon are explained. Following this chapter, OSL (Chapter 8) and RPL (Chapter 9) are introduced.

In Chapter 10, transparent ceramics for scintillators and dosimeter materials are introduced. Recently, R&D of transparent ceramics for radiation detector is ongoing, and it is one of the hottest topics in related fields.

Chapter 11 focuses on glass materials. Glasses have been used for many optical applications, including luminescent devices. This chapter explains some fundamental aspects of glasses and materials for radiation detectors.

Chapter 12 introduces ionizing radiation detectors using luminescent materials, and how to measure ionizing radiation by luminescent materials is introduced.

This book is suitable for undergraduate and graduate students, whose major is radiation detection using phosphors. The content of this book is also helpful to engineers and young researchers in the field of radiation detection. To fully understand the contents of this book, it is desirable that readers understand solid state physics, quantum mechanics, and inorganic chemistry. Some key concepts are explained in Chapters 1 and 2. We hope that this book will be useful to all its readers.

Takayuki Yanagida & Masanori Koshimizu
May 2021

Series Preface

Wiley Series in Materials for Electronic and Optoelectronic Applications

This book series is devoted to the rapidly developing class of materials used for electronic and optoelectronic applications. It is designed to provide much-needed information on the fundamental scientific principles of these materials, together with how these are employed in technological applications. The books are aimed at (postgraduate) students, researchers, and technologists, engaged in research, development, and the study of materials in electronics and photonics, and industrial scientists developing new materials, devices, and circuits for the electronic, optoelectronic and, communications industries.

The development of new electronic and optoelectronic materials depends not only on materials engineering at a practical level, but also on a clear understanding of the properties of materials, and the fundamental science behind these properties. It is the properties of a material that eventually determine its usefulness in an application. The series therefore also includes such titles as electrical conduction in solids, optical properties, thermal properties, and so on, all with applications and examples of materials in electronics and optoelectronics. The characterization of materials is also covered within the series in as much as it is impossible to develop new materials without the proper characterization of their structure and properties. Structure–property relationships have always been fundamentally and intrinsically important to materials science and engineering.

Materials science is well known for being one of the most interdisciplinary sciences. It is the interdisciplinary aspect of materials science that has led to many exciting discoveries, new materials, and new applications. It is not unusual to find scientists with a chemical engineering background working on materials projects with applications in electronics. In selecting titles for the series, we have tried to maintain the interdisciplinary aspect of the field, and hence its excitement to researchers in this field.

Arthur Willoughby
Peter Capper
Safa Kasap

1

Ionizing Radiation Induced Luminescence

Takayuki Yanagida
Nara Institute of Science and Technology, Ikoma, Japan

1.1 Introduction

Ionizing radiation was discovered more than one hundred years ago [1]. Ionizing radiation is defined as high energy quanta that ionize materials. For some years, the physical properties, and the uses and harmful effects of ionizing radiation, have been widely recognized. A typical property is its high penetration of materials, especially with high energy photons such as X- and γ-rays. Such properties makes it possible for us to investigate the inside of materials, including the human body, without damaging any of the internal structures. On the other hand, if the human body absorbs too much ionizing radiation, the radiation can cause harm such as a cancer. In order to merit the use of ionizing radiation, control of the amount and energy generated is necessary, for which accurate detection techniques are required. Ionizing radiation is invisible and odorless, and in order to detect it, we must first convert it into something which we can easily access. In most cases, we can use various tools to convert ionizing radiation into a current. From this current, we can easily gain the information desired by using common electronics. Here, the tools used to convert ionizing radiation into recognizable information are known as radiation detectors.

Figure 1.1 illustrates the classification of typical ionizing radiation detectors. Mainly, there are two types of solid materials that are most commonly used as radiation detectors. One is the semiconductor, and the other is in the form of luminescent materials known as scintillators and storage phosphors [2]. The former functions to absorb the energy of the ionizing radiation and converts it into a large number of carriers. The latter converts the ionizing radiation into a large number of photons, which can be detected by photodetectors such as the photomultiplier tube (PMT) and the photodiode (PD). The number of these carriers or

Phosphors for Radiation Detectors, First Edition. Edited by Takayuki Yanagida and Masanori Koshimizu.

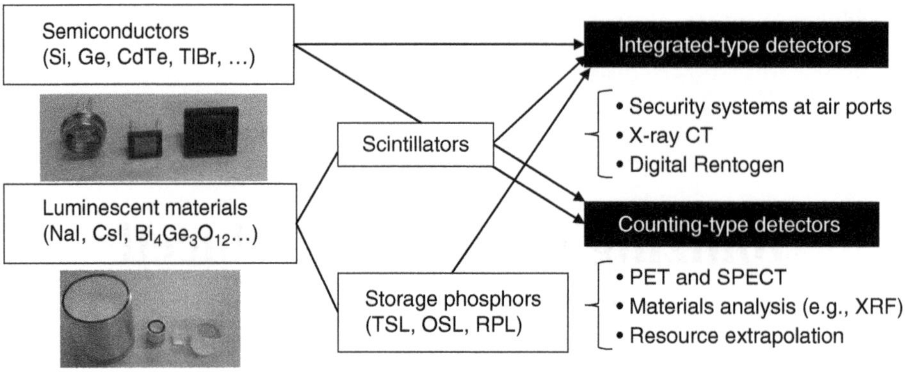

Figure 1.1 *Classification of solid state ionizing radiation detectors.*

photons is proportional to the quantity or energy of the incident ionizing radiation, and in this way, we can measure any type of ionizing radiation. Radiation detector types comprise two kinds of detection methodologies; photon counting-type and integration-type. In the photon counting-type detectors, because each radiation signal is processed event-by-event, a fast time-response typically in the order of ns to μs is important. In contrast, the integration-type detectors detect multiple events over a few ms, so a fast time-response is not required. If the counting rate of the target radiation is limited, we generally use the photon counting-type detectors, while we select the integration-type when the rate is high.

Semiconductors and scintillators can be applied to both types of detectors, while storage phosphors can only be applied to the integration-type detectors with a very long integration time (e.g., several weeks to months). These storage phosphors are mainly used for personnel protection dosimetry and imaging plates in the nondestructive study of humans (e.g., dental application) or objects. The storage phosphors used for such applications are mainly classified into three types by emission mechanisms, namely optically stimulated luminescence (OSL) [3], thermally stimulated luminescence (TSL) [4], and radiophotoluminescence (RPL) [5] materials. Hereafter, the two kinds of luminescent materials (scintillators and storage phosphors) are focused. Although the emission mechanisms of storage phosphors are different from those of scintillators, the readout system of dosimeters is similar to that of scintillation detectors. Their emissions are typically read out by PMT under photo- or thermal-stimulation (excitation). Details are described in later chapters.

Typically, luminescent materials for ionizing radiation measurements consist of the host material and the dopant for luminescence centers. For example, Tl-doped NaI is the most common scintillator [6], NaI is the host which is generally an insulator or semiconductor, and Tl is the dopant with a typically very low concentration. In these kind of materials, the host has a role to absorb the target ionizing radiation efficiently, and the dopant has a role to emit photons whose number is proportional to the incident radiation energy or amount. The combination of the host and the dopant is one of the more recent trends of R&D in this field. The same trend is true, not only for scintillators, but also for storage phosphors for dosimeters.

In addition to host-dopant type materials, undoped materials have also been investigated for the measurement of radiation. Although we can control the detection properties of luminescent materials by doping, some difficulties in the synthesis, such as a uniformity of dopant distribution, are always problematic. To avoid such problems, continuous effort has been paid to the study of undoped materials. The most common undoped materials for radiation measurement are semiconductor scintillators, and in addition to these, self-activated scintillators such as $Bi_4Ge_3O_{12}$ (BGO) [7] and Auger-free luminescence (AFL) scintillators (e.g., BaF_2 [8]) are also common. Undoped materials have been also investigated for dosimeter applications, and a common material used is BeO [9].

1.2 Interactions of Ionizing Radiation with Matter

Interaction probabilities of ionizing radiation depend on the species of ionizing radiation and matter under investigation. Figure 1.2 demonstrates a very common intuitive illustration of the interaction of ionizing radiation with matter. Although there are numerous species of ionizing radiations, the species included in Figure 1.2 are the most common.

α-rays (He nucleus) are common charged particle and have a high interaction probability with many materials, and a thin sheet of paper is enough to absorb these rays. In practical applications, a thin plate with a thickness of a few to several tens of μm scintillators are used as α-ray detectors, and the most common scintillator for this use is Ag-doped ZnS [10]. The weak point of the conventional Ag-doped ZnS powder is poor energy resolution without a clear full-energy deposited peak in pulse height spectrum. Generally, experimental environments of α-ray detection contain so much X- and γ-ray background, that detectors with Ag-doped ZnS have difficulty in discriminating the level of signal and noise. In order to solve this problem, some new approaches were recently proposed [11, 12].

β-rays are electrons that have higher penetrative power than α-rays, but they can be absorbed by thin metals such as aluminum plate. For β-rays detection, organic (plastic or liquid) scintillators are used [13], and organic scintillators are described in Chapter 2. In addition to organic scintillators, some inorganic scintillators are used by the electron detector in scanning electron microscopes (SEM) [14]. In β-ray detectors, a fast response (rise and decay times) is often required. Because charged particles such as α- and β-rays are easily stopped by detector components, dosimeters for these particles are not common. Recently,

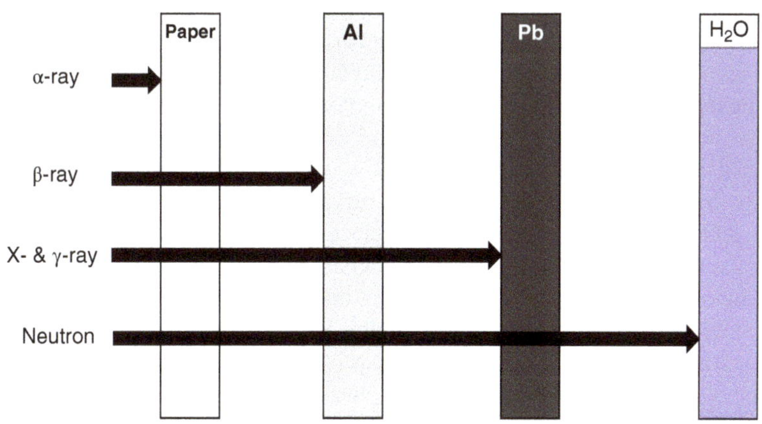

Figure 1.2 *Interaction of typical ionizing radiations with matter.*

the muon has been used for the non-destructive study of both artificial and natural objects [15], and plastic scintillators are used for muon detectors.

High energy photons, such as X- and γ-rays, are the most useful ionizing radiation in practical applications, and one of their characteristics is a high penetrative power. In order to absorb X- and γ-rays efficiently, we must prepare enough large and heavy materials such as Pb and Fe block. Generally, we use the density (ρ) and the effective atomic number (Z_{eff}) to evaluate the detection efficiency of X- and γ-ray detectors. Here, Z_{eff} is defined as

$$Z_{eff} = \sqrt[3\sim5]{\sum_i w_i Z_i^{3\sim5}} \tag{1.1}$$

where w_i is the weight ratio of the i-th element of the detector material, and Z_i is the atomic number (Z) of the i-th element. The power in the formula depends on its application. In the case of the scintillation detector, generally we evaluate the detection efficiency by Z_{eff}^4, since the interaction probability of the photoelectric absorption event is proportional to $\sim\rho\ Z_{eff}^4$, while we use the power of ~3 (sometimes 2.94) for individual dosimeter applications. It must be noted that Z_{eff} in scintillator and dosimeter fields can vary. In scintillators for X- and γ-ray detectors, generally, high ρ and Z_{eff} are preferable. On the other hand, light materials with Z_{eff} close to 7.1 are preferable for individual dosimeter applications, because Z_{eff} of the human body is around 7.

In the case of neutrons, the tendency is largely that of differently charged particles and photons. Although neutrons have a high penetrative power when applied to paper, thin metal plate, and heavy block, they have a high interaction probability with H, so H_2O is an effective material to interact with neutrons. In neutrons, some specific elements, such as H, ^3He, ^6Li, and ^{10}B, have a high interaction probability, and in order to detect neutrons efficiently, and detector materials should contain these elements. Especially, a ^3He-filled gas proportional counter is the standard detector for thermal neutron detectors.

1.3 Scintillation

1.3.1 Energy Conversion Mechanism

Figure 1.3 shows typical emission mechanisms of a scintillator assuming a combination of the host and the dopant for the luminescence center. When the incident ionizing radiation is absorbed by the ionizing radiation-induced luminescent materials, many secondary electrons (i.e., δ-rays, or carriers in a semiconductor field) are generated via electromagnetic interaction such as Coulomb scattering. In the case of X- or γ-ray irradiation, one primary electron is generated by one of the three interactions, including photoelectric absorption, Compton scattering, and pair creation, and the primary electron can excite many secondary electrons. In spite of the species of the incident ionizing radiation, many secondary electrons are generated, and the number of them is proportional to the incident energy of the ionizing radiation if we assume the irradiation of a single quantum. This first process is called the conversion process, and the typical time scale of this process is extremely fast (as ~fs). After the generation of a large number of secondary electrons, these electrons dissipate their kinetic energy via the interactions with each other or the host lattice. If some electrons can directly reach the dopant sites (luminescence centers), they can recombine with the holes, and we can

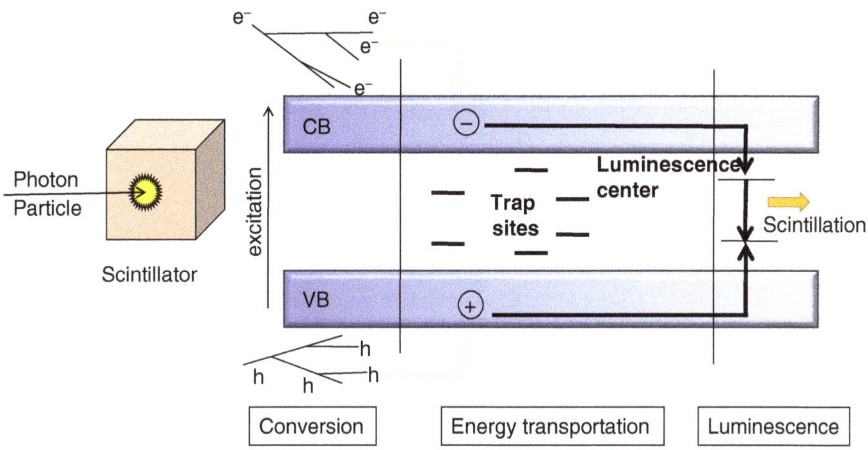

Figure 1.3 *Typical emission mechanisms of scintillation.*

observe an emission. This process is a diffusion process of secondary electrons, and is called an energy transportation process with a time scale of ps. In addition to the recombination, some secondary electrons directly excite the electron at the luminescence center, and we can also observe an emission by the relaxation of this excited electron at the luminescence center. Emission caused by these processes is called scintillation, and this final stage is known as the luminescence (or emission) process. The time scale of the luminescence process depends on the emission mechanism or dopant ion. For example, if the dopant shows a 5d-4f transition, a typical time scale is several tens to hundreds of ns.

The final process of the scintillation is the same with photoluminescence (PL), and we may have a question about the difference of the scintillation and PL. Although no standard definition has been put forward, we (the authors of this book) consider that the interactions of excited secondary electrons is necessary for scintillation. In PL, generally one UV or Vis photon can excite one electron in the outer orbital, and we observe an emission by the relaxation of this excited electron. In this case, the excited electron does not interact with other electrons. In the case of the direct band-gap excitation of semiconductor-type phosphor, the interaction of the excited electron with other electrons may be possible. But typical UV or Vis excitation energy is not so high, and the excited electron does not have enough energy to interact with the other electrons in most cases. Thus, the excitation of multiple electrons by quanta and interactions of these multiple electrons will be the main difference of scintillation with PL. In other words, if we consider typical insulator or semiconductor materials, at least several tens of eV of excitation energy will be required to cause scintillation, and excitation below this energy will be PL. Of course, the threshold energy of scintillation and PL depends on the materials used. In the case of small band-gap materials, the excitation energy of several eV may be enough to cause scintillation.

1.3.2 Emission Mechanism

Figure 1.4 summarizes typical emission mechanisms in a scintillator, and the same classification is also possible for dosimeter materials. There are two kinds of luminescence in scintillators, namely intrinsic and extrinsic emissions. The intrinsic luminescence represents

Luminescence without dopants (Intrinsic luminescence)

1. Free exciton luminescence (e.g., ZnO)
2. Self-trapped exciton (STE) luminescence (e.g., CaF_2)
3. Auger free luminescence (e.g., BaF_2)
4. Self-activation (e.g., $Bi_4Ge_3O_{12}$)

Luminescence with dopants (Extrinsic luminescence)

5. 1s ↔ 2p transition (e.g., F-center)
6. ns^2 ↔ nsnp transition (Ga^+, In^+, Tl^+, Ge^{2+}, Sn^{2+}, Pb^{2+}, Sb^{3+}, Bi^{3+}, ...)
7. 3d ↔ 3d, 4d ↔ 4f transition (transitional metals)
8. 4f ↔ 4f, 5f ↔ 5f transition (rare earth, Actinoid)
9. 4f ↔ 5d transition (Ce^{3+}, Eu^{2+}, ...)
10. charge transfer luminescence (Yb^{3+}, VO_4^{3-}, ...)

Figure 1.4 *Typical emission mechanisms of scintillation.*

free-exciton luminescence, self-trapped exciton (STE) luminescence, Auger free luminescence, and self-activation luminescence. The free exciton luminescence is observed in wide band-gap semiconductor materials having direct transitions. In this type, common examples are ZnO [16] and GaN [17]. Scintillation due to free exciton is characterized by fast scintillation decay (1–10 ns) with a sharp emission peak in the spectrum. Recently, Ga_2O_3 is found to have fast and very intense scintillation upon γ-ray irradiation [18]. Although Ga_2O_3 is still under discussion, whether it is a direct- or indirect-transition type of semiconductor, it shows an attractive scintillation performance.

STE is observed in wide-band-gap insulator materials, such as undoped alkali halides. Common properties of STE are summarized as follows: a broad emission peak in the emission spectrum, a large Stokes shift, a relatively high light yield in scintillation, a relatively fast decay time from several hundred ns to a few μs, and a large temperature dependence of the scintillation light yield. The most common materials are BaF_2 [8], SrF_2 [19], CaF_2 [20], and their mixed compounds.

AFL is observed in some materials which have a larger band-gap energy than the energy between the core and the valence bands [21]. AFL generally shows a very fast luminescence decay (~1 ns) with a short emission wavelength (VUV-UV). The common materials are BaF_2 [8], $BaMgF_4$ [22], CsF [23], and Cs_2ZnCl_4 [24]. Up to now, AFL has been observed in halide materials, and a search for other compounds (oxides and nitrides) is an interesting research topic. Conventionally, it was considered that they have no temperature dependence concerning the scintillation light yield [25], but we recently revealed that AFL materials also show temperature dependence [26]. The disadvantage of AFL scintillators is a relatively low scintillation light yield. The details of AFL are described in Chapter 4.

Self-activation type scintillators are composed only of the host, and this type can be classified as belonging to the intrinsic luminescence-type materials. The difference from other intrinsic-type materials is that the host material contains luminescent ions as a main element. Common scintillators of this type are BGO [7], CeF_3 [27], and $CeBr_3$ [28]. Scintillations from BGO are caused by ns^2 transitions of Bi^{3+}, and those of the latter two materials are due to the 5d-4f transitions of Ce^{3+}. Although the emission mechanism is not fully understood, recent results of Tl-based crystals, such as $TlMgCl_3$ and $(Tl_{1-x}A_x)MgCl_3$ (where A = alkali metal), show very high scintillation light yields [29, 30], and their emission will relate to Tl^+.

In these materials, the emission properties are governed by the luminescent ions of the host, but this class of materials do not suffer from concentration quenching.

Then, the topic moves to materials with extrinsic luminescence. The first one is defect-based luminescent materials, and the most common defect relating to luminescence is the F-center, which is characterized by one electron captured at an anion vacancy. Materials containing F-centers sometimes show scintillation. Common materials with an F-center are simple oxide materials, such as MgO [31] and Al_2O_3 [32]. In addition to F-centers, there are some other defect-based luminescence centers, for example, F^+- and F^--centers. Although defects are not exactly activators, defect-based luminescence is sometimes categorized as extrinsic emission.

Luminescence due to the $ns^2 \leftrightarrow nsnp$ transition is one of the common electron transitions involved in many scintillators, such as Tl-doped NaI [6] and Tl-doped CsI [33], which are very traditional scintillators. These traditional scintillators were discovered around 1940, and ns^2 ions have become a familiar dopant in this field. The spectral feature of the $ns^2 \leftrightarrow nsnp$ transition is characterized by a broad emission band, and the emission wavelength is from near UV to VIS. Because the decay time is from several hundred ns to several μs, it is acceptable for pulse-height-based scintillation detectors. In this type of luminescence, in addition to Tl^+ and Bi^{3+}, In^+ [34] and Sn^{2+} [35] have shown interesting scintillation properties in recent years.

In scintillators, transitional metal ions are sometimes activated to obtain a high luminescent efficiency for integration-type detectors. The spectrum of 3d-3d and 4d-4d (d-d) transitions has a broad luminescence feature, and the decay time is typically of the ms order. Among the transitional metal elements, Mn^{2+} [36] and Cr^{3+} [37] are sometimes selected for scintillation detectors due to a good spectral matching with Si-PD.

Electron transitions of the 4f-4f and 5f-5f levels are called f-f transitions, and some scintillators use the 4f-4f transitions of rare-earth elements because they typically exhibit a high emission intensity with acceptable decay time of several μs to few ms. The most common scintillator based on the 4f-4f transition is Pr^{3+}-doped Gd_2O_2S (GOS) ceramic [38], which is often part of the equipment used in medical X-ray CTs. The Pr-doped GOS shows a very high scintillation intensity with a decay time in the medium range (several μs). The luminescence due to the 4f-4f transitions of Pr^{3+} appears in yellow-green, which is well-matched with the spectral sensitivity of Si-PD. For practical uses, Ce^{3+} is co-doped to suppress the afterglow level. Recently, Eu^{3+}-doped $(Lu,Gd)_2O_3$ [39] has attracted much attention for X-ray CT. The emission intensity is higher than other materials, and the emission wavelength is in red where Si-PD has a high sensitivity. In addition to these materials, we have investigated doping with other rare-earth ions in order to develop near-infrared emitting scintillators. For this purpose, Nd^{3+}, Ho^{3+}, Er^{3+} Tm^{3+}, and Yb^{3+} have been selected as the emission center.

Emission due to 5d-4f transitions of trivalent and divalent rare-earth ions are very important in recent scintillation detectors because they show intense and fast emissions by the spin- and parity-allowed transitions. The most common emission center is Ce^{3+}, and examples of commercial scintillators are Ce-doped Lu_2SiO_5 (LSO) [40], $(Lu,Y)_2SiO_5$ (LYSO) [41], $Gd_3(Al,Ga)_5O_{12}$ (GAGG) [42], $Y_3Al_5O_{12}$ (YAG) [43], $YAlO_3$ (YAP) [44], $LaBr_3$ [45], and Cs_2LiYCl_6 [46]. Except for garnet materials, including GAGG, most scintillators have emission wavelengths of 300–400 nm with 30–60 ns decay times, and these properties are suitable for conventional PMT readouts. In addition to Ce^{3+}, Pr^{3+} can also show luminescence due to the 5d-4f transition in some host materials. The appearance of the 5d-4f transition depends on the relative positions between the lowest 5d and $^1S_0$ levels. Common Pr-doped scintillators

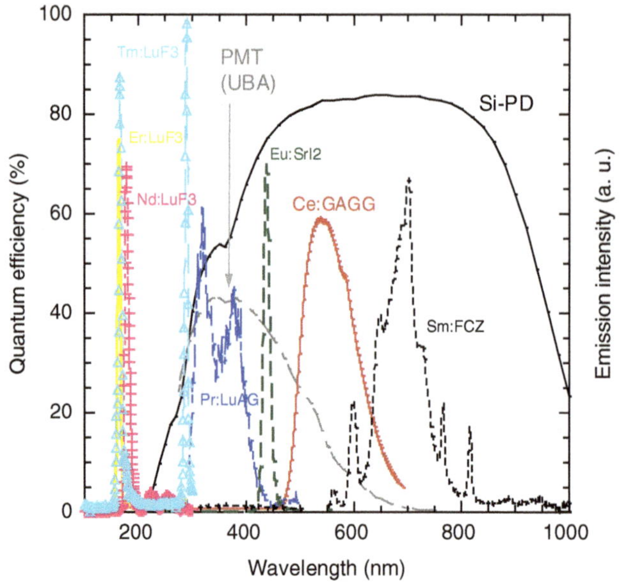

Figure 1.5 Emission spectra of scintillators under X-ray irradiation and typical quantum efficiencies of Si-PD and PMT.

are Pr-doped LuAG [47], $Lu_2Si_2O_7$ (LPS) [48], and YAP [49], which show light yields of 10 000–20 000 ph/MeV with a 20 ns decay time. Compared with Ce-doped scintillators, the emission of Pr-doped materials appears in the shorter wavelength range, typically at 250–350 nm. Some other trivalent rare-earth ions show 5d-4f transitions only in hosts with a wide-band gap energy, and the emission wavelength is in VUV. A common example is Nd-doped LaF_3, which has an emission wavelength at 175 nm with a few-ns decay time [50] due to the 5d-4f transition of Nd^{3+}. Divalent Eu also shows a very high scintillation light yield with a typically ~1 μs decay time, which is acceptable for the photon counting type detectors. Typical examples are Eu^{2+}-doped SrI_2 [51], CaF_2 [52], LiI [53], and $LiCaAlF_6$ [54]. Recently, luminescence due to the 5d-4f transitions of Sm^{2+} ions was also reported [55], and further investigations are in progress. Sm^{2+} shows emission in the near-infrared range, and has attracted much attention for use in Si-PD type photodetectors. Figure 1.5 shows typical scintillation spectra of such scintillators with various emission wavelengths superposed with typical quantum efficiency curves of common photodetectors. It should be noted that the emission wavelength by 5d-4f transitions strongly depends on the crystal field, which is specific to each host lattice; the presented data are only examples.

1.3.3 Scintillation Light Yield and Energy Resolution

In addition to the emission wavelength, light yield is the most important property for scintillators because it directly determines the signal to noise ratio (S/N) in all types of scintillation detectors. The scintillation light yield generally uses the unit of ph/MeV, which means a number of emitted scintillation photons with 1 MeV absorption of ionizing radiation. Typically, we measure it by X- and γ-ray irradiation. It must be noted that the light yields of one scintillator are different by species of irradiated ionizing radiation. If we irradiate

5.5 MeV α-ray from ^{241}Am and 662 keV γ-ray from ^{137}Cs to one sample, the observed light yields are different. The relative ratio of light yields under α-ray and γ(β)-ray irradiation is known as the α/γ- (α/β-) ratio. Although we do not have a universal theory for this ratio, empirically, the ratio of halide scintillators is close to 1, while that of oxide scintillators is ~0.2. In some fields of radiation physics and chemistry, the difference of energy conversion efficiency of different ionizing radiation species is recognized as the linear energy transfer (LET) effect.

Here, we will introduce the common explanation on the scintillation light yield. The semi-empirical approach was made in 1980 by Robbins [56] based on semiconductor physics. In the semiconductor radiation detector, empirical relation of ξ (average energy consumed per electron–hole pair) and E_g (band-gap energy) are connected by a parameter β as

$$\xi = \beta E_g. \tag{1.2}$$

In this approach, to consider ξ, the energy of electron–hole pairs, falls below the threshold energy for impact ionization:

$$\xi = E_i + E_{op} + 2E_f \tag{1.3}$$

where E_i, E_{op}, and E_f represent the threshold energy for impact ionization, energy emitted as optical phonons, and average residual energy of electron–hole pairs, respectively. Throughout this discussion, the unit of ξ (energy) is eV. Here, we consider the branching ratio of optical phonon emission with the probability of r and the impact ionization with (1-r) under the initial absorbed energy of E_0. If the energy after some processes, such as impact ionization and phonon emission, remains at $>E_i$, the impact ionization (excitation process) continues. In the ideal case, the limiting efficiency (Y) of the production of electron–hole pairs is

$$Y = \frac{Number\ of\ e-h\ pairs \times E_i}{E_0} \tag{1.4}$$

and by using this relation, the average energy per electron–hole pair is re-written as

$$\xi = \frac{E_i}{Y} = E_i + KE_i + 2L_f E_i \tag{1.5}$$

where $L_f = E_f/E_i$ and K means the ratio of rate of optical phonons rate of energy loss by ionization, expressed as

$$K = \frac{r}{1-r} \cdot \frac{\hbar\omega_{LO}}{E_i}. \tag{1.6}$$

In this equation, $\hbar\omega_{LO}$ means the energy of the longitudinal optical phonon. If we assume this energy and the optical phonon energy is constant, K can be expressed as

$$K = \frac{optical\ phonon\ generation\ rate}{ionization\ rate} \cdot \frac{\hbar\omega_{LO}}{E_i}. \tag{1.7}$$

Here, we assume special conditions of: (i) Ionization rate is constant for carrier energy; and (ii) $E_i = 1.5E_g$, and according to the avalanche multiplication data of Si, then K can be approximated to

$$K \sim \frac{optical\ phonon\ generation\ rate}{constant} \cdot \frac{\hbar\omega_{LO}}{1.5E_g}. \tag{1.8}$$

In order to proceed with the calculation, we assume a polaron model where a polaron accompanies $\alpha/2$ phonons, then α can be expressed as

$$\alpha = \frac{e^2}{h}\left(\frac{1}{K_\infty} - \frac{1}{K_0}\right)\sqrt{\frac{m}{2\hbar\omega_{LO}}} \tag{1.9}$$

where K_∞ and K_0 are static and high-frequency dielectric constants, respectively. Under this condition, the optical phonon generation rate is proportional to

$$\propto \frac{\alpha\omega_{LO}}{2} = constant\left(\frac{1}{K_\infty} - \frac{1}{K_0}\right)\sqrt{\hbar\omega_{LO}}. \tag{1.10}$$

Thus, K can be expressed as

$$K \sim constant \cdot \left(\frac{1}{K_\infty} - \frac{1}{K_0}\right)\frac{\left(\hbar\omega_{LO}\right)^{3/2}}{1.5E_g}. \tag{1.11}$$

By using these equations, we can estimate the scintillation emission efficiency semi-empirically. Following this first approach, the model was modified for actual use in daily experiments. In 1994 [57], the scintillation light output L per unit energy was expressed as

$$L = \frac{n_{e-h}}{E_\gamma} SQ = \frac{n_{\max}}{E_\gamma}\eta \tag{1.12}$$

where $n_{e\text{-}h}$ is the number of electron–hole pairs under γ-ray with energy of $E\gamma$ irradiation, n_{max} is the number of electron hole pairs which would be generated if there were no losses to optical phonons, S stands for transfer efficiency from the host to luminescence centers, Q is luminescence quantum efficiency at localized luminescence centers, and η is total scintillation efficiency.

Here, we will proceed with further consideration that this assumption is $E_i = 1.5E_g$. In the Robbins approach [56], the average energy used to produce one electron–hole pair under assumption of no optical phonon loss is $\xi_{min} = 2.3E_g$, and in this case, $n_{max} = 10^6E\gamma/2.3E_g$ where we use the energy unit of eV. Because $n_{e\text{-}h} = 10^6E\gamma/2.3\xi$ and $n_{max} = 10^6E\gamma/\xi_{min}$, we can deduce that $\beta_s = \xi_{min}/\xi$. In Equation (1.5), if we assume $K = 0$, we have $\xi_{min} = 1.5E_g(1+2L_f)$, and if we use the relation of $\xi_{min} = 2.3E_g$, $L_f = 0.27$. If we assume L_f does not depend on K, we can use $L_f = 0.27$ in Equation (1.5), and the average energy consumed per electron–hole pair can be expressed as $\xi = E_g(2.3 + 1.5K)$. Under this condition, the parameter β is approximated to

$$\beta_s = \frac{1}{1+0.65K}. \tag{1.13}$$

The number of electron–hole pairs is

$$n_{e-h} = \frac{10^6 E_\gamma}{2.3\xi} = \frac{10^6 E_\gamma}{2.3\beta_s E_g}. \tag{1.14}$$

Thus we can obtain

$$L = \frac{10^6 E_\gamma}{2.3\xi} = \frac{10^6 SQ}{2.3\beta_s E_g} = \frac{10^6 SQ}{2.3E_g}\left(1+0.65\mathrm{K}\right). \tag{1.15}$$

Equation (1.15) is a convenient form because it contains a parameter of optical phonon loss, and the evaluation of the model under various temperatures was made in 2010 [58].

Generally, experimental evaluations of effects of optical phonon (thermal) loss is difficult. For experimental research a very convenient model assuming $\beta_s = 2.5$ is proposed, based on the data of various scintillators [59], and most research uses this formula which is described as

$$L\left[ph/MeV\right] = \frac{10^6 E_\gamma}{2.3\xi} = \frac{10^6}{2.5E_g}SQ. \tag{1.16}$$

In this formula, we can determine L, E_g, and Q experimentally, and we can deduce S by using these values. L can be evaluated by the pulse height spectrum, E_g by optical absorption or reflection spectrum, and Q by PL quantum yield measurement. Although we cannot predict the scintillation light yield by these formulae, we can understand why halide and sulfide materials show higher light yields than oxide materials by E_g. In most cases, Q of commercial scintillators is close to 100%. As described in Chapter 11, in the case of the integration-type detector, a factor representing the absorption of ionizing radiation is multiplied to Equation (1.16).

Typical ways to determine L is to compare with the pulse height of a single photoelectron (~single photon) of photodetector with high quantum efficiency, and we can know the number of photoelectrons from the scintillator. In this manner, we can directly measure L after dividing the number of photoelectrons by the quantum efficiency of the photodetector at the emission wavelength of the scintillator. The other common way is to compare the pulse height of ^{55}Fe 5.9 keV X-ray measured directly by Si-based photodetectors. The photoabsorption peak due to 5.9 keV X-ray corresponds to ~1640 electron–hole pairs, and we can evaluate the number of photoelectrons generated by scintillation photons if we use the same experimental conditions. Another way is to compare relative pulse height with scintillators with known scintillation light yield. In these evaluation techniques, finally, we must divide the observed number of photoelectrons by the quantum efficiency of photodetectors at emission wavelength, and it contains a certain amount of error because the quantum efficiency of photodetectors has a wavelength dependence. Typically, the error for the estimation is 5–10% for experts of these kind of experiments. In some research, the scintillation light yield is calculated by the area intensity of the radioluminescence spectrum, and this is generally incorrect because radioluminescence intensity is not a quantitative but a qualitative value. The main reason is that we cannot correct the absorption probability of ionizing radiations in scintillators. For example, we have two samples with the same light yield, but one is light and the other is heavy. If we irradiated X-rays to these samples, the latter would show higher radioluminescence intensity. The other reason is the effect of TSL at room temperature, and we also cannot correct any effects from TSL. If we are to measure scintillation light yield quantitatively, pulse height measurements must be conducted. At present, we cannot measure the pulse height of scintillators with very slow decay (> 0.1 ms), and further technical development is required to measure pulse height.

The following topics are limited to photon counting-type detectors, since integration-type detectors cannot measure the energy of ionizing radiation, except in some special cases. The scintillation light yield is one of the most important properties of scintillation detectors because it directly relates to the energy resolution. Generally, the energy resolution obeys Poisson statistics. If we represent the quantum efficiency of the photodetector as q, the number of electron–hole pairs after photodetector output n is a product of q and the number of scintillation photons. The energy resolution under the absorbed energy of E is expressed as

$$\frac{\Delta E}{E} = \frac{\sqrt{n}}{n} = \frac{1}{\sqrt{n}}. \tag{1.17}$$

Therefore, we can obtain a better energy resolution in bright scintillators. In actual detectors, the energy resolution is different from the resolution derived from simple Poisson statistics, and the gap between actual dispersion of n and dispersion of n in Poisson distribution is called the Fano factor (F). By using the Fano factor, the limit of the energy resolution of actual detectors is expressed as

$$\frac{\Delta E}{E} = 2.35\sqrt{\frac{F}{n}}. \tag{1.18}$$

In semiconductor, gas, and scintillation detectors, F is ~0.1, 0.1–0.4, and 1, respectively. Therefore, semiconductor detectors such as Si, Ge, and CdTe are known to have a superior energy resolution compared to scintillation detectors, and the best energy resolution so far is ~2% at 662 keV [60, 61]. In practical detectors, energy resolution is not only affected by statistics but also by non-uniformity of the scintillator. Especially in luminescence center doped scintillators, because we generally use bulky larger material to interact with ionizing radiations effectively, non-uniform distribution of dopant ions cannot be avoided. The non-uniform distribution causes differences of light output at each point on the scintillator, and in such a case, the photoabsorption peak or some other features caused by ionizing radiation become, for example, a superposition of multiple Gaussian. Eventually, the shape of the peak in the pulse height spectrum becomes broad, and the energy resolution becomes worse. Such a non-uniformity is also observed in photodetectors, and the energy resolution observed in practical detectors depends on the non-uniform response of scintillators and photodetectors.

In order to evaluate the energy resolution of scintillators fundamentally, sometimes intrinsic energy resolution [62] is evaluated. Before the twenty-first century, the energy resolution was limited to 6–7% at 662 keV in common scintillators. After the invention of Ce-doped $LaCl_3$ [63] and $LaBr_3$ [45], some new scintillators with high energy resolution (2–4% at 662 keV) appeared. If we say $\Delta E/E = 0.01$ (1%), then n is 5590. If we use typical Si-based photodetectors, quantum efficiency at visible wavelength is ~80%, and 5590/0.8 ~7000 photons are required to achieve ~1% energy resolution. However, actual scintillation detectors show a large discrepancy with this simple calculation, and in order to explain such a discrepancy, the intrinsic energy resolution was introduced [64]. Although there are several expressions about the intrinsic energy resolution, logically, the energy resolution can be divided into several terms, for example

$$\left(\frac{\Delta E}{E}\right)^2 = \delta_{sc}^2 + \delta_{cir}^2 + \delta_{st}^2 \tag{1.19}$$

where δ_{sc}, δ_{cir}, and δ_{st} represent the intrinsic energy resolution, resolution due to circuit noise, and resolution expected by Poisson statistics, respectively. In pulse height spectrum, we observe $\Delta E/E$ directly, and we can estimate δ_{st} by the number of scintillation photons and the quantum efficiency of the photodetector. The circuit noise δ_{cir} can be directly estimated by the injection of a test pulse into the electrical circuit. Thus, we can calculate δ_{sc} by the subtraction of δ_{st} and δ_{cir} from $\Delta E/E$. Under most experimental conditions, estimation must consider a contribution not only by the scintillator but also by the photodetector, and the simplest measurement can be possible by Si-PD because it has no internal gain. When we measured and calculate the intrinsic energy resolution, that at 662 keV of Tl-doped NaI was 2.5% [64], Tl-doped CsI was 5% [65], Ce-doped $LaBr_3$ was 2.2% [66], Ce-doped LuAG was 4.5% [66], and Ce-doped LSO was 7.7% [66]. At present, the remaining problem is whether the intrinsic energy resolution we are observing is a fundamental limit or not, and whether the intrinsic energy resolution is the physical property of each material or not (detector property). The detailed explanations on energy resolution and intrinsic energy resolution are described in Chapter 12.

In an ideal scintillation detector, energy response (signal output intensity vs. incident radiation energy) should have a simple proportional relationship; however, the energy responses of actual detectors are not simple. The main reason is a non-proportional energy response of scintillators. Figure 1.6 (top) shows a non-proportional response plot, which shows a relationship between relative light output (pulse height) per unit energy and irradiated γ-ray energy. Although Eu-doped SrI_2, Ce-doped CLYC, and Pr-doped LuAG show a relatively flat response against γ-ray energy, BaF_2, Ce-doped GAGG, and Tl-doped NaI exhibit a large fluctuation against γ-ray energy. When we use the latter three types of scintillators as actual detectors, we must apply the gain correction function to measure the energy accurately. The main origin of the non-proportional response is considered as related to the K-edge of the main element of scintillator. This was first pointed out in Ce-doped GSO [67]. Now, we understand that the flat response against γ-ray energy is required to achieve high energy resolution. In practical scintillation detectors, we generally use a function between γ-ray energy and pulse height, as shown in the bottom of Figure 1.6. In a simple case, we fit the relationship by a linear function with the least squares method, and the bottom panel in the bottom of Figure 1.6 shows the residual from linear fitting at each energy. After we prepare

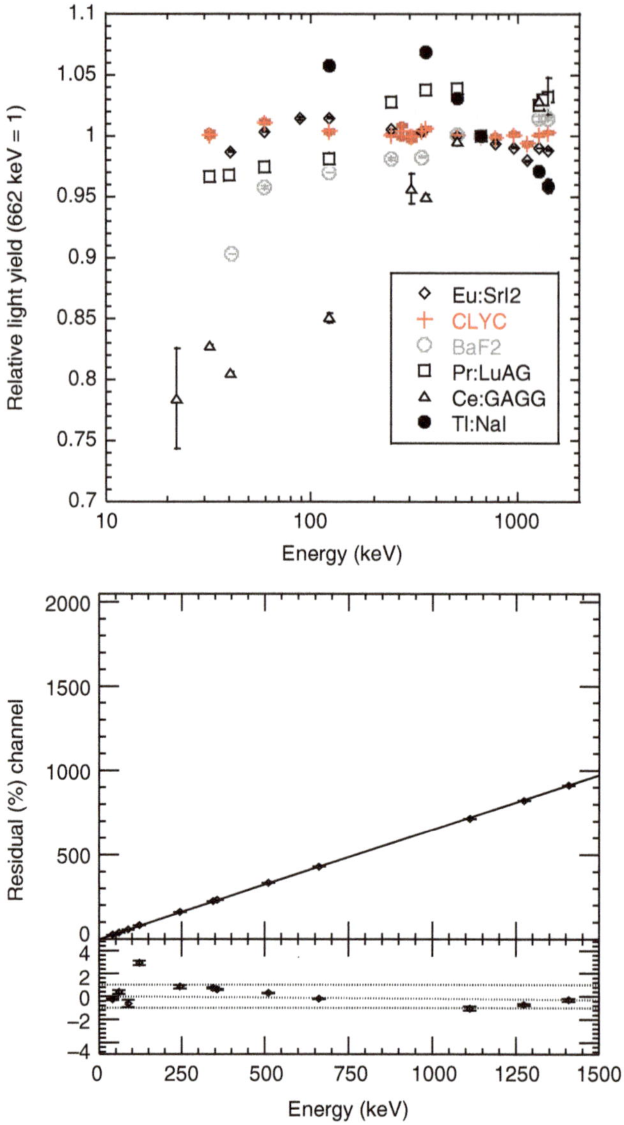

Figure 1.6 *(Top) Relationship between the scintillation decay time (ns) and emission wavelength (nm) and (bottom) relationship between γ-ray energy and photoabsorption peak channel.*

such a fitting function, we can convert pulse height channel to energy of ionizing radiation focused on each measurement. Although these two figures have the same physical meaning, people in basic science prefer a non-proportionality plot, and those using the actual detector prefer a linearity plot.

1.3.4 Timing Properties

Timing properties are also important for scintillation detectors, and depend on the rise and decay times of scintillators and photodetectors. Generally, we analyze the decay time profile by the sum of multi exponential functions, which can be understood as a typical rate equation. Here, we assume radiative and non-radiative rate constants of k_r and k_{nr}, respectively. $N(t)$ is the number of excited states, and it shows a relaxation with a rate of $-\Delta N_r$ during a very short time of Δt. This can be formulated as

$$-\Delta N_r = k_r N(t)\Delta t. \tag{1.20}$$

The number of the decrease of excited states $-\Delta N_{nr}$ is expressed as

$$-\Delta N_{nr} = k_{nr} N(t)\Delta t. \tag{1.21}$$

Totally, the decrease of the excited states $-\Delta N$ is a sum of Equations (1.20) and (1.21), and it can be expressed as

$$-\Delta N = -(\Delta N_r + \Delta N_{nr}) = (k_r + k_{nr})N(t)\Delta t. \tag{1.22}$$

Thus, we can obtain

$$-\frac{\Delta N}{N(t)} = (k_r + k_{nr})\Delta t \tag{1.23}$$

We can solve this differential equation as

$$-N(t) = N_0 e^{-(k_r + k_{nr})t} \tag{1.24}$$

where N_0 means the number of luminescence centers at $t = 0$. If we define $k_r + k_{nr} = 1/\tau$, we can obtain a typical decay time function for the analysis. Unlike the case of the decay time, the common confirmed analysis method for the rise time has not yet been constructed. One of the problems with the rise time is that it is too fast for typical photodetectors and irradiation sources to observe. Except for scintillators with very slow rise times, the typical rise time is in the order of ~ps, and it is technically difficult to detect such a fast phenomenon.

To date, unlike the situation of scintillation light yield, no widely accepted and confirmed model has been constructed. We apply a common model of PL to understand the scintillation phenomenon. The common relationship between scintillation decay time is expressed as

$$\Gamma = \frac{1}{\tau} \propto \frac{R}{\lambda^3}\left(\frac{R+2}{3}\right)^2 \left| < \psi_{excite} \middle| \mu \middle| \psi_{ground} > \right|^2 \tag{1.25}$$

where Γ means the radiative decay rate, τ is the decay time, R is the refractive index, λ is the emission wavelength, ψ_{excite} is the wave function of excited state, ψ_{excite} is the wave function of ground state, and μ is the electric dipole operator between the initial lowest energy of excited state and the final state of the configuration, respectively [68]. Based on this relationship, we understand that scintillators with shorter emission wavelengths show smaller decay

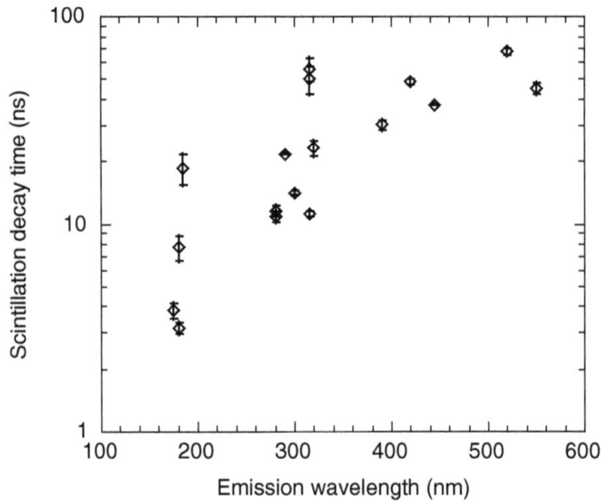

Figure 1.7 *Relationship between the scintillation decay time (ns) and emission wavelength (nm).*

times if we compare them with scintillators possessing the same emission mechanisms (e.g., comparison among d-f transitions). Figure 1.7 shows the relationship between the scintillation decay time and emission wavelength of some common materials. In this plot, we use Ce^{3+}, Pr^{3+}, and Nd^{3+} doped samples, and the emission mechanism of all the samples is 5d-4 transition. A clear proportional relationship based on Equation (1.25) is clearly observed. Typically, emission wavelength due to d-f transitions of Ce^{3+}, Pr^{3+}, and Nd^{3+} appear in the visible, ultraviolet, and vacuum ultra violet wavelengths, respectively. If we plot the same relationship on 4f-4f transitions of rare earth ions, a similar relation will be obtained, but it is technically difficult to measure the scintillation decay time of near infrared emission, which is the typical emission wavelength of 4f-4f transition of rare earth ions.

In some specific applications, a very fast response is required, and in such a case, timing resolution is important. The timing resolution is not a physical property of each material but a detector property depending on properties of detector components, including scintillators, photodetectors, and readout electronics. Although there is no clear numerical function form, the timing resolution is a function of scintillation light yield, rise, and decay times. In order to achieve a high timing resolution, fast rise and decay times accompanied with a high light yield are required. In addition, fast rise and decay times of photodetector are also required. The recent timing resolution is ~100 ps [69], and in time-of-flight positron emission tomography (TOF-PET), 100 ps timing resolution corresponds, for example, to 3 cm localization limitation of a tumor.

The other important property of the scintillation detector is an afterglow, which mainly affects integration-type detectors such as X-ray CT and security systems. If the afterglow is huge, the S/N and spatial resolution in the radiation image degrade. Because the main target is the human body, X-ray CT requires very low afterglow levels for scintillators, and typical afterglow levels are few, at the ppm level. Among commercial scintillators, BGO and $CdWO_4$ are known to show very low afterglow levels, and GOS also shows a relatively low afterglow level. Taking into account the balance of light yield and afterglow, most X-ray CT use GOS as a main device. In security systems, Tl-doped CsI is widely used, since it shows a high emission intensity with very low cost, although the afterglow level is the poorest among these commercial scintillators. Physically, the afterglow is TSL at room temperature, and scintillators for integration-type detector should not have TSL glow peaks below 100 °C. On the other hand, except for very huge afterglow, photon counting-type detectors can ignore the afterglow because it generally appears as a very small pulse with a high frequency, and is generally cut by lower threshold level of the electronics. Details of TSL is described in Chapter 7, and is a scientifically interesting topic deeply related to afterglow in long persistent luminescence for night illuminations [70].

1.3.5 Radiation Hardness

Some applications use a huge amount of irradiation of ionizing radiation, and to employ detectors in such an environment, radiation hardness (or radiation tolerance) is required. Ionizing radiations can cause ejections and ionizations of atoms. When we irradiate high energy photons, such as X- and γ-rays, ionization is a main phenomenon, and energetic particles mainly cause an ejection. In the case of e^-, p, and n, if their energy exceeds a threshold energy, generation of the Frenkel defect, which consists of hole and interstitial ion, starts. In such ejection-induced damage, generation of defects will occur, and will reduce scintillation light yield by trapping.

The most typical radiation damage is a degradation of transmittance, and coloration is often observed. If the transmittance decreases, emitted photons inside the scintillator cannot reach the photodetectors, and cause a reduction of signal output intensity from the photodetectors. In the case of $PbWO_4$, which is often used for experiments in high energy physics, p-irradiation with ~10^5 Gy causes a degradation of transmittance which induces a reduction of light yield [71]. When the coloration occurs due to exposure of ionizing radiation, it sometimes recovers by annealing (thermal annealing, thermal bleaching), UV-irradiation (UV-annealing), and keeping the sample under room temperature for a long time (spontaneous recovery). In typical interpretation, ejected ions stay near to the incident position, and by these perturbations of thermal or light energy, they can return to the incident position. On the other hand, damage by hadron beams are not or less recovered by these phenomena. It can be understood that the amount of damage is largely higher than that generated by X- or γ-rays.

There are several ways of evaluating the effects of radiation damage; one of the typical ways is to measure optical transmittance before ($T_0(\lambda)$) and after $T(\lambda)$ of the irradiation. By using the thickness of the sample d, the relationship before and after irradiation is

$$T(\lambda) = T_0(\lambda)e^{-\mu_{ir}d} \tag{1.26}$$

where μ_{ir} is an irradiation-induced absorption coefficient. By this μ_{ir}, we can evaluate the radiation damage quantitatively. In actual experiments of radiation hardness, after the irradiation, many small signals can be detected if you directly observe the output signal of scintillator + photodetector by oscilloscope, and these small signals are the afterglow induced by a huge amount of irradiation.

One of the relating phenomena is positive hysteresis (sometimes called radiation drift), which is a phenomenon in which the light yield of the scintillator is increased after exposure to ionizing radiation. From the viewpoint of solid-state physics, it is an interesting phenomenon, but from the engineering and operation of actual detectors, it is problematic because it causes a temporal change of the detector gain. The positive hysteresis was discovered in Tl-doped CsI [72] and Ce-doped GSO [73, 74]. After these pioneering works, we found Ce/Zr co-doped GSO [75] and Ce-doped GAGG [76] also exhibited the positive hysteresis. Up to now, several interpretations are proposed, and our interpretation is a generation of tentative excitation states [75, 76].

1.3.6 Temperature Dependence

As with other phosphors, scintillators show temperature dependence, especially on light yield and timing properties. When we consider the temperature dependence, population of excited states for luminescence and trap states and thermal quenching should be considered. The populations of excited states depend on the thermal vibration of the lattice, so if we increase the temperature, the light yield shows the highest value at the optimum temperature while the decay time decreases monotonically. The same is also true for trap states which deeply relate to TSL described in Section 1.4 and Chapter 7. If the environmental temperature for the sample is enough to excite trapped electrons to the bottom of the conduction band, trapped electrons can be released immediately after trapping, and contribute to scintillation. In this case, the scintillation light yield appears to have an optimum temperature. This phenomenon is called thermal activation. Thermal quenching is common for all the phosphors, and generally, higher light yield can be achieved at lower temperatures due to the suppression of thermal loss. In common thermal quenching, light yield and decay time decrease with the increase in temperature.

In most practical detectors, the operating guaranteed temperature is from −30 to 40 °C, but detectors in some special applications, such as oil-logging and monitoring the area around nuclear reactors, must work at several hundreds of °C. The temperature dependence of scintillators and photodetectors cannot be avoided, and we generally correct it with an analysis method or control the temperature by hard (cooling instrument). If we wish to measure the temperature dependence of the scintillator as in solid-state physics, it would be better to measure it by an X-ray induced radioluminescence spectrum because only the sample is cooled in this experiment. On the other hand, if we wish to measure the temperature dependence as a radiation detector, the device module (scintillator + photodetector) should be measured ([77]).

The temperature dependence relates to emission mechanisms, described in Section 1.3. Previously, AFL was considered to have no temperature dependence, since it was due to an electron transition between core and valance band, and it was experimentally confirmed from room temperature to ~90 °C [78]. However, recent research reveals that even AFL shows temperature dependence when it is investigated at higher temperatures than 100 °C [26].

1.4 Ionizing Radiation Induced Storage Luminescence

1.4.1 General Description

As explained in Section 1.3, after the absorption of ionizing radiations in a material, a large number of secondary electrons are generated, some of which contribute to scintillation. On the other hand, some of the other secondary electrons are trapped at trap sites which are often caused by lattice defects, and they sometimes become meta-stable. In this case, if we input the

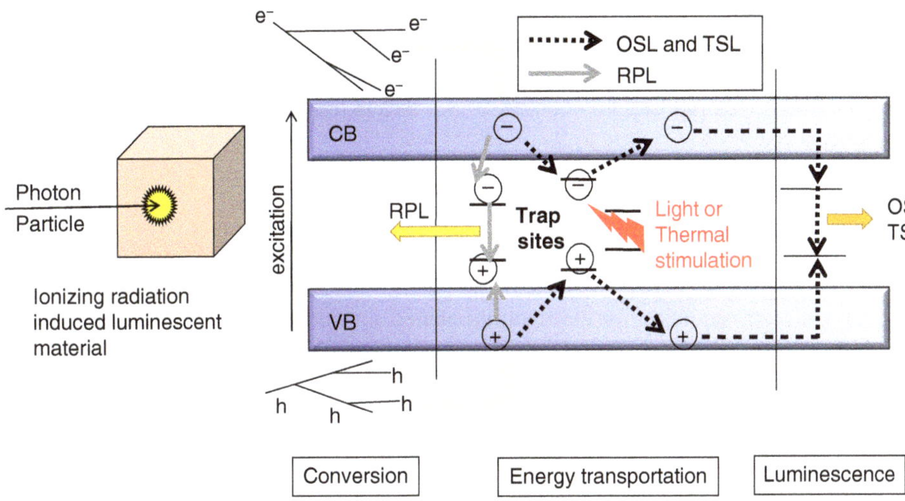

Figure 1.8 *Emission mechanisms of TSL, OSL, and RPL.*

energy corresponding to the energy difference between the bottom of the conduction band and the trap site which is considered as a trap depth, then these trapped electrons can be re-excited, and some of them can reach luminescence centers to emit photons. We refer to the input of the energy as a stimulation. When the stimulation is an illumination of light or heating, we call the emission phenomena an OSL or TSL, respectively. In the case where the trap depth is deeper than the energy of room temperature, we can store the information of the incident ionizing radiation as a form of the carrier trapping stably for a long time, and we can use such a storage phosphor with a deeper trap for the personal dosimeter. RPL can be considered as one of the special trapping phenomena. In the cases of OSL and TSL, the trapping site occupied with an electron or a hole does not show a PL property. But in a special case, the trapping site occupied by an electron or a hole can obtain a function of PL. In this case, we can observe a PL by such a newly generated emission center consisting of the trapping site and carrier, and we call this PL as RPL. These newly generated emission centers are called RPL centers. Figure 1.8 shows a similar drawing, but for TSL, OSL, and RPL, refer to Figure 1.3.

1.4.2 Analytical Description of TSL

There are several analysis methods of TSL, and here we introduce a model based on reaction kinetics. The following explanations on TSL analysis can be found in previous literature [79, 80]. In this section, we briefly introduce the common understanding of the analytic formula of TSL, while detailed explanations on practical applications appear in Chapter 7. If we assume E as the threshold of thermal ionization, the probability of thermal ionization is a typical Boltzmann distribution such as

$$p = se^{-\frac{E}{kT}} \tag{1.27}$$

where s, k, and T represent the frequency factor, the Boltzmann constant, and the temperature, respectively. We define n, m, η, and ζ as concentration of free electron, concentration of free hole, proportional constant of radiative recombination, and proportional constant of non-radiative recombination, respectively. Here, if N is the total number of the electron trapping centers, and n_1 is the number of the trapping centers occupied with an electron, then $N\text{-}n_1$ is a number of empty trapping centers. If we define α as a retrapping coefficient, the time dependence of n_1 is

$$\frac{dn_1}{dt} = -se^{-\frac{E}{kT}} n_1 + \gamma n(\mathrm{N} - n_1). \tag{1.28}$$

We can use the same equation for the trapping center occupied with a hole assuming M as the number of trapping centers for the hole, m_1 as the number of trapping centers occupied with the hole, and E' as an activation energy for the hole to the valence band, and the time dependence of the hole center is

$$\frac{dm_1}{dt} = -s' \exp\left(-\frac{E'}{kT}\right) m_1 + \gamma' m(\mathrm{M} - m_1) - \gamma_m n m_1 \tag{1.29}$$

where s' is a frequency factor for the hole, γ' is the retrapping coefficient of the free hole, and γ_m is the recombination coefficient, respectively. We assume J is a constant energy absorption of material (constant), and the number of free electrons and free holes are generated proportional to αJ, where α is a proportional constant. Then, the time dependence of concentrations of electrons and holes are

$$\frac{dn}{dt} = \alpha \mathrm{J} - (\eta + \zeta) nm - \frac{dn_1}{dt} - \gamma_m n m_1 \tag{1.30}$$

and

$$\frac{dm}{dt} = \alpha \mathrm{J} - (\eta + \zeta) nm - \frac{dm_1}{dt} - \gamma_m n m_1, \tag{1.31}$$

respectively. In these equations, dn_1/dt and dm_1/dt represent a charging (trapping) process, and $(\eta+\zeta)nm$ and $B_m nm_1$ represent dissipation processes. Let us consider the energy dissipation after stopping the irradiation at temperature T. In this case, if we assume $\mathrm{J} = 0$ in (1.27)–(1.30) at time $t = 0$, then TL from the electron centers by electron release is

$$I(t) = \gamma_m n m_1. \tag{1.32}$$

After this time, the time dependence of concentration of electron trapped centers is

$$\frac{dn_1}{dt} = -sn_1 \exp\left(-\frac{E}{kT}\right). \tag{1.33}$$

If we assume temperature $T_0 << E/k$ and the heating rate of β, the temperature is

$$T = T_0 + \beta t. \tag{1.34}$$

The simplest model of luminescence process under this condition is the Randall–Wilkins model. In this classical model, retrapping of electrons is not considered ($\gamma = 0$). If we assume $(\eta + \zeta) = 0$ in Equation (1.30), the TSL intensity can be written as

$$I(T) = B_m n m_1 = -\frac{\mathrm{d}n_1}{\mathrm{d}t}. \tag{1.35}$$

If we change a variable of Equation (1.33) by Equation (1.34), the temperature dependence of the electron concentration can be expressed as

$$\frac{\mathrm{d}n_1}{\mathrm{d}T} = -n_1(T)\frac{s}{\beta}\exp\left(-\frac{E}{\mathrm{k}T}\right) \tag{1.36}$$

and after the integration of Equation (1.35) by temperature T, we can obtain

$$n_1(T) = n_{10}\exp\left[-\int_{T_0}^{T}\frac{s}{\beta}\exp\left(-\frac{E}{\mathrm{k}T'}\right)\mathrm{d}T'\right]. \tag{1.37}$$

In this equation, n_{10} means an electron density at $T = T_0$, which equals the concentration of electron trapping centers generated by ionizing radiation exposure. If we combine Equation (1.36) with Equation (1.37), we obtain

$$I(T) = sn_{10}\exp\left(-\frac{E}{\mathrm{k}T}\right)\exp\left[-\frac{s}{\beta}\int_{T_0}^{T}\exp\left(-\frac{E}{\mathrm{k}T'}\right)\mathrm{d}T'\right]. \tag{1.38}$$

Equation (1.38) represents a first-order kinetics, and if the order is higher than 1, Equation (1.38) changes to

$$I(T) = sn_{10}\exp\left(-\frac{E}{\mathrm{k}T}\right)\left[1 + (b-1)\frac{s}{\beta}\int_{T_0}^{T}\exp\left(-\frac{E}{\mathrm{k}T'}\right)\mathrm{d}T'\right]^{-\frac{1}{b-1}} \tag{1.39}$$

where b is an arbitrary order. Generally, if we can assume $E \gg kT$ and $T_0 = 0$, then

$$\int_{T_0}^{T}\exp\left(-\frac{E}{\mathrm{k}T'}\right)\mathrm{d}T' \cong \frac{\mathrm{k}T^2}{E}\exp\left(-\frac{E}{\mathrm{k}T}\right)\left(1 - \frac{2\mathrm{k}T}{E}\right). \tag{1.40}$$

When we combine Equation (1.40) with Equation (1.38),

$$I(T) = sn_{10}\exp\left(-\frac{E}{kT}\right)\exp\left[-\frac{s}{\beta}\frac{kT^2}{E}\exp\left(-\frac{E}{kT}\right)\left(1-\frac{2kT}{E}\right)\right]. \tag{1.41}$$

When we differentiate Equation (1.38) and assume it is equal to 0, it means there is a glow peak in the TSL glow curve. By using the glow peak temperature T_m, we can obtain the relationship

$$\frac{\beta E}{kT_m^2} = s\exp\left(-\frac{E}{kT_m}\right) \tag{1.42}$$

$$s = \frac{\beta E}{kT_m^2}\exp\left(\frac{E}{kT_m}\right). \tag{1.43}$$

When we combine Equation (1.43) with Equation (1.41), we obtain the relationship

$$I(T) = n_{10}\frac{\beta E}{kT_m^2}\exp\left[\frac{E}{kT}\left(\frac{T-T_m}{T_m}\right)\right]\times\exp\left\{-\frac{T^2}{T_m^2}\exp\left[\frac{E}{kT}\left(\frac{T-T_m}{T_m}\right)\right](1-\Delta)\right\} \tag{1.44}$$

where $\Delta = 2kT/E$. If we combine Equation (1.42) with Equation (1.41), we obtain a maximum TSL intensity of

$$I_m = n_{10}\frac{\beta E}{kT_m^2}\exp\left[-(1-\Delta_m)\right] \tag{1.45}$$

and the equation of

$$n_{10}\frac{\beta E}{kT_m^2} = I_m\exp(1-\Delta_m) \tag{1.46}$$

where $\Delta_m = 2kT_m/E$. If we combine Equation (1.46) with Equation (1.44), we obtain

$$I(T) = I_m\exp(1-\Delta_m)\exp\left\{\frac{E}{kT}\left(\frac{T-T_m}{T_m}\right)\times\exp\left\{-\frac{T^2}{T_m^2}\exp\left[\frac{E}{kT}\left(\frac{T-T_m}{T_m}\right)\right]\right\}(1-\Delta_m)\right\} \tag{1.47}$$

and

$$I(T) = I_m\exp\left[1+\frac{E}{kT}\frac{T-T_m}{T_m}-\frac{T^2}{T_m^2}\times\exp\left(\frac{E}{kT}\frac{T-T_m}{T_m}\right)(1-\Delta)-\Delta_m\right]. \tag{1.48}$$

Equation (1.8) is a glow curve deconvolution function (GCD) of first-order kinetics. In the case of second-order kinetics (b = 2), Equation (1.39) becomes

$$I(T) = sn_{10}\exp\left(-\frac{E}{kT}\right)\left[\frac{s}{\beta}\frac{kT^2}{E}\exp\left(-\frac{E}{kT}\right)(1-\Delta)+1\right]^{-2}. \tag{1.49}$$

When $I(T)$ shows the maximum, we can obtain equations such as

$$\frac{\beta E}{kT_m^2} = (1+\Delta_m)s\exp\left(-\frac{E}{kT_m}\right) \tag{1.50}$$

$$s = \frac{\beta E}{kT_m^2}\frac{1}{1+\Delta_m}\exp\left(\frac{E}{kT_m}\right) \tag{1.51}$$

and

$$s\exp\left(-\frac{E}{kT_m}\right) = \frac{\beta E}{kT_m^2}\frac{1}{1+\Delta_m}. \tag{1.52}$$

If we combine Equation (1.51) with Equation (1.49), the TSL intensity can be expressed as

$$I(T) = n_{10}\frac{\beta E}{kT_m^2}\frac{1}{1+\Delta_m}\exp\left[\frac{E}{kT}\left(\frac{T-T_m}{T_m}\right)\right]\times\left\{\frac{T^2}{T_m^2}\frac{1-\Delta}{1+\Delta_m}\exp\left[\frac{E}{kT}\left(\frac{T-T_m}{T_m}\right)\right]+1\right\}^{-2}. \tag{1.53}$$

When the TSL intensity shows the maximum I_m, Equation (1.49) can be rewritten as

$$I_m = sn_{10}\exp\left(-\frac{E}{kT_m}\right)\left[\frac{s}{q}\frac{kT^2}{E}\exp\left(-\frac{E}{kT_m}\right)(1-\Delta_m)+1\right]^{-2}. \tag{1.54}$$

When we combine Equation (1.52) with Equation (1.54), the maximum intensity can be expressed as

$$I_m = n_{10}\frac{\beta E}{kT_m^2}\frac{1}{1+\Delta_m}\left(\frac{2}{1+\Delta_m}\right)^{-2} \tag{1.55}$$

and relationship of

$$n_{10}\frac{\beta E}{kT_m^2}\frac{1}{1+\Delta_m} = I_m\left(\frac{2}{1+\Delta_m}\right)^{2} \tag{1.56}$$

is deduced. Finally, if we combine Equation (1.56) with Equation (1.53), we obtain the equation on the second-order kinetics of

$$I(T) = 4I_{\mathrm{m}} \exp\left[\frac{E}{\mathrm{k}T}\left(\frac{T-T_{\mathrm{m}}}{T_{\mathrm{m}}}\right)\right] \times \left[\frac{T^2}{T_{\mathrm{m}}^2}(1-\Delta)\exp\left(\frac{E}{\mathrm{k}T}\frac{T-T_{\mathrm{m}}}{T_{\mathrm{m}}}\right) + 1 + \Delta_{\mathrm{m}}\right]^{-2}. \quad (1.57)$$

In addition to these standard analysis, analogical consideration of TSL efficiency with scintillation is considered as

$$I_{TSL} = \frac{h\nu}{\beta' E_g} \eta_{trap} S' Q \eta_{esc} \quad (1.58)$$

where η_{trap}, S', and η_{esc} are the trap efficiency of carriers at trapping centers, carrier transfer efficiency to luminescence centers, and a probability that emitted photons are not absorbed in TSL material [81]. Other parameters have the same meaning with scintillation. Some analogical relation is proposed to TSL and OSL [82], and they essentially have the same physical meaning that scintillation and storage luminescence should be treated as one theory.

1.4.3 Analytical Description of OSL

Here, we introduce a basic analytical treatment of OSL, and explanations on practical applications and common materials are described in Chapter 8. The concentration of the metastable state occupied with an electron or hole ($N_{\mathrm{OSL}}(t)$) can be expressed as

$$N_{OSL}(t) = \int_{\gamma_1}\int_{\gamma_2}\cdots\int_{\gamma_m} n(\gamma_1, \gamma_2, \ldots \gamma_m, t)\, d\gamma_1 d\gamma_2 \ldots d\gamma_m, \quad (1.59)$$

where $\gamma_1, \gamma_2, \ldots \gamma_{\mathrm{m}}$ mean the stability of the metastable state, that is they govern the probability per unit time in which the system will return to equilibrium, and $n(\gamma_1, \gamma_2, \ldots \gamma_{\mathrm{m}}, t)$ is a weighting function, or distribution, expressing the concentration of occupied states possessing the parameters $\gamma_1, \gamma_2, \ldots \gamma_{\mathrm{m}}, t$. Then, OSL intensity $I_{\mathrm{OSL}}(t)$ is written as

$$I_{OSL}(t) = \left|\frac{dN_{OSL}(t)}{dt}\right|. \quad (1.60)$$

If we assume that $P(t)$ is the probability per unit time of the decay of the metastable states $N_{\mathrm{OSL}}(t)$,

$$\frac{dN_{OSL}(t)}{dt} = -N_{OSL}(t)^l P(t). \quad (1.61)$$

In this formula, $l = 1$ means a first-order function. If each state $n(\gamma_1, \gamma_2, \ldots \gamma_{\mathrm{m}}, t)$ has its own probability function $p(\gamma_1, \gamma_2, \ldots \gamma_{\mathrm{m}})$ under the condition of $l = 1$, then

$$I_{OSL}(t)=\left|\frac{dN_{OSL}(t)}{dt}\right|=\int_{\gamma_1}\int_{\gamma_2}\cdots\int_{\gamma_m} n(\gamma_1,\gamma_2,\ldots\gamma_m,t)\,p(\gamma_1,\gamma_2,\ldots\gamma_m,)\,d\gamma_1 d\gamma_2\ldots d\gamma_m \quad (1.62)$$

where we assume that no interaction between states occur. This formula has no time dependence of t, and if we would like to treat the probability time dependently, $p(\gamma_1, \gamma_2, \ldots \gamma_m, t)$ should be used. The form of p depends on the stimulation methods such as TSL or OSL. For optical stimulation (OSL), we have

$$p(E_0)=\Phi\sigma(E_0) \quad (1.63)$$

where E_0, Φ, and $\sigma(E_0)$ are the threshold of optical stimulation energy, optical stimulation intensity, and photoionization cross-section, respectively. If $m = 1$, γ_1 equals to E_0. In previous works [83, 84], photoionization cross-section is expressed as

$$\sigma(E_0)\propto\frac{\sqrt{E_0}\,(h\nu-E_0)^{\frac{3}{2}}}{h\nu\left(h\nu+E_0\left(\frac{m_0}{m^*}-1\right)\right)^2} \quad (1.64)$$

where $h\nu$ is the energy of the incident photon of wavelength λ, m^* is the charge carrier effective mass, and m_0 is the rest of mass, respectively. There are several expressions of the photoionization cross-section, and the more simple form [85] is

$$\sigma(E_0)\propto\frac{(h\nu-E_0)^{\frac{3}{2}}}{(h\nu)^3}. \quad (1.65)$$

Photoionization is basically the same as the photoelectric (photoelectric absorption) effect, described in scintillation, but the energy assumed here is around visible photons (several eV).

Generally, stimulation intensity is a function of time, and can be expressed as

$$\Phi(t)=\Phi_0+\beta_\Phi t \quad (1.66)$$

where Φ_0 and $\beta\Phi$ are a constant of stimulation and a proportional constant of the stimulation intensity if we assume a linear time dependence. If $\Phi(t)$ is constant, it represents a continuous wave OSL (CW-OSL), and the OSL intensity becomes

$$I_{OSL}(t)=N_0 p(E_0)e^{-\frac{t}{\tau(E_0)}} \quad (1.67)$$

where $t(E_0)$ is $1/p(E_0)$. If $\Phi(t)$ is not constant, the situation is a linear modulation OSL (LM-OSL), and the OSL intensity is

$$I_{OSL}(t) = -\frac{dN_{OSL}(t)}{dt} = N_{OSL}(t)\sigma(E_0)\beta_\Phi t \tag{1.68}$$

where we assume $\Phi_0 = 0$. Generally, we measure OSL in typical PL machines where the environment is free from any other light, and $\Phi_0 = 0$ is a valid assumption. In this case, the OSL intensity becomes

$$I_{OSL}(t) = N_0\sigma(E_0)\beta_\Phi t e^{-\frac{\sigma(E_0)\beta_\Phi t^2}{2}} \tag{1.69}$$

Based on these analytical equations, OSL intensity shows exponential decay under stimulation, and this is an important property to distinguish PL and RPL from OSL. If readers would like to study OSL in more detail, it will be better to read specialized books on OSL (for example, [3]).

Sections 1.4.2 and 1.4.3 describe common confirmed formulations of TSL and OSL, but up to now, a widely accepted formulation of RPL has not been obtained. RPL is treated as PL after carrier trapping phenomena, and general treatment of PL can be applied.

1.5 Relationship of Scintillation and Storage Luminescence

Up to now, scintillation (1.3) and ionizing radiation induced storage luminescence (1.4) have been carried out in different fields and by different people. However, these phenomena occur at the same time, and Figure 1.9 describes the situation relating to luminescence by combining Figures 1.3 and 1.8.

In addition to the luminescence phenomena described in these figures, some of the secondary electrons lose their energy thermally, and in such a case, we obtain no emissions. In the model of scintillation described in Section 1.3, it sometimes contains thermal energy dissipation as in Equation (1.15), but it does not consider that a significant amount of carrier

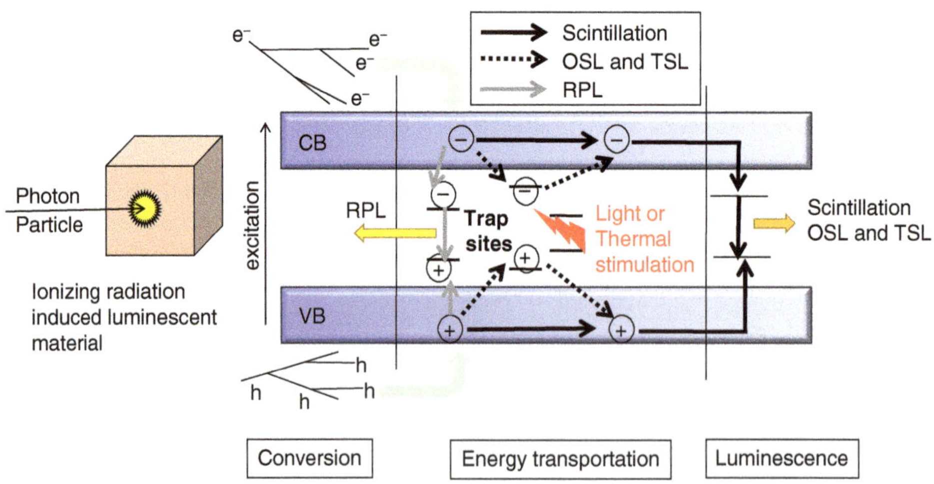

Figure 1.9 Typical emission mechanisms of scintillation, OSL, TSL, and RPL.

is stored at trapping centers. In storage luminescence, such as TSL and OSL explained in Section 1.4, analysis is done by the rate equation, which does not consider how many carriers can reach the trapping centers, and scintillation is never assumed. The rate equation is only considered after the trapping (or generally luminescence) sites are occupied, so it cannot be used before the occupation. As commented above, scintillation and storage luminescence occur at the same time after the absorption of ionizing radiation. Therefore, in ionizing radiation induced luminescent materials, branching ratio of scintillation, storage luminescence, and thermal loss is important, and the branching probability of these phenomena can be expressed as

$$P_{scintillation} + P_{storage} + P_{thermal} = 1 \tag{1.70}$$

where $P_{\mathrm{scintillation}}$, P_{storage}, and P_{thermal} are the branching probabilities of secondary electrons to scintillation, storage luminescence, and thermal loss, respectively. It should be noted that the thermal energy loss is a very rough concept, and contains energy loss during any diffusion processes and relaxation processes at localized trapping and luminescence centers. Although the above equation is quite natural if we assume energy conservation, such a relation has not been considered historically because scintillator and dosimeters with storage phosphors have been investigated in different scientific fields.

The relationship among scintillation, storage luminescence, and thermal loss has been experimentally investigated recently by the author [82, 86–88]. In materials recognized as a phosphor, the ratio of thermal loss is relatively small, and in such a case, scintillation and storage luminescence should show an inverse proportional relationship. In order to confirm the validity of our assumption, we investigated the relationship of scintillation and storage luminescence intensities of materials which have different dopant concentrations. Figure 1.10

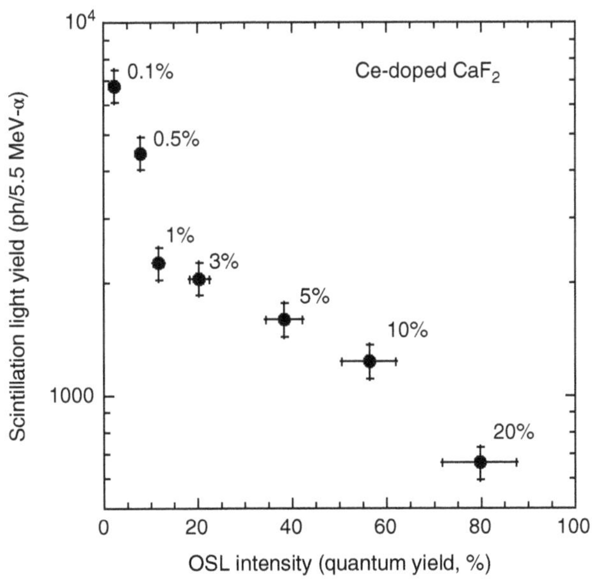

Figure 1.10 *The inverse proportional relationship of Ce differently doped* CaF_2 *single crystals on the plane of OSL intensity as a PL quantum yield (%) and scintillation light yield (ph/5.5 MeV-α). Source: The data taken from [86].*

shows a representative of the inverse proportional relationship of Ce differently doped CaF_2 on OSL vs. scintillation. A clear inverse proportionality is observed, and confirms the validity of the assumption. A similar relationship has been confirmed in many material forms, including single crystal, glass, and ceramics in scintillation vs. OSL/TSL. In most luminous materials, we can observe such a relationship, but in the case of Ce-doped $LiCaAlF_6$, we observed a positive proportional relationship [87]. Ce-doped $LiCaAlF_6$ is not luminous both in scintillation and storage luminescence, so we consider most of the absorbed energy will be converted to thermal loss. If we neglect the effect of thermal loss, a low temperature experiment will be helpful.

This experimental result presents some important problems to conventional understanding of ionizing radiation induced luminescence fields. For example, from the viewpoint of scintillation, the degradation of light yield in higher dopant concentration has been interpreted as concentration quenching. However, Figure 1.10 shows that most energy in highly Ce-doped samples is not converted to thermal loss but to energy storage. Another point is the ε-value, which is the average energy to generate one electron–hole pair in solid state materials. The most common example is the Si semiconductor detector, and the value of Si is known to be ~3.6 eV ($=\beta E_g$ of Si in conventional understanding based on Section 1.3), which also relates to the theoretical limit of Si solar cell of $<30\% \sim 1/3.6 \sim 28\%$. The ε-value is also defined for scintillators, and has been considered as 10–20 eV. This calculation is based on the scintillation light yield from a pulse height spectrum, and does not take into account storage luminescence. If we consider the ε-value from the definition, such an estimation in scintillators is not correct, because we do not count the contribution from the carrier storage.

As described above, we think ionizing radiation induced luminescence can be treated as one form of unified physics, and this is why we describe these topics in one book, although they have often been treated as different scientific fields. In this case, the base of the consideration is the energy conservation law, and it strongly assumes the integration of energy in infinite time, which is a standard strategy in astrophysics because the real-time (time-derivative) observation is impossible. The author (Yanagida) studied astrophysics, and the consideration depends on the field of the origin. Recently, another author (Koshimizu), whose field of origin is in solid state physics, proposed a real-time observation on the energy transportation (carrier diffusion) process, which is a key process to understanding S in the equations presented in the previous sections. Such an observation is enabled by transient absorption spectroscopy, i.e., optical absorption spectroscopy as a function of time after excitation by pulsed electron beams. Because the excited states are probed with optical absorption, their real-time dynamics prior to scintillation can be analyzed. Actually, slow decay of the transient absorption correlates with low scintillation intensity, and is consistent with observation results based on energy conservation. Such transient spectroscopy has long been used with pulsed light, from flash lamps to laser instruments, as excitation sources to elucidate the excited states dynamics. For ionizing radiation, pulsed electron beams can be used as excitation sources. Such a measurement technique based on pulsed electron beams has also long been used to analyze the chemical reaction dynamics in radiation chemistry and is called "pulse radiolysis." This technique can also be applied to ionizing radiation-induced luminescence materials and gives information on energy transfer, carrier trapping, and quenching. Thus, to understand S, both energy conservation and temporal analysis-based experiments have been used recently.

1.6 Common Characterization Techniques of Ionizing Radiation Induced Luminescence Properties

In this section, common characterization techniques for ionizing radiation induced luminescence phenomena are introduced. Transmittance and absorbance measurements are common characterization techniques for phosphors, because we can obtain the absorption bands of materials by this measurement. The PL spectrum is important when considering the emission origin. Figure 1.11a represents a typical methodology to measure transmittance and absorbance. At first, we measure the light transmission intensity of the sample holder, and we define its intensity as 100% transmittance at measured wavelength. Then, we place a sample on the holder, and measure the light transmission intensity of sample + holder. The ratio of the with-without sample is the transmittance. The transmittance can be understood as Equation (1.26), and if we can plot it in log-scale corrected by the thickness of the sample, we can draw the absorbance plot. The absorbance is in arbitrary units, and some literature is confused about the absorbance and absorption coefficient (cm^{-1}). It must be understood that if readers need to evaluate the absorption coefficient quantitatively, reflectance must be measured. Without the correction of reflectance, it is impossible physically to determine the absorption coefficient. In some special cases where the reflectance is close to zero, we can treat the absorbance with the rough absorption coefficient. In the case of undoped materials, the wavelength rapidly drops to transmittance to 0% (or rapid increase of absorbance to infinite value), and we can notice the wavelength (energy) as a band-gap of each material. Some literature use the term "band-gap" to absorption bands due to dopant in luminescence center doped materials. Such an expression is wrong, at least in solid state physics, although most readers can understand what authors of the paper mean. The transmittance obtained by Figure 1.11a is called in-line transmittance, and if we use an integration sphere to collect a non-straight line of light from the sample, we can measure diffuse transmittance. In ionizing radiation detector uses, diffuse transmittance is close to practical conditions, because we generally use reflectors to collect every photon generated by irradiation to photodetectors.

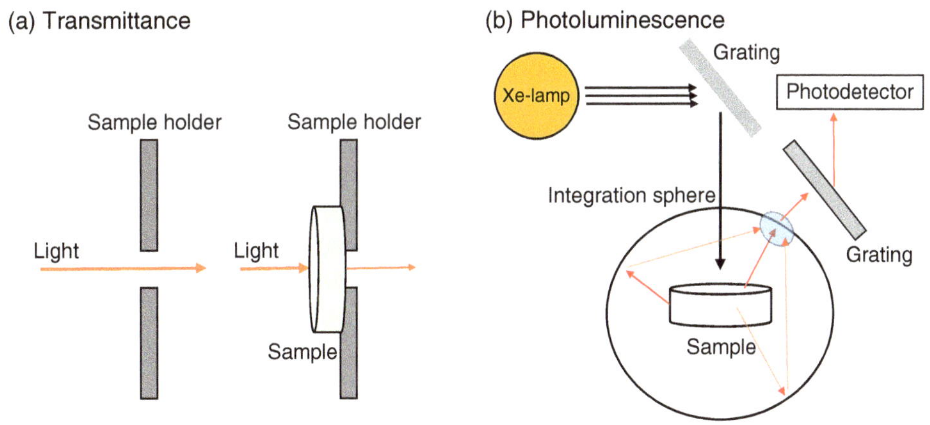

Figure 1.11 Schematic drawing of transmittance (absorption) (a) and PL measurement (b) setups.

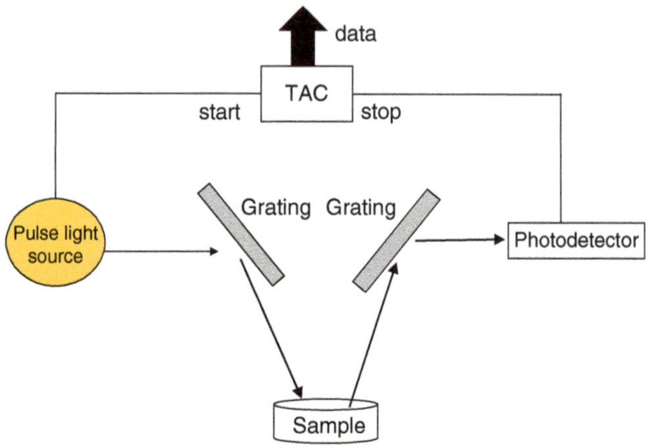

Figure 1.12 *Schematic drawing of PL decay setup.*

Figure 1.11b represents a setup of PL excitation and emission spectra by using the integration sphere. Generally, PL can be observed upon UV–VIS excitation from Xe-lamp, and we can select which luminescence center (or electron transition) can be induced by selecting the excitation wavelength using a grating. If we use the integration sphere, we can collect all the PL photons. If we define A and I as absorption, which is evaluated by the intensity of excitation light with and without the sample and PL emission intensity, respectively, we can deduce $Q = I/A$ experimentally. The detectors for PL photons are generally grating and photodetector. PL quantum yield is a quantitative value, but generally PL is a qualitative evaluation with an arbitrary unit, since PL intensity depends on the geometry of sample setting. In common PL measurement, we do not use the integration sphere.

Figure 1.12 represents a typical experimental configuration of PL decay measurement. The excitation light was emitted from a pulsed light source such as a lamp or an LED. In the lamp, a grating is used to select excitation wavelength, and selected excitation photons are irradiated to the sample. The timing of the excitation is injected as a start signal to time to amplitude converter (TAC). PL photons from the sample are also filtered by the grating to select the target emission, and then collected by the photodetector. The emitted PL photons are reduced to a single photon by filters, pinhole, etc., and this single photon signal is put into the TAC as a stop signal. A timing window is set for the target timing range (e.g., several tens ns for few ns phenomenon), and only the coincidence event within the timing window is recognized as a target signal. The TAC has a function to convert the timing difference of start and stop signals to a pulse signal. After accumulation of pulse signals many times, we can obtain a PL decay curve. Such a technique is called time correlated single photon counting (TCSPC). Generally, we analyze the PL decay curve by an approximation of sum of exponential functions as

$$I(t) = \sum_i A_i \exp(-t/\tau_i) \tag{1.71}$$

where A_i, t, are τ_i are intensity of each decay component, time, and decay time (lifetime) of each component, respectively. Sometimes, we deconvolute the excitation pulse (instrumental response function, IRF), and it should be noted that perfect deconvolution is generally difficult. After obtaining PL quantum yield Q and decay time τ, we can use convenient equations such as

$$k_f + k_{nr} = \frac{1}{\tau} \tag{1.72}$$

$$Q = \frac{k_f}{k_f + k_{nr}} \tag{1.73}$$

$$k_f = \frac{Q}{\tau} \tag{1.74}$$

where k_f and k_{nr} denote rate constants of radiative and non-radiative transitions, respectively. In this way, we can evaluate radiative and non-radiative rate constants quantitatively.

PL is a phenomenon at localized luminescence centers, and except for some cases like the semiconductor scintillator, we do not consider the movement and interactions of carriers. In other words, PL is a phenomenon within a band-gap. But scintillation is accompanied with interactions of carriers, and sometimes PL and scintillation offer very different emission spectra. As a basic scintillation property, we generally measure radioluminescence spectrum to obtain the scintillation emission intensity to select adequate photodetector for applications. A typical setup is illustrated in Figure 1.13. The most common setup uses an X-ray generator as an excitation source, and if readers have another radiation source, they can use that instead of the X-ray generator. We sometimes use α-rays instead of X-rays [89] in laboratory-level experiments, and sometimes use other radiation sources at large facilities [90] by taking our experimental setup there. Scintillation photons are generally collected via an optical fiber, since a direct hit of ionizing radiation causes radiation damage to grating and photodetector, and to avoid this, we generally set them apart from the radiation source. In some literature, light yield is calculated by an integrated area of radioluminescence emission spectrum. It must be noted that such an evaluation is not correct in most cases, as explained in Section 1.3.3. Figure 1.14 shows a geometry of transmission-type measurement, and in some experiments, we use a reflection type measurement similar to the PL configuration.

Scintillation decay time is generally measured by γ-ray or X-ray excitation. Figure 1.14 presents a typical setup for a γ-ray induced scintillation decay curve measurement. The basic concept is the same as that in PL decay curve measurements but the excitation source is

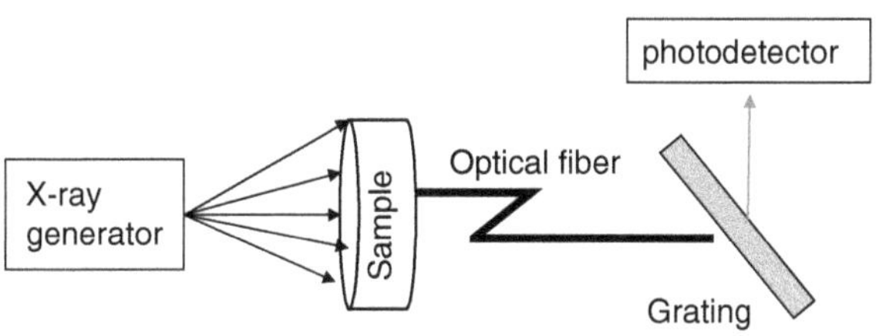

Figure 1.13 *Common setup of X-ray induced radioluminescence spectrum measurement.*

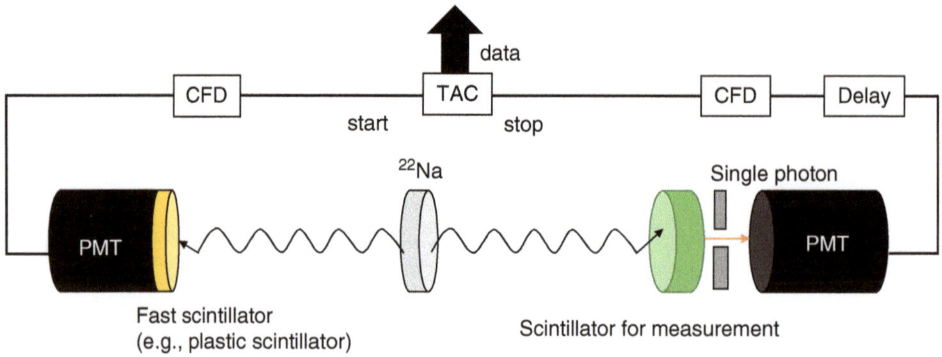

Figure 1.14 *Common setup of γ-ray induced scintillation decay curve measurement (delayed coincidence method).*

different. Generally, we use the ^{22}Na radioisotope as an excitation source, which emits two 511 keV γ-rays in 180° opposite directions. One of these two γ-rays hits a fast scintillator, and creates a start signal through a constant fraction discriminator (CFD). At the opposite side, the other γ-ray hits the target scintillator for measurement, and emits scintillation photons. Then, the number of these emitted scintillation photons are reduced to a single photon, and a signal pulse corresponding to the single photon is output from photodetector (typically, PMT). After some delay, the signal is injected into TAC as a stop signal through adequate delay and CFD. The process after TAC is the same with PL decay, and the analysis method is also the same. This typical experimental methodology is called delayed coincidence method (DCM). The difference with PL is that DCM generally does not resolve wavelength because of the low signal intensity and frequency. If we resolve the wavelength, a very long measurement time will be required.

Although DCM with a 511 keV γ-ray source is the most common way to evaluate scintillation decay time, it contains several technical disadvantages. Because the energy of γ-ray (511 keV) is high and has a high penetrative power of materials, detection efficiency is not high and requires a long time for the measurement to be made. In addition, low detection efficiency makes it difficult to measure a slow component, and generally, a component slower than several μs is difficult to measure. For example, measurements of emissions from Eu^{3+}, Tb^{3+} and most transition metal ions, are almost impossible, although they are used for integration-type detectors. The lack of the wavelength resolution is also a problem because identification of the scintillation emission origin is sometimes difficult. In most cases, we can guess the emission origin by PL decay, but decays of scintillation and PL sometimes show a large discrepancy.

In order to solve these problems, some groups have developed pulse X-ray-based instruments for scintillation decay measurement. Pioneering work was done by the Lawrence Berkeley Laboratory (USA) with the cooperation of Hamamatsu Photonics (Japan) [91]. The key technology is a stable pulse X-ray source which can be used at laboratory level, and if such a source is developed, we can use the same data processing system with PL decay and DCM. In this pulse X-ray tube, the origin of the excitation is a laser diode or LED, and emits visible photons with 10–10^6 Hz frequency, depending on the time range of measurement. Emitted visible photons hit a multi-alkali photocathode of the X-ray tube, and are converted into electrons. These electrons are accelerated by high voltage bias, and are lead to a tungsten target by a strong electric field. Then, bremsstrahlung X-rays are generated and irradiated to the sample through a Be window. In this way, we can make a pulse X-ray by the timing of laser diode or LED and use it as a start signal.

We now introduce some applications of this pulsed X-ray tube for scintillation decay measurement. The timing resolution of an X-ray tube is several tens ps, which is determined by the traveling time of accelerated electrons, and it is enough to measure the rise time of most scintillators [92]. Following these pioneering works, a pulse X-ray equipped streak camera system was developed, and enabled us to observe the wavelength resolved scintillation decay curve [93]. In this system, the detection part consists of a spectrometer chamber, streak camera, and two-dimensional CCD, and a two-dimensional image of wavelength vs. time can be observed. By the streak system, one of the problems of DCM, non-wavelength resolution, can be solved. Up until the development of the streak camera system, every study used a pulse laser diode as the root of the excitation source, and it was enough to measure fast-timing events. However, it is not sufficient to measure slow events, especially in the ms time range. One of the solutions for ms measurement is to replace the excitation source to common LED, which has higher power than the pulse-type laser diode but slower speed. The pulse LED type system was developed in 2014 [94]. This system enabled us to measure slow (ms) scintillation decay as well as the X-ray induced afterglow. Because the signal intensity per unit time is very low in slower scintillation, the system does not offer the function of wavelength resolution, and the detection part is one PMT. When we compare with typical measurements by DCM, the measurement time dramatically decreases to a few minutes for one sample. At present, no significant difference of excitation energy between 511 keV (DCM) and several tens keV (pulse X-ray tube) has been found.

In other ionizing radiations such as α-ray, measurement is generally difficult. One easy but rough way is to use an oscilloscope for direct observation of the signal output from a detector, consisting of scintillator and photodetector by a self-trigger mode. Although it may contain noise (e.g., thermal noise of photodetector), we can roughly observe scintillation decay time. Previously, most papers observed decay curves by such an oscilloscope measurement, but recently, a digitizer has been used to accumulate many decay curves. In such a case, we can accumulate a large amount of decay curves, and select correct decay curves by setting correct selection conditions of events. An accurate but technically difficult way is to use DCM by one scintillator sample and two photodetectors. If we take the case of α-ray irradiation as an example, let us assume one cylindrical scintillator optically coupled with two PMTs by edge surfaces, and α-ray is irradiated to the side of the scintillator. When the α-ray is absorbed by the center part of the scintillator, half of the scintillation photons are transported to one PMT and the other remaining half to another PMT. We can use one of the two output signals as a start signal, and if we can reduce the number of photons in another side to a single photon, we can use it as a stop signal. In this way, we can measure scintillation decay under particle excitation.

The pulse height spectrum is one of the most important properties of scintillation detectors, because most of measurements in other scientific fields use integration-type detectors, and pulse height (photon counting) type measurements is one of the original techniques in radiation and high energy physics. The detailed explanation is given in Chapter 12.

Typical experiments on dosimeters are described in Chapters 7 (TSL), 8 (OSL), and 9 (RPL). In most cases, dosimeter properties are examined by X-ray or γ-ray irradiation, since an accessibility of high energy photons are easier than other ionizing radiations. Especially, X-rays are mainly used for tests of individual dosimeters, because energy and amounts of X-rays can be controlled by bias voltage and tube current in a typical X-ray generator. In evaluations of dosimeters, the irradiated dose range should be adequate for the purpose. The lower detection limit of commercial individual dosimeters is 1–100 μGy, and experiments should be conducted from these lower detection limits to ~10 Gy. In some papers, several kGy irradiations are done, although the aim is an individual dosimeter. On the other hand, if the purpose is to monitor radiation therapy, tests of several hundreds Gy to several kGy irradiation will be valid. Another important aspect is fading. If the aim is individual dosimeters, fading should be tested at about one month, and if the aim is an imaging plate typically used in a dental clinic, three to four days' tests will be adequate. In addition to these general aspects, in evaluation of TSL properties, heating rate affects peak temperatures in the glow curve. In practical applications, the heating rate is around 10 °C/s, while 0.1–1 °C/s rate is common for studies of TSL as a physical property. In actual materials, we must consider time of heat conduction, and if the heating rate is too fast, responses of materials cannot follow. With a fast heating rate, we generally observe a relatively higher glow peak temperature than the true value, although we do not have a way to know the exact true value. When we measure TSL with a very slow heating rate, an observed glow peak temperature will be close to the true value. Another detailed explanation of the measurement technique of TSL can be found in Chapter 7.

In characterizations of OSL and RPL, we generally use common systems of PL spectrum. In OSL, after the irradiation of ionizing radiation, we put the sample in a PL machine and stimulate the sample under the adequate stimulation wavelength. In some spectrofluorometers, the automatic measurement function of PL excitation spectrum is installed, and by this function, we can easily investigate the stimulation spectrum. Unlike the PL excitation spectrum, which generally shows a clear excitation band due to electron transitions of dopant, OSL stimulation spectrum generally shows a broad unclear feature. For OSL and TSL measurements, an automation type instrument (Risø TL/OSL DA-20 reader) is widely used ([95]). Generally, this system equips β-ray (^{90}Sr) as an irradiation source, LED as a stimulation source for OSL having typically a 470 nm emission wavelength, and a heater for TSL measurement. One of the disadvantages of this automated system is a lack of wavelength resolution, which equals to non-determination capability on the emission origin, and some ideas are proposed to add a function of wavelength resolution [96].

In order to investigate OSL and RPL, some important aspects should be noted in actual experiments. Typically, the stimulation wavelength of OSL is longer than the emission wavelength, and resembles the situation of upconversion. In order to distinguish them, continuous stimulation is a simple way. If the phenomenon is OSL, the emission intensity decreases under the continuous stimulation, while in the case of upconversion, the intensity is constant. In addition, the stimulation wavelength of OSL is not necessarily longer than the emission wavelength, because we cannot deny the existence of a deep trap. In such a case, distinction of OSL and RPL becomes difficult. The continuous stimulation (excitation) is also a useful technique in this situation, and if the phenomenon is OSL, the emission intensity decreases with time, while that of RPL is constant against time. But in the special case where the trap depth and excitation band overlap, clear distinction is difficult. Other experimental aspects of OSL and RPL are introduced in Chapters 8 and 9, respectively.

References

1. Röntgen, W.C. (1895). On a new kind of rays. *Nature* 53 (1369): 274–276.
2. Knoll, G. (2000). *Radiation Detection and Measurement*. Hoboken, NJ: Wiley.
3. Yukihara, E.G. and McKeever, S.W.S. (2011). *Optically Stimulated Luminescence: Fundamentals and Applications*. New York: Wiley.
4. Mckeever, S.W.S. (1985). *Thermoluminescence of Solids*. Cambridge: Cambridge University Press.
5. Schulman, J.H., Ginther, R.J., and Klick, C.C. (1951). Dosimetry of X-rays and gamma-rays by radiophotoluminescence. *J. Appl. Phys.* 22: 1479–1487.
6. Hofstadter, R. (1948). Alkali halide scintillation counters. *Phys. Rev.* 74: 100–101.
7. Weber, M.J. and Monchamp, R.R. (1973). Luminescence of $Bi_4Ge_3O_{12}$: spectral and decay properties. *J. Appl. Phys.* 44: 5495–5499.
8. Laval, M., Moszyński, M., Allemand, R. et al. (1983). Barium fluoride – inorganic scintillator for subnanosecond timing. *Nucl. Instrum. Meth. Phys. Res.* 206: 169–176.
9. Jahn, A., Sommer, M., and Henniger, J. (2014). OSL efficiency for BeO OSL dosimeters. *Radiat. Meas.* 71: 104–107.
10. McElhaney, S.A., Ramsey, J.A., Bauer, M.L. et al. (1990). A ruggedized ZnS(Ag)/epoxy alpha scintillation detector. *Nucl. Instrum. Methods Phys. Res. A.* 299: 111–114.
11. Yanagida, T., Fujimoto, Y., Miyamoto, M. et al. (2014). Optical and scintillation properties of Cd doped ZnO film. *Jpn. J. Appl. Phys.* 53: 02BC13.
12. Kaneko, J.H., Izaki, K., Toui, K. et al. (2016). An alpha particle detector based on a GPS mosaic scintillator plate for continuous air monitoring in plutonium handling facilities. *Radiat. Meas.* 93: 13–19.
13. Maekawa, T., Sumita, A., and Makino, S. (1998). Thin Beta-ray detectors using plastic scintillator combined with wavelength-shifting fibers for surface contamination monitoring. *J. Nucl. Sci. Technol.* 35: 886–894.
14. Tzolov, M.B., Barbi, N.C., Bowser, C.T. et al. (2018). First-surface scintillator for low accelerating voltage scanning electron microscopy (SEM) imaging. *Microsc. Microanal.* 24: 488–496.
15. Tanaka, H.K.M. and Yokoyama, I. (2008). Muon radiography and deformation analysis of the lava dome formed by the 1944 eruption of Usu, Hokkaido – contact between high-energy physics and volcano physics. *Proc. Jpn. Acad. Ser. B.* 84: 107–116.
16. Yanagida, T., Fujimoto, Y., Yamanoi, K. et al. (2012). Optical and scintillation properties of bulk ZnO crystal. *Phys. Status Solidi C* 9: 2284–2287.
17. Yanagida, T., Fujimoto, Y., and Koshimizu, M. (2014). Evaluation of scintillation properties of GaN, e-J. *Surf. Sci. Nanotechnol.* 12: 396–399.
18. Yanagida, T., Okada, G., Kato, T. et al. (2016). Fast and high light yield scintillation in Ga_2O_3 semiconductor material. *Appl. Phys. Exp.* 9: 042601-1-042601-4.
19. Shendrik, R.Y., Radzhabov, A., and Nepomnyashchikh, A.I. (2013). Scintillation properties of SrF_2 and SrF_2-Ce^{3+} crystals. *Tech. Phys. Lett.* 39: 587–590.
20. Mikhailik, V.B., Kraus, H., Imber, J. et al. (2006). Scintillation properties of pure CaF_2. *Nucl. Instrum. Methods Phys. Res. A.* 566: 522–525.
21. Itoh, M. and Kamada, M. (2001). Comparative study of auger-free luminescence and valence-band photoemission in wide-gap materials. *J. Phys. Soc. Japan* 70: 3446–3451.
22. Yanagida, T., Kawaguchi, N., Fujimoto, Y. et al. (2010). Growth and scintillation properties of $BaMgF_4$. *Nucl. Instrum. Methods Phys. Res. A.* 621: 473–477.
23. Moszynski, M., Allemand, R., Odru, M.L.R. et al. (1983). Recent progress in fast timing with CsF scintillators in application to time-of-flight positron tomography in medicine. *Nucl. Instrum. Methods Phys. Res. Sect. A.* 205: 239–249.
24. Yahaba, N., Koshimizu, M., Yan, S. et al. (2014). X-ray detection capability of a Cs_2ZnCl_4 single-crystal scintillator. *Appl. Phys. Exp.* 7: 062602-1-062602-4.
25. Schotanus, P., van Eijk, C.W.E., Hollander, R.W. et al. (1985). Temperature dependence of BaF2 scintillation light yield. *Nucl. Instrum. Methods Phys. Res. A.* 238: 564–565.
26. Yanagida, T., Fujimoto, Y., Koshimizu, M. et al. (2015). Scintillation properties of CdF_2 crystal. *J. Lumin.* 157: 293–296.
27. Anderson, D.F. (1989). Properties of the high-density scintillator cerium fluoride. *IEEE Trans. Nucl. Sci.* 36: 137–140.

28. García-Toraño, E., Caro, B., Peyrés, V. et al. (2016). Characterization of a $CeBr_3$ detector and application to the measurement of some materials from steelworks. *Nucl. Instrum. Methods Phys. Res. A.* 837: 63–68.
29. Arai, M., Fujimoto, Y., Koshimizu, M. et al. (2020). Scintillation and photoluminescence properties of $(Tl_{1-x}A_x)MgCl_3$ (where a = alkali metal). *J. Alloys Compds* 823: 153871.
30. Fujimoto, Y., Koshimizu, M., Yanagida, T. et al. (2016). Thallium magnesium chloride: a high light yield, large effective atomic number, intrinsically activated crystalline scintillator for X-ray and gamma-ray detection. *Jpn. J. Appl. Phys.* 55: 090301-1-090301-3.
31. Kato, T., Okada, G., and Yanagida, T. (2016). Optical, scintillation and dosimeter properties of MgO transparent ceramic and single crystal. *Ceram. Int.* 42: 5617–5622.
32. Futami, Y., Yanagida, T., and Fujimoto, Y. (2014). Optical, dosimetric, and scintillation properties of pure sapphire crystals. *Jpn. J. Appl. Phys* 53: 02BC12.
33. Haas, J.T.M.d. and Dorenbos, P. (2008). Advances in yield calibration of scintillators. *IEEE Trans. Nucl. Sci.* 55: 1086–1092.
34. Masai, H., Yamada, Y., Okumura, S. et al. (2015). Photoluminescence of monovalent indium centres in phosphate glass. *Sci. Rep.* 5: 13646.
35. Masai, H., Yanagida, T., Fujimoto, Y. et al. (2012). Scintillation property of rare earth-free SnO-doped oxide glass. *Appl. Phys. Lett.* 101: 191906.
36. Kato, T., Okada, G., and Yanagida, T. (2016). Optical, scintillation and dosimeter properties of MgO transparent ceramic doped with Mn^{2+}. *J. Ceram. Soc. Jpn.* 124: 559–563.
37. Kato, T., Okada, G., and Yanagida, T. (2016). Optical, scintillation and dosimeter properties of MgO translucent ceramic doped with Cr^{3+}. *Opt. Mater.* 54: 134–138.
38. Grabmaier, B.C., Rossner, W., Berthold, T. et al. (eds.) (1996). *Inorganic Scintillators and their Application*, 29–35. Delft University Press.
39. Seferis, I., Michail, C., Valais, I. et al. (2014). Light emission efficiency and imaging performance of Lu_2O_3:Eu nanophosphor under X-ray radiography conditions: comparison with Gd_2O_2S:Eu. *J. Lumin.* 151: 229–234.
40. Melcher, C.L. and Schweitzer, J.S. (1992). Cerium-doped lutetium oxyorthosilicate – a fast, efficient new scintillator. *IEEE Trans. Nucl. Sci.* 39: 502–505.
41. Pidol, L., Kahn-Harari, A., Viana, B. et al. (2004). High efficiency of lutetium silicate scintillators, Ce-doped LPS, and LYSO crystals. *IEEE Trans. Nucl. Sci.* 51: 1084, 1087.
42. Kamada, K., Endo, T., Tsutumi, K. et al. (2011). Composition engineering in cerium-doped $(Lu,Gd)_3(Ga,Al)_5O_{12}$ single-crystal scintillators. *Cryst. Growth Des.* 11: 4484–4490.
43. Yanagida, T., Itoh, T., Takahashi, H. et al. (2007). Improvement of ceramic YAG(Ce) scintillators to $(YGd)_3Al_5O_{12}$(Ce) for gamma-ray detectors. *Nucl. Instrum. Methods Phys. Res. A.* 579: 23–26.
44. Baryshevsky, V.G., Korzhik, M.V., Moroz, V.I. et al. (1991). $YAlO_3$ – Ce-fast-acting scintillators for detection of ionizing-radiation. *Nucl. Instrum. Methods A* 58: 291–293.
45. van Loef, E.V.D., Dorenbos, P., van Eijk, C.W.E. et al. (2002). Scintillation properties of $LaBr_3$: Ce^{3+} crystals: fast, efficient and high-energy-resolution scintillators. *Nucl. Instrum. Methods A* 486: 254–258.
46. Combes, C.M., Dorenbos, P., van Eijk, C.W.E. et al. (1999). Optical and scintillation properties of pure and Ce^{3+}-doped Cs_2LiYCl_6 and Li_3YCl_6: Ce^{3+} crystals. *J. Lumin.* 82: 299–305.
47. Ogino, H., Yoshikawa, A., and Nikl, M. (2006). Scintillation characteristics of Pr-doped $Lu_3Al_5O_{12}$ single crystals. *J. Cryst. Growth* 292: 239–242.
48. Yanagida, T., Watanabe, K., Okada, G. et al. (2018). Optical, scintillation and radiation tolerance properties of Pr-doped pyrosilicate crystals. *Jpn. J. Appl. Phys.* 57: 106401.
49. Yanagida, T., Kamada, K., Fujimoto, Y. et al. (2010). Growth and scintillation properties of Pr doped YAP with different Pr concentrations. *Nucl. Instrum. Methods A* 623: 1020–1023.
50. Dorenbos, P., van Eijk, C.W.E., Hollander, R.W. et al. (1990). Scintillation properties of Nd^{3+} doped LaF_3 crystals. *IEEE Trans. Nucl. Sci.* 37: 119–123.
51. Sturm, B.W., Cherepy N.J., Drury, O.B. et al. (2011) Characteristics of undoped and europium-doped SrI_2 scintillator detectors. 2011 IEEE Nuclear Science Symposium and Medical Imaging Conference (SS/MIC 2011), 7
52. Shimizu, Y., Minowa, M., Suganuma, W. et al. (2006). Dark matter search experiment with CaF2(Eu) scintillator at Kamioka Observatory. *Phys. Lett. B.* 633: 195–200.
53. Murray, R.B. (1958). Use of Li^6I(Eu) as a scintillation detector and spectrometer for fast neutrons. *Nucl. Instrum. Methods* 2: 237–248.
54. Yanagida, T., Kawaguchi, N., and Fujimoto, Y. (2011). Basic study of Europium doped $LiCaAlF_6$ scintillator and its capability for thermal neutron imaging application. *Opt. Mater.* 33: 1243–1247.

55. Okada, G., Ueda, J., Tanabe, S. et al. (2014). Samarium-doped oxyfluoride glass-ceramic as a new fast erasable dosimetric detector material for microbeam radiation cancer therapy applications at the Canadian synchrotron. *J. Am. Ceram. Soc.* 97: 2147–2153.
56. Robbins, D.J. (1980). On predicting the maximum efficiency of phosphor systems excited by ionizing radiation. *J. Electrochem. Soc.* 127: 2694–2702.
57. Lempicki, A. and Wojtowicz, A.J. (1994). Fundamental limitations of scintillators. *J. Lumin.* 60 and 61: 942–947.
58. Mikhailik, V.B. and Kraus, H. (2010). Performance of scintillation materials at cryogenic temperatures. *Phys. Status Solidi B* 247: 1583–1599.
59. Dorenbos, P. (2002). Light output and energy resolution of Ce^{3+}-doped scintillators. *Nucl. Instrum. Methods Phys. Res. A* 486: 208–213.
60. Alekhin, M.S., de Haas, J.T.M., Khodyuk, I.V. et al. (2013). Improvement of ⊠-ray energy resolution of $LaBr_3$:Ce^{3+} scintillation detectors by Sr^{2+} and Ca^{2+} co-doping. *Appl. Phys. Lett.* 102: 161915.
61. Stand, L., Zhuravleva, M., Lindsey, A. et al. (2013) Potassium Strontium Iodide: A New High Light Yield Scintillator with 2.4% Energy Resolution, 2013 IEEE Nuclear Science Symposium and Medical Imaging Conference (2013 NSS/MIC), conference record
62. Moszyński, M., Zalipska, J., Balcerzyk, M. et al. (2002). Intrinsic energy resolution of NaI(Tl). *Nucl. Instrum. Methods Phys. Res. A.* 484: 459–469.
63. van Loef, E.V.D., Dorenbos, P., Kramer, K. et al. (2001). Scintillation properties of $LaCl_3$: Ce^{3+} crystals: fast, efficient, and high-energy resolution scintillators. *IEEE Trans. Nucl. Sci.* 48: 341–345.
64. Kelley, G.G., Bell, P.R., Davis, R.C. et al. (1956). Intrinsic scintillator resolution. *IRE Trans. Nucl. Sci.* 3: 57–58.
65. Syntfeld-Kazuch, A., Swiderski, L., Czarnacki, W. et al. (2006). Non-proportionality and Energy Resolution of CsI(Tl). *IEEE Nuclear Science Symposium Conference Record* 2006: N30–N134.
66. Chewpraditkula, W. and Moszynski, M. (2011). Scintillation properties of $Lu_3Al_5O_{12}$, Lu_2SiO_5 and $LaBr_3$ crystals activated with cerium. *Phys. Proc.* 22: 218–226.
67. Uchiyama, Y., Kouda, M., Tanihata, C. et al. (2001). Study of energy response of Gd_2SiO_5:Ce^{3+} scintillator for the ASTRO-E hard X-ray detector. *IEEE Trans. Nucl. Sci.* 48: 379–384.
68. Dorenbos, P. (2010). Fundamental limitations in the performance of Ce^{3+}-, Pr^{3+}-, and Eu^{2+}-activated scintillators. *IEEE Trans. Nucl. Sci.* 57: 1162–1167.
69. Gundacker, S., Turtos, R.M., Kratochwil, N. et al. (2020). Experimental time resolution limits of modern SiPMs and TOF-PET detectors exploring different scintillators and Cherenkov emission. *Phys. Med. Biol.* 65: 1–20.
70. Matsuzawa, T., Aoki, Y., Takeuchi, T. et al. (1996). A new long phosphorescent phosphor with high brightness, $SrAl_2O_4$:Eu^{2+},Dy^{3+}. *J. Electrochem. Soc.* 143: 2670–2673.
71. Nesshi-Tedaldi, F., Dissertori, G., Lecomte, P. et al. (2009). Studies of Cerium floride, LYSO and lead tungstate crystals expose to high hadron fluences, IEEE NSS MIC 2009 Conference record N32-3
72. Grescovich, G., Cusano, D., Hoffman, D. et al. (1992). Ceramic scintillators for advanced, medical X-ray-detectors. *Am. Ceram. Soc. Bull.* 71: 1120–1130.
73. Tanaka, M., Hara, K., Kim, S. et al. (2002). Applications of cerium-doped gadolinium silicate Gd2SiO5:Ce scintillator to calorimeters in high-radiation environment. *Nucl. Instrum. Methods Phys. Res. A.* 404: 283–294.
74. Kawade, K., Fukatsu, K., Itow, Y. et al. (2011). Study of radiation hardness of Gd_2SiO_5 scintillator for heavy ion beam. *JINST* 6: 1–12.
75. Yanagida, T., Fujimoto, Y., and Watanabe, K. (2014). Dopant concentration ependence on radiation induced positive hysteresis of Ce:GSO and Ce:GSOZ. *Radiat. Meas.* 61C: 16–20.
76. Yanagida, T., Fujimoto, Y., Koshimizu, M. et al. (2014). Positive hysteresis of Ce-doped GAGG scintillator. *Opt. Mater.* 36: 2016–2019.
77. Yanagida, T., Fujimoto, Y., Yamaji, A. et al. (2013). Study of the correlation of scintillation decay and emission wavelength. *Radiat. Meas.* 55: 99–102.
78. Melcher, C.L., Manente, R.A., and Schweitzer, J.S. (1989). Applicability of barium fluoride and cadmium tungstate scintillators for well logging. *IEEE Trans. Nucl. Sci.* 36: 1188–1192.
79. Kitis, G., Gomez-Ros, J.M., and Tuyn, J.W.N. (1998). Thermoluminescence glow-curve deconvolution functions for first, second and general orders of kinetics. *J. Phys. D Appl. Phys.* 31: 2636–2641.
80. Chen, R. and Winer, S.A.A. (1970). Effects of various heating rates on glow curves. *J. Appl. Phys.* 41: 5227–5232.
81. Bos, A.J.J. (2007). Theory of thermoluminescence. *Radiat. Meas.* 41: S45–S56.

82. Yanagida, T., Okada, G., and Kawaguchi, N. (2019). Ionizing-radiation-induced storage-luminescence for dosimetric applications. *J. Lumin.* 207: 14–21.
83. Grimmeis, H.G. and Ledebo, L.-A. (1974). Photo-ionization of deep impurity levels in semiconductors with non-parabolic bands. *J. Phys. C Sol. Stat. Phys.* 8: 2615–2626.
84. Grimmeis, H.G. and Ledebo, L.-A. (1975). Spectral distribution of photoionization cross sections by photoconductivity measurements. *J. Appl. Phys.* 46: 2155.
85. Lucovsky, G. (1965). On the photoionization of deep impurity centers in semiconductors. *Solid State Commun.* 3: 299–302.
86. Yanagida, T., Fujimoto, Y., Watanabe, K. et al. (2014). Scintillation and optical stimulated luminescence of Ce doped CaF_2. *Radiat. Meas.* 71: 162–165.
87. Yanagida, T. (2016). Ionizing radiation induced emission: scintillation and storage-type luminescence. *J. Lumin.* 169: 544–548.
88. Yanagida, T. (2018). Inorganic scintillating materials and scintillation detectors. *Proc. Japan Acad. B.* 94: 75–97.
89. Yanagida, T., Kamada, K., Fujimoto, Y. et al. (2011). Scintillation properties of transparent ceramic and single crystalline Nd:YAG scintillators. *Nucl. Instrum. Methods A* 631: 54–57.
90. Kawaguchi, N., Yanagida, T., Fujimoto, Y. et al. (2013). Neutron detection with $LiCaAlF_6$ scintillator doped with 3d-transition metal ions. *Radiat. Meas.* 55: 128–131.
91. Blankespoor, S.C., Derenzo, S.E., Moses, W.W. et al. (1994). Characterization of a pulsed X-ray source for fluorescent lifetime measurements. *IEEE Trans. Nucl. Sci.* 41: 698–702.
92. Derenzo, S.E., Weber, M.J., Moses, W.W. et al. (2000). Measurements of the intrinsic rise times of common inorganic scintillators. *IEEE Trans. Nucl. Sci. IEEE* 47: 860–864.
93. Yanagida, T., Fujimoto, Y., Yoshikawa, A. et al. (2010). Development and performance test of picosecond pulse X-ray excited streak camera system for scintillator characterization. *Appl. Phys. Express* 2: 056202.
94. Yanagida, T., Fujimoto, Y., Ito, T. et al. (2014). Development of X-ray induced afterglow characterization system. *Appl. Phys. Express* 7: 062401.
95. Wróbel, D., Bilski, P., Marczewska, B. et al. (2015). Characterization of the Risø TL/OSL DA-20 reader for application in TL dosimetry. *Radiat. Meas.* 74: 1–5.
96. Bos, A.J.J., Winkelman, A.J.M., Le Masson, N.J.M. et al. (2002). A TL/OSL emission spectrometer extension of the Risø reader, Radiat. Prot. *Dosimetry* 101: 111–114.

2

Organic Scintillators

Masanori Koshimizu
Graduate School of Engineering, Tohoku University, Sendai, Japan

2.1 Introduction

In this chapter, an overview of organic scintillators will be presented. In contrast to inorganic scintillators, the choice of elements is limited in organic scintillators. From the viewpoint of high-energy photon detection, the low effective atomic number is generally a shortcoming of organic scintillators, because they are mainly composed of elements with low atomic numbers, such as carbon, hydrogen, oxygen, nitrogen, etc. Nevertheless, organic scintillators have various advantages, such as the use of scintillators in a liquid phase enables the detection of α- or β-emitting radioactive isotopes dissolved in the liquid scintillators. Additionally, large-scale detector systems can easily be constructed on the basis of liquid scintillators. Most of organic scintillators exhibit fast scintillation decay, which enables their use in the applications in which excellent time resolution or high counting-rate detection capability are necessary. Plastic scintillators have high workability for their use in complicated shapes.

One of their important differences from inorganic scintillators is the occurrence of radiation-induced ionization of molecules. Ionization of the constituent molecules in organic scintillators often results in a chemical reaction of radical species. Such chemical reactions are one of the causes of relatively low scintillation light yields in organic scintillators. Also, such chemical reactions result in degradation of organic scintillators, which results in a vulnerability of organic scintillators to high radiation doses.

Another peculiarity of organic scintillators, or organic phosphors, is that they exhibit negligible storage-type luminescence, such as thermally or optically stimulated luminescence (TSL or OSL), which can be used for dosimetry. One of the possible reasons is considered to be the difficulty in creating trapping sites for electrons and holes: metastable trapping of

Phosphors for Radiation Detectors, First Edition. Edited by Takayuki Yanagida and Masanori Koshimizu.

electrons or holes (or metastable charge separation) at or above room temperature continues to be a challenge in organic phosphors [1].

The content of this chapter is as follows: the basic electronic processes responsible for scintillation in organic scintillators are described in Section 2.1. The electronic states in organic molecules are described in a different manner from that of inorganic solids. Also, energy transfer processes are introduced in this section: two representative energy transfer processes will be discussed. Some effect of ionization and triplet excited states are also introduced in relation with linear energy transfer (LET) effects. In Section 2.2, various liquid scintillators are introduced. The applications and the use of the liquid scintillators will be discussed. Organic crystal scintillators are introduced in Section 2.3. Plastic scintillators, which comprise of host organic polymers and fluorescence molecules, are introduced in Section 2.4. In Section 2.5, organic–inorganic hybrid scintillators, in which the scintillation originates mainly from the organic molecules, will be introduced. Plastic or liquid organic scintillators added or "loaded" with organometallic compounds will be introduced, and recent development of nanocomposite scintillators as plastic scintillators added with inorganic nanoparticles will be discussed.

2.2 Basic Electronic Processes in Organic Scintillators

Energy deposition by impinged radiation via ionization or electronic excitation, excitation energy transfer from host to guest molecules, and radiative transitions are described in this section. Understanding of these processes is important in the discussion of scintillation spectra, light yields, and temporal behavior. Firstly, electronic states in organic molecules are described by a quantum mechanical approach, which provides the basis of the scintillation processes in organic scintillators. Relaxation processes of excited states of organic molecules are also introduced. Secondly, energy transfer processes between organic molecules are discussed. These processes are important in liquid organic scintillators and plastic scintillators, in which excitation energy transfers from the host solvents or polymers to guest molecules. Finally, scintillation dynamics at high LETs will be discussed in relation to LET-dependent scintillation time profiles, which are useful for discrimination events of different radiations.

Most of the optical responses of organic scintillators is related to π electrons of organic molecules, which are involved in π bonds. They are formed in organic molecules via side-to-side overlap of 2p orbitals of neighboring carbon atoms. Among molecules having π electrons, π conjugated systems have mainly been used in organic scintillators. They are formally described as alternating single and multiple (in many cases, double) bonds, in which π electrons are delocalized. The delocalized π electrons are preferable from the viewpoint of optical responses, such as emission wavelengths appropriate to photon detectors or large oscillator strength which results in fast radiative decay.

2.2.1 Electronic States and Excited States Dynamics of Organic Molecules

Local electronic states in molecules or localized centers can be described in a configuration coordinate model in a framework of Born–Oppenheimer approximation: the energies of the electronic ground and excited states are described with parabolas having the same shape as adiabatic potential curves with the transverse axis of configuration coordinate, Q. The configuration coordinate is taken as the deformation of the molecules or the lattice structure

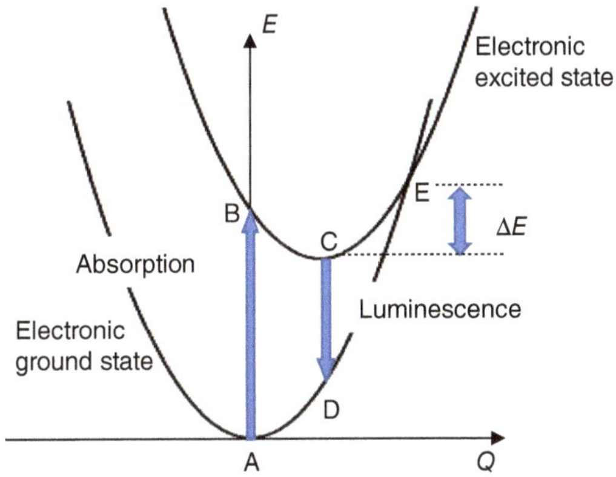

Figure 2.1 *Configuration coordinate model of localized electronic system.*

around the localized center along a particular normal vibration mode. The energy structure of the ground and the lowest excited states described in the configuration coordinate model is shown in Figure 2.1. The minimum of the excited state is displaced from that of the ground state, which indicates that the most stable molecular structure is different for the ground and excited states.

Electronic transitions are treated in the configuration coordinate model as vertical transitions, i.e., between electronic states with the same configuration coordinate Q, as shown in Figure 2.1. This is because the speed of the motion of electrons is significantly different to those of atoms or ions owing to their large difference in their masses: the motion of the atoms or ions can be treated as frozen during the electronic transitions. The electronic ground state at $T \rightarrow 0$ is at the origin of the configuration coordinate (point A) in Figure 2.1. Electronic excitation occurs from A to B as the vertical transition upon absorption of a photon. As for the excited state relaxation, we have to consider the competition between direct transition from B to A and relaxation within the electronic excited state, i.e. from B to C, the latter of which is generally much faster. Most of the excited electrons experience the relaxation to C, which is the minimum of the excited state, and subsequent electronic excitation to D via vertical transition, which corresponds to luminescence. In the case of finite temperature, the electronic ground and excited states are thermally populated around the minima at points A and C, which results in broad absorption and luminescence bands. Furthermore, the thermal population around C results in quenching: some electrons are populated at the crossing point of the parabolas of the electronic ground and excited states (point E) and experience the relaxation from E to A in the electronic ground state. In this case, the electrons undergo relaxation from the electronic excited state to the ground state without emission of a photon, i.e., thermal quenching. For the energy difference ΔE between E and C, the thermal quenching rate is written as $s \cdot \exp(-\Delta E/kT)$, where s is a frequency factor. Assuming that the radiative transition rate, A, does not depend on temperature, we can obtain the quantum efficiency of the emission, η, and the emission lifetime, τ, as functions of temperature as follows:

$$\eta = \frac{A}{A + s\exp\left(-\Delta E / kT\right)} \tag{2.1}$$

$$\tau = \frac{1}{A + s\exp(-\Delta E / kT)} \tag{2.2}$$

The adiabatic potential curves shown in Figure 2.1 assume the continuous potential energy. In some cases, where a small number of vibrational modes are involved, quantization of the vibration energy results in discrete energy levels for the electronic ground and excited states, as schematically shown in Figure 2.2. In this case, electronic transitions occur between the discrete levels of the electronic ground and excited states, which results in absorption or luminescence spectra having multiple peaks. Such structure of the absorption or luminescence spectra is often observed for organic molecules. The electronic ground state at $T \rightarrow 0$ is at the lowest level. Upon absorption of a photon, electrons are excited from the lowest level of the electronic ground state to the upper levels as well as the lowest level of the electronic excited state, as shown in Figure 2.2. The electronic transitions result in an absorption spectrum having multiple peaks corresponding to the vibrational levels of the electronic excited state. For the relaxation process from the upper levels of the electronic excited state, electronic transitions to the levels of the electronic ground state compete with the vibrational energy loss to the lowest level of the electronic excited state, the latter of which is generally much faster and known as internal conversion. In other words, the relaxation process of the excitation mostly occurs from the upper levels of the electronic excited state initially to the lowest level of the excited state and subsequently to one of the vibrational levels of the electronic ground state. Luminescent photons are emitted during the electronic transitions in the latter process. The luminescence spectrum is also composed of multiple peaks corresponding to the vibrational levels of the electronic ground state. At finite temperature, the upper levels are thermally populated in the electronic ground and excited states; however, at room temperature, the population is negligible in the case of organic molecules owing to their large vibrational energy.

Some excited states undergo a change in the spin state: from singlet excited states just after excitation to triplet excited states. This process is called intersystem crossing and is schematically depicted in Figure 2.3. Radiative transitions to the ground state, which is a spin singlet

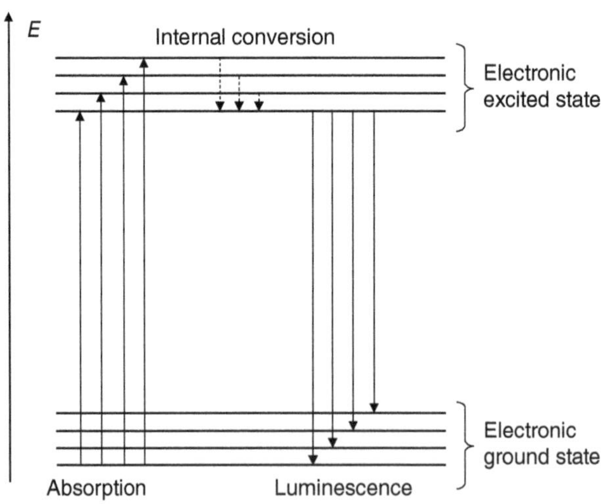

Figure 2.2 *Vibrational energy levels of organic molecules.*

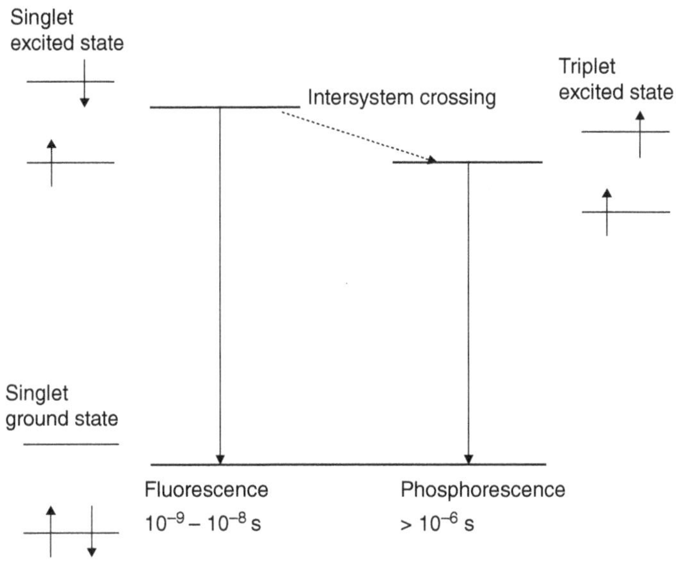

Figure 2.3 *Schematic diagram of singlet and triplet excited states and fluorescence and phosphorescence processes.*

state, from the singlet and triplet excited states are called fluorescence and phosphorescence, respectively. The time scale of the former spin-allowed process is 10^{-9} to 10^{-8} seconds, whereas the latter spin-forbidden process needs much longer, typically more than 10^{-6} seconds. In organic scintillators, triplet excited states do not contribute to the fast scintillation component, and contribute little to the total scintillation light yield. Their contribution is mainly in the slow component of approximately 10^{-6} seconds. The scintillation process in which the triplet excited states are involved is discussed in Sections 2.1.2 and 2.1.3.

2.2.2 Excitation Energy Transfer

In this subsection, we will consider the following energy transfer process: one molecule (or moiety in a macromolecule) in its excited state loses its energy and turns into the lower electronic state, and another molecule (or moiety) receives the lost energy in its excited state. The former and the latter are called a donor and an acceptor, respectively. Excitation energy transfer processes can roughly be divided into two: radiative and nonradiative processes. In the former process, a donor emits a photon via radiative transition from its excited state to a lower state (typically ground state), and the emitted photon is absorbed by an acceptor, which results in the acceptor in its excited state. On the contrary, in the nonradiative process, the same final state occurs from the same initial state as the radiative process without emission and absorption of a photon. The two energy transfer processes are schematically depicted in Figure 2.4.

The two energy transfer processes can clearly be distinguished with photoluminescence time profiles after pulsed excitation of donors: the kinetics of the energy transfer and photoluminescence are now discussed. The photoluminescence intensities of donors and acceptors as a function of time after excitation, $I_D(t)$ and $I_A(t)$, are proportional to the numbers of excited donors and acceptors, $N_D(t)$ and $N_A(t)$, respectively. We assume that only the donors are instantaneously excited at $t = 0$ in a system with both donors and acceptors, i.e., $N_D(0) = N_0$ and $N_A(0) = 0$.

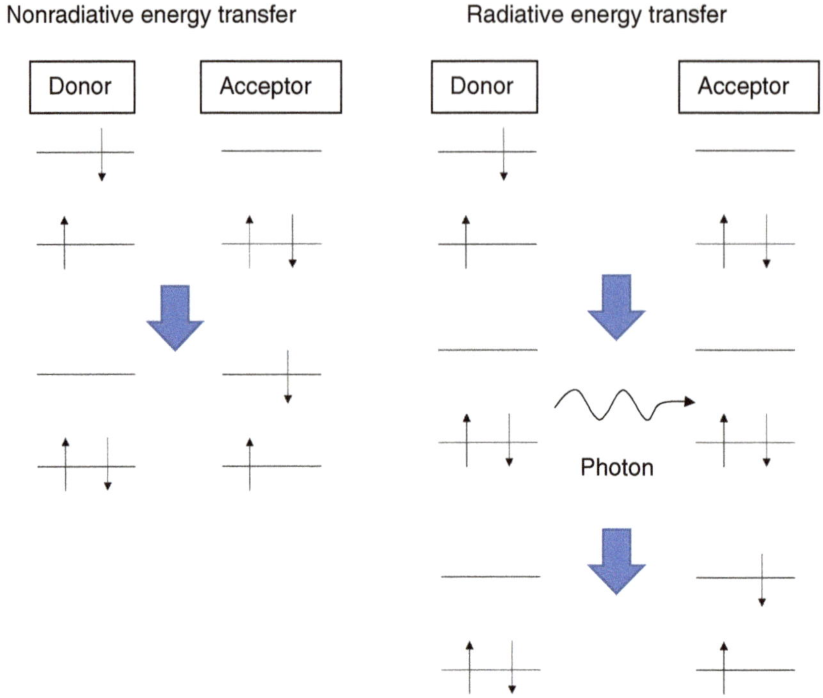

Figure 2.4 *Schematic representation of nonradiative and radiative energy transfer processes.*

Here, let us consider a simple case of nonradiative energy transfer with a single rate constant, k_{ET}. Assuming that the donors in their excited state exhibit photoluminescence with a single decay time constant, τ_D, with the acceptors, the differential equation for the number of the excited donors is

$$\frac{dN_D(t)}{dt} = -\left(\tau_D^{-1} + k_{ET}\right)N_D(t) \tag{2.3}$$

With $N_D(0) = N_0$, we can obtain the number of the excited donors as a function of time after excitation:

$$N_D(t) = N_0 \exp\left[-\left(\tau_D^{-1} + k_{ET}\right)t\right] \tag{2.4}$$

which results in

$$I_D(t) \propto \exp\left[-\left(\tau_D^{-1} + k_{ET}\right)t\right] \tag{2.5}$$

However, without the acceptors, the photoluminescence intensity of donors as a function of time after excitation is given by

$$I_D(t) \propto \exp\left(-t/\tau_D\right) \tag{2.6}$$

Comparing Equations (2.5) and (2.6), it is clearly indicated that the nonradiative energy transfer shortens the photoluminescence decay of the donors. Based on this simple competition model, the ratio of the numbers of donors experiencing the decay (return to the ground state) within the donors or the energy transfer is written with $\tau_D^{-1}:k_{ET}$, which results in an energy transfer efficiency of $\frac{k_{ET}}{\tau_D^{-1}+k_{ET}}$. If the photoluminescence decay of the acceptors is significantly slower than the energy transfer process, a slow rise corresponding to the energy transfer is observed in the photoluminescence time profile of the acceptors.

Contrary to the case of the nonradiative energy transfer, the photoluminescence decay kinetics of the donors are not altered by radiative energy transfer, because the probability of the absorption of the photons emitted by the donors, f, does not depend on time. Assuming that the acceptors in their excited state exhibit photoluminescence with a single decay time constant, τ_A, without donors, the differential equation for the number of the excited acceptors is

$$\frac{dN_A(t)}{dt} = f\tau_D^{-1}N_D(t) - \tau_A^{-1}N_A(t) = f\tau_D^{-1}N_0\exp(-t/\tau_D) - \tau_A^{-1}N_A(t) \tag{2.7}$$

Unfortunately, this differential equation cannot be solved analytically. The first term on the right-hand side of the equation corresponds to the excitation of the acceptors by absorption of the photons emitted by the donors. If τ_D is comparable or larger than τ_A, the decrease in the number of the excited acceptors is delayed, which results in an elongated decay of the acceptor photoluminescence. The absorption probability f is determined by the optical density, at the emission wavelength, from the donor emitted the photon to the surface of the sample. The radiative energy transfer in scintillators is important because they are commonly used as a bulk form with typical thickness of >1 mm and thus optical path length of the photons emitted by donors is relatively long. There are two important processes in scintillators in relation with the radiative energy transfer: self-absorption and wavelength shift. The former is the absorption of the scintillation photons by the scintillator itself. This process occurs for the emission and the absorption by the same molecules (moieties, or luminescent centers), as well as for the case of different molecules for the donors and acceptors. Because the quantum efficiency of the fluorescent molecules (moieties, or luminescent centers) is generally less than unity, the self-absorption results in a reduction in the scintillation light yield and an elongation of the scintillation decay. The self-absorption is frequently a problem for organic scintillators, since the Stokes shift of the fluorescent molecules used in many organic scintillators is relatively small. The other process is the wavelength shift, which is often used in liquid organic and plastic scintillators. Two kinds of fluorescence molecules are dissolved in solvents of liquid organic scintillators or added to host polymers of plastic scintillators: emission of primary fluorescent molecules is absorbed by secondary fluorescent molecules, and subsequently the secondary fluorescent molecules emit photons. Considering the Stokes shift of the primary and secondary molecules, this radiative energy transfer process shifts the emission toward the longer wavelength side. The shift in the emission wavelength is effective in avoiding a reduction in the scintillation light yield owing to the self-absorption by the primary fluorescence molecules: the addition of the secondary fluorescence molecules at low concentration results in low f. The wavelength shift is also used to match the emission wavelength with the sensitive wavelength region of photodetectors such as photomultiplier tubes.

In the following, the nonradiative energy transfer process is discussed in a quantum mechanical treatment [2]. Let us consider a closely located donor-acceptor pair in vacuum. We consider the electronic energy levels depicted in Figure 2.5. Because HOMO and LUMO of the

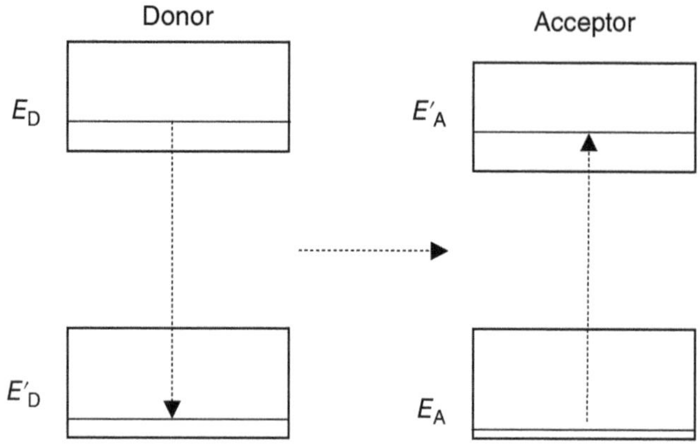

Figure 2.5 *Schematic representation of HOMO and LUMO bands of a donor and an acceptor.*

organic molecules comprise many vibrational levels, they are described as a broad band. Initially, a donor and an acceptor are assumed to be in their electronic excited and ground states, respectively. The wavefunction of the system in the initial state can be written as

$$\Psi_1 = \varphi'_D \varphi_A \Phi'_D(E'_D)\Phi_A(E_A) \tag{2.8}$$

where $\varphi'_{D\text{ or }A}$ denotes the electronic wavefunction of the donor or acceptor in their electronic excited states, and $\varphi_{D\text{ or }A}$ expresses the electronic wavefunction of the donor or acceptor in their electronic ground states. Likewise, $\Phi'_{D\text{ or }A}$ is the vibrational wavefunction of the donor or acceptor in their electronic excited states, and $\Phi_{D\text{ or }A}$ is the vibrational wavefunction of the donor or acceptor in their electronic ground states. After the nonradiative energy transfer, the donor and the acceptor are assumed to be in their electronic ground and excited states, respectively. The wavefunction of the system in the final state can be written as

$$\Psi_2 = \varphi_D \varphi'_A \Phi_D(E_D)\Phi'_A(E'_A) \tag{2.9}$$

Based on time-dependent perturbation theory of first order, the transition probability from the initial to the final state can be described with Fermi's golden rule:

$$dW = \frac{2\pi}{\hbar}|V_{12}|^2 \delta(\Delta E) \tag{2.10}$$

where $\hbar$ is Planck's constant and δ is the Dirac delta function. $\Delta E = E'_D - E_D - (E'_A - E_A)$ is the change in the total energy of the system during the energy transfer. The transition probability is nonzero only for $\Delta E = 0$, which means that transition satisfying energy conservation only occurs. $V_{12} = \langle \Psi_2 | \hat{V} | \Psi_1 \rangle$ is the transition matrix element, and $\hat{V}$ is the interaction operator.

Here we assume that the nonradiative energy transfer occurs via dipole–dipole interaction. This energy transfer process is called the Förster transfer and the most important one in many applications including scintillators. In this case, the transition matrix element is written as

$$\langle \Psi_1 | \hat{V} | \Psi_2 \rangle = M_{DA} S_D(E'_D, E_D) S_A(E'_A, E_A) \tag{2.11}$$

where $S_D(E'_D, E_D) = \langle \Phi'_D(E'_D) | \Phi_D(E_D) \rangle$ and $S_A(E'_A, E_A) = \langle \Phi'_A(E'_A) | \Phi_A(E_A) \rangle$ are the overlapping integrals of vibrational wavefunctions of the donor and the acceptor, respectively.

$M_{DA} = \chi |\boldsymbol{p}_D||\boldsymbol{p}_A|/R^3$, and R is the distance between the donor and the acceptor. $|\boldsymbol{p}_D|$ and $|\boldsymbol{p}_A|$ are the matrix element of the transition dipole of the donor and the acceptor, respectively. χ is a factor related to the relative direction of dipoles and the direction of the donor-acceptor pair. Now we consider the energy distribution in the initial state: $g'(E'_D)$ as the energy distribution of the donor in its excited state and $g(E_A)$ as the energy distribution of the acceptor in its ground state. The transition probability can be obtained via integration of Equation (2.10) over E'_D and E_A as

$$W = \frac{2\pi}{\hbar} \int M_{DA}^2 \left[\int g'(E'_D) S_D^2 (E'_D, E'_D - E) dE'_D \int g(E_A) S_A^2 (E_A, E_A - E) dE_A \right] dE \tag{2.12}$$

The variable of integration is changed and the equation is reorganized to be

$$W = \frac{2\pi\chi^2}{R^6} \int \left[p_D^2 \int g'(E'_D) S_D^2 (E'_D, E'_D - E) dE'_D \right] \left[p_A^2 \int g(E_A) S_A^2 (E_A, E_A - E) dE_A \right] d\omega \tag{2.13}$$

where the first and second factors of the integrand correspond to the emission spectral shape of the donor, $G_D(\omega)$, and the absorption spectral shape of the acceptor, $G_D(\omega)$, respectively, as follows:

$$\int g'(E'_D) S_D^2 (E'_D, E'_D - E) dE'_D = \frac{1}{\hbar} G_D(\omega) \tag{2.14}$$

$$\int g(E_A) S_A^2 (E_A, E_A - E) dE_A = \frac{1}{\hbar} G_A(\omega) \tag{2.15}$$

Here we consider the normalized spectral shape, i.e., $\int F_D(\omega) d\omega = 1$:

$$F_D(\omega) = \frac{4\omega^3}{3\hbar c^3} p_D^2 \tau_{R\_D} G_D(\omega) \tag{2.16}$$

where τ_{R_D} is the radiative lifetime of the excited donors and c is the velocity of light. Also, the absorption spectral shape of the donor is related with absorption coefficient of the acceptor, $\mu_A(\omega)$, as follows:

$$\mu_A(\omega) = \frac{4\pi^2\omega}{3\hbar c} N_A p_A^2 G_A(\omega) \tag{2.17}$$

where N_A is the concentration of the acceptor. Finally, we obtain

$$W = \frac{9\chi^2 c^4}{8\pi N_A \tau_{R\_D} R^6} \int F_D(\omega) \mu_A(\omega) \frac{d\omega}{\omega^4} \tag{2.18}$$

as the transition probability in a vacuum. In a medium of refractive index of n, the probability is

$$W = \frac{9\chi^2 n^4 c^4}{8\pi N_A \tau_{R\_D} R^6} \int F_D(\omega) \mu_A(\omega) \frac{d\omega}{\omega^4} \tag{2.19}$$

An important conclusion of this derivation is that the probability is inversely proportional to R^6, which indicates that the energy transfer occurs only for closely located donor-acceptor pairs. Another important conclusion is that the probability is proportional to the overlap integral of the normalized emission spectrum of the donor and the absorption coefficient spectrum of the acceptor. Based on the consideration of the kinetics, the energy transfer efficiency is 50% for the radius R_0:

$$R_0^6 = \frac{9\chi^2 n^4 c^4}{8\pi N_A} \int F_D(\omega) \mu_A(\omega) \frac{d\omega}{\omega^4} \tag{2.20}$$

R_0 is referred to as the Förster radius. A typical Förster radius for donor-acceptor pairs has a large overlap integral of the spectra is several nm.

In the case of forbidden dipole transition, other mechanisms make the largest contribution to energy transfer. One possible mechanism is a dipole–quadrupole interaction. In this case, the transition probability can be estimated in a similar manner as

$$W = \frac{135\pi\hbar c^8}{4\tau_D \tau_{R\_A} R^8} \int F_D(\omega) F_A(\omega) \frac{d\omega}{\omega^8} \tag{2.21}$$

where τ_{R_A} is the radiative lifetime of the acceptor and $F_A(\omega)$ is the normalized absorption spectrum of the acceptor. In this case, the probability is inversely proportional to R^8. Another possible mechanism is the exchange interaction, in which an excited electron of the donor is exchanged with a ground-state electron of the acceptor, as schematically shown in Figure 2.6. The transition probability can be expressed as

$$W = \frac{2\pi}{\hbar} Z^2 \int F_D(\omega) F_A(\omega) d\omega \tag{2.22}$$

where $Z^2 \propto \exp\left(-\frac{2R}{L}\right)$ represents the superposition of the wavefunctions of the excited electron of the donor and the ground-state electron of the acceptor, and L is the sum of the

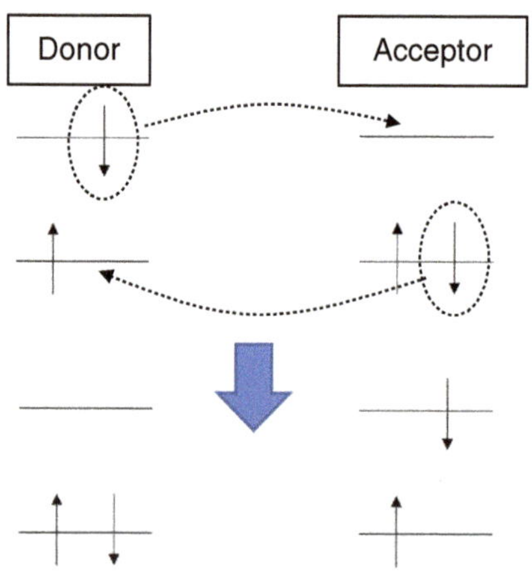

Figure 2.6 *Schematic representation of energy transfer via exchange interaction.*

radius of the donor and the acceptor (in spherical approximation). Also in this case, the energy transfer is effective only for closely-located donor–acceptor pairs.

Energy transfer process in organic scintillators has been analyzed since the 1950s. A simple kinetic consideration of the excitation energy transfer and competing quenching in organic scintillators was reported as early as 1952 [3]. It should be noted that the electronic processes are similar in the cases of liquid and solid scintillators; however, diffusion (or Brownian motion) of the molecules in liquid scintillators contribute to excitation energy migration in solvents, whereas it makes no contribution to solid scintillators. A set of rate equations of radiative or nonradiative energy transfer, quenching, and fluorescence have been proposed [4]. Subsequently, excimer formation and diffusion of molecules are taken into account [5]. At high solute concentration, nonradiative energy transfer from the solvent to the solute before internal conversion has also been reported [6]. Also, such a model was extended to plastic and organic crystalline scintillators to elucidate their pulse shape (scintillation time profile) [7]. From the viewpoint of experimental analysis of the energy transfer process, solvents of molecules having π electrons were diluted with aliphatic molecules such as n-heptane [5], cyclohexane [8–13], or paraffine [8–11, 13] to suppress the excited state migration among the solvent molecules or sometimes enhance the viscosity to suppress the diffusion motion of the solvent molecules. Also, some quenchers, such as bromobenzene [12–16], aniline [17, 18], or CCl_4 [18, 19], were added to scintillators to analyze the competition between the energy transfer and quenching. In addition, the effects of temperature on the energy transfer process were also analyzed [8–17]. Furthermore, the energy transfer process from primary to secondary fluorescence molecules was analyzed [20].

Contrary to the case of organic liquid scintillators, whose energy transfer processes have attracted considerable attention from the viewpoint of physical chemistry of excited states, energy transfer processes in plastic scintillators have been analyzed with an emphasis on the resulting scintillation properties, such as light yield or detection signal shape resembling the scintillation time profiles. Some researchers have proposed an energy transfer model, which is similar to those of organic liquid scintillators apart from the diffusion of solvent molecules [21]. The scintillation time profiles of plastic scintillators have been analyzed in terms of energy transfer process, because the plastic scintillators have been used for applications in which fast response of scintillation detectors is necessary [22–25]. It has been reported that excimer formation in the host has an effect on the scintillation properties [26].

In the case of liquid organic and plastic scintillators, they are composed of solvents or host polymers with at least one kind of fluorescent molecule because the scintillation efficiency of the solvents or host polymers is low: efficient energy transfer from the solvents or host polymers to the fluorescent molecule is necessary. Typical concentration of the fluorescent molecule or "primary" molecule is from less than 1 wt% up to several wt%, depending on the fluorescent molecule species. It should be noted that addition of primary fluorescent molecules at high concentrations sometimes results in concentration quenching [27]. Because the primary fluorescent molecules generally have a small Stokes shift, such binary organic scintillators may suffer the loss of scintillation photons via self-absorption when the volume of the scintillators is large. The self-absorption can be avoided with the addition of secondary fluorescent molecules with longer emission wavelengths at much lower concentration.

The contribution of triplet excited states has also been analyzed [28]. Slow components from the singlet excited state of fluorescence molecules in organic scintillators originate from two processes: one is the interaction of two molecules of their triplet state, which results in a molecule in the singlet excited state and the other in the singlet ground state. The fluorescent molecules in their triplet excited state are mainly formed via energy transfer from the solvent molecules in their triplet excited state. As for the other process, thermally activated delayed fluorescence, in which singlet excited state is formed via thermal activation from the triplet excited state within one molecule, has been recently applied to organic scintillators [29].

2.2.3 Scintillation Dynamics in Organic Scintillators at High Linear Energy Transfer

Linear Energy Transfer (LET) is the energy density deposited by an ionizing radiation in matter per unit length of its trajectory. High-energy electrons deposit their energy at low LET, and high-energy photons (X-rays or gamma rays) also have low LET, because the energy is finally deposited by high-energy electrons such as photoelectrons and secondary electrons. However, high-energy protons or heavier particles have high LET. A significant different in LET generally results in a difference in the scintillation light yield and the time profiles. The scintillation light yield dependent on LET was formalized by Birks as early as 1951 [30] and the formalization has long been used to account for the LET effect in spectroscopy. Also, the scintillation time profiles dependent on LET has a technological importance: the pulse shape discrimination (PSD) technique used for neutron or other high-LET particle detection in the background of gamma rays [31]. Because the detection signal shape of a scintillation detector resembles the scintillation time profile of the scintillator, a significant difference in the scintillation time profiles results in different shapes of detection signals. Hence, radiations having significantly different LETs can be discriminated on the basis of the difference in the shape of the detection signals by using a scintillator having LET-dependent scintillation time profiles. Neutrons are often detected via conversion to energetic protons or heavy charged particles, which have high LETs. Based on the high LETs of the reaction products of neutrons and scintillators having an LET-dependent scintillation time profile enables PSD of detection events of neutrons and gamma rays. PSD techniques and neutron detection in general will be introduced in Chapter 12. In this chapter, neutron detection using organic scintillators based on PSD technique will be introduced.

A neutron has no electric charge and cannot induce ionizations and electronic excitations by itself. To detect neutrons using scintillation detectors, they have to be converted into high-energy charged particles. Neutrons are often categorized into two groups according to their energies: thermal and fast neutrons, for which effective methods of the conversion to high-energy charged particles are different. Fast neutrons are detected via energy deposition to nuclei via elastic collisions, and subsequently the recoil nuclei as high-energy charged particles induce ionizations and excitations. Among the nuclei, the kinetic energy of fast neutrons is most effectively transferred to protons. Hence, scintillators having a large amount of hydrogen atoms are advantageous for fast neutron detection. Organic scintillators are widely used for fast neutron detection for this reason, and relatively high LET of the recoil protons enables PSD of the fast neutrons and gamma rays. However, thermal neutrons do not have sufficient kinetic energies to be transferred to nuclei to generate high-energy charged particles. Nuclear reactions of thermal neutrons with several nuclei, such as ^{6}Li, ^{10}B, or Gd isotopes, are used to detect thermal neutrons using scintillation detectors. The nuclear reactions with ^{6}Li or ^{10}B generates energetic α-rays and ^{3}H or ^{7}Li, respectively, whereas the Gd isotopes emit gamma rays upon absorption of thermal neutrons. In the case of scintillators having ^{6}Li or ^{10}B as the constituent, scintillation induced by high-LET charged particles is expected, which enables PSD of the thermal neutrons and gamma rays. Fast neutron detection using organic liquid, single crystalline, and plastic scintillators will be introduced in Sections 2.2, 2.3, and 2.4, respectively. Thermal neutron detection needs the inclusion of ^{6}Li, ^{10}B, or Gd isotopes and hence will be introduced in Section 2.5.

The LET-dependent scintillation time profiles are explained in terms of two different scintillation components having different susceptibilities to LET: one is the fast emission process without the intersystem crossing, which contributes to the fast decay component of scintillation. The other is the interaction between the fluorescent molecules of their triplet excited state, which results in a delayed emission from the fluorescent molecules of their singlet state, as was explained in the Subsection 2.2.2. Another fast process competing with internal conversion becomes important at high LET. This process is called ionization quenching, in which an excited molecule in its singlet states is lost owing to its ionization [32]. Two processes are proposed for the ionization quenching: one is that of a bimolecular reaction of closely located excited molecules of their singlet states, which results in the deexcitation of one molecule and ionization of the other molecule. This process is efficient at high LET because the concentration of excited molecules is high. The other process is the ionization of an excited molecule in its higher singlet state by an electric field supplied by closely located ionized molecules. This process is also efficient at high LET, for which ionizations and electronic excitations occur at high concentration. These quenching processes explain the lower light yield at higher LET in organic scintillators. However, the slow process is less susceptible to the fast quenching process, and the yield depends little on LET [33]. As a result, the contribution of the slow component becomes larger at higher LET, which enables PSD. To obtain better PSD capability, deuterated liquid organic scintillators have also been developed to exploit higher LET by recoil deutrons [34].

Several compositions of liquid organic and plastic scintillators having good PSD characteristics are commercially available to be used for fast and thermal neutron detection. Some organic molecular crystals have also been applied to fast neutron detection using PSD. Technical aspects of the PSD technique in general will be introduced in Chapter 12.

2.3 Liquid Scintillators

Liquid organic scintillators have several uses: analysis of radioactive isotopes emitting α- or β-rays [35], detection of fast neutrons via recoil protons, or construction for large-scale detection systems. Several kinds of liquid scintillators with different compositions, properties, and applications are commercially available. Table 2.1 summarizes the properties and suitable applications of the commercially available liquid organic scintillators including loaded scintillators. The emission wavelength is generally optimized to the wavelength region of high detection efficiency of photomultiplier tubes (PMTs). Scintillation light yields are limited up to 15 000 photons/MeV compared to the scintillation light yields of inorganic scintillators. The decay time constants of the main scintillation component are generally less than several ns, accompanied with weak long components. Some liquid scintillators have PSD capability. The scintillation properties of liquid organic scintillators depend on the combination of the solvent and the fluorescent molecules as the solutes.

For the analysis of radioactive isotopes, the use of liquid scintillators is advantageous over solid scintillators because the radioactive isotopes as the samples are surrounded with the liquid scintillators if they can be included homogeneously. High scintillation light yield is necessary for the liquid scintillators to detect the radioactive isotopes at high efficiency. In particular, to detect efficiently radioactive isotopes emitting low-energy β-rays, such as ^{3}H,

Table 2.1 *Scintillation properties of commercially available liquid organic scintillators.*

Eljen	Saint-Gobain	Nuclear enterprise	Emission wavelength (nm)	Scintillation light yield (photons/MeV)	Decay time constants (ns)	Density (g/cm^3)	Note
EJ-301	BC-501A	NE-213	425	12 000	3.16, 32.3, 270	0.874	PSD
EJ-309			424	12 300	3.5	0.959	PSD High flash point
EJ-309B			424	8800–10 600	3.5	0.96	B-loaded 1–5 wt%
EJ-313	BC-509	NE-226	425	3060	3	1.619	Hydrogen-free Neutron detection
EJ-315	BC-537	NE-230	425	9200	3.5	0.954	Deuterated
EJ-321L	BC-517L	NE-235L	425	6000	2.0	0.86	Mineral oil-based Cosmic
EJ-321H	BC-517H	NE-235H	425	8000	2.0	0.86	Mineral oil-based PSD
EJ-321P	BC-517P		425	4300	2.2	0.85	Mineral oil-based High flash point
EJ-321S	BC-517S		425	10 200	2.0	0.87	Mineral oil-based High light yield
EJ-325A	BC-519	NE-235C	425	9500	3.5	0.954	Mineral oil-based PSD
EJ-331	BC-521	NE-323	424	10 500		0.90	Gd-loaded High light yield
EJ-335	BC-525		424	8500		0.89	Gd-loaded Mineral oil-based
EJ-339	BC-523A	NE-321A	425	10 000–10 800		0.92	^{10}B-loaded (1–5 wt%), PSD
EJ-351	BC-573	NE-220	425	10 000	3.8	1.036	Dioxane-based Aqueous samples
	BC-505		425	12 300	2.5	0.877	High light yield High flash point
	BC-533		425	7800	3	0.80	High flash point Low cost

high scintillation light yield, is very important. Also, fast scintillation decay is preferable to discriminate the detection signals effectively from the noise. In addition, appropriate solvent should be used to contain the radioactive samples homogeneously. The solvent is chosen in some cases from a viewpoint of miscibility with water. Toluene, xylene, dioxane, linear alkyl benzene (frequently abbreviated as LAB), etc. have been frequently used as the solvent. Various kinds of organic phosphors have been used, for example, 2,5-diphenyloxazole (PPO), 1,4-bis-2-(5-phenyloxazolyl)-benzene(POPOP), 1,4-bis-2(4-methyl-5-phenyloxazolyl)-benzene(methylPOPOP), 2-phenyl-5-(4-biphenyl)-1,3,4-oxadiazole(PBD), 2-(4-t-butylphenyl)-5-(4-biphenylyl)-1,3,4-oxadiazole (butyl-PBD), 9,10-diphenylanthracene (DPA), etc. [36]. They are used as primary or secondary fluorescent molecules. An important difference with the use of organic liquid scintillators for the analysis of the radioactive isotopes with the other cases is the inclusion of the samples inside the scintillators, which often cause quenching and the lowering of the scintillation light yield: therefore, the organic scintillators without the samples should have high scintillation light yields to accommodate the lowering in the light yield in the organic scintillators including the samples by the quenching. The lowered scintillation light yields result in small detection signals. PMTs with high efficiency and low noise are necessary to achieve the detection of the radioactive isotopes at high efficiency [37]. Recently, a new method which enables the correction for the quenching have been proposed and implemented in commercially available measurement systems [38].

As for the detection of β-rays from the solutes as the samples, search for neutrinoless double β-decay events also follows this usage, although liquid scintillators for this application are often classified as loaded organic scintillators. In particle physics, detection of neutrinoless double β-decay events is an important experimental issue to confirm the presence of the neutrino mass and its Majorana nature (i.e., a neutrino is its own antiparticle), which requires modifications of the Standard Model of particle physics. In normal double β-decay events, the energy released by the β-decay is divided to two electrons as β-rays and two neutrinos, whereas all released energy is given to the two electrons in neutrinoless double β-decay events. The two decay modes can be discriminated on the basis of the spectroscopy of the β-rays because the energy deposition by the neutrinos in the detected volume is negligible. One of detectors, which have been proposed for this detection. is the liquid organic scintillator dissolved or dispersed with compounds comprising the candidate isotopes for the neutrinoless double β-decay [39]. To accommodate the extremely long half-lives of the candidate isotopes of $>10^{25}$ years, i.e., extremely low radioactivity, large-scale detectors comprising the candidate isotopes at high concentration and with extremely low undesired radioactivity are necessary. Several liquid scintillators dissolved with compounds comprising the candidate isotopes, e.g., ^{96}Zr [40] or ^{100}Mo [41], have been proposed; however, a significant lowering of the light yield with the addition of the compounds comprising the candidate isotopes cannot be avoided.

Liquid organic scintillators within transparent container have also been used for detection of ionizing radiations from outside the liquid scintillators. As was mentioned in Section 2.1.3, liquid organic scintillators have long been used for detection of fast and thermal neutrons: as for fast neutrons, ionizations and electronic excitations are induced mainly by recoil protons, which obtain a part of their kinetic energy from fast neutrons via elastic collision. However, thermal neutrons are detected via nuclear reactions with ^{6}Li, ^{10}B, or Gd isotopes because they cannot induce electronic excitations or ionization by themselves. In the latter case, inclusion of such isotopes is necessary to detect thermal neutron. Such liquid organic scintillators are called loaded scintillators and will be introduced in Section 2.5. In many applications of fast neutron detection, discrimination of the detection events of fast neutrons and gamma rays is necessary to suppress the detection signals of background gamma rays. The discrimination has been achieved using difference in the shape of the detection signals for fast neutrons and gamma rays, such as the PSD technique, as was introduced in Section 2.1.2.

Liquid organic scintillators with PSD capability have been widely used for detection of radioactive sources [42, 43], fissile materials [44, 45], or radioxenon in the atmosphere [46] for nuclear security, for detection of neutrons from unstable nuclei at accelerator facilities in nuclear physics experiments [47–49], or to estimate double differential cross-section of spallation neutron production [50], for spectroscopy of neutrons for plasma diagnosis in fusion reactors [51, 52], for monitoring of secondary neutron production during proton therapy [53], or fast neutron tomography [54]. PSD technique has also been used for the detection of radioactive isotopes emitting α-rays by incorporating them into the liquid scintillators [55, 56]. They are also used for large-scale detection systems of solar neutrino [57] or solar neutrons [58] while rejecting background signals with the PSD capability in astrophysics.

2.4 Organic Crystalline Scintillators

Compared to the case of inorganic scintillators, for which single-crystalline scintillators with various compounds are commercially available and a huge number of papers have been published on the development of single-crystalline scintillators, the number of crystalline scintillators of organic molecules is limited. Among them, anthracene crystals are the most famous example. They have long been used as a standard for the scintillation light yield for organic scintillators. The scintillation light yield of anthracene crystals was estimated to be slightly less than 20 000 photons/MeV [59], and recently a consistent result of an estimated light yield of 20 100 photons/MeV was reported based on a pulse height spectrum for gamma rays from ^{137}Cs [60]. In addition to anthracene, stilbene and p-terphenyl crystals are commercially available as organic crystalline scintillators. The light yields of these crystals are 16 000 and 19 400 photons/MeV, respectively [60]. In contrast to anthracene, whose scintillation decay time constants are 27.6 and 224.1 ns, stilbene and p-terphenyl exhibit fast scintillation decay. Table 2.2 summarizes the scintillation properties of organic crystals. As for stilbene, it has recently been reported that crystals grown from solutions exhibit significantly higher scintillation light yield than those of crystals grown from the melt [67]. As for p-terphenyl crystals, excellent time resolution was achieved in scintillation detectors coupled with SiPM [68].

Among the commercially available organic crystalline scintillators, stilbene has frequently been used. One of the applications of stilbene crystals is neutron monitoring in fusion reactor facilities. Stilbene-based scintillation detectors have been used to monitor the neutron emission profiles from Tokamak-type fusion reactors, JT-60U [69], FTU [70], and H2-LA [71]. They have also been used in nuclear physics [72] or for environmental monitoring of radioactive isotopes emitting α-rays in β-rays the background [73]. Furthermore, they have been used to detect β-emitting radioxenon in a background of radon as α-emitter [74] or to detect neutrons in differential die-away analysis [75] for nuclear security.

In addition to those commercially available, novel organic crystalline scintillators have been developed. The main focus has been on crystals of dyes used in organic liquid or plastic scintillators as fluorescent molecules. Solution growth technique has been frequently used as the growth technique [76, 77]. Single crystals with several organic molecules having phenyl rings have been grown from the solutions, and the light yields for α- and β-rays and the PSD capability of the crystals have been characterized [61]. Single crystals of 9,10-diphenylanthracene

Table 2.2 Scintillation properties of organic crystals.

	Emission wavelength (nm)	Light yield (photons/MeV)	Decay time constants (ns)	References
Anthracene	445	20100	27.6, 224.1	[60]
		19550		[61]
Stilbene	390	16000	3.4, 19.4	[60]
		19552		[61]
p-Terphenyl	420	19400	1.9, 14.1	[60]
9,10-diphenylanthracene	465	16000	21, 200	[62]
		12750		[61]
Phenylazaboranaphthalene	380, 520	<1000	20, 500	[63]
Ammonium Salicylate		11960	NA	[61]
Salicylamide		21540	NA	[61]
Triphenyl benzene		7360	NA	[61]
Tetraphenyl butadiene		22490	NA	[61]
Bibenzyl doped with			2.46	[64]
stilbene			2.58, 40.7, 1998	[65]
9-phenylcarbazole (pure)		<5000	NA	[66]
9-phenylcarbazole (doped)		25000	NA	[66]

(DPA) grown from solution exhibited a light yield of 16000 photons/MeV and had decay time constants of 21 and 200 ns [62]. PSD capability of these crystals has also been reported [78]. As for DPA, some researchers have reported that the crystals grown from the melt had a higher scintillation light yield than that of solution-grown crystals [79]. Crystals of boron-containing organic molecules have also been grown from these solutions, and among them, crystals of phenylazaboranaphthalene exhibited a light yield of less than 1000 photons/MeV [63]. Bibenzyl crystals doped with stilbene have also been developed as scintillators having PSD capability, and their scintillation decay time constant of the main component of less than 3 ns have been reported [64, 65]. Fabrication of 9-phenylcarbazole-based pure and doped crystals have been reported [66]. For the pure crystal, the scintillation light yield was estimated to be less than 5000 photons/MeV. For the doped one, the scintillation light yield of 25 000 photons/MeV was estimated, although the dopant was not clarified. As for PSD capabilities, some researchers showed that PSD capability can be enhanced via the control of the symmetry of the crystal structure for the crystals of various organic molecules [80].

2.5 Plastic Scintillators

Plastic scintillators in general are composed of host polymers and one or two kinds of fluorescent molecules. As for the host polymers, polyvinyltoluene (PVT) or polystyrene have commonly been used, because an efficient energy transfer from the polymers to the fluorescent molecules can be expected. Commercially available plastic scintillators are also based on these polymers. Many researchers have attempted to use other host polymers to achieve higher scintillation light yield; however, the improvement have been limited [81]. As for the polymers often used for optical materials, polymethyl methacrylate (PMMA) is a common choice for its transparency. Actually, various kinds of plastic scintillators based on PMMA have been developed using different combinations of added fluorescent molecules [82]. The use of naphthalene or 1,1,3-trimethyl-3-phenylindan as the primary fluorescent molecules was successful in achieving high scintillation light yields. To achieve high radiation tolerance and high mechanical strength, polysiloxanes have also been used [83]. They are composed of Si-O-Si main chains and the main constituent of silicone rubber. Thus far, it has been difficult to obtain high scintillation efficiency owing to a low energy transfer efficiency from the host polymers to the primary fluorescent molecules and their low solubilities. Some polyesters have also been used for the host to achieve scintillation light yields comparable to commercially available ones [84]. Recently, carbazole-based polymers, which are frequently used in the development of organic LED devices, have been used as the host of the plastic scintillators [85]. As for the fluorescent molecules, luminescent dyes used in liquid organic scintillators have also been used in plastic scintillators, because the energy transfer process is considered to have a similarity in the cases of liquid organic and plastic scintillators.

Plastic scintillators are generally fabricated from the solution of fluorescent molecules in the monomer to be polymerized, such as methylstyrene or styrene, as the solvent. It should be noted that the dissolved oxygen in the monomer solution should be removed before the polymerization because it acts as an efficient quencher. The fabrication is performed via radical polymerization with, e.g., azobis(isobutyronitrile) as the initiator for the polymerization. Some researchers have used drying processes of the solution containing the host polymers and fluorescent molecules as the solutes to obtain plastic scintillators. In this case, thick scintillators are difficult to fabricate because of the viscosity of the solutions and formation of bubbles during the evaporation of the solvents. In applications in which a large number of plastic scintillators are necessary in high-energy physics, injection molding technique has also been used [86]. In contrast to the cases of the above fabrication methods, the scintillators have to be melted at high temperatures before the molding. Hence, thermal stability is necessary for the constituents of the plastic scintillators.

Table 2.3 summarizes the properties of commercially available plastic scintillators, including loaded scintillators. In general, plastic scintillators have scintillation light yields from several thousands up to slightly larger than 10000 photons/MeV. Emission wavelengths are generally in the UV or blue regions, i.e., 350–500 nm. Scintillation decay time constants of the main component are generally less than 5 ns, which is suitable for applications in which excellent time resolution or detection capability at high counting rate are necessary. Also, some plastic scintillators have PSD capability. Compared to inorganic scintillators, tolerance at high temperature and high dose is limited. Also, it should be noted that plastic scintillators have lower chemical resistance toward organic solvents than those of inorganic scintillators.

The applications of plastic scintillators can be grouped into four: one is applications in which a large-scale detector system is necessary. They are widely used in high-energy physics [87]. For example, in the OPERA neutrino project, plastic scintillators are used as a part of the large detector to provide real-time detection signals of charged particles [88]. Plastic scintillators are also used in the ATLAS detector at the Large Hadron Collider of CERN [89]. The radiation tolerance of the plastic scintillators is a shortcoming to be overcome in this application. They are also used in radiation portal monitors for border screening owing to their cost-effectiveness [90]. Secondly, plastic scintillators are applied as fiber sensors to monitor spatial distribution of radiations. Such fibers are commercially available from several suppliers with core of polystyrene-based plastic scintillators and cladding made of PMMA. Detectors comprising the plastic scintillator fibers have been used for three-dimensional particle tracking in high-energy physics [91], neutron distribution imaging in nuclear fusion

Table 2.3 *Scintillation properties of commercially available plastic scintillators.*

Eljen	Saint-Gobain	Nuclear enterprise	Emission wavelength (nm)	Scintillation light yield (photons/MeV)	Decay time constants (ns)	Density (g/cm^3)	Note
EJ-200	BC-408	Pilot F	425	10 000	2.1	1.023	Good general properties
EJ-204	BC-404	NE-104	408	10 400	1.8	1.023	Good general properties
EJ-208	BC-412	NE-110	435	9200	3.3	1.023	Good general properties
EJ-212	BC-400	NE-102A	423	10 000	2.4	1.023	Good general properties Thin films
EJ-214			435	9000	2	1.02	Ultra-thin films
EJ-228	BC-418	Pilot U	391	10 200	1.4	1.023	Fast timing Small sizes
EJ-230	BC-420	Pilot U2	391	9700	1.5	1.023	Fast timing
EJ-232	BC-422	NE-111A	370	8400	1.6	1.023	Fast timing
EJ-232Q	BC-422Q		370	460–2900	0.7	1.023	Very fast timing
EJ-240	BC-444	NE-115	430	6300	285	1.023	Long decay
EJ-244	BC-440		434	8600	3.3	1.023	Elevated temperature
EJ-244M	BC-440M		434	8600	3.3	1.023	Elevated temperature Crosslinked
EJ-248	BC-448		425	9200	2.1	1.023	Elevated temperature

(*Continued*)

Table 2.3 *(Continued)*

Eljen	Saint-Gobain	Nuclear enterprise	Emission wavelength (nm)	Scintillation light yield (photons/MeV)	Decay time constants (ns)	Density (g/cm^3)	Note
EJ-248M	BC-448M		425	9200	2.1	1.023	Elevated temperature Crosslinked
EJ-254	BC-454		425	7500–9200	1.51	1.02	B-loaded (1–5 wt%)
EJ-256	BC-452	NE-142	425	5200–7700	2.1	1.037–1.081	Pb-loaded (1.5–5 wt%)
EJ-260	BC-428	NE-103	490	9200	9.2	1.023	Green-emitting
EJ-262			481	8700	2.1	1.023	Green-emitting
EJ-276			425	8600	13, 35, 270	1.096	PSD
EJ-290	BC-490	NE-120	423	9000	3	1.02	Casting resin
EJ-296	BC-498		435	9000	2.5	1.02	Paint
	BC-416		434	5800	4	1.023	Low cost
	BC-430		580	6900	16.8	1.032	Red-emitting

facility [92], or dose distribution estimation during radiotherapy [93]. Third, is the use for *in vivo* dosimetry during radiotherapy [94, 95]. Modern radiotherapy delivers radiation to patients with complicated dose distribution to maximize and minimize the doses in tumors and healthy tissues, respectively. The *in vivo* dosimetry is necessary to confirm that correct dose distribution is achieved inside the body of the patient. A small plastic scintillator coupled with an optical fiber can be used for real-time dosimetry at any point inside the body of the patient. The low effective atomic number of plastic scintillators is advantageous in this application because of their tissue-equivalent absorption characteristics of ionizing radiations, which enables dosimetry inside the body without disturbing the dose distribution. Lastly, plastic scintillators with PSD capability are also applied to fast neutron detection. Similar to the case of fast neutron detection using liquid organic scintillators, recoil protons are detected and discriminated from gamma rays based on the PSD technique. They have been applied to detect fast neutrons from plutonium, to distinguish in its metal or oxide states for security [96], measurement of time profiles of fast neutron production from pulsed neutron sources [97], or monitoring fast neutrons generated during proton therapy [98].

2.6 Organic–Inorganic Hybrid Scintillators

Liquid organic or plastic scintillators have often been modified by incorporation of molecules or nanoparticles, which include inorganic elements to bring some properties to the scintillators, e.g., enhanced detection efficiency to high-energy photons or thermal neutrons. To enhance the sensitivity to high-energy photons, heavy elements should be included in scintillators at high concentrations. However, to enhance the detection efficiency for thermal neutrons, some nuclei with which thermal neutrons experience nuclear reactions, such as ^{6}Li or ^{10}B, should be included. Such elements or nuclei can be included as compound molecules or nanoparticles to form loaded scintillators or organic–inorganic composite scintillators, respectively.

In this section, two approaches are introduced for the inclusion of the inorganic elements in organic scintillators: one is the dissolution of molecules for the inclusion of inorganic elements. In this case, typically, organometallic compounds, which have high solubility in organic substances, are dissolved in organic scintillators. The other is to include inorganic nanoparticles stably dispersed in organic scintillators to obtain organic–inorganic nanocomposite scintillators. In the latter case, surface modification of the inorganic nanoparticles is necessary to achieve stable dispersion.

2.6.1 Loaded Organic Scintillators

For high-energy photon detection, loading of heavy elements is beneficial to enhance the interaction probability of the scintillators and the high-energy photons [81, 99]. There are two approaches to develop loaded organic scintillators: if the organic scintillators are used for their fast response to be exploited, loading of organometallic compounds without emission is appropriate. For example, Pb-loaded plastic scintillators are commercially available, as shown in Table 2.3. Also, organometallic compounds of heavy metal element have been included in loaded organic scintillators by many researchers. Based on the approach, Bi- [100–102] or Pb- [103] loaded plastic scintillators have been developed. Significant enhancement in the detection efficiency of high-energy photons has successfully been achieved; however, the loading of the heavy elements significantly lowered the scintillation light yield owing to the quenching by the loaded compounds. The other approach is to exploit the other characteristics of plastic scintillators, such as scalability, cost effectiveness, or workability. In this case, organometallic compounds with highly emissive properties can be used to enhance both the interaction probability with high-energy photons and the scintillation light yield. It has been reported that loading of Ir organometallic complex, which has been used in organic LED, in combination with triphenyl bismuth significantly enhanced the scintillation light yield in polyvinylcarbazole (PVK)-based plastic scintillators [104].

Loaded organic scintillators have also been developed for thermal neutron detection. Since thermal neutrons are detected via nuclear reactions with several nuclei having a high cross-section of the nuclear reactions, loading of ^{10}B, ^{6}Li, or Gd is effective. For the case of ^{10}B and ^{6}Li, plastic scintillator hosts having PSD capability enables the discrimination of detection events of thermal neutrons and gamma rays. Boron-loaded plastic and liquid organic scintillators are commercially available. as shown in Tables 2.1 and 2.3. Also, Li- [102, 105–107], B- [107–110], and Gd- [111–113] loaded plastic scintillators have been developed.

2.6.2 Organic–Inorganic Nanocomposite Scintillators

Figure 2.7 shows the schematic representation of the scintillation process in the organic–inorganic nanocomposite scintillators. In the case of high-energy photon detection such as X-rays and gamma rays, inorganic nanoparticles with heavy elements as the main constituent have been included. High-energy photons impinging to the nanocomposite scintillators interact mainly with the nanoparticles via photoelectric absorption or Compton scattering. In both cases, photoelectrons or scattered electrons are emitted via the interaction and they produce numerous ionizations and electronic excitations along their trajectories. For the detection of thermal neutrons, inorganic nanoparticles with ^{10}B or ^{6}Li as the main constituent have been included. In this case, high-energy α-rays or other charged particles (^{3}H or ^{7}Li) are generated as the products of the nuclear reactions of thermal neutrons and ^{10}B or ^{6}Li, and they produce many ionizations and electronic excitations along their trajectories. In both cases, the nanocomposite scintillators have to be uniform within the range of the produced charged particles via interaction of the impinged radiation with the inorganic nanoparticles. At present, no nanocomposite scintillators are commercially available.

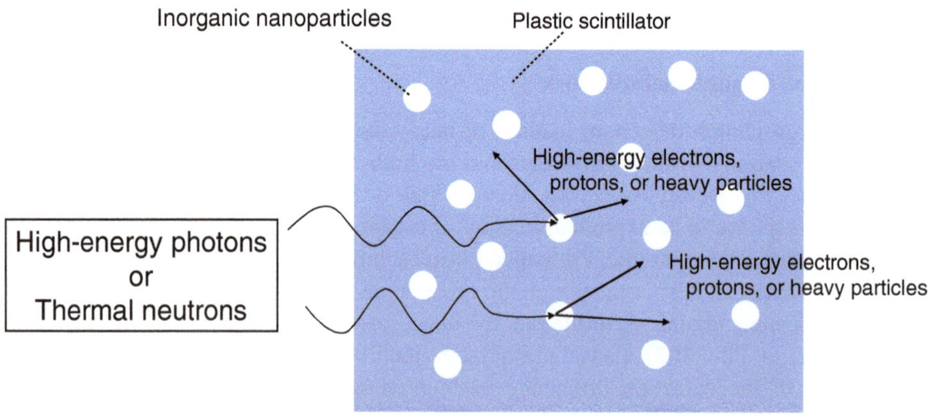

Figure 2.7 Schematic representation of scintillation process in nanocomposite scintillators.

Similar to the case of loaded organic scintillators, there have been two approaches to the fabrication of nanocomposite scintillators: one is the addition of inorganic nanoparticles without emission to exploit the fast response of the plastic scintillators. In this case, the interaction probability with the radiations to be detected is enhanced. The other is the addition of highly emissive inorganic nanoparticles. In this case, the interaction probability with the radiations to be detected and the scintillation light yield are enhanced, whereas the fast response of the plastic scintillators is generally degraded.

As for detection of high-energy photons, nanoparticles of heavy element oxides or fluorides have been added to the plastic scintillators in the former approach. To fabricate transparent nanocomposites, surface modification with organic molecules is important to disperse the inorganic nanoparticles in the polymer matrices. Nanoparticles of hafnia-silica [114, 115], HfO_2 [116, 117], Gd_2O_3 [118], YbF_3 [119], or Bi_2O_3 [120] have been added to plastic scintillators. Some scintillators have been applied to detect high-energy X-rays at synchrotron radiation facilities [121, 122]. Highly emissive nanoparticles have also been introduced in plastic scintillators. Among them, addition of semiconductor quantum dots have enhanced the scintillation light yield and the detection efficiency for high-energy photons, and the degradation of the fast response of the plastic scintillators was limited owing to the fast emission from the quantum dots [123, 124]. Insulator-based emissive nanoparticles have also been introduced in plastic scintillators. Interestingly, scintillation from insulator-based nanoparticles of small sizes was not observed in the nanocomposite scintillators, whereas it was observed from the nanoparticles only. This result strongly suggests that nonradiative energy transfer occurs from the inorganic nanoparticles to the host polymers before the scintillation [125–128].

Nanocomposite scintillators for thermal neutrons have also been developed. Nanocomposite scintillators based on borosilicate glass and polyethylene glycol have been developed [129]. Plastic scintillators introduced with nanoparticles of LiF [130] or lithium salicylate [131] have also been developed.

References

1. Kabe, R. and Adachi, C. (2017). Organic long persistent luminescence. *Nature* 550: 384.
2. Agranovich, V.M. and Galanin, M.D. (1982). *Electronic Excitation Energy Transfer in Condensed Matter*. Amsterdam: North-Holland Publishing Co.
3. Reid, C. (1952). Scintillators and energy transfer processes. *Physiol. Rev.* 88: 422.
4. Melhuish, W.H. (1964). Excitation energy transfer from toluene to some organic scintillators. *J. Chem. Phys.* 40: 1369.
5. Birks, J.B., Conte, J.M.C., and Walker, G. (1966). The influence of excimer formation on solvent-solute energy transfer in organic liquid scintillators. *IEEE Trans. Nucl. Sci.* 13: 148.
6. Horrocks, D.L. (1970). The influence of excimer formation on solvent-solute energy transfer in organic liquid scintillators. *J. Chem. Phys.* 52: 1566.
7. Birks, J.B. (1968). Energy transfer in organic systems. VI: Fluorescence response functions and scintillation pulse shapes. *J. Phys. B: At. Mol. Opt. Phys.* 1: 946.
8. Mathad, R.D. (1985). Environmental effects on electronic excitation energy transfer in organic liquid scintillation. *Pramana J. Phys.* 25: 95.
9. Mathad, R.D., Chikkur, G.C., and Umakantha, N. (1986). Electronic excitation energy transfer in organic Liquid scintillators. *J. Chem. Phys.* 84: 2187.
10. Mathad, R.D. and Umakantha, N. (1986). Effects of dilution, temperature and viscosity on intermolecular energy transfer in scintillators. *J. Phys. D: Appl. Phys.* 19: 1381.
11. Hanagodimath, S.M., Chikkur, G.C., and Gadaginmath, G.S. (1990). Environmental effects on the energy migration coefficient, the effective energy transfer and quenching distances in an organic liquid scintillator. *Chem. Phys.* 148: 347.

12. Hanagodimath, S.M., Gadaginmath, G.S., and Chikkur, G.C. (1992). On the mechanism of excitation energy transfer involving long and short range interaction in dilute organic liquid scintillator systems. *Acta. Phys. Polonica. A* 81: 361.
13. Math, B.G., Chikkur, G.C., and Malimath, G.H. (1992). The role of diffusion, migration and long-range interaction in energy transfer and quenching processes in an organic liquid scintillator. *Appl. Radiat. Isot.* 43: 1349.
14. Umarjee, D.M. and Chikkur, G.C. (1985). Temperature dependence of energy transfer and quenching studies in an organic liquid scintillator. *Phys. Lett. A* 112: 57.
15. Hanagodimath, S.M., Chikkur, G.C., and Gadaginmath, G.S. (1991). A unique method of determining the excitation energy migration coefficient in organic liquid scintillators. *Pramana J. Phys.* 37: 153.
16. Malimath, G.H. and Chikkur, G.C. (1994). Role of energy migration in an organic liquid scintillator system in the 20–70°c temperature range. *Appl. Radiat. Isot.* 45: 143.
17. Umarjee, D.M. and Chikkur, G.C. (1987). Effect of dilution and temperature on the energy transfer and quenching in an organic liquid scintillator. *Phys. Lett. A* 123: 23.
18. Giraddi, T.P., Kadadevarmath, J.S., and Malimath, et al. (1996). Effect of solvent on the fluorescence quenching of organic liquid scintillators by aniline and carbon tetrachloride. *Appl. Radiat. Isot.* 47: 461.
19. Hariharan, C. and Mishra, A.K. (2000). Quenching of liquid scintillator fluorescence by chloroalkanes and chloroalkenes. *Radiat. Meas.* 32: 113.
20. Malimath, G.H., Chikkur, G.C., Pal, H. et al. (1997). Role of internal mechanisms in energy transfer processes in organic liquid scintillators. *Appl. Radiat. Isot.* 48: 359.
21. Hirayama, F., Basile, L.J., and Kikuchi, C. (1968). Energy transfer and quenching in plastic scintillators. *Mol. Cryst.* 4: 83.
22. Bengston, B. and Moszyński, M. (1974). Energy-transfer and light-collection characteristics for different types of plastic scintillators. *Nucl. Instrum. Methods* 117: 227.
23. Bengston, B. and Moszyński, M. (1978). Study of primary energy transfer process in ultrafast plastic scintillators. *Nucl. Instrum. Methods* 155: 221.
24. Hallam, A. and Birks, J.B. (1978). Energy transfer in organic systems. XIII: Plastic scintillators. *J. Phys. B: At. Mol. Opt. Phys.* 11: 3273.
25. Huijun, Y., Xiangtuo, W., Yuying, J. et al. (1984). Energy transfer in multi component plastic scintillators. *J. Lumin.* 31–32: 833.
26. Polacki, Z. (1984). Influence of polystyrene excimers on the excitation energy transfer in solid scintillating solutions. *J. Polym. Sci. Polym. Phys.* 22: 2275.
27. Burshtein, A.I. (1984). Concentration quenching of noncoherent excitation in solutions. *Sov. Phys. Usp.* 27: 579.
28. Birks, J.B. (1970). Triplet-triplet interactions in organic scintillators. *Chem. Phys. Lett.* 7: 293.
29. Hajagos, T.J., Garcia, E., Kishpaugh, D. et al. (2019). Plastic scintillators based on thermally activated delayed fluorescence dyes. *Nucl. Instrum. Methods Phys. Res. A.* 940: 185.
30. Birks, J.B. (1951). Scintillations from organic crystals: Specific fluorescence and relative response to different radiations. *Proc. Phys. Soc. A.* 64: 874.
31. Ciéslak, M.J., Gamage, K.A.A., and Glover, R. (2019). Critical review of scintillating crystals for neutron detection. *Crystals* 9: 480.
32. Laustriat, G. (1968). The luminescence decay of organic scintillators. *Mol. Cryst.* 4: 127.
33. Brooks, F. (1979). Development of organic scintillators. *Nucl. Instrum. Methods* 162: 477.
34. Bildstein, V., Garrett, P.E., Wong, J. et al. (2013). Comparison of deuterated and normal liquid scintillators for fast-neutron detection. *Nucl. Instrum. Methods Phys. Res. A.* 729: 188.
35. Hou, X. (2018). Liquid scintillation counting for determination of radionuclides in environmental and nuclear application. *J. Radioanal. Nucl. Chem.* 318: 1597.
36. Parmentier, J.H. and Ten Haaf, F.E.L. (1969). Developments in liquid scintillation counting since 1963. *Int. J. Appl. Radiat. Isot.* 20: 305.
37. Vaninbroukx, R. and Stanef, I. (1973). Present status in the field of precision liquid scintillation counting. *Nucl. Instrum. Methods* 112: 111.
38. Broda, R. (2003). A review of the triple-to-double coincidence ratio (TDCR) method for standardizing radionuclides. *Appl. Radiat. Isot.* 58: 585.
39. Shimizu, I. and Chen, M. (2019). Double beta decay experiments with loaded liquid scintillator. *Front. Physiol.* 7: 33.
40. Fukuda, Y., Moriyama, S., and Ogawa, I. (2013). Development of liquid scintillator containing a zirconium complex for neutrinoless double beta decay experiment. *Nucl. Instrum. Methods Phys. Res. A.* 732: 397.

41. Gehman, V.M., Doe, P.J., Robertson, R.G.H. et al. (2010). Solubility, light output and energy resolution studies of molybdenum-loaded liquid scintillators. *Nucl. Instrum. Methods Phys. Res. A.* 622: 602.
42. Flaska, M. and Pozzi, S.A. (2007). Identification of shielded neutron sources with the liquid scintillator bc-501a using a digital pulse shape discrimination method. *Nucl. Instrum. Methods Phys. Res. A.* 577: 654.
43. Poitrasson-Rivière, A., Hamel, M.C., Polack, J.K. et al. (2014). Dual-particle imaging system based on simultaneous detection of photon and neutron collision events. *Nucl. Instrum. Methods Phys. Res. A.* 760: 40.
44. Pozzi, S.A., Clarke, S.D., Flaska, M. et al. (2009). Pulse-height distributions of neutron and gamma rays from plutonium-oxide samples. *Nucl. Instrum. Methods Phys. Res. A.* 608: 310.
45. Dolan, J.L., Flaska, M., Poitrasson-Riviere, A. et al. (2014). Plutonium measurements with a fast-neutron multiplicity counter for nuclear safeguards applications. *Nucl. Instrum. Methods Phys. Res. A.* 763: 565.
46. Russ, W.R., Stuenkel, D.O., Valentine, J.D. et al. (1999). An evaluation of radioxenon detection techniques for use with a fluid-based concentration system. *IEEE Trans. Nucl. Sci.* 46: 296.
47. Arnell, S.E., Roth, H.A., Skeppstedt, Ö. et al. (1991). A 2π neutron and γ-ray multiplicity filter for the NORDBALL detection system. *Nucl. Instrum. Methods Phys. Res. A.* 300: 303.
48. Laurent, H., Lefort, H., Beaumel, D. et al. (1993). EDEN: a neutron time-of-flight multidetector for decay studies of giant states. *Nucl. Instrum. Methods Phys. Res. A.* 326: 517.
49. Nilsson, B., Adler, J.-O., Andersson, B.-E. et al. (2007). Measurement of the $^4He(\gamma,n)$ reaction from $23 < E\gamma < 70$ MeV. *Phys. Rev. C.* 75: 014007.
50. Nakamoto, T., Ishibashi, K., Matsufuji, N. et al. (1995). Spallation neutron measurement by the time-of-flight method with a short flight path. *J. Nucl. Sci. Technol.* 32: 827.
51. Giacomelli, L., Zimbal, A., Tittelmeier, K. et al. and ASDEX Upgrade Team. (2011). The compact neutron spectrometer at ASDEX upgrade. *Rev. Sci. Instrum.* 82: 123504.
52. Belli, F., Conroy, S., Esposito, B. et al. and JET EFDA contributors. (2012). Conceptual design, development and preliminary tests of a compact neutron spectrometer for the JET experiment, IEEE Trans. *Nucl. Sci.* 59: 2512.
53. Clarke, S.D., Pryser, E., Wieger, B.M. et al. (2016). A scintillator-based approach to monitor secondary neutron production during proton therapy. *Med. Phys.* 43: 5915.
54. Joyce, M.J., Agar, S., Aspinall, M.D. et al. (2016). Fast neutron tomography with real-time pulse-shape discrimination in organic scintillation detectors. *Nucl. Instrum. Methodss Phys. Res. A.* 834: 36.
55. Yang, D., Zhu, Y., and Jiao, R. (1994). Determination of Np, Pu and Am in high level radioactive waste with extraction-liquid scintillation counting. *J. Radioanal. Nucl. Chem.* 183: 245.
56. Tinker, R.A. and Smith, J.D. (1996). Simultaneous measurement of 226Ra and 133Ba using liquid scintillation counting with pulse shape discrimination. *Anal. Chim. Acta* 332: 291.
57. Borexino Collaboration (2002). Science and technology of Borexino: A real-time detector for low energy solar neutrinos. *Astropart. Phys.* 16: 205.
58. Pirard, B., Woolf, R.S., Bravar, U. et al. (2009). Test and simulation of a fast neutron imaging telescope. *Nucl. Instrum. Methods Phys. Res. A.* 603: 406.
59. Wright, G.T. (1959). Absolute scintillation efficiency of anthracene crystals. *Proc. Phys. Soc. B.* 68: 929.
60. Yanagida, T., Watanabe, K., and Fujimoto, Y. (2015). Comparative study of neutron and gamma-ray pulse shape discrimination of anthracene, stilbene, and p-terphenyl. *Nucl. Instrum. Methods Phys. Res. A.* 784: 111.
61. Hull, G., Zaitseva, N.P., Cherepy, N.J. et al. (2009). New organic crystals for pulse shape discrimination. *IEEE Trans. Nucl. Sci.* 56: 899.
62. van Loef, E.V., Mukhopadhyay, S., Zaitseva, N. et al. (2012). Crystal growth and characterization of 9,10-diphenylanthracene. *J. Cryst. Growth* 352: 103.
63. van Loef, E.V., Marshall, J., Zakharov, L.N. et al. (2013). Novel boron-containing organic scintillators: azaboranaphthalene and phenylazaboranaphthalene. *IEEE Trans. Nucl. Sci.* 60: 952.
64. Brown, J.A., Goldblum, B.L., Bernstein, L.A. et al. (2014). Relative light yield and temporal response of a stilbene-doped bibenzyl organic scintillator for neutron detection. *J. Appl. Phys.* 115: 193504.
65. Hatarik, R., Bernstein, L.A., Caggiano, J.A. et al. (2012). Characterizing time decay of bibenzyl scintillator using time correlated single photon counting. *Rev. Sci. Instrum.* 83: 10D911.
66. van Loef, E.V., Markosyan, G., Shirwadkar, U. et al. (2017). Solution growth and scintillation properties of 9-phenylcarbazole. *J. Cryst. Growth* 480: 145.
67. Zaitseva, N., Glenn, A., Carman, L. et al. (2015). Scintillation properties of solution-grown trans-stilbene single crystals. *Nucl. Instrum. Methods Phys. Res. A* 789: 8.
68. De Gerone, M., Biasotti, M., Ceriale, V. et al. (2016). Properties of single crystal para-terphenyl as medium for high resolution TOF detector. *Nucl. Instrum. Methods Phys. Res. A.* 824: 192.

69. Ishikawa, M., Nishitani, T., Morioka, A. et al. (2002). First measurement of neutron emission profile on JT-60U using Stilbene neutron detector with neutron-gamma discrimination. *Rev. Sci. Instrum.* 73: 4237.
70. Esposito, B., Kaschuck, Y., Rizzo, A. et al. (2004). Digital pulse shape discrimination in organic scintillators for fusion applications. *Nucl. Instrum. Methods Phys. Res. A.* 518: 626.
71. Zhang, X., Yuan, X., Xie, X. et al. (2012). A digital delay-line-shaping method for pulse shape discrimination in stilbene neutron detector and application to fusion neutron measurement at HL-2A tokamak. *Nucl. Instrum. Methods Phys. Res. A.* 687: 7.
72. Steinetz, B.M., Benyo, T.L., Chait, A. et al. (2020). Novel nuclear reactions observed in Bremsstrahlung-irradiated deuterated metals. *Phys. Rev. C.* 101: 044610.
73. Morishita, Y., Di Fulvio, A., Clarke, S.D. et al. (2019). Organic scintillator-based alpha/beta detector for radiological decontamination. *Nucl. Instrum. Methods Phys. Res. A.* 935: 207.
74. Sivels, C.B., McIntyre, J.I., Prinke, A.M. et al. (2019). Stilbene cell development to improve radioxenon detection. *Nucl. Instrum. Methods Phys. Res. A.* 923: 72.
75. Miller, C.A., Peters, W.A., Odeh, F.Y. et al. (2020). Sub-critical assembly die-away analysis with organic scintillators. *Nucl. Instrum. Methods Phys. Res. A.* 959: 163598.
76. Zaitseva, N., Carman, L., Glenn, A. et al. (2011). Application of solution techniques for rapid growth of organic crystals. *J. Cryst. Growth* 314: 163.
77. Fraboni, B., Fraleoni-Morgera, A., and Zaitseva, N. (2016). Ionizing radiation detectors based on solution-grown organic single crystals. *Adv. Funct. Mater.* 26: 2276.
78. Whitney, C., Johnson, E., Chen, X.J. et al. (2013). DPA-based fast neutron dosimeter for the space environment. *IEEE Trans. Nucl. Sci.* 60: 830.
79. Zhu, T., Liu, X., Qin, J. et al. (2019). Studies on the luminescence and scintillation properties of 9,10-diphenylanthracene crystals grown by two methods. *Cryst. Growth Des.* 19: 6855.
80. Feng, P.L. and Foster, M.E. (2013). Pulse-shape discrimination in high-symmetry organic scintillators. *IEEE Trans. Nucl. Sci.* 60: 3142.
81. Bertrand, G.H., Hamel, M., and Sguerra, F. (2014). Current status on plastic scintillators modifications. *Chem. A. Eur. J.* 20: 15660.
82. Salimgareeva, V.N. and Kolesov, S.V. (2005). Plastic scintillators based on polymethyl methacrylate: A review. *Instrum. Exp. Tech.* 48: 273.
83. Quaranta, A., Carturan, S., Cinausero, M. et al. (2013). Characterization of polysiloxane organic scintillators produced with different phenyl containing blends. *Mater. Chem. Phys.* 137: 951.
84. Nakamura, H., Shirakawa, Y., Takahashi, S. et al. (2011). Evidence of deep-blue photon emission at high efficiency by common plastic. *EPL* 95: 22001.
85. Montbarbon, E., Sguerra, F., Bertrand, G.H.V. et al. (2017). Structural variation of carbazole derivatives for plastic scintillation applications. *Chem. Photo. Chem.* 1: 451.
86. Yoshimura, Y., Inagaki, T., Morimoto, T. et al. (1998). Plastic scintillator produced by the injection-molding technique. *Nucl. Instrum. Methods Phys. Res. A.* 406: 435.
87. Kharzheev, Y.N. (2015). Scintillation counters in modern high-energy physics experiments (review). *Phys. Part. Nucl.* 46: 678.
88. Acquafredda, R., Adam, T., Agafonova, N. et al. (2009). The OPERA experiment in the CERN to Gran Sasso neutrino beam. *J. Instrum.* 4: P04018.
89. Jivan, H., Sideras-Haddad, E., Erasmus, R. et al. (2015). Radiation hardness of plastic scintillators for the Tile Calorimeter of the ATLAS detector. *J. Phys. Conf. Ser.* 645: 012019.
90. Siciliano, E.R., Ely, J.H., Kouzes, R.T. et al. (2005). Comparison of PVT and NaI(Tl) scintillators for vehicle portal monitor applications. *Nucl. Instrum. Methods Phys. Res. A.* 550: 647.
91. E224 Collaboration (1996). Search for the H dibaryon in (K^-, K^+) reaction with scintillating fiber active target. *Phys. Lett. B* 378: 53.
92. Ress, D., Lerche, R.A., Ellis, R.J. et al. (1995). High-sensitivity scintillating-fiber imaging detector for high-energy neutrons. *Rev. Sci. Instrum.* 66: 4943.
93. Shin, S.H., Yoo, W.J., Jang, K.W. et al. (2016). Development of an all-in-one phantom and scintillator radiation sensor for real-time monitoring of source position and dose distribution in high-dose-rate brachytherapy. *Sensors Mater.* 28: 619.
94. O'Keeffe, S., Maccarthy, D., Woulfe, P. et al. (2015). A review of recent advances in optical fibre sensors for *in vivo* dosimetry during radiotherapy. *Br. J. Radiol.* 88: 20140702.
95. Kron, T., Lehmann, J., and Greer, P.B. (2016). Dosimetry of ionising radiation in modern radiation oncology. *Phys. Med. Biol.* 61: R167.

96. Pozzi, S.A., Bourne, M.M., Dolan, J.L. et al. (2014). Plutonium metal vs. oxide determination with the pulse-shape-discrimination-capable plastic scintillator EJ-299-33. *Nucl. Instrum. Methods Phys. Res. A.* 767: 188.
97. Preston, R.M., Eberhardt, J.E., and Tickner, J.R. (2013). Neutron generator burst timing measured using a pulse shape discrimination plastic scintillator with silicon photomultiplier readout. *J. Instrum.* 8: P12005.
98. Chung, S., Kacperek, A., Speller, R. et al. (2017). Optimisation of pulse shape discrimination using EJ299-33 for high energy neutron detection in proton beam therapy. *J. Instrum.* 12: C11033.
99. Hajagos, T.J., Liu, C., Cherepy, N.J. et al. (2018). High-Z sensitized plastic scintillators: A review. *Adv. Mater.* 30: 1706956.
100. Bertrand, G.H.V., Sguerra, F., Dehé-Pittance, C. et al. (2014). Influence of bismuth loading in polystyrene-based plastic scintillators for low energy gamma spectroscopy. *J. Mater. Chem. C.* 2: 7304.
101. Koshimizu, M., Bertrand, G.H.V., Hamel, M. et al. (2015). X-ray detection capability of bismuth-loaded plastic scintillators. *Jpn. J. Appl. Phys.* 54: 102202.
102. Cherepy, N.J., Sanner, R.D., Beck, P.R. et al. (2015). Bismuth- and lithium-loaded plastic scintillators for gamma and neutron detection. *Nucl. Instrum. Methods Phys. Res. A.* 778: 126.
103. Hamel, M., Turk, G., Rousseau, A. et al. (2011). Preparation and characterization of highly lead-loaded red plastic scintillators under low energy X-rays. *Nucl. Instrum. Methods Phys. Res. A.* 660: 57.
104. Rupert, B.L., Cherepy, N.J., Sturm, B.W. et al. (2012). Bismuth-loaded plastic scintillators for gamma-ray spectroscopy. *EPL* 97: 22002.
105. Sen, I., Penumadu, D., Williamson, M. et al. (2011). Thermal neutron scintillator detectors based on poly (2-vinylnaphthalene) composite films. *IEEE Trans. Nucl. Sci.* 58: 1386.
106. Mabe, A.N., Auxier, J.D. II, Urffer, M.J. et al. (2013). Transparent lithiated polymer films for thermal neutron detection. *Nucl. Instrum. Methods Phys. Res. A* 722: 29.
107. Britvich, I.G., Vasil'chenko, V.G., Kirichenko, V.N. et al. (2002). New polystyrene-based scintillators. *Instrum. Exp. Tech.* 45: 644.
108. Normand, S., Mouanda, B., Haan, S. et al. (2002). Discrimination methods between neutron and gamma rays for boron loaded plastic scintillators. *Nucl. Instrum. Methods Phys. Res. A.* 484: 342.
109. Normand, S., Mouanda, B., Haan, S. et al. (2002). Study of a new boron loaded plastic scintillator (revised). *IEEE Trans. Nucl. Sci.* 49: 1603.
110. Britvich, G.I., Vasil'chenko, V.G., Lishin, V.A. et al. (2001). New scintillation materials for particle detectors. *Instrum. Exp. Tech.* 44: 472.
111. Czirr, J.B. (1973). Gd-loaded plastic scintillator. *Nucl. Instrum. Methods* 108: 613.
112. Velmozhnaya, E.S., Bedrik, A.I., Zhmurin, P.N. et al. (2015). Mixed-ligand complexes of gadolinium carboxylates containing unsaturated bonds in plastic scintillators. *Funct. Mater.* 22: 274.
113. Dumazert, J., Coulon, R., Hamel, M. et al. (2016). Gadolinium-loaded plastic scintillators for thermal neutron detection using compensation. *IEEE Trans. Nucl. Sci.* 63: 1551.
114. Sun, Y., Koshimizu, M., Yahaba, N. et al. (2014). High-energy X-ray detection by hafnium-doped organic-inorganic hybrid scintillators prepared by sol-gel method. *Appl. Phys. Lett.* 104: 174104.
115. Kagami, K., Koshimizu, M., Fujimoto, Y. et al. (2020). High-energy X-ray detection capabilities of Hf-loaded plastic scintillators synthesized by sol-gel method. *J. Mater. Sci. Mater. Electron.* 31: 896.
116. Liu, C., Hajagos, T.J., Kishpaugh, D. et al. (2015). Facile single-precursor synthesis and surface modification of hafnium oxide nanoparticles for nanocomposite γ-ray scintillators. *Adv. Funct. Mater.* 25: 4607.
117. Hiyama, F., Noguchi, T., Koshimizu, M. et al. (2018). X-ray detection capabilities of plastic scintillators incorporated with hafnium oxide nanoparticles surface-modified with phenyl propionic acid. *Jpn. J. Appl. Phys.* 57: 012601.
118. Cai, W., Chen, Q., Cherepy, N. et al. (2013). Synthesis of bulk-size transparent gadolinium oxide–polymer nanocomposites for gamma ray spectroscopy. *J. Mater. Chem. C.* 1: 1970.
119. Jin, Y., Kishpaugh, D., Liu, C. et al. (2016). Partial ligand exchange as a critical approach to the synthesis of transparent ytterbium fluoride–polymer nanocomposite monoliths for gamma ray scintillation. *J. Mater. Chem. C.* 4: 3654.
120. Hiyama, F., Noguchi, T., Koshimizu, M. et al. (2018). X-ray detection properties of plastic scintillators containing surface-modified Bi_2O_3 nanoparticles. *Jpn. J. Appl. Phys.* 57: 052203.
121. Inoue, K., Koshimizu, M., Hiyama, F. et al. (2018). Measurement of nuclear resonant scattering on 61Ni with fast scintillation detector using proportional-mode silicon avalanche photodiode. *IEEE Trans. Nucl. Sci.* 65: 1012.
122. Kishimoto, S., Koshimizu, M., Hiyama, F. et al. (2020). Measurements of 73-keV X-ray time spectrum with avalanche-photodiode scintillation detector using Bi_2O_3-nanoparticle-doped plastic scintillator. *Nucl. Instrum. Methods Phys. Res. A.* 968: 163908.

123. Liu, C., Li, Z., Hajagos, T.J. et al. (2017). Transparent ultra-high-loading quantum dot/polymer nanocomposite monolith for gamma scintillation. *ACS Nano.* 11: 6422.
124. Gandini, M., Villa, I., Beretta, M. et al. (2020). Efficient, fast and reabsorption-free perovskite nanocrystal-based sensitized plastic scintillators. *Nat. Nanotechnol.* 15: 462.
125. Sahi, S., Chen, W., and Jiang, K. (2015). Luminescence enhancement of PPO/PVT scintillators by CeF_3 nanoparticles. *J. Lumin.* 159: 105.
126. Demkiv, T.M., Vistovskyy, V.V., Halyatkin, O.O. et al. (2018). Luminescence of polystyrene composites loaded with CeF_3 nanoparticles. *Nucl. Instrum. Methods Phys. Res. A.* 908: 309.
127. Demkiv, T.M., Halyatkin, O.O., Vistovskyy, V.V. et al. (2016). X-ray excited luminescence of polystyrene-based scintillator loaded with $LaPO_4$–Pr nanoparticles. *J. Appl. Phys.* 120: 144301.
128. Demkiv, T.M., Halyatkin, O.O., Vistovskyy, V.V. et al. (2016). Luminescent and kinetic properties of the polystyrene composites based on BaF_2 nanoparticles. *Nucl. Instrum. Methods Phys. Res. A.* 810: 1.
129. Koshimizu, M., Kitajima, H., Iwai, T. et al. (2008). Organic-inorganic hybrid scintillator for neutron detection fabricated by Sol-Gel method. *Jpn. J. Appl. Phys.* 47: 5717.
130. Sen, I., Urffer, M., Penumadu, D. et al. (2012). Polyester composite thermal neutron scintillation films. *IEEE Trans. Nucl. Sci.* 59: 1781.
131. Young, S.A., Sen, I., and Penumadu, D. (2012). Investigation of ^{6}Li enriched particle dispersion in fluorescent electrospun polymer nanofibers to be used as thermal neutron scintillators. *J. Eng. Mater. Technol.* 134: 010908.

3

Inorganic Oxide Scintillators

Daisuke Nakauchi, Noriaki Kawaguchi, and Takayuki Yanagida
Nara Institute of Science and Technology, Ikoma, Japan

3.1 Introduction

A wide variety of inorganic scintillators such as single crystals, polycrystalline ceramics, and glasses have been developed depending on their application. Following the discovery of X-rays, ZnS and $CaWO_4$ powders have been used as photographic film-based radiation detectors for decades. However, since the application of medical diagnoses such as gamma cameras began, bulk single crystals have been adopted as the material form of inorganic scintillators in most practical applications. Compared with ceramics or glasses, a single crystal exhibits high transmittance at its own emission wavelength, and also has high density and homogeneity, so that high detection efficiency can be achieved in a large bulk body. Many oxide scintillators exhibit a highly effective atomic number (Z_{eff}), non-deliquescence, and low afterglow, compared to halide scintillators such as Tl:NaI. For this reason, extensive research has been conducted on oxide scintillators for medical and security applications [1]. In general, oxides have a larger band-gap energy (E_g) than halides such as bromide and iodide, so that the *LY* tends to be relatively low [2]. However, several oxide scintillators, including well-organized garnets [3] and silicates [4], showing *LY* comparable with Tl:NaI have been recently discovered. This chapter briefly describes synthesis and properties of oxide scintillators.

3.2 Crystal Growth

Most of the practical single crystalline scintillators are grown by the Czochralski (CZ) method or the Bridgman–Stockberger method. Figure 3.1 shows a schematic diagram of the

Phosphors for Radiation Detectors, First Edition. Edited by Takayuki Yanagida and Masanori Koshimizu.

CZ furnace. The raw powder is placed into a crucible and heated to a temperature above the melting point. After the raw materials are completely molten, a seed crystal is lowered and slowly dipped into the melt surface, until the bottom part of the dipped seed becomes molten. Then, the dipped seed is pulled away from the melt, and the melt then slowly starts to crystallize from the contact interface. To produce the desired diameter, the rotation speed, pulling rate, and heating power are adjusted to produce and maintain the desired diameter. With this method, since the single crystal can be produced without directly contacting the crucible and a mechanical distortion due to the crucible, a most nearly perfect and homogeneous crystals can be obtained. Therefore, the CZ method is the most typical method for bulk crystal growth, and most of oxide single crystalline scintillators have been synthesized in this way. In some cases, multiple heaters are installed in the furnace to reduce the temperature gradient. As shown in Table 3.1, it is necessary to select the most compatible crucible for each material, and crucibles for high melting point materials such as Mo and W must be performed in an inert atmosphere because they are easily oxidized and volatilized.

The Bridgman–Stockberger technique is also a popular method in which a single crystal is grown from a melt by progressively freezing from one end to the other in a crucible. This method is conducted by moving the raw materials placed in a crucible in a horizontal or vertical direction, and is called the horizontal (HB) method or vertical Bridgman (VB) method, respectively. Figure 3.2 shows a schematic diagram of the VB method, in which the furnace has one or two heated zones to create a temperature gradient to the growth direction.

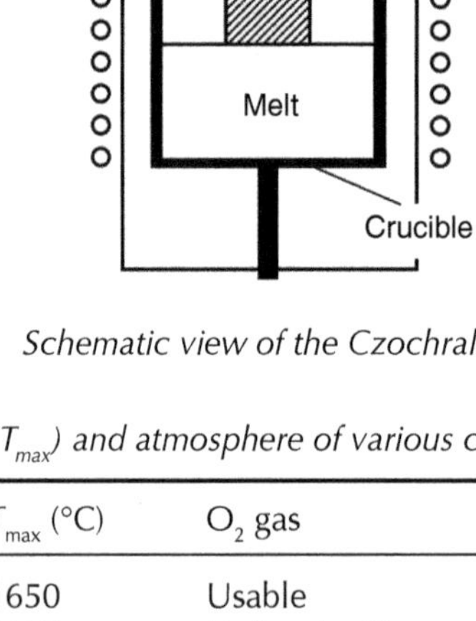

Figure 3.1 *Schematic view of the Czochralski method.*

Table 3.1 *Usable temperature (T_{max}) and atmosphere of various crucibles for melt growth.*

Crucible	m.p. (°C)	T_{max} (°C)	O_2 gas	Inert gas	H_2 gas
Pt	1772	1650	Usable	Usable	Unusable
Ir	2453	2150	Only a few%	Usable	Usable
Mo	2620	2300	Unusable	Usable	Usable
W	3370	2800	Unusable	Usable	Usable
SiO_2	1710	1500	Usable	Usable	Usable
Graphite	3500	3700	Unusable	Usable	Usable

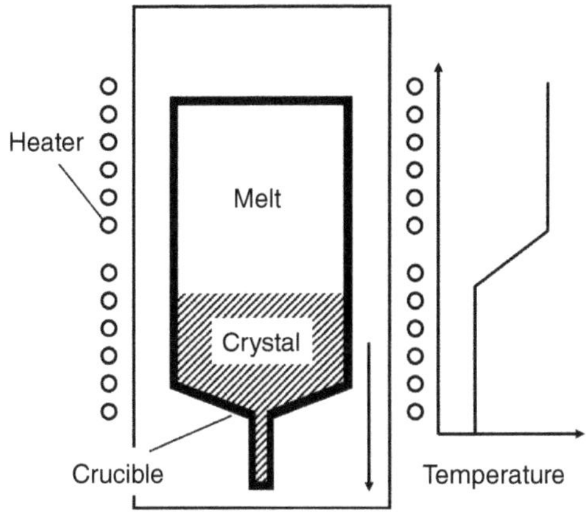

Figure 3.2 *Schematic view of the vertical Bridgman-Stockberger method.*

A seed crystal is placed at the cold end of a temperature gradient furnace such as one end of a crucible boat or the bottom of a cylindrical tube. Then, the crystal is directionally solidified from the seed crystal by shifting the crucible, thereby moving the heat zone and the growth interface along with it. This method requires little supervision owing to the relatively simple technology. In both the CZ and Bridgman–Stockberger methods, a typical running time to obtain one ingot is from one to several weeks.

For primary tests searching for new scintillator materials, the micro-pulling down (μ-PD) method and the floating zone (FZ) method, which can easily produce small crystals of about several mm in diameter, have been used. Figure 3.3 illustrates a schematic view of the μ-PD furnace. The μ-PD method involves downward pulling of a fiber crystal through a micro-nozzle incorporated into the bottom of the crucible [5–7]. The shape of the crystal can be controlled by changing the shape of the nozzle. The prepared crystals are homogeneous with almost no variation in diameter and composition throughout the entire region. In rare-earth-doped garnet crystals grown by the μ-PD method, the dopant distribution along the radial direction demonstrated concave or flat profiles with a dopant segregation coefficiency of about 1 [8]. The merit of the μ-PD method is rapid crystal growth with acceptable crystalline quality for all applications, and typically, we can obtain one crystal rod of several mm × several cm length within one day (8–12 hours). After crystal growth, the crucible should be washed with acids, which takes one day (8–12 hours). Thus, it should be noted that the typical running speed is one crystal every two days if the user has one crucible.

The FZ method is one of the zone-melting techniques, which has been developed as a method for refining crystalline materials [9–11]. A crystalline rod is melted and solidified from one end, and the impurities are discharged from the crystal and segregated out at the other end. A schematic diagram of the optical FZ technique is shown in Figure 3.4. In order to melt the raw material, light from lamps was reflected by a mirror surface and focused at the center. A narrow region of a feed rod was melted, and this molten zone was slowly moved downward along the rod axis. Since the FZ furnace equipped with Xe arc lamps enables us to grow a single crystal with a melting point higher than 2800 °C [12], Hf- and Zr-based oxides have been studied as candidate scintillators in recent years [13, 14]. However, optical FZ growth is limited to materials that are not volatile and absorb the wavelength of the lamp. Furthermore, since stability of the molten zone strongly depends on the surface tension of

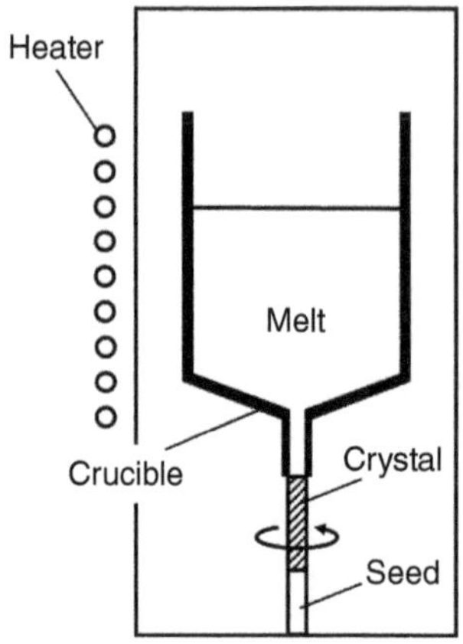

Figure 3.3 Schematic view of the micro-pulling down method.

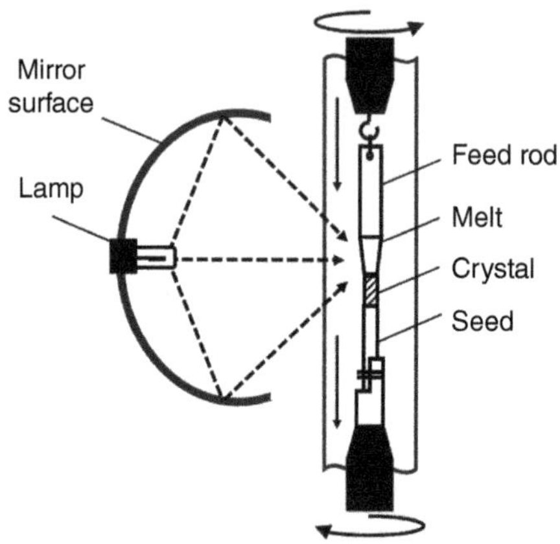

Figure 3.4 Schematic view of the optical floating zone method.

the melt, the FZ method is not suitable for growing large-diameter crystals. Since the melt zone is maintained in a floating state by surface tension, crystal growth is performed without a crucible. This feature avoids impurities from a crucible and provides various choices of growing atmosphere such as oxygen under high pressure [15]. A typical running time is a total of two days, one day (8–12 hours) for the preparation of the feed rod and one day (4–12 hours) to grow one crystal rod of several mm × several cm length. If users can prepare several feed rods at the same time, two to three samples can be grown each day with one machine. The main purpose of the μ-PD and FZ methods is material screening, and the selection of these two methods depends on the preference of each user.

3.3 Outlines of Oxide Scintillators

At the beginning of the research and development history of oxide scintillators, natural ore-based phosphors such as $CaWO_4$ were investigated following the invention of photodetectors.

$CdWO_4$ was discovered as an alternative to natural ore-based phosphors [16], and the first oxide bulk single crystal scintillators such as $CaWO_4$ and $CdWO_4$ were introduced by Gillette in 1950 [17]. In this era, intrinsic-type phosphors without dopant have mainly attracted attention. The characteristics of scintillators showing intrinsic luminescence are listed in Table 3.2, and most of them show a decay time as long as that of Tl:NaI.

Inorganic crystal scintillators have been widely used in medical diagnoses since the 1950s: X- or γ-ray detectors such as gamma cameras since 1958 [35], X-ray computed tomography since 1973 [36], and positron emission tomography (PET) since 1975, [37] have been used. As medical applications have progressed, $Bi_4Ge_3O_{12}$ (BGO) was discovered by Weber et al. in 1973 [38]. BGO is one of the most outstanding scintillating crystals established after the discovery of Tl:NaI, and has taken over from Tl:NaI in many applications. After entering the development stage of PET, which requires a short decay time of tens of nanoseconds, for these demands, the allowed 5d-4f transitions in rare-earth ions offered an attractive possibility to create new scintillators with a fast response. Following the discovery of $Ce:Gd_2SiO_5$ (Ce:GSO) in 1983 [39], the research and development for Ce-doped scintillators became intense. Table 3.3 shows developed oxide scintillators doped with Ce. The different emission wavelengths of 5d-4f transitions can be explained by the following. A crystal field having a low symmetry at Ce^{3+} sites splits a 5d level into five sublevels. A strong crystal field and a low symmetry make the lowest 5d level close to the 4f ground state. The 5d-4f emission

Table 3.2 *Oxide scintillators with intrinsic luminescence.*

	Z_{eff}	Emission (nm)	Decay time (ns)	*LY* (photons/MeV)	Refs.
$CsLiB_6O_{10}$	43	530	687, 3130	64 000	[18]
$Li_2B_4O_7$	7	310	287	277	[19, 20]
$LiAlO_2$	11	320	2200	5900	[21, 22]
$LiGaO_2$	28	330	460, 2800	3500	[21, 22]
LaB_3O_6	48	342	>10 000	500	[20]
Y_2O_3	37	335, 400	34	9300	[23]
Lu_2O_3	69	335, 400	–	500	[23]
$Y_3Al_5O_{12}$	32	310	460	14 300	[24]
$Lu_3Al_5O_{12}$	63	310, 390	25, 970	8600	[24]
$Lu_3Sc_2Al_3O_{12}$	62	340	850	5000	[25]
$YTaO_4$	63	420	700, 3800	15 200	[26]
$LuTaO_4$	69	330	700, 1400, 3100	9500	[26]
Li_3TaO_4	67	400	6100 + slow	17 400	[26]
$Mg_4Ta_2O_9$	64	350	3900 + slow	19 900	[26]
$Zn_3Ta_2O_8$	62	400	200, 1100	26 000	[26]
$MgWO_4$	67	470	36 000	15 000	[27]
$CaMoO_4$	35	520	20 000	2100	[28]
$CaWO_4$	66	430	8200	15 800	[29]
$ZnMoO_4$	35	600	~140 000	3550 (at 77 K)	[30]
$ZnWO_4$	65	500	20 000	9300	[31]
$CdWO_4$	64	480	12 000	26 000	[31, 32]
$PbWO_4$	76	420	6.6	200	[33]
$Bi_4Si_3O_{12}$	77	480	100	1700	[34]
$Bi_4Ge_3O_{12}$	75	480	300	8200	[31]

Table 3.3 *Ce-doped oxide scintillators.*

	Z_{eff}	Emission (nm)	Decay time (ns)	*LY* (photons/MeV)	Refs.
$Ce:Y_2SiO_5$	35	420	37	26300	[40, 41]
$Ce:Gd_2SiO_5$	59	440	41	8000	[39, 42]
$Ce:Lu_2SiO_5$	66	420	40	30000	[43]
$Ce,Ca:Lu_2SiO_5$	66	420	37	39000	[44]
$Ce:Lu_{1.8}Gd_{0.2}SiO_5$	66	420	40	36000	[45]
$Ce:Lu_{0.4}Gd_{1.6}SiO_5$	61	430	50	17200	[45]
$Ce:Lu_{1.8}Y_{0.2}SiO_5$	65	400	41	33800	[46]
$Ce:La_2Si_2O_7$	51	370	29	5400	[47]
$Ce:Gd_2Si_2O_7$	57	390	47	20000	[48, 49]
$Ce:La_{0.2}Gd_{1.8}Si_2O_7$	57	390	46, 346	36000	[50]
$Ce:Lu_2Si_2O_7$	65	380	38	26300	[46]
$Ce:Y_3Al_5O_{12}$	32	520	88, 302	20300	[51]
$Ce:Y_3Al_2Ga_3O_{12}$	33	510	55	32900	[52]
$Ce:Gd_3Al_2Ga_3O_{12}$	54	520	88	46000	[53]
$Ce:Tb_3Al_5O_{12}$	57	560	44 + slow	57000	[54]
$Ce:Tb_3Al_2Ga_3O_{12}$	55	540	31 + slow	10000	[55]
$Ce:Lu_3Al_5O_{12}$	63	510	58	12500	[56, 57]
$Ce:Lu_3Ga_2Al_3O_{12}$	62	490	36, 93	20400	[52]
$Ce:YAlO_3$	34	370	25	21400	[42, 58]
$Ce:GdAlO_3$	58	360	30	9000	[59]
$Ce:LuAlO_3$	65	360	19	11800	[60]
$Ce:Gd_{0.6}Y_{0.4}AlO_3$	53	390	15, 139	10000	[61]
$Ce:Lu_{0.3}Y_{0.7}AlO_3$	53	370	15–20	14200	[62]
$Ce:LuBO_3$	66	380	24	11000	[63]
$Ce:CaHfO_3$	65	440	35	8100	[64]
$Ce:SrHfO_3$	63	390	40	20000	[65]
$Ce:BaHfO_3$	64	400	25	10000	[66]

occurs by relaxation from the lowest 5d state to the 4f ground state, and the emission wavelengths change correspondingly.

Other rare-earth ions showing luminescence due to allowed 5d-4f or 4f-4f transitions were also focused on as a luminescence center. as listed in Table 3.4. Here, most of scintillators listed appear to be aluminates or silicates. This fact suggests that a rigid crystal lattice is likely to lead to high light yield (LY) in extrinsic-type scintillators. Some Eu-doped aluminates and silicates are also potential candidates for scintillators with a high *LY* of 40000–50000 photon/MeV, but the decay times are not as short as Ce-doped ones. In recent years, for applications requiring time resolution, a scintillation decay time shorter than Ce-doped scintillators has been demanded. Due to such demands, scintillators exhibiting luminescence due to 5d-4f transitions of Pr with a decay time constant of about 20 ns have been investigated: $Pr:Lu_3Al_5O_{12}$ (Pr:LuAG) has both short decay time and high *LY* [82]. In addition, scintillators based on charge transfer luminescence of Yb^{3+} with a decay time constant of a few ns have also attracted attention, but quenching at room temperature results in lower *LY* [83]. Very recent literature has also found Ce-doped Hf-based scintillators with higher density, higher *LY*, and shorter decay time than BGO [64].

Table 3.4 Other rare-earth-doped oxide scintillators.

	Z_{eff}	Emission (nm)	Decay time (ns)	*LY* (photons/MeV)	Refs.
$Nd:Y_3Al_5O_{12}$	32	4f-4f	3160	11 000	[67]
$Nd:Lu_3Al_5O_{12}$	63	4f-4f	762, 5060	11 000	[68]
$Nd:Lu_{0.1}Y_{0.9}AlO_3$	44	4f-4f	3800	6000	[68]
$Pr:Y_2SiO_5$	35	280, 310	31	3200	[69]
$Pr:Lu_2SiO_5$	66	290, 320	26	2400	[69]
$Pr:Y_2Si_2O_7$	34	380	8	8300	[70]
$Pr:Gd_2Si_2O_7$	57	350	23	1800	[70]
$Pr:Lu_2Si_2O_7$	65	380	11	9700	[70]
$Pr:YAlO_3$	34	250	14	7000	[71]
$Pr:LuAlO_3$	65	250, 300	9	4600	[72]
$Pr:Y_3Al_5O_{12}$	32	320	18	8000	[73]
$Pr:Lu_3Al_5O_{12}$	63	300	20	20 000	[74]
$Eu:Ba_2MgSi_2O_7$	44	505	280, 795	10 000	[75]
$Eu:Sr_2MgSi_2O_7$	32	465	186, 435	4000	[75]
$Eu:Ca_2MgSi_2O_7$	16	530	185, 487	11 000	[76]
$Eu:Ba_2Si_3O_8$	49	505	355, 1400	35 000	[75]
$Eu:Ba_2SiO_4$	52	505	242, 668	40 000	[75]
$Eu:BaSi_2O_5$	47	520	2800	30 000	[75]
$Eu:Ba_5Si_8O_{21}$	48	453, 511	305, 852	20 000	[75]
$Eu:BaSiO_3$	50	550	285	8100	[77]
$Eu:CaAl_2O_4$	15	440	166	12 000	[78]
$Eu:SrAl_2O_4$	31	510	447	46 000	[79]
$Eu:BaAl_2O_4$	48	500	438	28 000	
$Ho:Lu_3Al_5O_{12}$	63	4f-4f	590, 3100	3100	[68]
$Er:Lu_3Al_5O_{12}$	63	4f-4f	810, 5800	6300	[68]
$Tm:YAlO_3$	34	4f-4f	18 000	15 000	[68]
$Tm:Lu_3Al_5O_{12}$	63	4f-4f	720, 1180	12 000	[68]
$Yb:Lu_2O_3$	69	350, 460	1.3	500	[80]
$Yb:Y_3Al_5O_{12}$	32	330	930	930	[81]

3.4 Silicate Materials

3.4.1 $Ce:Gd_2SiO_5$ (Ce:GSO)

Ce:GSO was developed by Takagi et al. in 1983 [39], and is the first scintillator to achieve a short decay time by using Ce^{3+} as an emission center for PET application. The discovery of GSO is important because the trend of the activation of Ce^{3+} starts from here. Ce:GSO is an exceptionally good performance scintillator with high radiation resistance, short decay time of 30–60 ns, *LY* of about 20% of Tl:NaI, and energy resolution of 7–8% [84]. However, there is the problem that high dose irradiation increases the *LY* (positive hysteresis). Figure 3.5 shows change in the *LY* of 0.5% and 1.5% Ce-doped GSO scintillators when ^{60}Co is γ-ray irradiated at a dose rate of 45 krad/h [85]. The *LY* of 0.5% Ce:GSO increases about 1.3 times compared with before irradiation. The positive hysteresis is a problem for practical detectors since it equals a fluctuation in gain. Ce:GSO has good

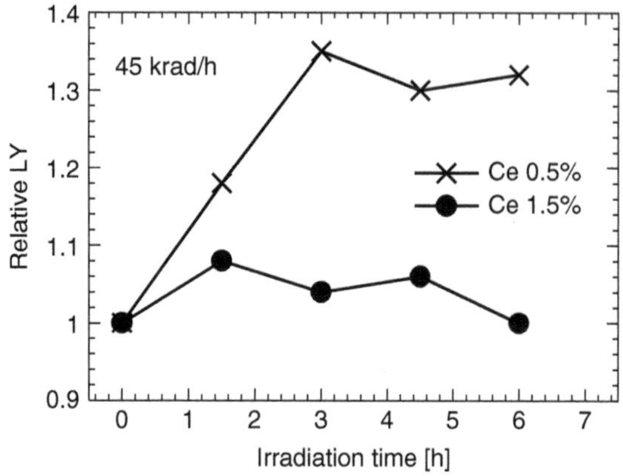

Figure 3.5 *Changes of LY (positive hysteresis) in Ce:GSO with a function of irradiation time.*

temperature dependence on the scintillation output, and the scintillation output at 150 °C is approximately 60% of the value at room temperature [86]. Therefore, Ce:GSO are practically used in well-logging for thermal neutron capture measurements [87].

GSO only has a congruent melting composition in a Gd_2O_3-SiO_2 binary system with a melting point of 1900 °C [88]. GSO has a monoclinic symmetry with a space group of $P2_{1/C}$ and a cleavage plane (100), and the thermal expansion along the [010] axis is two to three times as large as the other axes [89]. Therefore, a crystal crack due to cleavage may occur when growing a single crystal. In addition, it is necessary to perform cutting and polishing in consideration of the cleavage surface.

Ce:GSO exhibits scintillation with a broad emission peak at 430 nm due to 5d-4f transition of Ce^{3+} under X-ray exposure [90]. In PL, excitation peaks appear at 250, 280, and 330 nm. Excitation peaks at 250 and 280 nm were also observed in undoped GSO, due to the 4f-4f transition of Gd^{3+} ($^8S_{7/2} - {}^6I_J$ and $^6D_J - {}^6I_J$). Excited electrons relax with non-radiative transitions to ^6P_J, and energy transfer to Ce occurs owing to the overlap of the $^6P_J - {}^8S_{7/2}$ transitions and the 4f-5d transitions of Ce^{3+}. The decay curve of 0.5% Ce:GSO is approximated by the sum of two exponential decay components. The decay time constants of the main component (~80%) and the slow second component (~20%) are 30–60 ns and 300–600 ns, respectively. *LY* is reduced to around 80% by 1.5% Ce-doping compared with 0.5% Ce:GSO, but also reduces the decay time constant by about half [89]. As shown in Figure 3.6, Ce:GSO clearly shows the slow rise time (14 ns) of the light pulse, resulting in limit of the measured time resolution [91].

3.4.2 Ce:Lu_2SiO_5 (Ce:LSO)

Ce:LSO, which was developed by Melcher et al. in 1992, is an excellent blue-emitting scintillator doped with Ce^{3+} as an emission center [4, 43]. Ce:LSO has outstanding properties such as high density of 7.39 g/cm^3, short decay time of 40–50 ns, and three times as high *LY* as that of a BGO scintillator, so it has been intensely used as a scintillator for PET applications. However, the presence of Lu leads to high background radiation caused by self-irradiation because of ^{176}Lu isotopes with a half-life of 3.78×10^{10} years and 2.59% natural abundance. For this reason, the energy resolution of Lu-based silicates is not good (~9%) despite the high *LY* [92]. The crystal structure of LSO is monoclinic, but LSO belongs to a

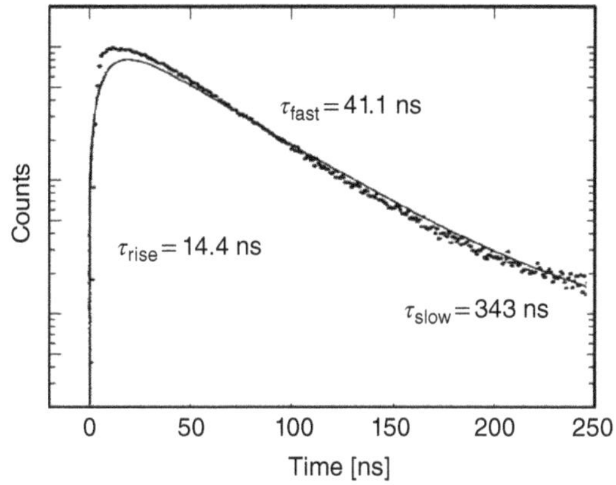

Figure 3.6 *Scintillation decay curve of Ce:GSO. Source: Reproduced from M. Moszyński et al. [91].*

different space group (*C*2/c) from GSO due to smaller radii of Lu ions than those of Gd ions. As shown in Figure 3.7, LSO has a congruent melting composition [93], and the bulk crystals can be produced by the CZ method. However, due to the extremely high melting point of 2150 °C and high anisotropy of linear expansion coefficient, there are also industrial problems about manufacturing and processing costs are high.

LSO has two inequivalent Lu^{3+} sites in the crystal structure, which are coordinated by seven and six oxygen atoms. An emission due to Ce1 substituted at one Lu^{3+} site is observed at 393 and 427 nm, and Ce_2 substituted at another site exhibits a peak at 460 nm. Unlike GSO, there is no migration process relate to Gd, and the rise time is fast. The temperature dependence of the scintillation output has also been investigated in the literature [94]. Unlike Ce:GSO, the output of Ce:LSO drastically degraded above 50 °C and dropped to 20% at 100 °C. Thus, Ce:LSO could not be applied to well-logging applications.

LYSO and LGSO are compounds having mixed crystal compositions of two rare-earth orthosilicates. Both have an LSO-type crystal structure, and the melting points are about

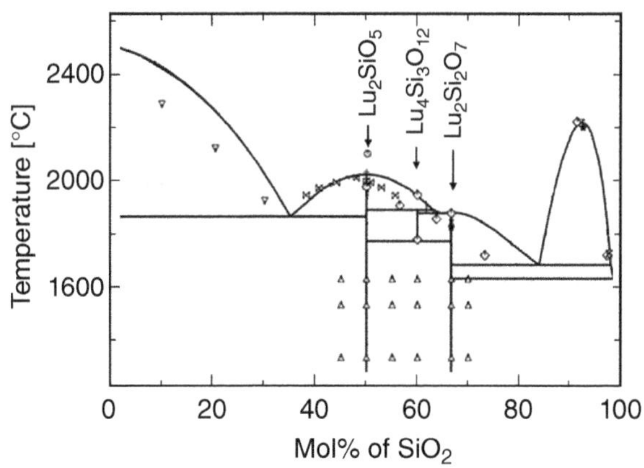

Figure 3.7 *Phase diagram of Lu_2O_3-SiO_2 binary system. Source: Reproduced from X. Ye et al. [93].*

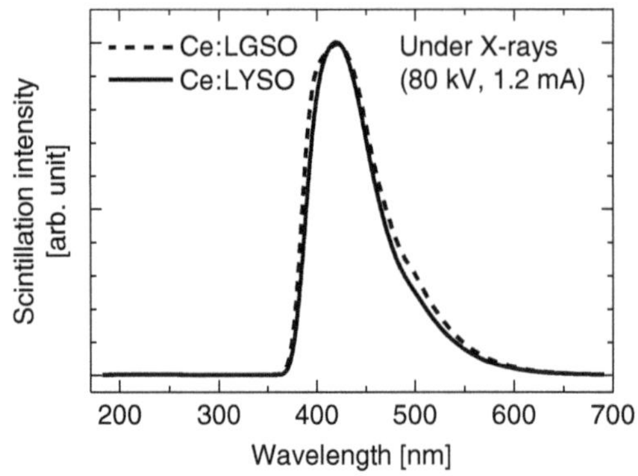

Figure 3.8 *X-ray-induced scintillation spectra of Ce:LGSO and Ce:LYSO crystals.*

2100 °C higher than that of GSO. As shown in Figure 3.8, they exhibit scintillation with a broad peak at 410–430 nm. The decay time constants are 43–45 ns which is shorter than Ce:GSO, and Ce:LYSO and Ce:LGSO do not show a slow rise time. The characteristics are similar to the LSO mentioned above, but the *LY* was slightly higher than that of Ce:LSO. In addition to improvements of the scintillation properties, low Lu content decreases the production costs of raw material and radioactivity of ^{176}Lu.

3.4.3 Ce:$Gd_2Si_2O_7$ (Ce:GPS)

Ce:GPS has attracted attention as a new high-luminance scintillator with no self-radioactivity and ρZ_{eff}^4 about twice that of Tl:NaI. Since Ce:GPS polycrystalline powders showed a scintillation output higher than that of Ce:GSO crystal, in 2009, Kawamura et al. performed crystal growth of Ce:GPS by the FZ method [48]. Although it is difficult to grow a pure GPS crystal, 10% Ce-substituted GPS is a congruent melting composition with triclinic symmetry and can be successfully synthesized by the FZ method. The triclinic Ce:GPS achieved a *LY* comparable with Ce:LSO and a high energy resolution of 6–8% for 662 keV γ-rays from ^{137}Cs [95].

The crystal structure of the high-temperature phase is orthorhombic, while the low-temperature phase displays a triclinic crystal structure with a density of 6.0 g/cm^3. The melting point is 1800 °C, and it is difficult to grow GPS from the melt because GPS has an incongruent melting composition. Rare-earth pyrosilicates ($RE_2Si_2O_7$) including GPS show various phases depending on the average ionic radius of the rare-earth sites. Figure 3.9 shows the correlation between the average ion radius rare-earth of pyrosilicates and the polymorphism, as revealed by Felshe [96]. Based on this knowledge, crystal growth of 10% La-admixed GPS (La-GPS) was attempted, and a high-quality single crystal was obtained [50, 97].

Figure 3.10 shows the X-ray-induced scintillation spectra of Ce:GPS and Ce:La-GPS. Both Ce:GPS and Ce:La-GPS exhibit scintillation with a broad emission band at 390 nm [97]. The primary decay time constant is tens of nanoseconds, and a slow component is significantly observed in Ce:La-GPS. From pulse height analyses, Ce:La-GPS exhibits significantly higher *LY* (30 000–40 000 photons/MeV [50]) than that of triclinic Ce:GPS. Both Ce:GPS and Ce:La-GPS shows good temperature dependence, and the scintillation output is still stable above 150 °C [98]. Therefore, they can be expected to contribute to well-logging applications.

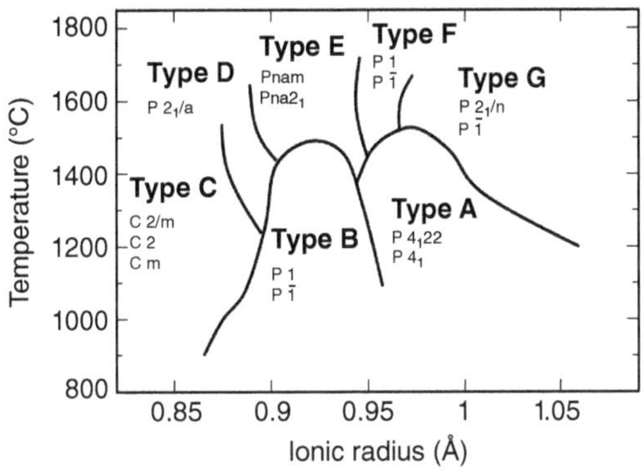

Figure 3.9 The unit-cell geometry of the observed structure types of pyrosilicates.

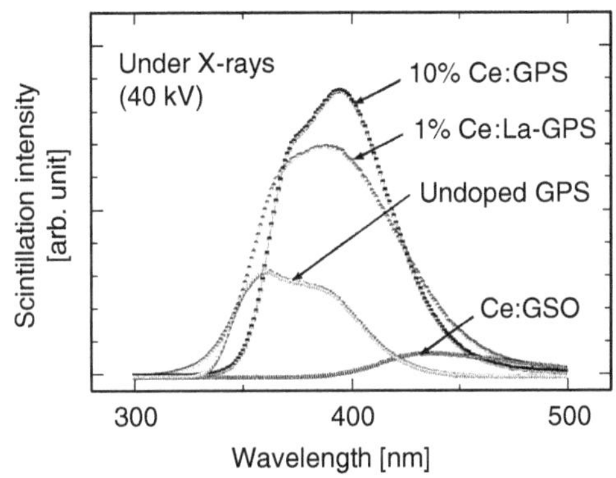

Figure 3.10 X-ray-induced scintillation spectra of GPS and GSO scintillators. Source: Reproduced from I. Gerasymov et al. [97].

3.4.4 LPS

Ce:LPS is a high-performance scintillator developed in 2000 [99], having excellent scintillation properties with a high density (6.2 g/cm^3), high *LY* comparable with Ce:LSO [46], short decay time (38 ns), and low afterglow; therefore, Ce:LPS is a promising candidate for X- and γ-ray detection. Ce:LPS shows the *LY* of 26 300 photons/MeV with energy resolution of approximately 11% [46]. Ce:LPS can be used for well-logging applications because of its good temperature dependence, in which the *LY* is stable up to 450 K [46, 100]. The crystal structure has monoclinic symmetry with the thortveitite structure belonging to space group of *C*2/m and one crystallographic Lu-site [101]. Ce:LPS exhibits scintillation with a broad emission band peaking at 380 nm as well as a shoulder peak around 405 nm due to the 5d-4f transitions of Ce^{3+}.

3.5 Garnet Materials

3.5.1 Ce:$Y_3Al_5O_{12}$ (Ce:YAG)

Ce:YAG has excellent mechanical and chemical stability and exhibits short decay time (60–80 ns), high *LY*, and emission peak at 520–550 nm [51], which makes it suitable for uses combined with Si-based photodetectors. Since Ce:YAG has been intensely studied as a common phosphor, the luminescence characteristics under ionizing radiations have been reported for a long time.

The crystal structure is cubic and consists of four types of unit cells. Since it congruently melts at 1970 °C [102], YAG crystals can be successfully grown from the melt. A partially substituted mixed garnet crystal is easily formed, but most garnets have a lattice defect such as anti-site defects and oxygen vacancies [103]. In grown crystals, Y^{3+} ions often substitute Al^{3+} sites to form anti-site defects (AD) Y^{3+}_{Al}, and polycrystals have less defect and show higher transmittance than garnet single crystals [24].

Figure 3.11 shows the scintillation spectra of Ce:YAG under X-ray irradiation, and a broad emission peak due to the 5d-4f transitions of Ce^{3+} is observed at 520 nm with a high LY of 14 300 photons/MeV [24]. The emission wavelength is longer than other Ce-doped phosphors because Ce ions are introduced into eight-coordinated Y sites, and the 5d excited levels approach the 4f ground level due to a strong crystal field. As shown in Figure 3.12, the decay time under γ-ray irradiation is 80–90 ns and 300–400 ns, while those under α-ray irradiation become shorter. Therefore, the pulse-shape discrimination (PSD) of α and γ signals in Ce:YAG has been studied by using different pulse shapes [51].

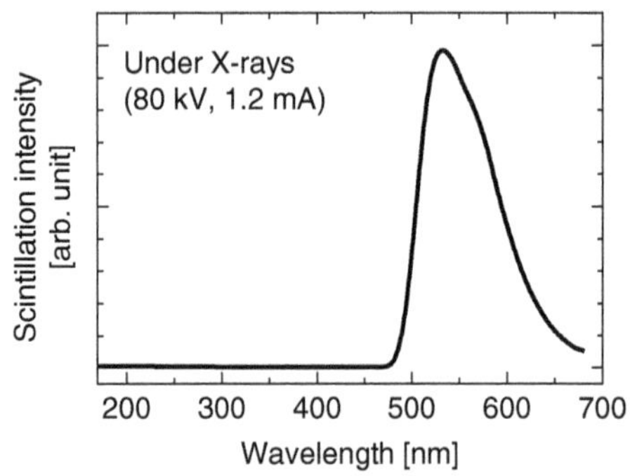

Figure 3.11 *X-ray-induced scintillation spectra of Ce:YAG.*

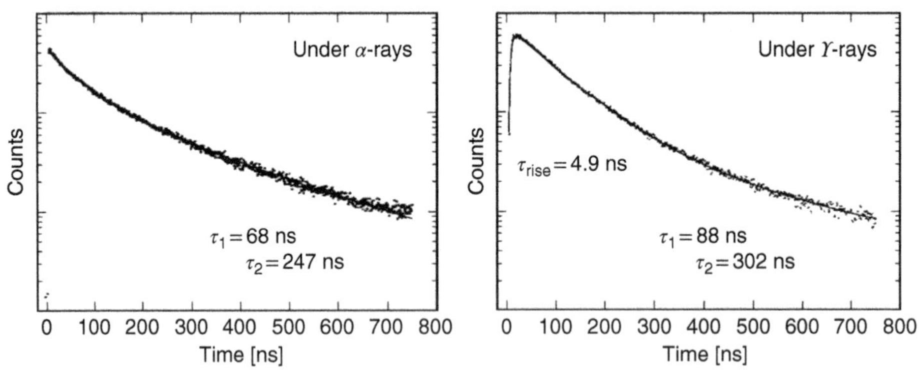

Figure 3.12 *Scintillation decay curves of Ce:YAG under α and γ-ray irradiation. Source: Reproduced from M. Moszynski et al. [51].*

3.5.2 Ce:$Lu_3Al_5O_{12}$ (Ce:LuAG), Pr:$Lu_3Al_5O_{12}$ (Pr:LuAG)

Ce- or Pr-doped LuAG have been widely studied owing to their excellent scintillation properties and a high density (6.73 g/cm^3) suitable for X- and γ-ray detection. Pr:LuAG show a high *LY* of 20 000 photons/MeV and a short decay time constant of approximately 20 ns [74], so it has been already utilized for positron emission mammography [104]. In contrast, Ce:LuAG is also a good performing scintillator with *LY* of 12 500 photons/MeV [56] and a short decay time, and co-doping of Ga achieves an *LY* higher than 20 000 photons/MeV [3] and higher density. However, LuAG also has intrinsic radioactivity because of containing ^{176}Lu. LuAG has a cubic crystal structure and congruently melts at 1950 °C. Because of high mechanical and chemical stability and no cleavage, there is no problem in processing. Ce and Pr have a large ionic radius as compared with Lu, so that the segregation coefficient is small. Figure 3.13 shows the X-ray-induced scintillation spectra of Ce- and Pr-doped LuAG. Ce:LuAG exhibits scintillation with a broad emission peak due to the 5d-4f transition of Ce observed at 510 nm, which is slightly shorter than that of Ce:YAG [51]. The rise time is 1 ns, which is several times faster than Ce:YAG. Pr:LuAG shows emissions from the 5d-4f transition of Pr^{3+} appearing at 310 and 375 nm, which are attributable to 5d-4f (^3H_j, ^3F_j) transitions, respectively. Several sharp peaks are also observed in the range longer than 480 nm and belong to the 4f-4f transitions of Pr^{3+}. The pulse height spectrum of ^{137}Cs γ-rays demonstrates that Pr:LuAG shows the *LY* of 19 000 photons/MeV with energy resolution of 4.3% [105]. The Pr 5d-4f emission shows good temperature dependence, and the scintillation *LY* is 70% at 150 °C [106].

3.5.3 Ce:$Gd_3Al_2Ga_3O_{12}$ (Ce:GAGG)

Ce:GAGG shows an outstanding *LY* of approximately 46 000 photons/MeV with a good resolution of 5–6% [107, 108], which is one of the oxide scintillators showing the highest *LY*. Ce:GAGG has no intrinsic radioactivity, has a high density, and is not deliquescent. Thus, it has practical use for a scintillator mounted on the product such as a survey meter for γ-ray detection. Despite the high *LY*, the temperature dependence is very poor, with *LY* decreasing to about 10% at 150 °C [106]. GAGG has a cubic crystal structure, and congruently melts at 1850 °C, so that the crystal growth can be performed by the CZ method.

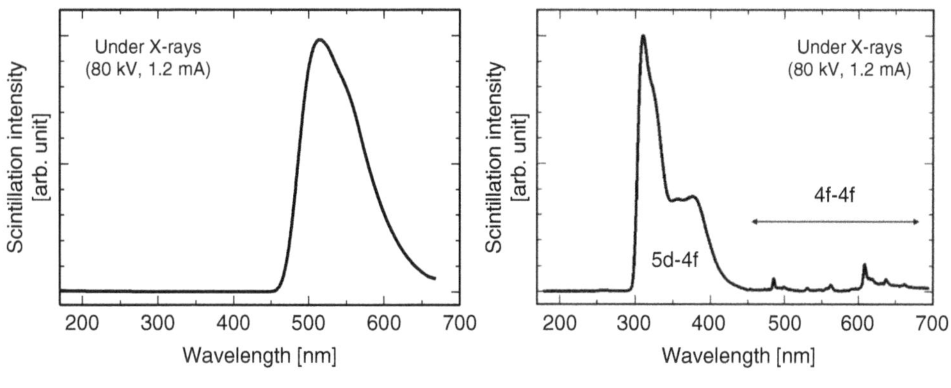

Figure 3.13 *X-ray-induced scintillation spectra of (left) Ce:LuAG and (right) Pr:LuAG crystals.*

$Gd_3Al_5O_{12}$ is also an attractive host having high density (6.2 g/cm^3) and no intrinsic radioactivity. However, $Gd_3Al_5O_{12}$ single crystals cannot be synthesized by the melt growth, because the ionic radius of the rare-earth site is large, and $Gd_3Al_5O_{12}$ has an incongruent melting composition, [88] as shown in Figure 3.14. $Gd_3Ga_5O_{12}$ also congruently melts, but the emission from Ce:$Gd_3Ga_5O_{12}$ is hardly observed [3]. The quench can be interpreted by the 5d level buried in the conduction band.

Figure 3.15 shows the scintillation spectrum, and Ce:GAGG shows scintillation with a broad peak at 530 nm, with decay time constants of 80–100 and 200–300 ns [53]. As well as Ce:YAG, the rise time under γ-ray exposure is slow (4.6 ns), and the scintillation pulses of Ce:GAGG have significantly different decay and rise times for γ- and α-rays, [110] as shown in Figure 3.16. The decay and rise times for γ-rays are shorter than those for α-rays. Ce:GAGG may have an advantage in particle detectors, where discrimination between γ- and α-rays is required.

3.5.4 Ce:$Tb_3Al_5O_{12}$ (Ce:TAG)

Ce:TAG has been investigated as a Lu-free garnet scintillator as well as Gd-based garnets. Ce:TAG shows efficient energy transfer from Tb to Ce. In recent years, it has been reported

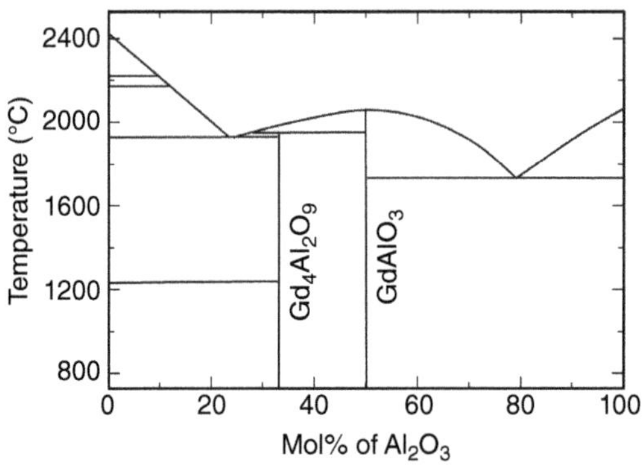

Figure 3.14 *Phase diagram of Gd_2O_3-Al_2O_3 binary system. Source: Reproduced from S. Lakiza [109].*

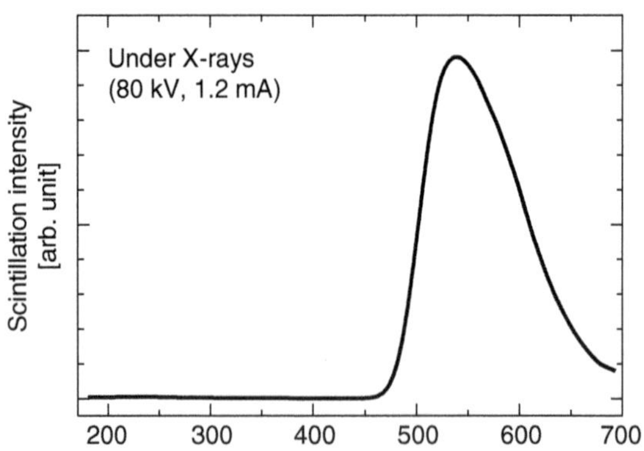

Figure 3.15 *X-ray-induced scintillation spectra of Ce:GAGG crystal.*

that Ce:TAG shows high *LY* [54, 111]. The crystal structure is cubic symmetry, and the melting point is 1970 °C. As shown in Figure 3.17, TAG has an incongruent melting composition, but congruent melting can be performed on the Al_2O_3-rich side of TAG [112, 113]. However, it is still difficult to synthesize a high-quality single crystal owing to contamination of perovskite phases.

The X-ray-induced scintillation spectra of Ce:TAG are shown in Figure 3.18. An intense broad emission band appears at 560 nm, and sharp peaks due to 4f–4f transitions of Tb^{3+} ($^5D_4 \rightarrow {}^7F_4$) are also observed. The emission wavelength is slightly longer than other Ce-doped garnet scintillators because of the modification of the local crystal field. The decay time constant of the first component is several tens of nanoseconds, but slow components in microseconds cannot be ignored, as shown in Figure 3.19. Ce:TAG exhibits a high *LY* of 57 000 photons/MeV [54], which is comparable with Ce:GAGG. The *LY* decreases as the Al site is replaced with Ga [55], which is a similar behavior to $Pr:Lu_3(Ga, Al)_5O_{12}$. These garnet materials can be synthesized in the transparent ceramic form described in Chapter 4.

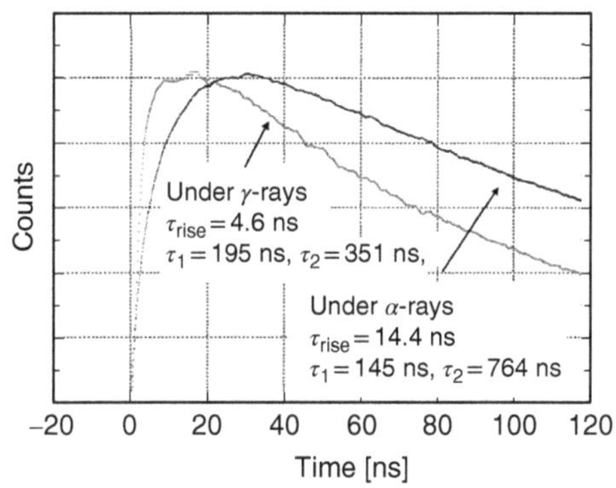

Figure 3.16 *Scintillation decay curves of Ce:GAGG under α- and γ-ray irradiation. Source: Reproduced from M. Kobayashi et al. [110].*

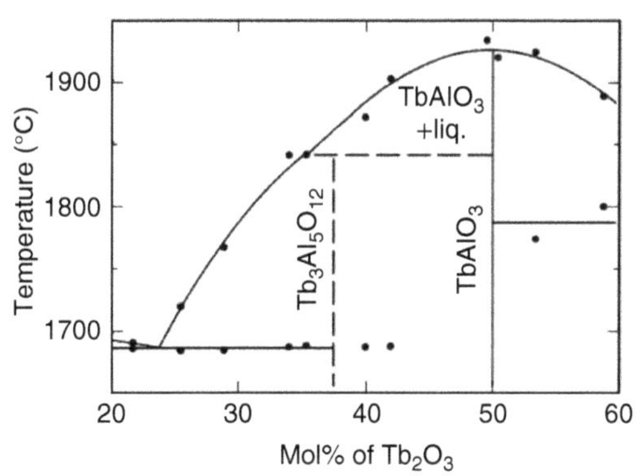

Figure 3.17 *Phase diagram of Tb_2O_3-Al_2O_3 binary system. Source: Reproduced from M. Geho et al. [112].*

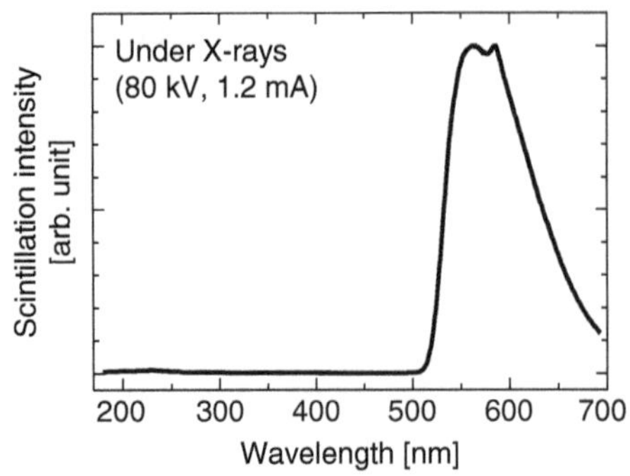

Figure 3.18 *X-ray-induced scintillation spectra of Ce:TAG crystal.*

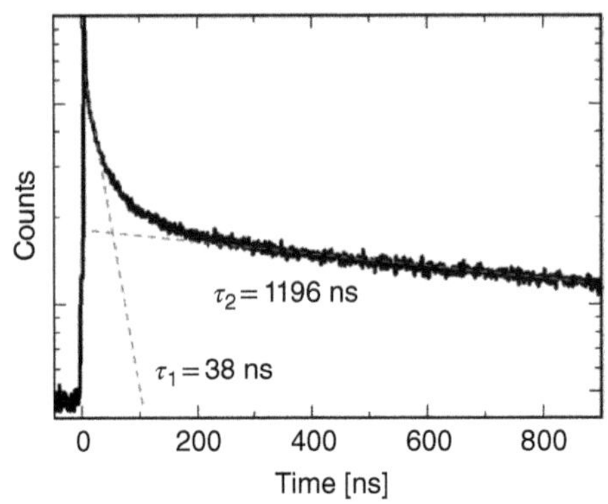

Figure 3.19 *Scintillation decay curve of Ce:TAG crystal.*

3.6 Perovskite Materials

3.6.1 Ce:$YAlO_3$ (Ce:YAP)

Ce:YAP crystals have a short decay time and mechanical stability ideal for use in the detection of low energy X- and γ-rays. Due to the short decay time, scintillation detectors based on Ce:YAP can be operated at a count rate of a few MHz. YAP has an orthorhombic crystal structure and congruently melts at 1875 °C. It is difficult to grow a large volume of single crystal because the conditions for stable growth are severe, and the crystal can often become twinned. Figure 3.20 shows scintillation spectrum of Ce:YAP under X-ray irradiation. Ce:YAP shows scintillation with a broad peak at 350 nm [42], and the emission wavelength is shorter than those of other Ce-doped garnets. The decay time constant is 24 ns, and the rise time is very short (0.9 ns). The *LY* increases from low to room temperature, slowly decreases up to 200 °C, and drops sharply at high temperatures [42, 114]. Figure 3.21 illustrates the pulse height spectra of ^{137}Cs γ-rays measured using Ce:YAP crystal, which shows *LY* of 21 400 photons/MeV with excellent energy resolution of 4.38% [58].

3.6.2 Ce:$LuAlO_3$ (Ce:LuAP)

Ce:LuAP has been attracting attention owing to its higher density (8.34 g/cm^3) than BGO. The linear attenuation coefficient against γ-rays is the highest among all commercial scintillators.

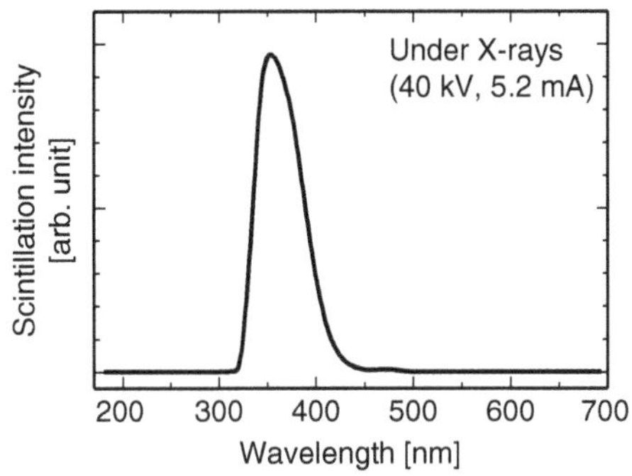

Figure 3.20 *X-ray-induced scintillation spectra of Ce:YAP crystal.*

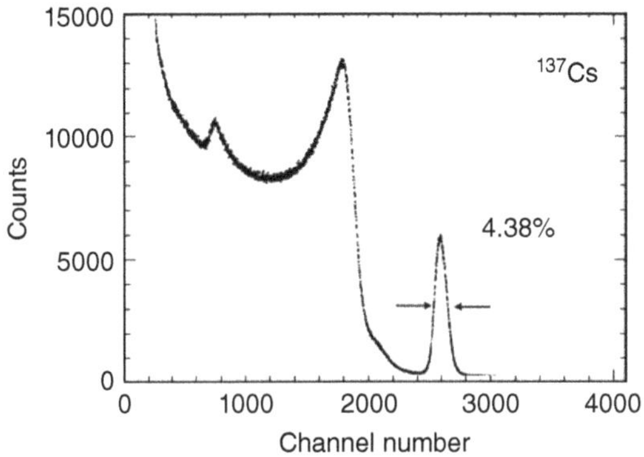

Figure 3.21 *Pulse height spectra of ^{137}Cs γ-rays measured using Ce:YAP crystal.*
Source: Reproduced from M. Kapusta et al. [58].

The crystal structure has the same orthorhombic perovskite structure as YAP. LuAP has an incongruent melting composition with the melting point of 1900 °C [115], so that the contaminated garnet phase causes variations in *LY* and decay characteristics. To solve the instability, mixed crystals such as LuYAP and LuGAP, in which a part of Lu is replaced by Y or Gd having a large ionic radius, have been developed [62]. Scintillation properties of Ce:LuYAP at high temperatures were reported in some literature, and the scintillation output at high temperature is higher than 60% compared with the value at room temperature [94, 116] Ce:LuAP exhibits scintillation with a broad peak at 360 nm due to 5d-4f transitions of Ce with a very short decay time of about 20 ns [60]. The *LY* has 11 800 photons/MeV with energy resolution of 6.8% [117].

3.7 Materials with Intrinsic Luminescence

3.7.1 $CdWO_4$

$CdWO_4$ is a well-known scintillator due to its useful features such as high density (7.9 g/cm^3), large Z_{eff} (61.2), and high scintillation output (~15 000 photons/MeV [31]). In addition, $CdWO_4$ has no deliquescence and low afterglow and is advantageous compared with $CaWO_4$. Therefore, $CdWO_4$ has been used with Si-PD for X-ray detectors for security applications [118]. $CdWO_4$

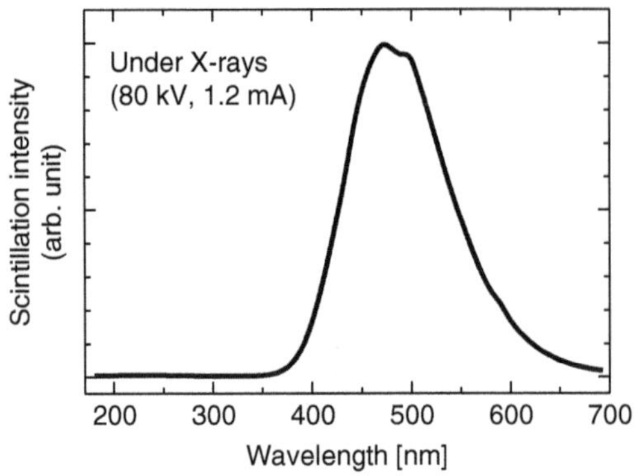

Figure 3.22 *X-ray-induced scintillation spectra of $CdWO_4$ crystal.*

has a monoclinic wolframite-type crystal structure and has a congruently melting composition whose melting point is 1325 °C [119], and $CdWO_4$ scintillators are synthesized by the CZ method in general.

As shown in Figure 3.22, $CdWO_4$ exhibits a broad emission peak at 470 nm due to $(WO_6)^{-6}$ complexion, and shows a relatively long decay time of several μs [120]. The *LY* is about 26 000 photons/MeV with energy resolution of 6.6% [32]. Although $CdWO_4$ exhibits totally high performance, long decay time constants prevent use in applications requiring high counting rates.

3.7.2 $Bi_4Ge_3O_{12}$ (BGO)

BGO has a highly effective atomic number (75.2) and a high density (7.13 g/cm^3), resulting in high detection efficiency. Therefore, BGO has been widely used in X- and γ-ray measurement applications such as medicine and security, since the scintillation properties of BGO was first reported in 1973 [38]. The crystal structure is cubic symmetry, and BGO has a congruent melt composition with a melting point of 1044 °C [121]. In 1965, Nitsche first reported on the crystal growth of BGO by the CZ method [122].

Figure 3.23 shows X-ray-induced scintillation spectrum of BGO, and a broad emission peak is observed around 480 nm with a decay time constant of about 300 ns [31]. Generally, two luminescence mechanisms have been proposed [123]. One is emission due to excitation and relaxation of 6sp-6s^2 transitions of Bi^{3+}. The other is that absorbed energy is transferred

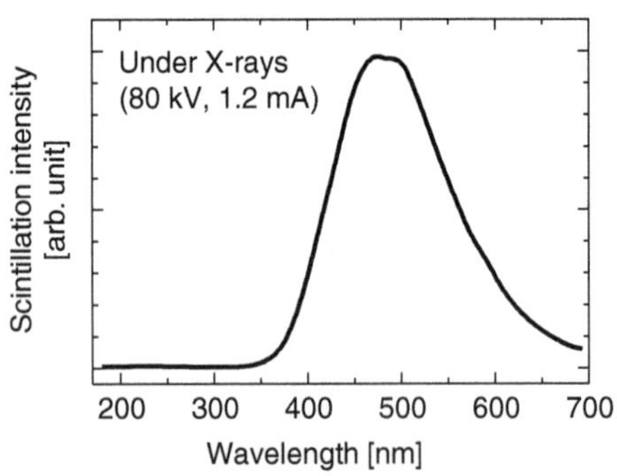

Figure 3.23 *X-ray-induced scintillation spectra of BGO crystal.*

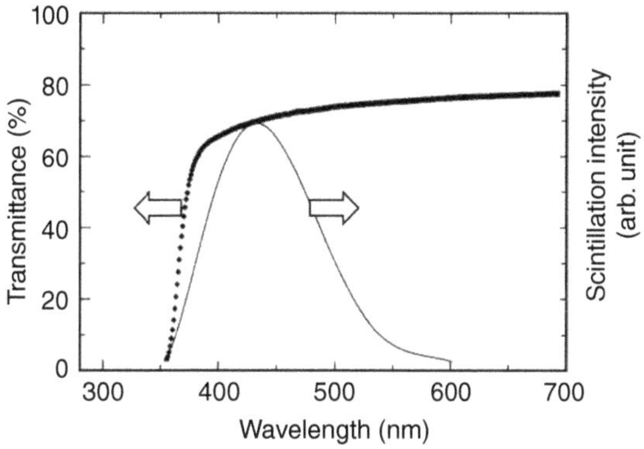

Figure 3.24 *X-ray-induced scintillation spectra of $PbWO_4$ crystal. Source: Reproduced from A. Annenkov et al. [33].*

to Bi or trap levels via excitons, and then recombines with holes near the top of the valence band. BGO exhibits a *LY* of about 8000 photons/MeV, with energy resolution of 9–10% [124].

3.7.3 $PbWO_4$

$PbWO_4$ has been studied as a fast and high-dense scintillator (8.28 g/cm^3) with short radiation length, so it has been applied to calorimeters in high energy experiments. The undoped crystals had disadvantages of low transmittance and low radiation resistance, so did not attract much attention for a long time. After being rediscovered in 1992, the transmittance at short wavelengths and radiation hardness were dramatically improved, and slow components were suppressed by removing Mo impurities and doping trivalent rare-earth ions such as La, Gd, and Y [125, 126]. The crystal structure has a tetragonal symmetry with a celite-type structure. $PbWO_4$ has a congruent melting composition with a melting point of 930 °C [127].

Figure 3.24 illustrates the X-ray-induced scintillation spectra of $PbWO_4$ crystal, which shows a broad emission peak at 430 nm [33]. The emission origin is a $(WO_4)^{2-}$ complex ion, which emits via radiative transition involving charge transfer between W^{6+} and O^{2-}. The decay time is very short (6 ns), which accounts for more than 80% of the decay curve [128].

References

1. van Eijk, C.W.E. (2003). Inorganic scintillators in medical imaging detectors. *Nucl. Instrum. Methods Phys. Res. A.* 509: 17–25. 10.1016/S0168-9002(03)01542-0.
2. Dorenbos, P. (2002). Light output and energy resolution of Ce^{3+}-doped scintillators. *Nucl. Instrum. Methods Phys. Res. A.* 486: 208–213. 10.1016/S0168-9002(02)00704-0.
3. Kamada, K., Endo, T., Tsutumi, K. et al. (2011). Composition engineering in cerium-doped $(Lu,Gd)_3(Ga,Al)_5O_{12}$ single-crystal scintillators. *Cryst. Growth Des.* 11: 4484–4490. 10.1021/cg200694a.
4. Melcher, C.L. and Schweitzer, J.S. (1992). A promising new scintillator: cerium-doped lutetium oxyorthosilicate. *Nucl. Instrum. Methods Phys. Res. A.* 314: 212–214. 10.1016/0168-9002(92)90517-8.
5. Yoon, D.-H., Yonenaga, I., Fukuda, T. et al. (1994). Crystal growth of dislocation-free $LiNbO_3$ single crystals by micro pulling down method. *J. Cryst. Growth* 142: 339–343. 10.1016/0022-0248(94)90342-5.
6. Tatarchenko, V.A. (1977). Capillary shaping in crystal growth from melts. *J. Cryst. Growth* 37: 272–284. 10.1016/0022-0248(77)90120-8.

7. Miyazawa, S. (1982). Shaped melt lowering (SML): a novel growth technique permitting keff = 1. *J. Cryst. Growth* 60: 331–337. 10.1016/0022-0248(82)90107-5.
8. Sugiyama, M., Yokota, Y., Fujimoto, Y. et al. (2012). Dopant segregation in rare earth doped lutetium aluminum garnet single crystals grown by the micro-pulling down method. *J. Cryst. Growth* 352: 110–114. 10.1016/j.jcrysgro.2011.12.039.
9. Pfann, W.G. (1962). Zone melting: this technique offers unique advantages in purification and in control of composition in various substances. *Science* 135: 1101–1109. 10.1126/science.135.3509.1101.
10. Pfann, W.G. (1966). *Zone Melting*, 2e. New York: Wiley.
11. Pfann, W.G. (1952). Principles of zone-melting. *JOM* 4: 747–753. 10.1007/BF03398137.
12. Poplawsky, R.P. and Thomas, J.E. (1960). Floating zone crystals using an arc image furnace. *Rev. Sci. Instrum.* 31: 1303–1308. 10.1063/1.1716879.
13. Nakauchi, D., Okada, G., Kawaguchi, N. et al. (2018). Scintillation properties of $RE_2Hf_2O_7$ (RE=La, Gd, Lu) single crystals prepared by xenon arc floating zone furnace. *Jpn. J. Appl. Phys.* 57: 100307. 10.7567/JJAP.57.100307.
14. Akatsuka, M., Nakauchi, D., Kawaguchi, N. et al. (2019). Scintillation properties of yttrium-stabilized zirconia crystals synthesized by the floating-zone method. *Sens. Mater.* 31: 1289–1295. 10.18494/SAM.2019.2189.
15. Balbashov, A.M. and Egorov, S.K. (1981). Apparatus for growth of single crystals of oxide compounds by floating zone melting with radiation heating. *J. Cryst. Growth* 52: 498–504. 10.1016/0022-0248(81)90328-6.
16. Kröger, F.A. and Urbach, F. (1948). Some aspects of the luminescence of solids. *Phys. Today* 1: 24–25. 10.1063/1.3066188.
17. Gillette, R.H. (1950). Calcium and cadmium tungstate as scintillation counter crystals for gamma-ray detection. *Rev. Sci. Instrum.* 21: 294–301. 10.1063/1.1745567.
18. Yanagida, T., Sakairi, M., Kato, T. et al. (2020). Scintillation detector properties of $CsLiB_6O_{10}$ (CLBO) crystal. *Appl. Phys. Express.* 13: 016001. 10.7567/1882-0786/ab5ffa.
19. Kobayashi, M., Ishii, M., and Senguttuvan, N. (2015) Scintillation characteristics of undoped and Cu+ doped $Li_2B_4O_7$ single crystals, ArXiv Prepr. ArXiv1503.03759. http://arxiv.org/abs/1503.03759.
20. Chernikov, V.V., Dubovik, M.F., Gavrylyuk, V.P. et al. (2003). Peculiarities of scintillation parameters of some complex composition borate single crystals. *Nucl. Instrum. Methods Phys. Res. A.* 498: 424–429. 10.1016/S0168-9002(02)02051-X.
21. Yanagida, T., Fujimoto, Y., Koshimizu, M. et al. (2017). Comparative studies of optical and scintillation properties between $LiGaO_2$ and $LiAlO_2$ crystals. *J. Physical Soc. Japan* 86: 094201. 10.7566/JPSJ.86.094201.
22. Yanagida, T., Watanabe, K., Okada, G. et al. (2019). Neutron and gamma-ray pulse shape discrimination of $LiAlO_2$ and $LiGaO_2$ crystals. *Nucl. Instrum. Methods Phys. Res. A.* 919: 64–67. 10.1016/j.nima.2018.11.135.
23. Futami, Y., Yanagida, T., Fujimoto, Y. et al. (2013). Optical and scintillation properties of Sc_2O_3, Y_2O_3 and Lu_2O_3 transparent ceramics synthesized by SPS method. *Radiat. Meas.* 55: 136–140. 10.1016/j.radmeas.2013.01.014.
24. Fujimoto, Y., Yanagida, T., Yagi, H. et al. (2014). Comparative study of intrinsic luminescence in undoped transparent ceramic and single crystal garnet scintillators. *Opt. Mater.* 36: 1926–1929. 10.1016/j.optmat.2014.06.019.
25. Ryskin, N.N., Dorenbos, P., Van Eijk, C.W.E. et al. (1995). Scintillation properties of $Lu_3Al_{5-x}Sc_xO_{12}$ crystals. *Radiat. Effects Defects Solids* 135: 311–313. 10.1080/10420159508229857.
26. Bourret, E.D., Smiadak, D.M., Borade, R.B. et al. (2018). Scintillation of tantalate compounds. *J. Lumin.* 202: 332–338. 10.1016/j.jlumin.2018.05.044.
27. Danevich, F.A., Chernyak, D.M., Dubovik, A.M. et al. (2009)). $MgWO_4$ – a new crystal scintillator. *Nucl. Instrum. Methods Phys. Res. A.* 608: 107–115. 10.1016/j.nima.2009.06.040.
28. Bavykina, I., Angloher, G., Hauff, D. et al. (2009). Development of cryogenic phonon detectors based on $CaMoO_4$ and $ZnWO_4$ scintillating crystals for direct dark matter search experiments. *Opt. Mater.* 31: 1382–1387. 10.1016/j.optmat.2008.09.018.
29. Moszyński, M., Balcerzyk, M., Czarnacki, W. et al. (2005). Characterization of $CaWO_4$ scintillator at room and liquid nitrogen temperatures. *Nucl. Instrum. Methods Phys. Res. A.* 553: 578–591. 10.1016/j.nima.2005.07.052.
30. Chernyak, D.M., Danevich, F.A., Degoda, V.Y. et al. (2015). Effect of tungsten doping on $ZnMoO_4$ scintillating bolometer performance. *Opt. Mater.* 49: 67–74. 10.1016/j.optmat.2015.08.017.

31. Holl, I., Lorenz, E., and Mageras, G. (1988). A measurement of the light yield of common inorganic scintillators. *IEEE Trans. Nucl. Sci.* 35: 105–109. 10.1109/23.12684.
32. Moszynski, M., Balcerzyk, M., Kapusta, M. et al. (2004). $CdWO_4$ crystal in gamma-ray spectrometry. *IEEE Symp. Conf. Rec. Nucl. Sci.* 2: 798–803. 10.1109/NSSMIC.2004.1462330.
33. Annenkov, A., Korzhik, M., and Lecoq, P. (2002). Lead tungstate scintillation material. *Nucl. Instrum. Methods Phys. Res. A.* 490: 30–50. 10.1016/S0168-9002(02)00916-6.
34. Kobayashi, M., Morimoto, K., Yoshida, H. et al. (1983). Bismuth silicate as a scintillating material for electromagnetic shower detectors. *Nucl. Instrum. Methods Phys. Res., Sect. A.* 205: 133–136. 10.1016/0167-5087(83)90181-3.
35. Anger, H.O. (1958). Scintillation camera. *Rev. Sci. Instrum.* 29: 27–33. 10.1063/1.1715998.
36. Hounsfield, G.N. (1973). Computerized transverse axial scanning (tomography): part 1. Description of system. *Br. J. Radiol.* 46: 1016–1022. 10.1259/0007-1285-46-552-1016.
37. Ter Pogossian, M.M., Phelps, M.E., Hoffman, E.J.N.A. et al. (1975). A positron emission transaxial tomograph for nuclear imaging (PETT). *Radiology* 114: 89–98. 10.1148/114.1.89.
38. Weber, M.J. and Monchamp, R.R. (1973). Luminescence of $Bi_4Ge_3O_{12}$: spectral and decay properties. *J. Appl. Phys.* 44: 5495–5499. 10.1063/1.1662183.
39. Takagi, K. and Fukazawa, T. (1983). Cerium-activated Gd_2SiO_5 single crystal scintillator. *Appl. Phys. Lett.* 42: 43–45. 10.1063/1.93760.
40. Wanarak, C., Phunpueok, A., and Chewpraditkul, W. (2012). Scintillation response of $Lu_{1.95}Y_{0.05}SiO_5$:Ce and Y_2SiO_5:Ce single crystal scintillators. *Nucl. Instrum. Methods Phys. Res., Sect. B* 286: 72–75. 10.1016/j.nimb.2011.11.028.
41. Tombrello, T.A., Melcher, C.L., and Schweitzer, J.S. (1992). UV and gamma-ray excited luminescence of cerium-doped rare-earth oxyorthosilicates. *Nucl. Instrum. Methods Phys. Res. A.* 320: 263–272. 10.1016/0168-9002(92)90784-2.
42. Baccaro, S., Blaẑek, K., de Notaristefani, F. et al. (1995). Scintillation properties of YAP:Ce. *Nucl. Instrum. Methods Phys. Res. A.* 361: 209–215. 10.1016/0168-9002(95)00016-X.
43. Melcher, C.L. and Schweitzer, J.S. (1992). Cerium-doped lutetium oxyorthosilicate: a fast, efficient new scintillator. *IEEE Trans. Nucl. Sci.* 39: 502–505.
44. Spurrier, M.A., Szupryczynski, P., Yang, K. et al. (2008). Effects of Ca^{2+} co-doping on the scintillation properties of LSO:Ce. *IEEE Trans. Nucl. Sci.* 55: 1178–1182. 10.1109/TNS.2007.913486.
45. Kurata, Y., Usui, T., Shimura, N. et al. (2009). LGSO scintillators for PET. *Med. Imaging Technol.* 27: 13–17. 10.11409/mit.27.13.
46. Pidol, L., Kahn-Harari, A., Viana, B. et al. (2004). High efficiency of lutetium silicate scintillators, Ce-doped LPS, and LYSO crystals. *IEEE Trans. Nucl. Sci.* 51: 1084–1087. 10.1109/TNS.2004.829542.
47. Wan, H., Wang, Y., Zhuang, L. et al. (2018). Photoluminescence, scintillation properties and trap states of $La_2Si_2O_7$ Ce single crystal. *Mater. Res. Express.* 5: 086202. 10.1088/2053-1591/aad1dd.
48. Kawamura, S., Kaneko, J.H., Higuchi, M. et al. (2007). Floating zone growth and scintillation characteristics of cerium-doped gadolinium pyrosilicate single crystals. *IEEE Trans. Nucl. Sci.* 54: 1383–1386. 10.1109/TNS.2007.902372.
49. Kawamura, S., Kaneko, J.H., Higuchi, M. et al. (2009). Scintillation characteristics of Ce:$GGGd_2Si_2O_7$ (Ce 2.530 mol%) single crystals prepared by the floating zone method. *IEEE Trans. Nucl. Sci.* 56: 328–330. 10.1109/TNS.2008.2010043.
50. Suzuki, A., Kurosawa, S., Shishido, T. et al. (2012). Fast and high-energy-resolution oxide scintillator: Ce-doped $(La,Gd)_2Si_2O_7$. *Appl. Phys. Express.* 5: 102601. 10.1143/APEX.5.102601.
51. Moszynski, M., Ludziehewski, T., Wolski, D. et al. (1994). Properties of the YAG:Ce scintillator. *Nucl. Instrum. Methods Phys. Res. A.* 345: 461–467.
52. Kamada, K., Yanagida, T., Pejchal, J. et al. (2011). Scintillator-oriented combinatorial search in Ce-doped $(Y,Gd)_3(Ga,Al)$ $5O_{12}$ multicomponent garnet compounds. *J. Phys. D Appl. Phys.* 44 10.1088/0022-3727/44/50/505104.
53. Yanagida, T., Kamada, K., Fujimoto, Y. et al. (2013). Comparative study of ceramic and single crystal Ce:GAGG scintillator. *Opt. Mater.* 35: 2480–2485. 10.1016/j.optmat.2013.07.002.
54. Nakauchi, D., Okada, G., Kawano, N. et al. (2017). Luminescent and scintillation properties of Ce-doped $Tb_3Al_5O_{12}$ crystal grown from Al-rich composition. *Appl. Phys. Express.* 10: 072601. 10.7567/APEX.10.072601.
55. Nakauchi, D., Okada, G., Kawano, N. et al. (2018). Effects of Ga substitution in Ce:$Tb_3Ga_xAl_{5-x}O_{12}$ single crystals for scintillator applications. *Jpn. J. Appl. Phys.* 57 10.7567/JJAP.57.02CB02.

56. Mares, J.A., Beitlerova, A., Nikl, M. et al. (2004). Scintillation response of Ce-doped or intrinsic scintillating crystals in the range up to 1 MeV. *Radiat. Meas.* 38: 353–357. 10.1016/j.radmeas.2004.04.004.
57. Nikl, M., Mares, J.A., Solovieva, N. et al. (2004). Energy transfer to the Ce3+ centers in $Lu_3Al_5O_{12}$:Ce scintillator. *Phys. Status Solidi* 201: R41–R44. 10.1002/pssa.200409041.
58. Kapusta, M., Balcerzyk, M., Moszyński, M. et al. (1999). A high-energy resolution observed from a YAP:Ce scintillator. *Nucl. Instrum. Methods Phys. Res. A.* 421: 610–613. 10.1016/S0168-9002(98)01232-7.
59. Dorenbos, P., Bougrine, E., De Haas, J.T.M. et al. (1995). Scintillation properties of $GdAlO_3$:Ce crystals. *Radiat. Effects Defects Solids* 135: 321–323. 10.1080/10420159508229859.
60. Lempicki, A. and Glodo, J. (1998). Ce-doped scintillators: LSO and LuAP. *Nucl. Instrum. Methods Phys. Res. A.* 416: 333–344. 10.1016/S0168-9002(98)00689-5.
61. Kamada, K., Endo, T., Tsutsumi, K. et al. (2012). Crystal growth and scintillation properties of Ce doped (Gd,Y)AlO_3 perovskite single crystals. *Phys. Status Solidi Curr. Top. Solid State Phys.* 9: 2263–2266. 10.1002/pssc.201200264.
62. Chval, J., Clément, D., Giba, J. et al. (2000). Development of new mixed lux(RE^{3+})1-xAP:Ce scintillators ($RE^{3+} = Y^{3+}$ or Gd^{3+}): comparison with other Ce-doped or intrinsic scintillating crystals. *Nucl. Instrum. Methods Phys. Res. A.* 443: 331–341. 10.1016/S0168-9002(99)01066-9.
63. Zhang, L., Pédrini, C., Madej, C. et al. (1999). Fast fluorescence and scintillation properties of cerium and praseodymium doped lutetium orthoborates. *Radiat. Effects Defects Solids* 150: 47–52. 10.1080/10420159908226206.
64. Fukushima, H., Nakauchi, D., Okada, G. et al. (2018, 2018). Synthesis and scintillation properties of Ce-doped $CaHfO_3$ crystals. *J. Mater. Sci. Mater. Electron.* 29: 21033–21039. 10.1007/s10854-018-0249-9.
65. van Eijk, C.W.E. (2002). Inorganic scintillators in medical imaging. *Phys. Med. Biol.* 47: R85–R106. 10.1088/0031-9155/47/8/201.
66. Greskovich, C. and Duclos, S. (1997). Ceramic scintillators. *Annu. Rev. Mater. Sci.* 27: 69–88. 10.1146/annurev.matsci.27.1.69.
67. Yanagida, T. and Sato, T. (2014). Optical and scintillation properties of Nd-doped complex garnet. *Opt. Mater.* 38: 174–178. 10.1016/j.optmat.2014.10.020.
68. Yanagida, T. (2013). Study of rare-earth-doped scintillators. *Opt. Mater.* 35: 1987–1992. 10.1016/j.optmat.2012.11.002.
69. Pejchal, J., Nikl, M., Mihókov, E. et al. (2009). Pr^{3+}-doped complex oxide single crystal scintillators. *J. Phys. D Appl. Phys.* 42 10.1088/0022-3727/42/5/055117.
70. Yanagida, T., Watanabe, K., Okada, G. et al. (2018). Optical, scintillation and radiation tolerance properties of Pr-doped pyrosilicate crystals. *Jpn. J. Appl. Phys.* 57: 106401. 10.7567/JJAP.57.106401.
71. van Eijk, C.W.E., Dorenbos, P., and Visser, R. (1994). Nd^{3+} and Pr^{3+} doped inorganic scintillators. *IEEE Trans. Nucl. Sci.* 41: 738–741. 10.1109/23.322798.
72. Drozdowski, W., Wojtowicz, A.J., Wiśniewski, D. et al. (2006). Scintillation properties of Pr-activated $LuAlO_3$. *Opt. Mater.* 28: 102–105. 10.1016/j.optmat.2004.09.031.
73. Dorenbos, P., Visser, R., Khaidukov, N.M. et al. (1993). Scintillation properties of some Ce^{3+} and Pr^{3+} doped inorganic crystals. *IEEE Trans Nucl. Sci.* 40: 388–394.
74. Ogino, H., Yoshikawa, A., Nikl, M. et al. (2006). Scintillation characteristics of Pr-doped $Lu_3Al_5O_{12}$ single crystals. *J. Cryst. Growth* 292: 239–242. 10.1016/j.jcrysgro.2006.04.021.
75. Derenzo, S., Bizarri, G., Borade, R. et al. (2011). New scintillators discovered by high-throughput screening. *Nucl. Instrum. Methods Phys. Res. A.* 652: 247–250. 10.1016/j.nima.2010.09.156.
76. Ogawa, T., Nakauchi, D., Okada, G. et al. (2019). Scintillation properties of Ce- and Eu-doped $Ca_2MgSi_2O_7$ crystals. *Opt. Mater.* 89: 63–67. 10.1016/j.optmat.2018.12.056.
77. Nakauchi, D., Okada, G., Kawaguchi, N. et al. (2017). Scintillation properties of $BaSiO_3$:Eu single crystals prepared by the floating zone method. *J. Mater. Sci. Mater. Electron.* 28: 6972–6976. 10.1007/s10854-016-5752-2.
78. Nakauchi, D., Kawaguchi, N., and Yanagida, T. (2019). Luminescence and scintillation properties of Eu-doped $CaAl_2O_4$ bulk crystals. *Sens. Mater.* 31: 1249–1255. 10.18494/SAM.2019.2184.
79. Nakauchi, D., Okada, G., Koshimizu, M. et al. (2016). Storage luminescence and scintillation properties of Eu-doped $SrAl_2O_4$ crystals. *J. Lumin.* 176: 342–346. 10.1016/j.jlumin.2016.04.008.
80. Yanagida, T., Fujimoto, Y., Yagi, H. et al. (2014). Optical and scintillation properties of transparent ceramic Yb:Lu_2O_3 with different Yb concentrations. *Opt. Mater.* 36: 1044–1048. 10.1016/j.optmat.2014.01.022.

81. Antonini, P., Bressi, G., Carugno, G. et al. (2001). Scintillation properties of YAG:Yb crystals. *Nucl. Instrum. Methods Phys. Res. A.* 460: 469–471. 10.1016/S0168-9002(00)01066-4.
82. Nikl, M., Ogino, H., Krasnikov, A. et al. (2005). Photo- and radioluminescence of Pr-doped $Lu_3Al_5O_{12}$ single crystal. *Phys. Status Solidi Appl. Mater. Sci.* 202: 202–204. 10.1002/pssa.200409079.
83. Guerassimova, N., Garnier, N., Dujardin, C. et al. (2001). X-ray excited charge transfer luminescence of ytterbium-containing aluminium garnets. *Chem. Phys. Lett.* 339: 197–202. 10.1016/S0009-2614(01)00331-1.
84. Kokubun, M., Abe, K., Ezoe, Y. et al. (2004). Improvements of the astro-E2 hard X-ray detector (HXD-II). *IEEE Trans. Nucl. Sci.* 51: 1991–1996. 10.1109/TNS.2004.832921.
85. Tanaka, M., Hara, K., Kim, S. et al. (1998). Applications of cerium-doped gadolinium silicate Gd_2SiO_5:Ce scintillator to calorimeters in high-radiation environment. *Nucl. Instrum. Methods Phys. Res. A.* 404: 283–294. 10.1016/S0168-9002(97)01142-X.
86. Melcher, C.L., Schweitzer, S., Manente, R.S. et al. (1991). Applicability of GSO scintillators for well logging. *IEEE Trans. Nucl. Sci.* 38: 506–509. 10.1109/23.289349.
87. Nikitin, A. and Bliven, S. (2010) Needs of well logging industry in new nuclear detectors, in: IEEE Nucl. Sci. Symp. Med. Imaging Conf., IEEE, 2010: pp. 1214–1219. doi:10.1109/NSSMIC.2010.5873961.
88. Smolin, Y.I. and Tkachev, S.P. (1969). Determination of the structure of gadolinium oxyorthosilicate Gd_2O_3-SiO_2. *Sov. Phys. Crystal.* 14: 14–16.
89. Ishibashi, H., Kurashige, K., Kurata, Y. et al. (1998). Scintillation performance of large Ce-doped Gd_2SiO_5 (GSO) single crystal. *IEEE Trans. Nucl. Sci.* 45: 518–521.
90. Melcher, C.L., Schweitzer, J.S., Peterson, C.A. et al. (1996) Crystal growth and scintillation properties of the rare earth oxyorthosilicates, in: Proceeding Int. Conf. Inorg. Scintill. Their Appl., Delft University Press, pp. 309–315.
91. Moszyński, M., Ludziejewski, T., Wolski, D. et al. (1996). Timing properties of GSO, LSO and other Ce doped scintillators. *Nucl. Instrum. Methods Phys. Res. A.* 372: 51–58. 10.1016/0168-9002(95)01244-3.
92. Balcerzyk, M., Moszynski, M., Kapusta, M. et al. (2000). YSO, LSO, GSO and LGSO. A study of energy resolution and nonproportionality. *IEEE Trans. Nucl. Sci.* 47: 1319–1323. 10.1109/23.872971.
93. Ye, X., Luo, Y., Liu, S. et al. (2017). Experimental study and thermodynamic calculation of Lu_2O_3-SiO_2 binary system. *J. Rare Earths* 35: 927–933. 10.1016/S1002-0721(17)60996-7.
94. Fedorov, A., Korzhik, M., Lobko, A. et al. (2005)). Light yield temperature dependence of lutetium-based scintillation crystals. *Nucl. Instrum. Methods Phys. Res. A.* 537: 276–278. 10.1016/j.nima.2004.08.025.
95. Tsubota, Y., Kaneko, J.H., Higuchi, M. et al. (2014). Crystal growth and scintillation properties of $La_{0.2}Ce_{0.05}Gd_{1.75}Si_2O_7$ (La-GPS) single crystal with triclinic structure. *Opt. Mater.* 36: 665–669. 10.1016/j.optmat.2013.11.007.
96. Felsche, J. (1970). Polymorphism and crystal data of the rare-earth disilicates of type R.E.$2Si_2O_7$. *J. Less Common Met.* 21: 1–14. 10.1016/0022-5088(70)90159-1.
97. Gerasymov, I., Sidletskiy, O., Neicheva, S. et al. (2011). Growth of bulk gadolinium pyrosilicate single crystals for scintillators. *J. Cryst. Growth* 318: 805–808. 10.1016/j.jcrysgro.2010.11.047.
98. Tsubota, Y., Kaneko, J.H., Higuchi, M. et al. (2015). High-temperature scintillation properties of orthorhombic $Gd_2Si_2O_7$ aiming at well logging. *Appl. Phys. Express* 8: 062602. 10.7567/APEX.8.062602.
99. Pauwels, D., Le Masson, N., Viana, B. et al. (2000). A novel inorganic scintillator: $Lu_2Si_2O_7$:Ce_{3+} (LPS). *IEEE Trans. Nucl. Sci.* 47: 1787–1790. 10.1109/23.914446.
100. Pidol, L., Khan-Harari, A., Viana, B. et al. (2003). Scintillation properties of $Lu_2Si_2O_7$:Ce^{3+}, a fast and efficient scintillator crystal. *J. Phys. Condens. Matter* 15: 2091–2102. 10.1088/0953-8984/15/12/326.
101. Bretheau-Raynal, F., Lance, M., and Charpin, P. (1981). Crystal data for $Lu_2Si_2O_7$. *J. Appl. Cryst.* 14: 349–350. 10.1107/S0021889881009527.
102. Fabrichnaya, O., Seifert, H.J., Ludwig, T. et al. (2001). The assessment of thermodynamic parameters in the Al_2O_3-Y_2O_3 system and phase relations in the Y-Al-O system. *Scand. J. Metall.* 30: 175–183. 10.1034/j.1600-0692.2001.300308.x.
103. Zorenko, Y., Voloshinovskii, A., Savchyn, V. et al. (2007). Exciton and antisite defect-relateduminescence in $Lu_3Al_5O_{12}$ and $Y_3Al_5O_{12}$ garnets. *Phys. Status Solidi Basic Res.* 244: 2180–2189. 10.1002/pssb.200642431.
104. Yanagida, T., Yoshikawa, A., Yokota, Y. et al. (2010). Development of Pr:LuAG scintillator array and assembly for positron emission mammography. *IEEE Trans. Nucl. Sci.* 57: 1492–1495. 10.1109/TNS.2009.2032265.
105. Yanagida, T., Yoshikawa, A., Ikesue, A. et al. (2009). Basic properties of ceramic Pr:LuAG scintillators. *IEEE Trans. Nucl. Sci.* 56: 2955–2959. 10.1109/TNS.2009.2026475.

106. Yanagida, T., Fujimoto, Y., Kurosawa, S. et al. (2013). Temperature dependence of scintillation properties of bright oxide scintillators for well-logging. *Jpn. J. Appl. Phys.* 52: 076401. 10.7567/JJAP.52.076401.
107. Kamada, K., Yanagida, T., Endo, T. et al. (2012). 2 inch diameter single crystal growth and scintillation properties of Ce:$Gd_3Al_2Ga_3O_{12}$. *J. Cryst. Growth* 352: 88–90. 10.1016/j.jcrysgro.2011.11.085.
108. Iwanowska, J., Swiderski, L., Szczesniak, T. et al. (2013). Performance of cerium-doped $Gd_3Al_2Ga_3O_{12}$ (GAGG:Ce) scintillator in gamma-ray spectrometry. *Nucl. Instrum. Methods Phys. Res. A.* 712: 34–40. 10.1016/j.nima.2013.01.064.
109. Lakiza, S., Fabrichnaya, O., Wang, C. et al. (2006). Phase diagram of the ZrO_2-Gd_2O_3-Al_2O_3 system. *J. Eur. Ceram. Soc.* 26: 233–246. 10.1016/j.jeurceramsoc.2004.11.011.
110. Kobayashi, M., Tamagawa, Y., Tomita, S. et al. (2012). Significantly different pulse shapes for ⊠- and ⊠-rays in $Gd_3Al_2Ga_3O$ 12:Ce_3 scintillating crystals. *Nucl. Instrum. Methods Phys. Res. A.* 694: 91–94. 10.1016/j.nima.2012.07.055.
111. Zorenko, Y., Douissard, P.A., Martin, T. et al. (2017). Scintillating screens based on the LPE grown $Tb_3Al_5O1_2$:Ce single crystalline films. *Opt. Mater.* 65: 73–81. 10.1016/j.optmat.2016.09.066.
112. Geho, M., Sekijima, T., and Fujii, T. (2004). Growth of terbium aluminum garnet ($Tb_3Al_5O_{12}$; TAG) single crystals by the hybrid laser floating zone machine. *J. Cryst. Growth* 267: 188–193. 10.1016/j.jcrysgro.2004.03.068.
113. Ganschow, S., Klimm, D., Reiche, P. et al. (1999). On the crystallization of terbium aluminium garnet. *Cryst. Res. Technol.* 34: 615–619. 10.1016/0022-2836(78)90206-1.
114. Glodo, J. and Wojtowicz, A. (2000). Thermoluminescence and scintillation properties of LuAP and YAP. *J. Alloys Compd.* 300–301: 289–294. 10.1016/S0925-8388(99)00726-4.
115. Petrosyan, A.G., Popova, V.F., Gusarov, V.V. et al. (2006). The Lu_2O_3-Al_2O_3 system: relationships for equilibrium-phase and supercooled states. *J. Cryst. Growth* 293: 74–77. 10.1016/j.jcrysgro.2006.05.017.
116. Baberdin, A., Dutova, A., Fedorov, A. et al. (2008). (Lu-Y)AlO_3:Ce scintillator for well logging. *IEEE Trans. Nucl. Sci.* 55: 1170–1173. 10.1109/TNS.2008.919261.
117. Balcerzyk, M., Moszyński, M., Galazka, Z. et al. (2005). Perspectives for high resolution and high light output LuAP:Ce crystals. *IEEE Trans. Nucl. Sci.* 52: 1823–1829. 10.1109/TNS.2005.856744.
118. Fritz, S.L. and Cook, L.T. (1987). High-resolution digital X-ray detector utilizing a discrete array of $CdWO_4$ scintillators and a self-scanned photodiode array. *Med. Phys.* 14: 244–248. 10.1118/1.596077.
119. Nestor, O.H. (1982). Scintillator material growth. *MRS Proc.* 16: 77. 10.1557/PROC-16-77.
120. Melcher, C.L., Manente, R.A., and Schweitzer, J.S. (1989). Applicability of barium fluoride and cadmium tungstate for scintillators for well logging. *IEEE Trans. Nucl. Sci.* 36: 1188–1192.
121. Voszka, R., Gévay, G., Földvári, I. et al. (1982). Growth and characterization of $Bi_4Ge_3O_{12}$ single crystals. *Acta Phys. Acad. Sci. Hungaricae* 53: 7–13. 10.1007/BF03156171.
122. Nitsche, R. (1965). Crystal growth and electro-optic effect of bismuth Germanate, Bi_4 $(GeO_4)_3$. *J. Appl. Phys.* 36: 2358–2360. 10.1063/1.1714490.
123. Rogemond, F., Pedrini, C., Moine, B. et al. (1985). Fluorescence properties of $Bi_4Ge_3O_{12}$(BGO) single crystals under laser excitation: excited state dynamics and saturation effects. *J. Lumin.* 33: 455–486. 10.1016/0022-2313(85)90114-0.
124. Moszyński, M., Balcerzyk, M., Czarnacki, W. et al. (2003). Intrinsic energy resolution and light yield non-proportionality of BGO. *IEEE Nucl. Sci. Symp. Conf. Rec.* 1: 92–96. 10.1109/nssmic.2003.1352005.
125. Kobayashi, M., Usuki, Y., Ishii, M. et al. (1999). Significant improvement of $PbWO_4$ scintillating crystals by doping with trivalent ions. *Nucl. Instrum. Methods Phys. Res. A.* 434: 412–423. 10.1016/S0168-9002(99)00550-1.
126. Kobayashi, M., Usuki, Y., Ishii, M. et al. (1997). Improvement in transmittance and decay time of $PbWO_4$ scintillating crystals by La-doping. *Nucl. Instrum. Methods Phys. Res. A.* 399: 261–268. 10.1016/S0168-9002(97)00929-7.
127. Chang, L.L.Y. (1971). Phase relations in the system PbO-WO_3. *J. Am. Ceram. Soc.* 54: 357–358. 10.1111/j.1151-2916.1971.tb12316.x.
128. Derenzo, S.E., Moses, W.W., Cahoon, J.L. et al. (1990). Prospects for new inorganic scintillators. *IEEE Trans. Nucl. Sci.* 37: 203–208. 10.1109/23.106619.

4

Inorganic Fluoride Scintillators

Noriaki Kawaguchi, Hiromi Kimura, Daisuke Nakauchi, Takumi Kato, and Takayuki Yanagida
Nara Institute of Science and Technology, Ikoma, Japan

4.1 Introduction

Since the ZnS:Ag scintillator's use for radiation detection experiments over 100 years ago [1], various scintillators have since been developed, including oxides, iodides, bromides, and fluorides. Figure 4.1 shows decay times and light yields of commercially available γ-ray scintillators. The horizontal axis is the light yield, the vertical axis is the decay time, and the information on "hygroscopic," "slightly hygroscopic," or "non-hygroscopic" are indicated by the symbols. From the figure, it can be seen that the fluoride scintillators are characterized by fast decay times, low hygroscopicities, and relatively lower light yields than iodide scintillators.

The light yield *LY* (photons/MeV) and the number of emitted photons N_{ph} (photons) under γ-ray excitation are expressed by the following equations [7]:

$$LY = \frac{N_{\text{ph}}}{E_\gamma} \tag{4.1}$$

$$N_{ph} = \frac{E_\gamma \cdot 10^6}{\beta \cdot E_{\text{gap}}} \cdot S \cdot Q \tag{4.2}$$

where $E\gamma$ is the γ-ray energy absorbed by the scintillator (MeV), and $\beta \cdot E_{\text{gap}}$ is the energy required to generate one electron–hole pair. The β is a constant, depending on the composition

Phosphors for Radiation Detectors, First Edition. Edited by Takayuki Yanagida and Masanori Koshimizu.

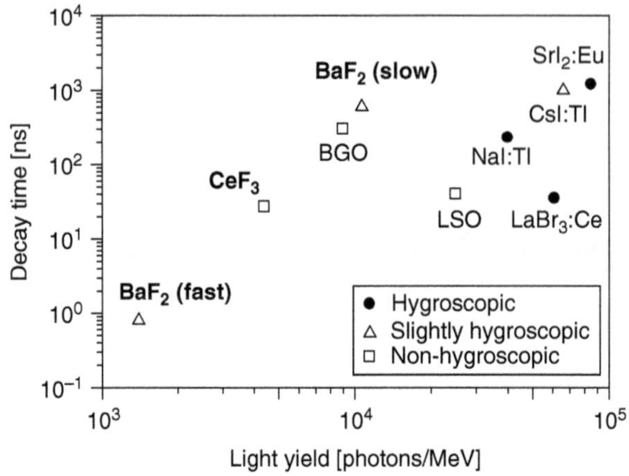

Figure 4.1 *Decay times and light yields of commercial scintillators for γ-ray detection [2–6].*

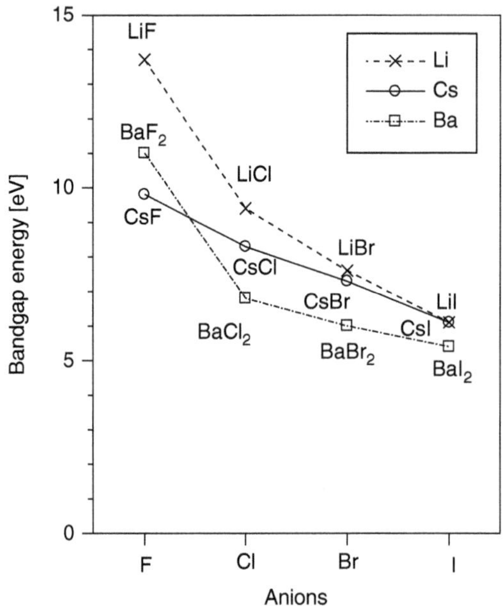

Figure 4.2 *Band-gap energies of selected halides. The values of Li- and Cs-compounds are from [8]. The values of Ba-compounds are from [9].*

of the scintillator, which typically has a value of approximately 2–3 [7]. The E_{gap} is the band-gap energy (eV), S is the transport efficiency of electron–hole pair energy to the luminescent center, and Q is the quantum yield in the luminescence center. As clearly seen, these equations are basically identical to Equation (1.15) in the case of $E\gamma = 1\,\text{MeV}$ and $\beta = 2.5$. According to these equations, the light yield LY tends to be low when the band-gap E_{gap} has a large value. This formula shows that scintillators with a wider band-gap tend to have a lower light yield. Figure 4.2 shows the band-gap energy of the halide. In this figure, it can be seen that the band-gap is wider in the order of fluoride, chloride, bromide, and iodide. In general, it is regarded that fluorides have the widest band-gap among all inorganic materials.

As described in Equation (1.24) in Section 1.3.4, the decay time tends to be fast when the emission wavelength is short [10]. The emission wavelength with the 5d-4f transition of Ce^{3+} in fluoride crystals is short due to the wide band-gaps of fluorides. Therefore, the decay times tend to be fast in the fluoride crystal. In addition, core-valence luminescence, which will be described later, has an extremely fast decay time constant, but is observed only in materials with a wide band-gap. This is also the reason why many fluorides have a fast decay time. Such extremely fast decay times are important in nuclear medicine and high energy physics applications. Such a fast decay time is also one of the advantage of fluorides.

In addition, fluorides have the advantage that many of their compounds have low hygroscopicity as well as those of oxides. Figure 4.3 shows the solubility of halides in water at 25 °C [11]. It can be seen that the solubility of the Li compound and the Ba compound is higher in the order of iodide, bromide, chloride, and fluoride, and that fluoride is the lowest. This tendency is seen in most elements, but there are exceptions. For example, the Cs compound is one of them. In the case of Cs compounds, iodide has a lower solubility than fluoride, and CsI has been applied to scintillators as a relatively stable material in air. However, most fluorides have a lower hygroscopicity than other halides, and like oxides, many fluoride compounds are stable in air.

Some fluorides show lower hygroscopicities than oxides. Lithium fluoride has a lower solubility than that of lithium oxide and it is one of the most stable lithium compounds, so it is used as a neutron converter that converts neutrons into charged particles by a $^{6}Li(n, \alpha)^{3}H$ reaction. A neutron converter can form a neutron scintillator by combining with a phosphor, and LiF/ZnS:Ag is particularly widely used as such a neutron scintillator [12]. This scintillator is a mixture of fluoride and sulfide, and the sulfide part emits light. However, it can be regarded as one of the typical examples of use of fluorides in scintillators.

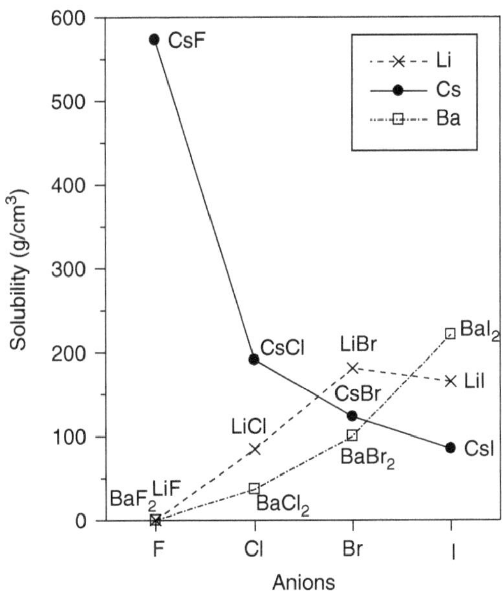

Figure 4.3 *Solubilities of selected halides (in water, 25 °C) [11].*

Fluorides have a hygroscopicity as low as oxides and a lower melting point than oxides. Figure 4.4 shows a plot of the melting points of common oxides and fluorides [13] against the atomic numbers of cations. From the figure, it can be seen that most fluorides have a lower melting point than oxides with the same cations. Generally, if the melting point is too high, it is difficult to grow a single crystal. However, when the melting point of a metal oxide (MO) is too low (such as B_2O_3), the fluoride may be in a gas state at room temperature (such as BF_3), and in this case, it is difficult to use fluoride as scintillators. In addition, Na, K, Rb, and Cs oxides have relatively low melting points. Since these oxides react violently with water to form hydroxides, carbonates are generally used as raw materials when preparing oxide materials containing Na, K, Rb, and Cs. These raw materials are also often used in the production of oxide glasses, which require a melt quenching process, and can achieve a relatively low melting point. Unlike oxides, Na, K, Rb, and Cs fluorides do not react with water, and NaF, KF, RbF, and CsF are used as raw materials for growth of single crystals.

As described above, fluoride has the advantage of a low melting point, but has the disadvantage that the atmosphere must be strictly controlled during single crystal growth. In order to grow high-purity fluoride single crystals, it is necessary to remove water and oxygen, which will be explained in detail in Section 4.2.2.

In summary, fluorides show fast decay times and low hygroscopicities. Compared to non-hygroscopic oxides, fluorides tend to show lower melting points. They can be of advantage for fluoride scintillators in terms of synthesis. In the following sections, methods for growing fluoride single crystals, a summary of scintillation properties, and the examples of fluoride scintillators for γ-ray, neutron, and charged particle detection are explained.

4.2 Crystal Growth of Fluorides

4.2.1 Classification of Methods for Crystal Growth

This section introduces a method for growing fluoride single crystals. It is also possible to obtain transparent ceramics of fluorides, and the solid-phase reaction (sintering) will be introduced in Chapter 10. Figure 4.5 shows single crystal growth methods for inorganic

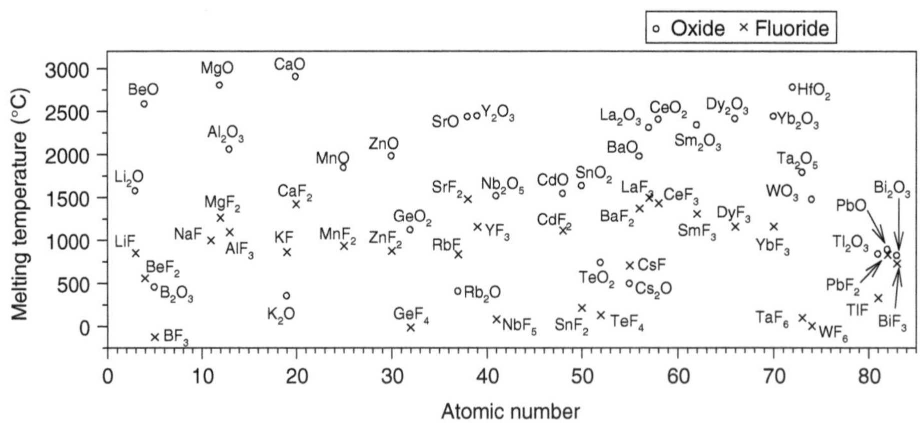

Figure 4.4 Melting temperatures of fluorides and oxides [13].

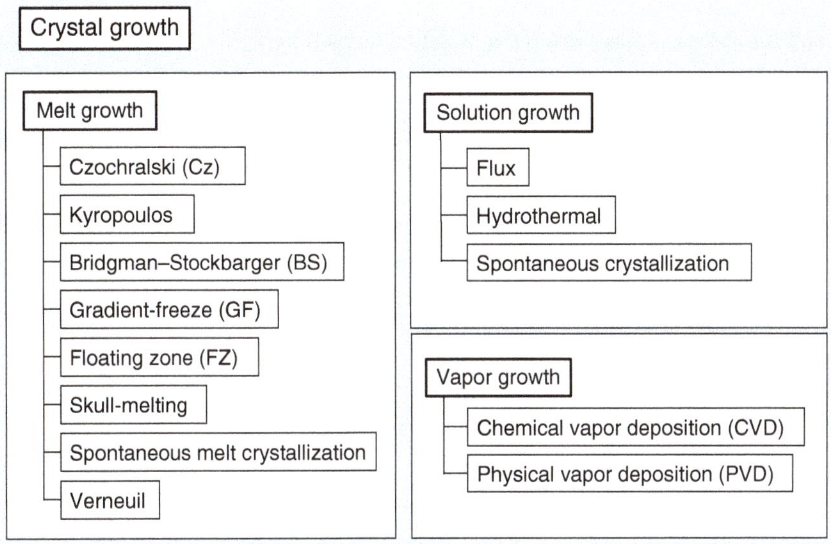

Figure 4.5 *Methods for inorganic crystal growth.*

materials. The single crystal growth methods can be classified into melt growth, solution growth, and vapor growth. Melt growth is a crystal growth method in which a single crystal is obtained by supercooling from a liquid phase having the same chemical composition as the grown single crystal. In solution growth, the solute is dissolved in the solvent crystallizes by slow cooling of the solution or supersaturation due to evaporation of the solvent. Vapor growth is a crystallization technique from supersaturated vapors.

The most industrially used growth method for fluoride crystals is the melt growth method. In this method, fluoride single crystals are obtained by gradually solidifying the melt of the raw fluoride materials. The Czochralski method and the Bridgman–Stockbarger method are often used for fluoride crystal growth. The Czochralski method is where a seed crystal is contacted with a melt in a crucible and gradually pulled up to a low temperature zone to solidify in one direction to obtain a single crystal. The Bridgman–Stockbarger method is where the crucible is gradually moved to a low temperature zone to solidify the melt in the crucible in one direction to grow a single crystal. The melt is crystallized outside the crucible in the Czochralski method, while the melt is crystallized inside the crucible in the Bridgman–Stockbarger method.

4.2.2 Furnace Materials, Atmosphere, and Scavengers for Fluoride Crystal Growth

In melt growth, it is important to select the right material for the crucible. Fluoride single crystals need to be grown in an inert gas atmosphere to prevent oxidation. The crucible material needs to be stable up to a high temperature in an inert gas atmosphere. Furthermore, it is convenient for the crucible to have a low wettability with fluorides because they can be easily taken off after the crystal growth. Quartz can be used for the crystal growth of halides other than fluoride, but quartz is not suitable with fluoride because it reacts with hydrofluoric acid. High-purity carbon is a suitable material for fluoride crucibles. Carbon can be used as a material for both heaters and heat insulators for fluoride crystal growth. Carbon heaters can be applied for both resistance heating and induction heating.

Furthermore, in order to prepare an inert gas atmosphere in the furnace, it is preferable to place the crucible and the heater into a metal vacuum chamber. In general, metals are corroded by acid, but some metals with relatively strong corrosion resistance to hydrofluoric acid, such as SUS316 (type 316 stainless steel), can be used.

In crystal growth, it is important to remove water and oxygen contained in the raw material powder and the vacuum chamber, and a scavenger is used for this purpose. PbF_2 and ZnF_2 are known as solid scavengers [14, 15]. They remove oxygen and water by the following reactions. As an example, the reaction between PbF_2 and divalent MO is shown.

$$MO + PbF_2 \rightarrow MF_2 + PbO \tag{4.3}$$

$$H_2O + PbF_2 \rightarrow PbO + 2HF \tag{4.4}$$

$$PbO + C \rightarrow Pb + CO \tag{4.5}$$

The method using a gas scavenger is excellent in that cations do not remain in the crystal. HF and CF_4 are known as gas scavengers [16, 17], but CF_4 is easier to use because it is nontoxic. It is considered that water is removed by the following reaction:

$$CF_4 + 2H_2O \rightarrow CO_2 + 4HF \tag{4.6}$$

Another gas scavenger candidate is the use of trace amounts of carbonyl fluoride (COF_2). The COF_2 gas can be used at a relatively lower temperature (400–900 K) than the CF_4 gas, and oxygen and water are removed by the following reactions [18]:

$$2COF_2 + O_2 \rightarrow 2CO_2 + 2F_2 \tag{4.7}$$

$$COF_2 + H_2O \rightarrow CO_2 + 2HF \tag{4.8}$$

4.2.3 Fluoride Crystal Growth Methods by Pulling Out from the Melt

Examples of the method of crystallizing while pulling out from a crucible, include the Czochralski method, the edge-defined film-fed growth (EFG) method, and the micro-pulling down method.

The Czochralski method is one of the most important fluoride single crystal growth methods. In this method, a seed crystal is contacted with the melted raw material in the crucible and is gradually pulled up to solidify it. As described in the previous section, a high-quality large single crystal can be obtained by controlling the atmosphere and by using a high-purity carbon crucible. Examples of fluoride single crystals grown by the Czochralski method are shown in the following. Figure 4.6 shows the LaF_3:Nd single crystal [19] by the Czochralski method. It can be seen that a transparent single crystal showing a purple color due to the light absorption characteristics of Nd^{3+} ions is obtained. The Nd^{3+} ion has the same valence as La^{3+} ion and has a similar ionic radius, so it can be doped homogeneously in the single crystal ingot. Therefore, the color of the single crystal ingot is also homogeneous. Figure 4.7 shows the CaF_2 single crystal [20] by the Czochralski method. In the growth

Figure 4.6 *The 2 in. diameter LaF_3:Nd single crystal by the Czochralski method.*

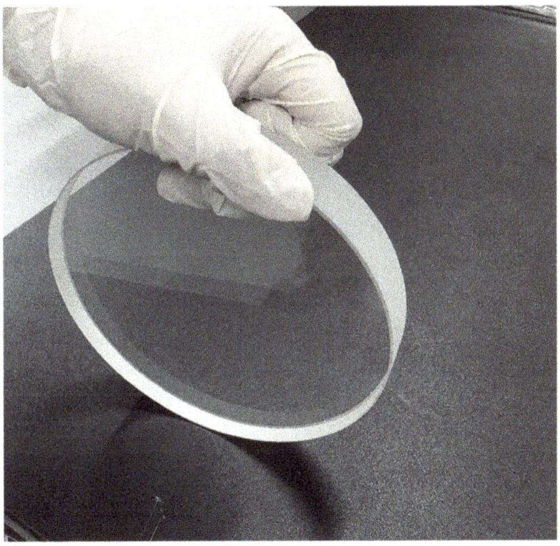

Figure 4.7 *The CaF_2 single crystal by the Czochralski method.*

of this CaF_2 single crystal, CF_4 gas is used for removing oxygen and water to prevent a decrease in transmittance in the short wavelength region. It is also possible to perform flux growth using the Czochralski furnace. The LuF_3 single crystal, which has a structural phase transition at high temperature, is grown by flux at a lower temperature using LiF as a solvent (flux) [21].

The EFG method can be regarded as an advanced Czochralski growth method. In the EFG method, a slit called a die is inserted at the upper side of the crucible, and the melt through the die is pulled up and crystallized. It is advantageous in that an arbitrary cross-sectional shape can be obtained and a single crystal with a shape close to the final product can be manufactured. Although it has been applied as a growth method for oxide scintillators [22], it is not common as a growth method for a fluoride scintillator. There is a report on single crystal growth of CaF_2 by the EFG method, but the obtained single crystal seems to have many defects [23].

In the micro-pulling down method [24], single crystals are grown by solidifying melts pulled out from the pinhole at the bottom of the crucible. A Pt wire or a seed crystal are contacted with melts and then pulled down from the crucible. There are many examples of crystal growth by the micro-pulling down method, mainly for research purposes [24]. Figure 4.8 shows the LiF single crystal grown by the micro-pulling-down method. Such rod-shaped single crystals can be obtained by this method. The diameter of the LiF crystal in the figure is approximately 2 mm, and one crystal with this size can be prepared in one day; therefore, it is suitable for material research.

4.2.4 Fluoride Crystal Growth Methods by Solidifying the Melt in the Crucible

The Bridgman–Stockbarger method is one of the most important growth techniques for growing crystals in a crucible. In this method, single crystals are obtained by melting the raw material in the crucible and gradually solidifying the melt by moving the crucible out of the hot zone. By placing a seed crystal in the bottom of the crucible and melting only a part of the seed crystal, it is possible to obtain a high-quality single crystal. In the case of the self-seeding Bridgman–Stockbarger method, no seed crystals are used. Large-diameter single crystals of CaF_2 for optical material applications are often grown in a vacuum by this method [25].

In the Kyropoulos method, a part of the melt is crystallized by pulling up, as in the Czochralski method, and then a single crystal is obtained in the crucible grown from the crystallized part by controlling the temperature. It can be applied for fluoride; for example, LiF and LiF:Pb single crystals have been grown by the Kyropoulos method [26].

The spontaneous melt crystallization is a very simple solidification method. Crystals are grown in the crucible in the same way as the Bridgman–Stockbarger method, but crystals are obtained by gradually cooling the crucible instead of unidirectional solidification by moving the crucible. This method is self-seeding crystallization using no seed crystals. Figure 4.9 shows examples of crystals obtained by this method. In this method, transparent crystals can be easily obtained in the case of a cubic crystal system having high symmetry; however, it is difficult to obtain transparent crystals in the case of the other crystal systems. Figure 4.9f shows an example of a hexagonal crystal. Scattering centers can be visually observed in this crystal. Since multiple crucibles can be used at the same time, the production efficiency is higher than that of the micro-pulling-down method, and it is possible to obtain dozens of samples a day. Even in the case of a cubic crystal, crystals by this method may not be used for optical material applications because the crystallinity is not good enough. However, it has the potential to be used as a scintillator, at least for research purposes. There are several examples that have actually been used in the study of fluoride scintillators [27–30].

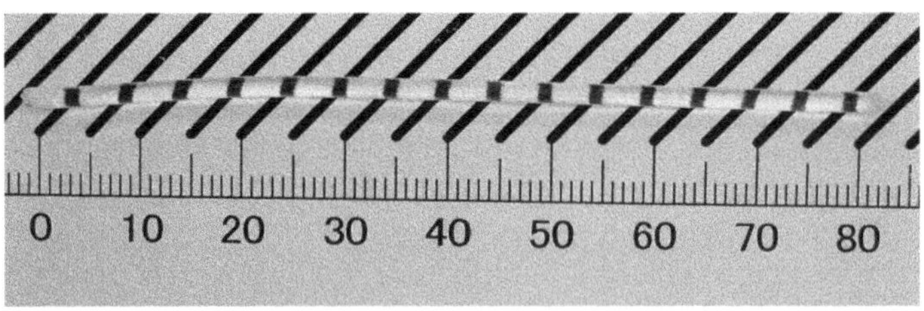

Figure 4.8 The LiF single crystal grown by the micro-pulling-down method.

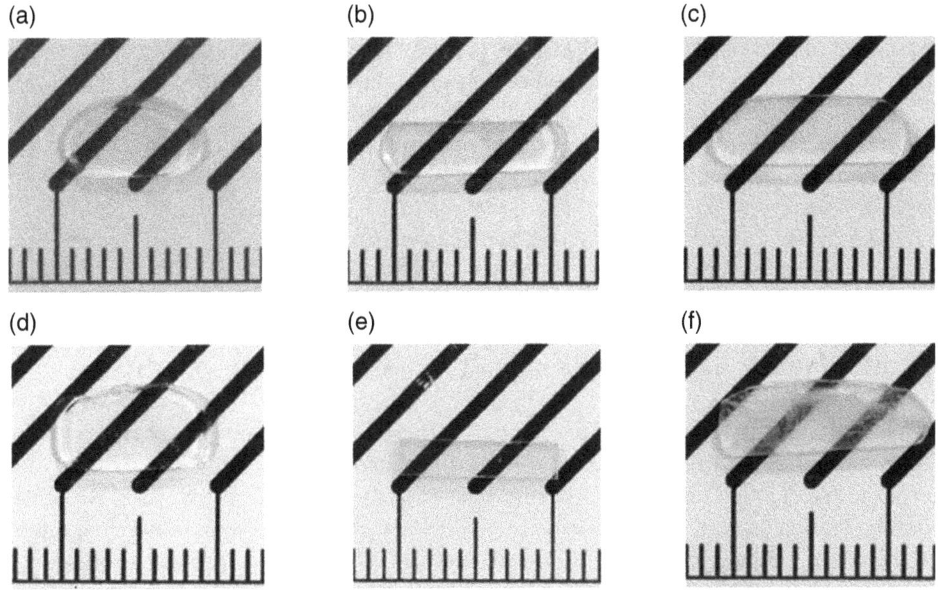

Figure 4.9 CaF_2 (a), SrF_2 (b), BaF_2 (c), LiF (d), PbF_2 (e), and $LiCaAlF_6$ (f) crystals by the spontaneous melt crystallization method. Scattering centers can be seen only in the hexagonal $LiCaAlF_6$ crystal and cannot be seen in all the other cubic crystals.

4.2.5 Fluoride Crystal Growth Methods Without Using Crucibles

As for crystal growth methods that do not use crucibles, there are the skull melting (crucible) method, the solution growth method, the vapor growth method, and the floating zone method.

The skull melting process is a technique in which the raw material itself is used as a crucible, and only the central portion is melted and crystallized. In this technique, impurities are not mixed in the melt from the crucible. There is the study of fluoride scintillators using the skull melting process. The LiF single crystal by the skull melting process [31] has been grown. The disadvantage of the skull melting process is a huge waste of target material. As explained above, only a central part of a large amount of raw material can be grown, and remaining parts are unusable.

As for the solution growth method, there are reports on the spontaneous crystallization of fluorides from a solution. The solution growth of LiF [32] and CaF_2 [33] crystals have been reported. These studies have not been performed for the purpose of scintillator applications, but this method has been used for research on iodide scintillators [34].

There are reports on crystal films of CeF_3 [35], NdF_3 [36], and $KMgF_3$ [37] prepared in vapor growth, but these studies are for the purpose of UV photosensors or vacuum ultra violet (VUV) light sources, not for scintillators, because scintillator applications of thin film shape are limited. The floating zone method is often used for growing oxide single crystals, but is not often used for fluoride due to a difficulty in controlling the atmosphere. In the case of oxide growth shown in Section 3.2, light from lamps is reflected by a mirror surface and focused at the center, and the central part is generally surrounded by quartz cover which is transparent to reflected light from the mirror. If we apply this method to fluoride materials, quartz cannot be used because it reacts with HF. Thus, we must prepare a cover which is transparent to reflection light and has less reactivity against HF gas, and generally, such a material for the cover is difficult to approach under common experimental requirements such as low cost, enough reusability easy for handling, etc.

Even though the traditional Czochralski and Bridgman–Stockbarger methods can be easily used for growing fluoride single crystals, there is very little research into methods that do not use crucibles. In general, the Czochralski method or the Bridgman–Stockbarger method may be selected as the first choice for the production of fluoride single crystal crystal scintillators.

4.3 Outline of Fluoride Scintillators

In this section, the scintillation properties of fluoride scintillators are reviewed using Tables 4.1–4.4. Fluoride scintillators were classified into four types. Table 4.1 shows fluoride scintillators based on luminescence from 5d-4f transitions of Ce^{3+} ions. Table 4.2 shows fluoride scintillators based on core-valence luminescence. Table 4.3 shows fluoride scintillators based on luminescence from 5d-4f transitions of Nd^{3+} ions and Eu^{2+} ions for γ-ray or charged particles detection. Table 4.4 shows fluoride scintillators for thermal neutron detection. In these tables, the values of Z_{eff} are shown for the host crystals.

Table 4.1 *Fluoride scintillators based on luminescence from 5d-4f transitions of Ce^{3+} ions.*

	Z_{eff}	Emission wavelength (nm)	Decay time (ns)	*LY* (photons/MeV)	References
BaF_2:0.2% Ce	52.7	310, 325	60, ~600	7000	[3]
BaF_2:4.5% Ce	52.7	340, 365	50	2200	[3]
CaF_2:0.1% Ce	17.1	320, 335	~350	2400	[3]
CaF_2:5% Ce	17.1	370	50	550	[3]
CeF_3	53.3	284, 300	4 (4%), 26.5 (94%), 135 (2%)	4400	[38]
LaF_3:10% Ce	52.3	300, 350	3 (10%), 26.5 (86%) 185 (4%)	2200	[38]
BaY_2F_8:9.9% Ce	44.1	330, 340	46 ± 1	780	[39]
$BaLu_2F_8$:1% Ce	63.3	320	35 ± 2	760	[39]
CsY_2F_7:5% Ce	43.8	333	32 ± 1	1400 ± 300	[40]
$CsGd_2F_7$:Ce	57.8	340	20 ± 5 (22%), 3000 ± 1000	7300 ± 700	[40]
$LiYF_4$:0.1% Ce	33.1	305, 325	94	1210	[41]
$LiLuF_4$:0.1% Ce	64.4	310, 335	320	3600 ± 400	[42]
LuF_3:Ce	66.2	310	28	8000	[43]

Table 4.2 Fluoride scintillators based on core-valence luminescence.

	Z_{eff}	Emission wavelength (nm)	Decay time (ns)	*LY* (photons/MeV)	References
BaF_2 (fast component)	52.7	195, 220	0.8	1400	[3]
BaF_2 (slow component)	52.7	310	600	10 660	[3]
BaY_2F_8	44.1	290	~3	–	[44, 45]
$BaLu_2F_8$	63.3	300	0.86	200	[39, 46]
$BaMgF_4$	48.8	205	0.57 ± 0.01 (70%), 2.2 ± 0.31 (30%)	1300 ± 100	[47]
$LiBaF_3$ (fast component)	50.9	190, 230	0.8	1400	[48]
KF	17.3	156	<2	–	[49]
$KMgF_3$	14.9	140–190	1.3	1400	[48]
$KCaF_3$	17.2	140–190	< 2	1400	[48]
KYF_4:Rb	31.9	140–200	1.9 ± 0.4	1800 ± 300	[50]
K_2YF_5	30.1	170	1.3	300	[48]
$KLuF_4$	62.6	170–200	1.3	~200	[48]
KLu_2F_7	64.3	165	1.3	~200	[48]
RbF	35.2	203, 234	1.3	1700	[48]
CsF	53.2	390	2.9	2000	[48]

Table 4.3 Fluoride scintillators based on luminescence from 5d-4f transitions of Nd^{3+} ions and Eu^{2+} ions for γ-ray or charged particles detection.

	Z_{eff}	Emission wavelength [nm]	Decay time (ns)	*LY* (photons/MeV)	References
LaF_3:8% Nd	52.3	173	7.6	140 ± 30	[19]
$LuLiF_4$:1% Nd	64.4	180	12	300	[51]
LuF_3:0.5% Nd	66.2	180	7.63 ± 1.05 (82%), 96.37 ± 86.5 (18%)	1200 ± 120	[52]
CaF_2:Eu	17.1	435	940	19 000	[53]

4.4 Fluoride Scintillators for γ-Ray Detection

4.4.1 Fluoride Scintillators Based on Luminescence from 5d-4f Transitions of Ce^{3+} Ions

Although the light yields of Ce-doped fluoride scintillators tend to be lower than those of Ce-doped oxide ones, in contrast, the decay times tend to be faster, as the fast Ce-doped fluoride scintillators, BaF_2:Ce and CeF_3 are well known.

The decay time of BaF_2:Ce depends on the Ce concentration. When the Ce concentration is 1% or more, the component with the 5d-4f transitions of Ce^{3+} ions is dominant [63].

Table 4.4 *Fluoride scintillators for thermal neutron detection.*

	Z_{eff}	Emission wavelength (nm)	Decay time (ns)	*LY* (photons/neutron)	References
$LiBaF_3$:Ce, Rb	50.9	190–330	1, 34, 2400	3600	[12]
$LiCaAlF_6$:Eu	14.3	370	1150	30000	[54, 55]
$LiCaAlF_6$:Eu, Na	14.3	370	1490	40000	[54, 56]
$LiCaAlF_6$:Ce	14.3	285, 310	46.7	3000	[54, 57]
$LiCaAlF_6$:Ce, Na	14.3	285, 310	38.9	5000	[54, 57]
LiF/CaF_2:Eu eutectic	8.3/17.1	435	700	10000	[58]
LiF/CaF_2:Ce eutectic	8.3/17.1	–	–	3000	[59]
LiF/SrF_2:Eu eutectic	8.3/17.1	430	410–611	5000–7000	[60]
LiF/SrF_2:Ce eutectic	8.3/17.1	270, 310, 330	120–160, 600–900	4500–5500	[61]
LiF/CaF_2 eutectic	8.3/17.1	300	260	4700	[62]
LiF/SrF_2 eutectic	8.3/17.1	300	90	9400	[62]

Radiation resistance of BaF_2:Ce is significantly improved by Ce-doping. When the Ce concentration is 27%, BaF_2:Ce shows a high radiation resistance and a fast decay time of 38 ns [64].

The CeF_3 single crystal is one of the promising scintillators in high energy physics experiments due to its fast decay time, non-hygroscopicity, and high radiation resistance. The light yield is 4400 photons/MeV, and the main component of the decay times is 26.5 ns [38]. CeF_3 single crystals with few defects can show emission peaks at around 280 and 300 nm. They are ascribed to the two ground states $^2F_{5/2}$ and $^2F_{7/2}$ of the 5d-4f transitions of Ce^{3+} ions. When the γ-ray dose is over 10^4 Gy, the effect of radiation damage can be observed in the CeF_3 single crystal [65].

4.4.2 Fluoride Scintillators Based on Core-Valence Luminescence

Figure 4.10 shows the emission mechanism of core-valence luminescence [66]. Core-valence luminescence is emission generated by the recombination of holes and electrons between the valence band and the core band. Core-valence luminescence is also called Auger-free luminescence or cross-luminescence [48]. When the band-gap (E_{bg}) is smaller than the energy between the conduction band and the core band (E_{cv}), as shown in Figure 4.10 (left), Auger electrons are emitted. On the other hand, in the case of $E_{bg'} > E_{cv'}$, as shown in Figure 4.10 (right), Auger electrons cannot be emitted and core-valence luminescence is also emitted [66].

Core-valence luminescence shows an extremely fast decay time and it is observed in some types of halides with a wide band-gap. Such materials, BaF_2, $BaLu_2F_8$, $BaMgF_4$, $LiBaF_3$, KF, CsF, RbF, and $KMgF_3$ are well known.

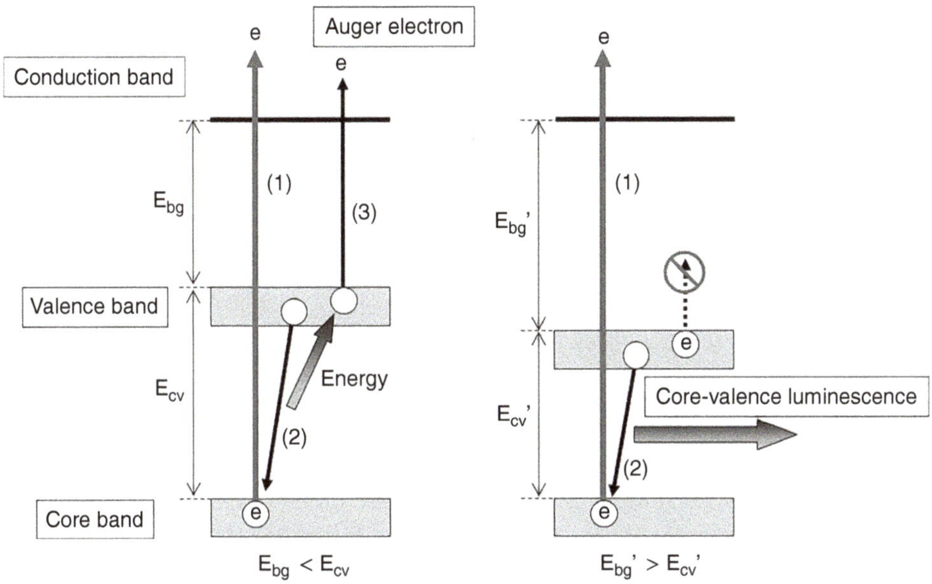

Figure 4.10 *The emission mechanism of core-valence luminescence.*

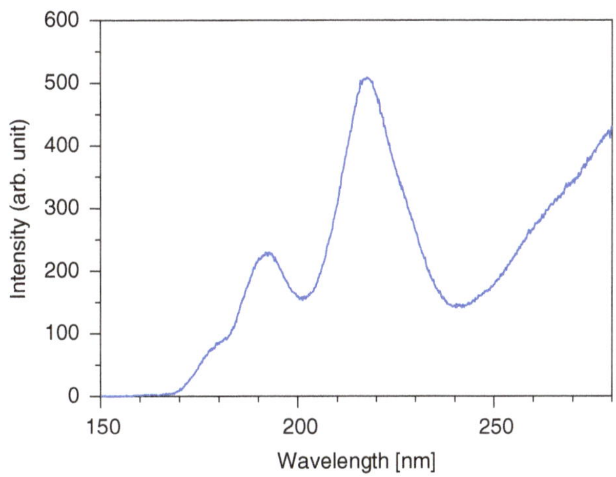

Figure 4.11 *The scintillation spectrum of BaF_2 under X-ray excitation measured in the N_2 gas atmosphere.*

The BaF_2 single crystal has both fast and slow emissions. Figure 4.11 shows the scintillation spectrum of BaF_2 under X-ray excitation measured in an N_2 gas atmosphere. The UV emission over 250 nm is caused by the self-trapped exciton (STE). It has a relatively high light yield (~10 660 photons/MeV) and a slow decay time (600 ns) [3]. The fast emission of BaF_2 is caused by core-valence luminescence [66]. Its decay time is extremely fast (0.8 ns), but the emission wavelength is in the VUV region (195 and 220 nm) and the light yield is not high (1400 photons/MeV) [3].

Since STE emission is significantly slower than core-valence luminescence, impurity ions doped BaF_2 crystals have been studied to suppress the STE emission, because some applications require only the fast decay component. For example, La^{3+} doping has been investigated and the STE emission of BaF_2 has successfully decreased [67]. The problem with this approach is a degradation of light output of core-valence luminescence as well as STE.

To develop the Ba-based core-valence luminescence scintillators with no slow components, some materials have been studied. Among them, $BaLu_2F_8$ shows core-valence luminescence; however, its light yield is low (200 photons/MeV) [39, 46].

Although $BaMgF_4$ has lower Z_{eff} than BaF_2, it shows only fast core-valence luminescence [47]. Figure 4.12 shows the scintillation decay curves of BaF_2 and $BaMgF_4$ single crystals under γ-ray excitation using the ^{137}Cs sealed source and the VUV sensitive photomultiplier tube (R8778, Hamamatsu). It is confirmed that the $BaMgF_4$ shows a faster decay curve than BaF_2. Its light yield is 1300 ± 100 photons/MeV.

The alkali-metal fluorides, KF, RbF, and CsF have been reported to exhibit fast core-valence luminescence. Among them, CsF has a higher effective atomic number than NaI:Tl; therefore, it attracted attention as a fast scintillator for time-of-flight positron emission tomography (TOF-PET). Its emission wavelengths are at 310 and 390 nm and they match the sensitivity of typical photomultiplier tubes. The light yield is low (2000 photons/MeV), but the decay time is fast (2.9 ns) [48]. In addition, there are no components due to slow STE emissions. However, the actual performance as a positron emission tomography (PET) scintillator was lower than that of BGO, and it has not been in practical use.

KF-based fluorides such as $KMgF_3$, $KCaF_3$, KYF_4:Rb, K_2YF_5, $KLuF_4$, and KLu_2F_7 have been reported as scintillators showing core-valence luminescence. Figure 4.13 shows the scintillation spectrum of a $KMgF_3$ single crystal under X-ray excitation measured in the N_2 gas atmosphere. The $KMgF_3$ single crystal shows the broad VUV emission peaking at 150 and 170 nm. It has been reported that the decay time and light yield of $KMgF_3$ are 1.3 ns and 1400 photons/MeV, respectively [48]. Because its emission wavelength is too short, KF-based fluoride scintillators are not put into practical use.

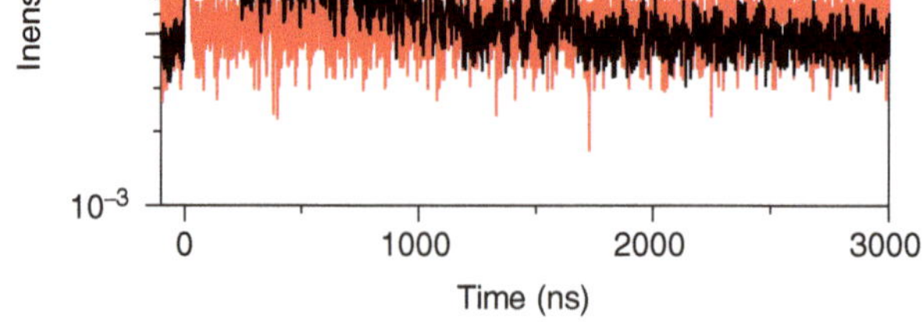

Figure 4.12 *The scintillation decay curves of BaF_2 and $BaMgF_4$ single crystals under γ-ray excitation using the ^{137}Cs sealed source and the VUV sensitive photomultiplier tube (R8778, Hamamatsu).*

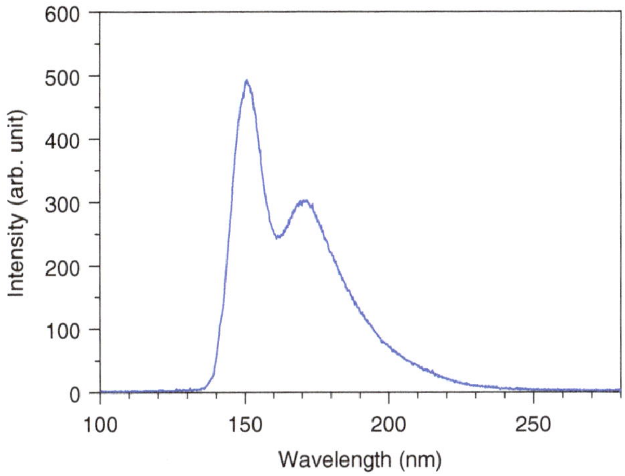

Figure 4.13 *The scintillation spectrum of $KMgF_3$ under X-ray excitation measured in the N_2 gas atmosphere.*

4.4.3 VUV Emitting Fluoride Scintillators Doped with Nd^{3+}, Er^{3+}, and Tm^{3+} Ions

VUV emitting scintillators coupled with VUV sensitive gas counters have been intensely studied as novel detection systems. The main advantage to using gas counters as photodetectors is the possibility to produce a large area detector at low cost. When gas counters are filled with photosensitive gases, such as triethylamine or tetrakis-dimethylamine-ethylene, they are sensitive to VUV photons. For such purposes, VUV emitting scintillators have been developed. Typical examples of VUV emitting scintillators are Nd-doped fluoride single crystals, which show scintillation in the VUV region due to 5d-4f transitions of Nd^{3+} ions [19, 51, 52]. Because conventional scintillators such as oxide and sulfide generally have an optical absorption edge in the near-ultraviolet region, they cannot show transparency in the VUV region. On the other hand, fluoride single crystals are a promising host material as VUV-emitting scintillators, since the scintillators used for this purpose are required to have a large band-gap and transparency in the VUV region. In addition to Nd^{3+} ions, Er^{3+} and Tm^{3+} ions can also exhibit emissions with 5d-4f transitions in the VUV region [52], as shown in Figure 1.5. However, scintillation decay times with emissions by Er^{3+} and Tm^{3+} ions are not fast.

LaF_3:Nd is one of the best materials as the VUV scintillator due to its short emission wavelength at 173 nm. Moreover, LaF_3 has a sufficient stopping power for γ-rays due to its large atomic number and high density. The transmittance is about 80% at 173 nm, which contributes to showing VUV luminescence from 5d-4f transition of Nd^{3+} ions [19]. The scintillation decay time is 7.6 ns, and the light yield of LaF_3:8%Nd is 140 ± 30 photons/MeV [19].

$LuLiF_4$:Nd has a wide band-gap and shows scintillation in the VUV region due to 5d-4f transitions of Nd^{3+} ions. 5d-4f absorption lines in $LuLiF_4$:Nd is observed at approximately 140, 160, and 175 nm, and radiation irradiation results in three 5d-4f emission peaks at approximately 180, 230, and 260 nm [51]. The decay time constant is approximately 12 ns as measured with α-ray irradiation, and the light yield is 300 ph/MeV for 1% Nd-doped $LuLiF_4$ [51].

LuF_3:Nd has a very high density of 8.3 g/cm^3 and a high effective atomic number (Z_{eff} = 66.2). LuF_3:Nd shows a deep absorption below 175 nm due to 4f-5d transitions of Nd^{3+} ions, and exhibits emission peaks due to 5d-4f transitions of Nd^{3+} ions at 180, 230, and 250 nm ascribed to $4f^25d \rightarrow {}^4I_{9/2}$, ${}^4F_x$, and ${}^4G_x$, respectively [52]. The scintillation decay times under X-ray irradiation are 7.63 (82%) and 96.3 (18%) ns, and the light yield of LuF_3:0.5%Nd is 1200 ph/MeV [52]. LuF_3:Nd shows the highest light yield among the Nd-doped VUV scintillators.

Although Nd-doped VUV scintillators have been intensively studied, high light yield materials have not yet been developed. In order to realize the radiation detectors with the scintillator coupled with VUV sensitive gas counters, further improvements are required.

4.5 Fluoride Scintillators for Neutron Detection

4.5.1 Review for Neutron Scintillators

The neutron measurement technique has applications such as non-destructive inspection, space exploration, well logging for oil and gas exploration, and border security inspection. In these applications, ^3He gas proportional counters have been used. Natural abundance of ^3He is 0.000137% and ^3He gas is mainly supplied by the USA. Because the demand of ^3He has increased for homeland security in USA, a shortage of ^3He has become a problem since around 2010 [68]. Therefore, solid neutron scintillators have been intensively studied as an alternative to ^3He gas.

Neutrons have different names depending on their energy, such as fast neutrons, intermediate neutrons, slow neutrons, epithermal neutrons, thermal neutrons, and cold neutrons. Fast neutrons have high energy and are detected by measuring recoil protons resulting from $^1H(n, n)^1H$ reactions of hydrogen atoms in plastic scintillators. Thermal neutrons, that have lower energy than fast neutrons, can easily interact with scintillators; therefore, fast neutrons are often detected using a neutron moderator made of polyethylene containing hydrogen atoms. ^3He, ^6Li, and ^{10}B can react with thermal neutrons and produce particle radiation as follows [12]:

$$^3\mathrm{He} + \mathrm{n} \rightarrow \mathrm{H}(0.573\mathrm{MeV}) + {}^3\mathrm{H}(0.191\mathrm{MeV}) \tag{4.9}$$

$$^6\mathrm{Li} + \mathrm{n} \rightarrow {}^3\mathrm{H}(2.73\,\mathrm{MeV}) + {}^4\mathrm{He}(2.05\,\mathrm{MeV}) \tag{4.10}$$

$$^{10}\mathrm{B} + \mathrm{n} \rightarrow {}^7\mathrm{Li}(0.84\mathrm{MeV}) + {}^4\mathrm{He}(1.47\mathrm{MeV}) + \gamma(0.48\mathrm{MeV})\ [93\%] \tag{4.11}$$

$$^{10}\mathrm{B} + \mathrm{n} \rightarrow {}^7\mathrm{Li}(1.01\mathrm{MeV}) + {}^4\mathrm{He}(1.78\mathrm{MeV})\ [7\%] \tag{4.12}$$

Here, the reaction between ^3He and thermal neutrons is used in the ^3He gas proportional counter described above. Helium is in a gas state at normal temperature and pressure. In contrast, some lithium compounds and boron compounds are solid at room temperature and can be used for neutron scintillators. Natural abundance of ^{10}B is about 20%, and is the most resource-rich and has a reaction cross-section for thermal neutrons about four times that of ^6Li, but the energy of the nuclear reaction (Q-value) is lower than ^6Li, and γ-rays are also generated at the same time. Therefore, ^6Li-based neutron scintillators are mainly used. Here, the Q-value of the nuclear reaction can be calculated using the mass defects in the nuclear reaction and the equation of $E = mc^2$. When two particles are generated, the particle with the lower mass can obtain the higher energy. The energy can be calculated from the value obtained with dividing the mass of another particle by the sum of the masses of the particles.

The neutron scintillators emit light when they are excited by high-energy particle radiation such as α-rays generated by these nuclear reactions of ^{6}Li or ^{10}B. In order to increase the probability of interaction with neutrons, it is necessary to contain ^{6}Li or ^{10}B at a high density. ^{6}Li has a lower abundance ratio in natural Li than ^{7}Li, which is only about 7.6%. In neutron scintillators, ^{6}Li is often enriched (up to 95% ^{6}Li enriched product is available). In addition, neutrons scintillators should not contain the elements that do not contribute to emissions but have a high neutron capture cross-section. Such elements, for example, Gd and Cd, are known. The Gd element can be considered as a neutron converter in a neutron scintillator, such as ^{6}Li and ^{10}B, but Gd is not suitable for the neutron converter in scintillators because the low energy γ-rays are mainly generated by nuclear reactions between Gd and neutrons. Generally, background γ-rays are sufficient in the environment of neutron measurement, and distinction of signal neutron from Gd-based reaction and background γ-rays is technically difficult. However, an X-ray imaging plate with a Gd converter can act as a neutron imaging plate, which is used for neutron radiography (it is not a scintillator application). The Cd element also generates the low energy γ-rays and should not be contained in the scintillator, but Cd is excellent as a neutron shielding material. In addition, the Eu element also has a high neutron capture cross-section, but Eu is often used as the emission center of scintillators for neutron measurement. This is because the Eu concentration of typical neutron scintillators is sufficiently low. Furthermore, it is also not desirable to contain both ^{6}Li and ^{10}B, as it results in multiple peaks in photon counting [69]. In order to obtain a single peak with such a composition, it is necessary to enrich both ^{6}Li and ^{11}B or to enrich both ^{7}Li and ^{10}B.

Since neutron scintillators are required to have low sensitivity to γ-rays which cause measurement noise, they are composed of light elements. This becomes more important, especially in a measurement field where the γ-ray dose is too high because the signal level of γ-ray noise may increase due to pile-up (overlapping of multiple pulses). For the same purpose, it is ideal for neutron scintillators to have a significantly high light yield only in the case of neutron irradiation, that is, neutron scintillators having a high n/γ ratio is preferable. It is easy to perform the pulse height discrimination using such neutron scintillators. As another method, in the case of materials having different decay times during neutron irradiation and γ-ray irradiation, it is also possible to remove the signals due to γ-rays by the pulse shape discrimination. Although a small difference in decay time between neutron and γ-ray irradiation can occur, due to the difference in interaction for many scintillators, a big difference in decay time is achieved by core-valence luminescence. Core-valence luminescence is generated only by γ-rays and β-rays and has an extremely short decay time (easy to distinguish from other emissions). Therefore, neutron scintillators with core-valence luminescence show excellent performance in the pulse height discrimination.

Widely used neutron scintillators, the LiF/ZnS:Ag composite, the LiI:Eu single crystal, and the Ce-doped lithium silicate glass (for example, GS20, Scintacor, Cambridge, UK) are known [12]. Among them, LiI:Eu shows an excellent performance due to its high light yield, high n/γ ratio, and good transparency. Figure 4.14 shows the pulse height spectrum of LiI:Eu under thermal neutron irradiation from ^{252}Cf surrounded by polystyrene blocks.

Figure 4.14 indicates that LiI:Eu is suitable for the pulse height discrimination. Although LiI:Eu has such excellent properties, its hygroscopicity can be its weak point. Even though it is sealed, LiI:Eu can absorb moisture. It may depend on the storage conditions and the sealing method, but even traditionally sealed NaI:Tl still causes a decrease in the light yield, typically

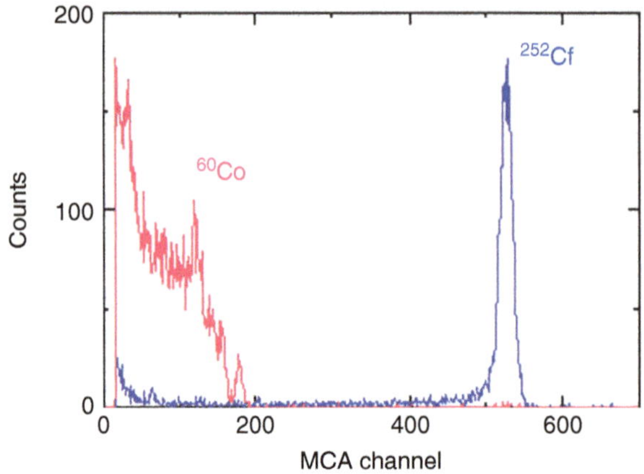

Figure 4.14 *Pulse height spectra of the LiI:Eu single crystal under γ-ray irradiation by ^{60}Co and thermal neutron irradiation by ^{252}Cf and the polyethylene modulator.*

within several years due to moisture. Therefore, it may be difficult to avoid LiI:Eu absorbing moisture in the long term. The relatively new material Cs_2LiYCl_6:Ce (CLYC) [12] has the similar excellent scintillation properties of LiI:Eu, and also shows core-valence luminescence, which enables pulse shape discrimination. In addition, the novel material $Na_{1-x}Li_xI$:Tl (NaIL) [70] can be produced as easily as NaI:Tl and also has excellent properties. However, these materials are also highly hygroscopic, so can be regarded as similar materials to LiI:Eu.

LiF/ZnS:Ag is not hygroscopic and shows excellent properties. This scintillator is a composite of LiF and ZnS:Ag particles. LiF is one of the lowest hygroscopic lithium compounds, and ZnS:Ag is an excellent particle scintillator with a high α/β (n/γ) ratio. Although luminescence is obtained from ZnS:Ag (sulfide), this composite material can be regarded as one of the most important scintillator applications of fluorides. Among solid neutron scintillators, LiF/ZnS:Ag is one of the lowest γ-ray sensitive materials due to its high n/γ ratio. However, LiF/ZnS:Ag also has disadvantages due to its opacity. A sharp peak cannot be obtained in the pulse height spectrum. The thickness cannot be increased because it is not transparent. As a result, its neutron detection efficiency is not high. The Ce-doped lithium silicate glass is a material that does not have such problems due to its high transparency and its non-hygroscopicity. However, the Ce-doped lithium glass has a lower n/γ ratio than LiF/ZnS:Ag and a relatively lower light yield (6000–7000 photons/neutron).

As a fluoride neutron scintillator other than these traditional materials, $LiBaF_3$ based materials have been studied [12]. $LiBaF_3$ shows a fast emission due to core-valence luminescence together with a slower STE emission. Therefore, $LiBaF_3$ can be applied for pulse shape discrimination. However, the emission wavelength of its core-valence luminescence is in the VUV region, which is difficult to detect. In addition, light yields of its STE and core-valence luminescence are not high. Currently, $LiBaF_3$ is not regarded as one of the standard materials, such as LiF/ZnS:Ag and Ce-doped lithium silicate glass.

4.5.2 $LiCaAlF_6$ Single Crystals

$LiCaAlF_6$ (LiCAF):Eu is a recently developed non-hygroscopic fluoride neutron scintillator. This material was developed around 2008 [71] and was commercially available from Tokuyama Corporation in Japan. When the Eu concentration is high and Na is co-doped, the light yield of LiCAF:Eu is maximized. The LiCAF:Eu, Na can show a light yield of 40 000 photons/neutron [56], which is approximately seven times that of Ce-doped lithium silicate glass GS20. If the Eu concentration is too high, the single crystal ingot will become cloudy due to the heterogeneous phase derived from Eu, and its production yield is decreased. This is a common phenomenon to single crystal materials with a low segregation coefficient of the dopant. By controlling the segregation of the dopant, a transparent single crystal can be obtained, as shown in Figure 4.15.

Figure 4.16 shows the pulse height spectra of LiCAF:Eu and LiF/ZnS:Ag under thermal neutron irradiation by a ^{252}Cf sealed source and a polyethylene modulator. The figure

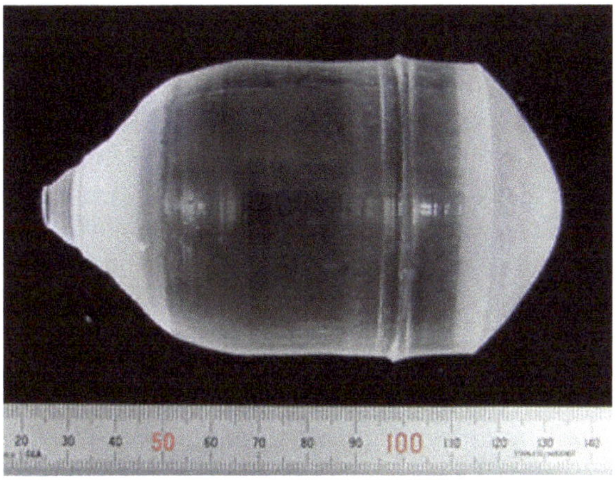

Figure 4.15 *The $LiCaAlF_6$:Eu single crystal by the Czochralski method.*

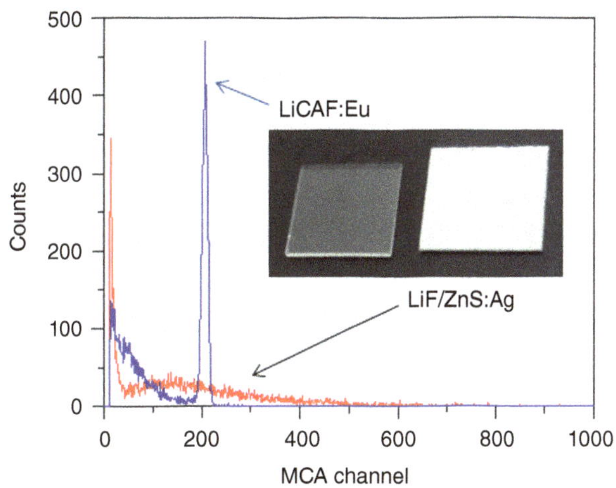

Figure 4.16 *Pulse height spectra of the LiCAF:Eu single crystal ($9 \times 9 \times 0.5\,mm^3$) and LiF/ZnS:Ag ($10 \times 10 \times 0.4\,mm^3$) under thermal neutron irradiation by ^{252}Cf and a polyethylene modulator with the same measurement time and experimental setup.*

indicates that LiCAF:Eu has a higher light yield than the Ce-doped lithium silicate glass GS20. Although LiCAF:Eu and LiF/ZnS:Ag samples are of similar size in this experiment, the peak height of LiCAF:Eu is significantly higher than that of LiF/ZnS:Ag in the same measurement time. Because LiF/ZnS:Ag is an opaque material, it is difficult to obtain a sharp peak as with LiCAF:Eu, and the detection efficiency tends to be lower than transparent single crystal materials. In contrast, the disadvantage of LiCAF:Eu is its high sensitivity to γ-rays, and in order to eliminate the influence of γ-ray noise, it is necessary to shield it with lead or to use the phoswich detector technique. Although the n/γ ratio of LiCAF:Ce is superior to that of LiCAF:Eu, its light yield is significantly lower. LiCAF:Ce shows a lower light yield and a faster decay time than the Ce-doped lithium silicate glass GS20 [12].

Figure 4.17 shows the experimental results of neutron imaging using the LiCAF scintillator plate. This experiment was carried out using the MUSASI (Multi-Purpose Thermal Neutron Application and Science) neutron beam port in the Japan Research Reactor No. 3 (JRR-3) as the thermal neutron beam source. Metallic cadmium (Cd) was used as the image object, and a position-sensitive photomultiplier tube was used as the photodetector. The results of imaging indicate that thermal neutron transmission images of the metallic Cd objects are successfully obtained. In addition, the imaging experiments of 1 mm wires and pinholes have also been successful. The results indicate that the LiCAF scintillator can be used for non-destructive inspection using thermal neutron beams. The non-hygroscopic scintillators are suitable for such applications, because it is easy to process into the desired shape.

The non-hygroscopic scintillators are also suitable for use in harsh environments due to their excellent durability. LiCAF:Eu was used for the investigation of the high-altitude neutron environment. Albedo neutrons are generated when the Earth's atmosphere is irradiated

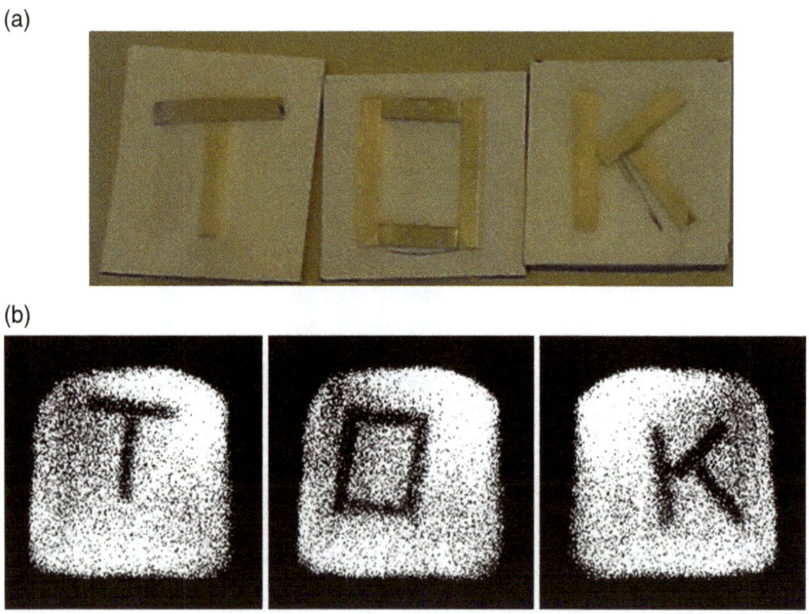

Figure 4.17 *Metallic Cd objects attached on papers (a) and the results of neutron imaging (b) using the LiCAF scintillator plate.*

by high-energy particles from outer space (cosmic-rays). The phoswich detector consisting of LiCAF:Eu and BGO single crystals was used for this study [72].

LiCAF:Eu has also been studied as a neutron scintillator for well logging to prospect for oil and gas. Neutron detectors are used in geological surveys in resource exploration, but since measurements are taken while drilling, the temperature of the detector becomes approximately 200 °C. Therefore, the high temperature properties of LiCAF:Eu have been investigated. As a result, LiCAF:Eu has shown an acceptable light yield in high temperature environments [54].

Although the non-hygroscopicity of LiCAF:Eu is the advantage, its high γ-ray sensitivity can be the big disadvantage. To decrease the γ-ray sensitivity of LiCAF:Eu, a composite material has been developed by dispersing LiCAF:Eu particles in a silicone resin [73]. This inorganic–organic composite material is called "TRUST (Transparent RUbber SheeT) LiCAF" [74]. It shows a significantly lower γ-ray sensitivity than the LiCAF:Eu single crystal. In addition to the low sensitivity to γ-rays, its excellent formability can be an advantage.

4.5.3 LiF/CaF$_2$ Eutectic Composites

LiCAF:Eu is one of the highest light-yielding materials among non-hygroscopic neutron scintillators, but its low dopant segregation coefficient increases the manufacturing cost. The LiF/CaF$_2$:Eu eutectic composite was developed to achieve a low manufacturing cost and similar performance [75]. In the eutectic system, a single liquid phase changes to a homogeneous mixture of multiple solid phases after solidification. Eutectic composite materials can have a homogeneous microstructure. In the eutectic materials, a lamellar structure is typically observed, but rod-like, globular, and acicular structures can be also obtained. Eutectic compositions show relatively lower melting points than the similar compositions. As with LiF/ZnS:Ag, this material is an inorganic–inorganic composite neutron scintillator. The big difference is its translucency. Figure 4.18 shows a LiF/CaF$_2$:Eu eutectic composite. It can be visually confirmed that the LiF/CaF$_2$:Eu eutectic composite has good translucency.

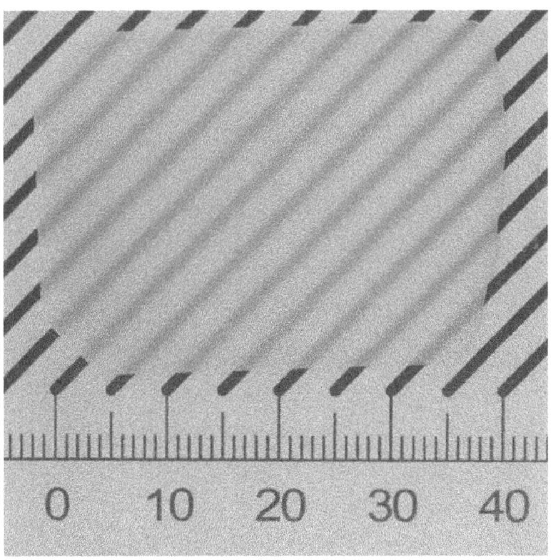

Figure 4.18 *The LiF/CaF$_2$:Eu eutectic composite.*

Figure 4.19 shows the scanning electron microscope (SEM) image of the non-doped LiF/CaF_2 eutectic composite, in which the lamellar microstructure can be observed.

In contrast to LiF/ZnS:Ag, since the LiF/CaF_2:Eu eutectic composite has translucency, a neutron detection peak can be observed in its pulse height spectrum. Figure 4.20 shows an example of the pulse height spectra. It can be seen that the peak is not as sharp as that of a transparent material such as the Ce-doped lithium silicate glass, but a clearer peak is obtained compared to LiF/ZnS:Ag. The light yield changes depending on Eu concentration, and when the Eu concentration is 0.005%, the light yield is approximately 10 000 photons/neutron. In addition, the fast component of decay time has been estimated as 700 ns [58]. Although the LiF/CaF_2:Eu eutectic composite is expected to show a low manufacturing cost, it also can achieve a higher light yield than Ce-doped lithium silicate glass GS20. The n/γ ratio of the LiF/CaF_2:Eu eutectic composite is not high, being comparable with that of LiCAF:Eu.

Figure 4.19 *The SEM image of the LiF/CaF_2 eutectic composite.*

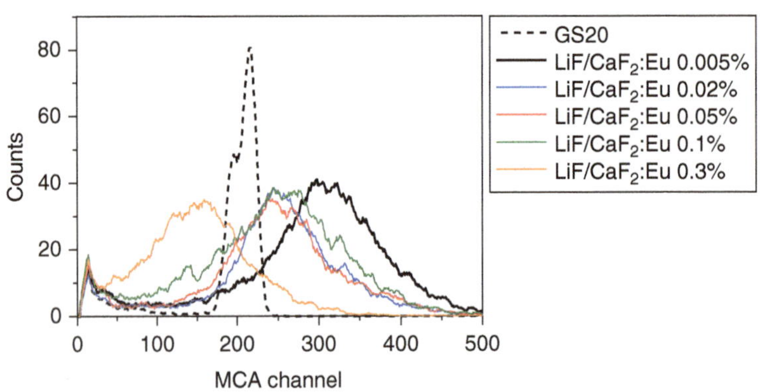

Figure 4.20 *Pulse height spectra of the LiF/CaF_2:Eu eutectic composites and the Ce-doped lithium silicate glass GS20.*

Table 4.5 *Properties of the LiCAF:Eu, Na single crystal, the LiF/CaF$_2$:Eu eutectic composite, and traditional neutron scintillators.*

	Hygro-scopicity	Light yield (photons/neutron)	Decay time (ns)	Z_{eff}	n/γ ratio
LiI:Eu	Huge	50 000 [77]	1400 [77]	52.3	0.87 [77]
Cs_2LiYCl_6:Ce (CLYC)	Huge	70 000 [77]	50, 1000 [77]	46.4	0.73 [77]
LiF/ZnS:Ag	Non	160 000 [77]	75 000 [77]	8.3/27.4	0.44 [77]
GS20	Non	6000 [12]–7000 [77]	70 [77]	25.2 [78]	0.35 [77]
LiCAF:Eu, Na	Non	40 000 [56]	1490 [56]	14.3	~0.2
LiF/CaF$_2$:Eu eutectic	Non	10 000 [58]	~700 [58]	8.3/17.1	~0.2

The LiF/CaF$_2$:Eu eutectic composite has been adopted by the neutron survey meter (NSN4 scintillation-based neutron survey meter [76], Fuji Electric, Tokyo, Japan).

Table 4.5 summarizes the properties of the LiCAF:Eu single crystal, the LiF/CaF$_2$:Eu eutectic composite, and traditional neutron scintillators. Among non-hygroscopic materials, LiF/ZnS:Ag, the LiCAF:Eu single crystal, and the LiF/CaF$_2$:Eu eutectic composite shows higher light yields. In addition, in contrast to opaque LiF/ZnS:Ag, the LiCAF:Eu single crystal and the LiF/CaF$_2$:Eu eutectic composite have translucent or transparent bodies and they can show neutron detection peaks in the pulse height spectra. Thus, the neutrons scintillator is a suitable application for fluoride materials.

Although demand for solid neutron scintillators is increasing from the viewpoint of resource saving, the ^{3}He gas proportional counter still have advantages in performance. Further research and development of solid neutron scintillators are required. However, the solid neutron scintillators are at least even currently superior in terms of neutron detection efficiency.

4.6 Fluoride Scintillators for Charged Particle Detection

4.6.1 Methods for Charged Particle Detection

Particle radiation is classified into uncharged particles and charged particles. Uncharged particles include neutrons and neutrinos. Fluoride scintillators can be used to detect neutrons, but are not used to detect neutrinos. Neutrinos have a low probability of interaction with matter, because they are uncharged and have an extremely low mass. Kamiokande and Super-Kamiokande are well-known facilities for neutrino detection. They use a large volume of water to detect Cherenkov radiation emitted when electrons and muons travel through water that are rarely generated by interaction with neutrinos. On the other hand, Kamioka Liquid Scintillator Anti-Neutrino Detector (KamLAND), which is the facility for detecting anti-neutrinos, uses a liquid organic scintillator, and is one of the few examples in which scintillators are used to detect neutrinos.

Unlike uncharged particles, charged particles can interact with matter. α-rays and β-rays are well-known examples of charged particles. α-rays easily interact with matter and lose all their energy when traveling short distances. α-rays travel while losing energy in matter, and finally they drastically lose energy (Bragg curve). In contrast, β-rays travel a longer distance than α-rays while generating braking X-rays, and gradually lose energy. Due to such a difference in interactions with matter, α-rays generate lower-energy electron–hole pairs at a higher density than β-rays. Although there are differences in the types of such interactions, charged particles such as α-rays and β-rays can be classified as radiation that easily interacts with matter, and can be directly detected by semiconductor detectors and gas detectors. Measurement using a scintillator is also widely performed.

α-rays are helium nuclei ($^4He^{2+}$) emitted by α-decay and have a positive charge. ZnS:Ag, which has a high α/β ratio, or a plastic scintillator, which is inexpensive, can be used for a large sized scintillator, are selected. Proton beams and heavy particle beams also have properties similar to α-rays and can be measured using similar scintillators. In fact, research on proton beam computed tomography (CT) has been conducted using scintillators. However, in radiation dose control in radiotherapy, since the radiation dose to be measured is too high, it seems better to use passive detectors made of thermo-luminescence materials, optically stimulated luminescence materials, or radio-photoluminescent materials.

β-rays are electrons or positrons that are emitted at high speed by β-decay. Electrons are emitted during β^--decay, and positrons are emitted during β^+-decay. The energy of β-decay is the sum of the kinetic energies of electrons and neutrinos, and since their ratio is not constant, the energy of β-rays has a continuous distribution. Usually, β-rays refer to electrons due to β^--decay. However, the internal conversion electron is an electron that jumps out of the orbital energy by giving energy to the orbital electron without emitting γ-rays when the excited nucleus returns to the ground state. The distribution is a line spectrum, which is different from β-rays. As scintillators for β-ray detection, plastic scintillators are inexpensive and can be used for a large size, or CaF_2:Eu having a low α/β ratio is utilized [53]. Compared to X-rays and γ-rays, particle radiation such as α-rays and β-rays often penetrates light-shielding materials, so it is necessary to achieve both light-shielding and detection. For example, it would be necessary to deposit a metal thin film on the surface that allows charged particles to pass through but does not allow the passage of light.

Muons generated by the interaction of radiation from space and atoms in the atmosphere have properties similar to β-rays. They have a high interaction probability and gradually lose energy in matter, but since they have significantly high energy, they can pass through matter while losing energy. Research has been conducted on the muon transmission images (muography) of volcanoes and pyramids that make use of their penetrating power, and plastic scintillators are used to detect them [79] in some of the studies. However, in many other studies, nuclear emulsions with the photosensitive silver halides are used. On the other hand, the detection method of positrons is different from that of other particle radiations. As described in the previous sections, positrons are rapidly annihilated by surrounding electrons and emit γ-rays, so when using a scintillator, γ-rays are the measurement target. Applications of positrons include evaluation of vacant defects in material science and PET in nuclear medicine, but both applications are for indirect measurement by detecting 511 keV annihilation γ-rays so, of course, γ-ray scintillators are used in these applications.

As mentioned above, scintillators that directly detect charged particles can be replaced with other measurement methods using no scintillators. However, a scintillation counter is one of the good candidates for charged particles detection. Therefore, some charged particles scintillators are commercially available. Among fluoride materials, the CaF_2:Eu single crystal is an example of a practical use of a scintillator for charged particle detection.

4.6.2 CaF_2 Based Scintillators for Charged Particle Detection

Scintillators for charged particles detection are mainly required to have the following characteristics:

1. High probability of interaction with the charged particles to be measured.
2. Low probability of interaction with radiation (for example, γ-rays and neutrons) other than the charged particles to be measured.
3. High light yield under charged particles excitation.
4. Fast decay time.
5. Emission wavelength suitable for photodetectors.
6. High radiation resistance.
7. Low hygroscopicity and high chemical durability.

Since the interaction probability between charged particles and matter is high, (1) is usually satisfied. However, hygroscopic scintillators such as NaI:Tl require strict packaging. In this case, charged particles cannot reach the inside of the packaging. That is, (1) is related to (7). (4)–(6) are general required characteristics for scintillators. (3) is achieved by the low α/β ratio. (2) is related to elements contained in the scintillators. Scintillators should be composed of light elements other than Li and B.

These conditions limit the fluoride candidates that can be used for charged particle detection. First, solid fluorides composed of atoms with an atomic number of 20 or less are LiF, BeF_2, NaF, MgF_2, AlF_3, KF, and CaF_2. Among them, BeF_2 and KF are hugely hygroscopic. Figure 4.21 shows the neutron absorption cross-sections of atoms. As can be seen from the figure, LiF has the problem of nuclear reaction with neutrons. In addition, since NaF, MgF_2, and AlF_3 are not easy to substitute rare earth elements, it is difficult to satisfy (3)–(5). After all, CaF_2, $KMgF_3$, $KCaF_3$, and $NaMgF_3$ can be the candidates, but among them, CaF_2 has the simplest composition and is easy to prepare. Therefore, CaF_2 is most suitable as a host material for charged particles scintillators.

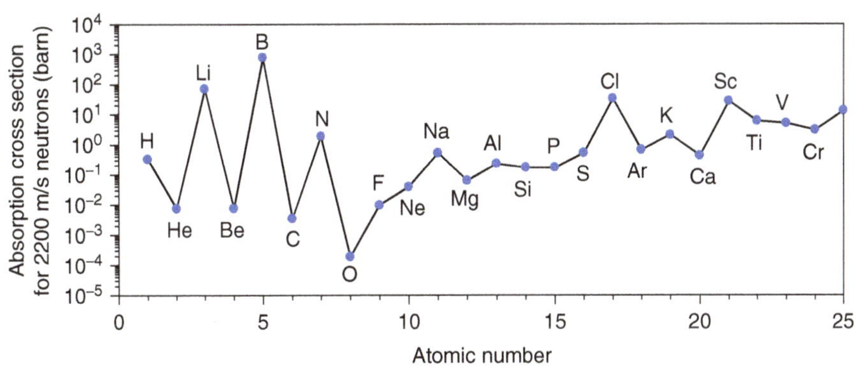

Figure 4.21 *Neutron absorption cross-sections of atoms Source: data from [80].*

According to the product data sheet [53] (Saint-Gobain Crystals, Hiram, Ohio, USA), the light yield of CaF_2:Eu is 19000 photons/MeV under γ-ray excitation and the decay time is 940 ns. The full energy peak under 5.48 MeV α-ray excitation appears at an equivalent γ-ray energy of approximately 1 MeV. The α/β ratio is calculated as approximately 0.2, which is a low value compared to those of typical halide and oxide scintillators.

Although CaF_2:Eu is one of the best scintillators for β-ray detection, its decay time can be improved by changing the emission center from Eu^{2+} to Ce^{3+}. The main component of the decay time of CaF_2:20% Ce is reported as 33.6 ns [81]. However, its light yield is not high compared to CaF_2:Eu. When the decay time of CaF_2:Ce is approximately 33.6 ns (Ce concentration is 20%), its light yield is approximately 800 photons/MeV under γ-ray excitation [81].

The decay time of the 5d-4f transition of Pr^{3+} ions can be faster than that of Ce^{3+} ions. Scintillation properties of CaF_2:Pr have also been investigated [30]. The CaF_2:Pr crystals were prepared by the spontaneous melt crystallization method, as shown in Figure 4.22. Scintillation properties have been studied under α-ray irradiation from ^{241}Am. Figure 4.23 shows scintillation decay curves of CaF_2:Pr and CaF_2:Eu crystals. When Pr concentration is higher, the scintillation decay time is faster. Light yields are shown in Figure 4.24. The light yield significantly decreases with increasing Pr concentration.

In summary, CaF_2 based scintillators with a faster decay time than CaF_2:Eu can be obtained by Ce or Pr doping. However, their light yields may not be sufficient for practical use. CaF_2:Eu has non-hygroscopicity and relatively lower α/β ratio among inorganic scintillators. As mentioned above, CaF_2:Eu is suitable for β-ray detection, even when compared with other CaF_2 based scintillators.

Other than inorganic materials, plastic scintillators also have a low α/β ratio and they can be used for β-ray detection. However, CaF_2:Eu is one of the most important scintillators for β-ray detection, because inorganic materials generally have the advantage of radiation resistance. Thus, the β-ray scintillator is a suitable application for fluoride materials.

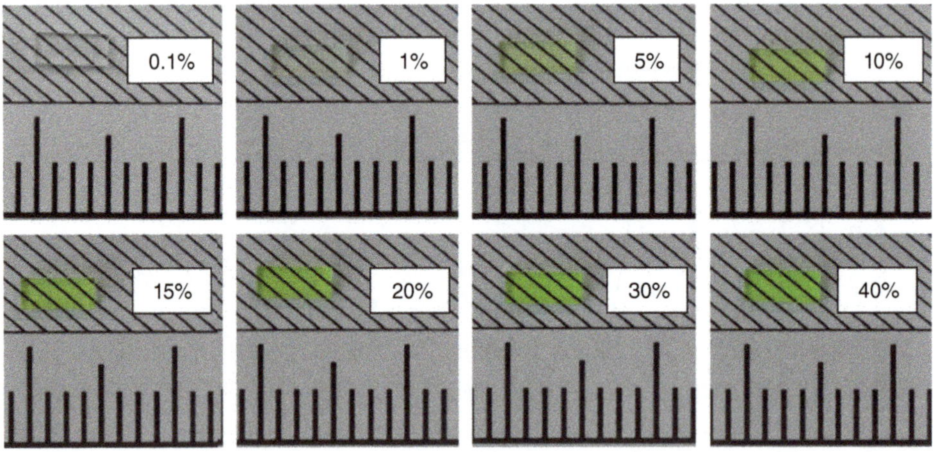

Figure 4.22 *The 0.1–40% Pr-doped CaF_2 crystals by the spontaneous melt crystallization method.*

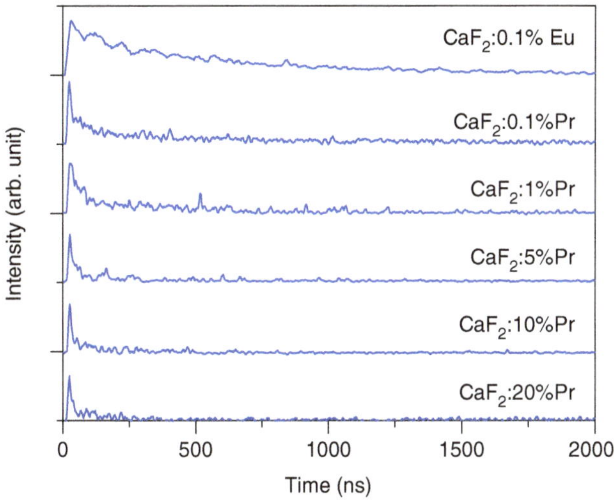

Figure 4.23 *Scintillation decay curves of* CaF_2*:Pr and* CaF_2*:Eu crystals under α-ray irradiation from* ^{241}Am*.*

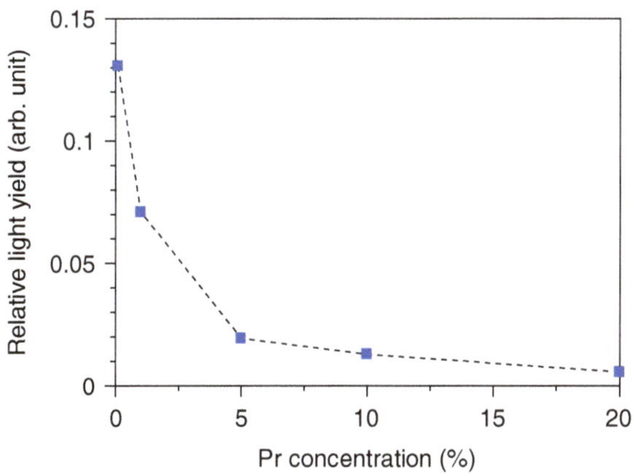

Figure 4.24 *Relative light yields of* CaF_2*:Pr crystals under α-ray irradiation from* ^{241}Am *when that of the* CaF_2*:Eu crystal is 1.*

References

1. Crookes, W. (1903). The emanations of radium. *Proc. Roy. Soc. Lond. A.* 71: 405–408. 10.1098/rspl.1902.0116.
2. Moses, W.W. and Derenzo, S.E. (1989). Cerium fluoride, a new fast, heavy scintillator. *IEEE Trans. Nucl. Sci.* 36: 173–176. 10.1109/23.34428.
3. van Eijk, C.W.E. (1993). Fast scintillators and their applications. *Nucl. Tracks Radiat. Meas.* 21: 5–10. 10.1016/1359-0189(93)90035-8.
4. van Eijk, C.W.E. (2001). Inorganic-scintillator development. *Nucl. Instrum. Methods A.* 460: 1–14. 10.1016/S0168-9002(00)01088-3.
5. van Loef, E.V.D., Dorenbos, P., van Eijk, C.W.E. et al. (2001). High-energy-resolution scintillator: Ce^{3+} activated $LaBr_3$. *Appl. Phys. Lett.* 79: 1573–1575. 10.1063/1.1385342.

6. Cherepy, N.J., Hull, G., Drobshoff, A.D. et al. (2008). Strontium and barium iodide high light yield scintillators. *Appl. Phys. Lett.* 92: 083508. 10.1063/1.2885728.
7. Rodnyi, P.A., Dorenbos, P., and van Eijk, C.W.E. (1995). Energy loss in inorganic scintillators. *Phys. Status. Solidi. B.* 187: 15–29. http://doi.org/10.1002/pssb.2221870102.
8. Song, K. and Williams, R.T. (1993). *Self-Trapped Excitons*. Springer 10.1007/978-3-642-97432-8.
9. Aguado, A., Ayuela, A., López, J.M. et al. (1999). *J. Phys. Soc. Jpn.* 68: 2829–2835. 10.1143/JPSJ.68.2829.
10. Weber, M.J. (2002). Inorganic scintillators: today and tomorrow. *J. Lumin.* 100: 35–45. 10.1016/S0022-2313(02)00423-4.
11. Lide, D.R. (ed.) (2009). *CRC Handbook of Chemistry and Physics*, vol. 90. CRC Press.
12. van Eijk, C.W.E. (2004). Inorganic thermal-neutron scintillators. *Nucl. Instrum. Methods Phys. Res. Sect. A.* 529: 260–267. 10.1016/j.nima.2004.04.163.
13. Dean, J.A. (ed.) (1998). *Lange's Handbook of Chemistry*, vol. 15. New York: McGraw-Hill Book Co.
14. Yonezawa, T., Matsuo, K., Nakayama, J. et al. (2003). Behaviors of metal-oxide impurities in CaF_2 and BaF_2 single-crystals grown with PbF_2 scavenger by Stockbarger's method. *J. Cryst. Growth* 258: 385–393. 10.1016/S0022-0248(03)01567-7.
15. Ko, J.M., Tozawa, S., Yoshikawa, A. et al. (2001). Czochralski growth of UV-grade CaF_2 single crystals using ZnF_2 additive as scavenger. *J. Cryst. Growth* 222: 243–248. 10.1016/S0022-0248(00)00928-3.
16. Guggenheim, H. (1963). Growth of highly perfect fluoride single crystals for optical lasers. *J. Appl. Phys.* 34: 2482–2485. 10.1063/1.1702768.
17. Pastor, R.C. and Arita, K. (1976). Crystal growth of alkaline earth fluorides in a reactive atmosphere. Part II. *Mater. Res. Bull.* 11: 1037–1047. 10.1016/0025-5408(76)90182-3.
18. Ishizu, S., Sekiya, A., Fukuda, K. et al., (2008). Pretreated metal fluorides and process for production of fluoride crystals. Patent WO 2009/139473 A1 filed 16 May 2008 and issued 19 November 2009.
19. Fukuda, K., Kawaguchi, N., Ishizu, S. et al. (2010). Crystal growth and scintillation characteristics of the Nd^{3+} doped LaF_3 single crystal. *Opt. Mater.* 32: 1142–1145. 10.1016/j.optmat.2010.03.021.
20. Yanagi, H., Nawata, T. Yoji, I. et al. (2004) Properties of large CaF_2 crystals grown by CZ method for lens materials, Proc. SPIE 5377, Opt. Microlith. XVII 1886–1893. 10.1117/12.556614
21. Fukuda, K., Ishizu, S., Kawaguchi, N. et al. (2011). Crystal growth and optical properties of the Nd^{3+} doped LuF_3 single crystals. *Opt. Mater.* 33: 1143–1147. 10.1016/j.optmat.2010.10.054.
22. Tatartchenko, V., Zanella, S.A., Locher, J.W. et al. (2008). Scintillator crystals and methods of forming. US Patent 2009/0136731 A1, filed 22 October 2008 and issued 28 May 2009.
23. Nicolov, M. (2002). Defect structure of CaF_2 crystals. *J. Optoelectron. Adv. Mater.* 4: 155–158.
24. Fukuda, T. and Chani, V. *Shaped Crystals: Growth by Micro-Pulling-Down Technique*. Springer-Verlag. ISBN: 978-3-540-71294-7.
25. Nicoar , D. and Nicoar , I. An improved Bridgman–Stockbarger crystal-growth system. *Mater. Sci. Eng., A* 102: L1–L4. 10.1016/0025-5416(88)90584-8.
26. Somma, F., Aloe, P., d'Acapito, F. et al. (2010). Optical and X-ray absorption spectroscopy in lead doped lithium fluoride crystals. IOP Conf. Ser.: Mater. Sci. Eng. 15: 012035. 10.1088/1757-899X/15/1/012035
27. Kawaguchi, N., Fukuda, K., Suyama, T. et al. (2010). Nd concentration dependence on the optical and scintillation properties of Nd doped BaF_2. *Opt. Mater.* 32: 1325–1328. 10.1016/j.optmat.2010.04.019.
28. Tanaka, H., Furuya, Y., Kawaguchi, N. et al. (2011). Crystal growth and scintillation properties of Nd doped CaF_2 single crystal. *Opt. Mater.* 33: 284–287. 10.1016/j.optmat.2010.08.031.
29. Kawaguchi, N., Yanagida, T., Futami, Y. et al. (2013). Doping concentration dependence on VUV luminescence of Tm:CaF_2. *Opt. Mater.* 35: 1898–1901. 10.1016/j.optmat.2013.01.026.
30. Kawaguchi, N., Kimura, H., Akatsuka, M. et al. (2018). Scintillation characteristics of Pr:CaF_2 crystals for charged-particle detection. *Sens. Mater.* 30: 1585–1590. 10.18494/SAM.2018.1926.
31. Taranyuk, V., Gektin, A., Shiran, N. et al. (2013). Modification of oxygen content in LiF crystals grown by skull method. *J. Cryst. Growth* 380: 205–208. 10.1016/crystalline/j.jcrysgro.2013.06.025.
32. Jindra, I.J. (1969). The preparation of lithium fluoride by controlled precipitation. *Cryst. Res. Technol.* 4: 69–75. 10.1002/crat.19690040109.
33. Markovic, M., Takagi, S., Chow, L.C. et al. (2009). Calcium fluoride precipitation and deposition from 12 mmol/L fluoride solutions with different calcium addition rates. *J. Res. Nat. Inst. Stand. Technol.* 114: 293–301. 10.6028/jres.114.021.
34. Wang, W., Qi, H., Liu, F. et al. (2020). Approaching the theoretical light yield limit in CsI (Tl) scintillator single crystals by a low-temperature solution method. *Cryst. Growth Des.* 20: 3474–3481. 10.1021/acs.cgd.0c00256.
35. Ieda, M., Ishimaru, T., Ono, S. et al. (2012). Filterless ultraviolet detector based on cerium fluoride thin film grown by pulsed laser deposition. Jpn. J. Appl. Phys. 51: 062202-1–3. 10.1143/JJAP.51.062202.

36. Ieda, M., Ishimaru, T., Ono, S. et al. (2011) Evaluation of NdF_3 thin films as a vacuum ultraviolet photoconductive detector, Conference on Lasers and Electro-Optics/Pacific Rim, CLEOPR 2011, 1225–1226. 10.1109/IQEC-CLEO.2011.6193789
37. Ishimaru, T., Ieda, M., Kamisaka, H. et al. (2011) Vacuum ultraviolet field emission lamp based on a $KMgF_3$ thin film phosphor, Conference on Lasers and Electro-Optics/Pacific Rim, CLEOPR 2011, 1117–1118.10.1109/IQEC-CLEO.2011.6193788
38. Moses, W.W. and Derenzo, S.E. (1990). The scintillation properties of cerium-doped lanthanum fluoride. *Nucl. Instrum. Methods Phys. Res. A.* 299: 51–56. 10.1016/0168-9002(90)90746-S.
39. van't Spijker, J.C., Dorenbos, P., van Eijk, C.W.E. et al. (1999). Luminescence and scintillation properties of BaY_2F_8:Ce^{3+}, $BaLu_2F_8$ and $BaLu_2F_8$:Ce^{3+}. *J. Lumin.* 85: 11–19. 10.1016/S0022-2313(99)00154-4.
40. Schaart, D.R., Dorenbos, P., van Eijk, C.W.E. et al. (1995). Properties and mechanisms of scintillation in $CsGd_2F_7$:Ce^{3+} and CsY_2F_7:Ce^{3+} crystals. *J. Phys. Condens. Matter* 7: 3063–3088. 10.1088/0953-8984/7/15/013.
41. Combes, C.M., Dorenbos, P., van Eijk, C.W.E. et al. (1997). Optical and scintillation properties of Ce^{3+} doped $LiYF_4$ and $LiLuF_4$ crystals. *J. Lumin.* 71: 65–70. 10.1016/S0022-2313(96)00118-4.
42. Yanagida, T., Fujimoto, Y., Kawaguchi, N. et al. (2011). Scintillation properties of Ce-doped $LuLiF_4$ and $LuScBO_3$. *Nucl. Instrum. Methods Phys. Res. A.* 652: 251–255. 10.1016/j.nima.2010.08.115.
43. Birowosuto, M.D. and Dorenbos, P. (2009). Novel γ- and X-ray scintillator research: on the emission wavelength, light yield and time response of Ce^{3+} doped halide scintillators. *Phys. Status Solidi.* 206: 9–20. 10.1002/pssa.200723669.
44. Aleksandrov, Y.M., Kuusmann, I.L., Makhov, V.N. et al. (1991). Intrinsic and impurity cross-luminescence in three-component barium-containing compounds. *Nucl. Instrum. Methods* 308: 208–210. 10.1016/0168-9002(91)90628-4.
45. Kirikova, N.Y., Klimenko, V.E., Kozlov, V.A. et al. (1995). Cross-luminescence of several complex fluorides excited by synchrotron radiation. *Nucl. Instrum. Methods A.* 359: 351–353. 10.1016/0168-9002(94)01383-7.
46. Weber, M J., Derenzo, S E., Dujardin, C. et al. Dense Ce^{3+}-Activated Scintillator Materials SCINT '95, Delft, Netherlands.
47. Yanagida, T., Kawaguchi, N., Fujimoto, Y. et al. (2010). Growth and scintillation properties of $BaMgF_4$. *Nucl. Instrum. Methods A.* 621: 473–477. 10.1016/j.nima.2010.06.218.
48. van Eijk, C.W.E. (1994). Cross-luminescence. *J. Lumin.* 60–61: 936–941. 10.1016/0022-2313(94)90316-6.
49. Jansons, J.L., Krumins, V.J., Rachko, Z.A. et al. (1988). Crossluminescence of KF and related compounds. *Solid State Commun.* 67: 183–185. 10.1016/0038-1098(88)90960-X.
50. Dorenbos, P., Visser, R., Andriessen, J. et al. (1993). Scintillation properties of possible cross-luminescence materials. *Nucl. Tracks Radiat. Meas.* 21: 101–103. 10.1016/1359-0189(93)90052-B.
51. Yanagida, T., Kawaguchi, N., Ishizu, S. et al. (2010). Crystal growth and evaluations of Nd:$LuLiF_4$ scintillators for different Nd concentration. *IEEE Trans. Nucl. Sci.* 57: 1312–1315. 10.1109/TNS.2010.2042965.
52. Yanagida, T., Kawaguchi, N., Fukuda, K. et al. (2011). Scintillation properties of Nd^{3+}, Tm^{3+}, and Er^{3+} doped LuF_3 scintillators in the vacuum ultra violet region. *Nucl. Instrum. Methods A.* 659: 258–261.
53. Saint-Gobain Crystals (2021) CaF_2(Eu) Calcium Fluoride Scintillation Material. Available at: https://www.crystals.saint-gobain.com/sites/imdf.crystals.com/files/documents/calcium-fluoride-data-sheet_69768.pdf (Accessed: 15 April 2021).
54. Kawaguchi, N., Okada, G., Fukuda, K. et al. (2020). Temperature dependence of scintillation responses in rare-earth-ions-doped $LiCaAlF_6$ single crystals. *Nucl. Instrum. Methods A.* 954: 161518. 10.1016/j.nima.2018.10.196.
55. Yanagida, T., Kawaguchi, N., Fujimoto, Y. et al. (2011). Basic study of europium doped $LiCaAlF_6$ scintillator and its capability for thermal neutron imaging application. *Opt. Mater.* 33: 1243–1247. 10.1016/j.optmat.2011.02.016.
56. Yanagida, T., Yamaji, A., Kawaguchi, N. et al. (2011). Europium and sodium codoped $LiCaAlF_6$ scintillator for neutron detection. *Appl. Phys Express* 4: 106401. 10.1143/APEX.4.106401.
57. Yokota, Y., Fujimoto, Y., Yanagida, T. et al. (2011). Crystal growth of Na-co-doped Ce:$LiCaAlF_6$ single crystals and their optical, scintillation, and physical properties. *Cryst. Growth Des.* 11: 4775–4779. 10.1016/j.optmat.2011.02.016.
58. Masai, H., Yanagida, T., Mizoguchi, T. et al. (2015). Local coordination state of rare earth in eutectic scintillators for neutron detector applications. *Sci. Rep.* 5: 13332. 10.1038/srep13332.
59. Kawaguchi, N., Fukuda, K., Yanagida, T. et al. (2010) Neutron Response of Rare-Earth-Doped ^{6}LiF/CaF_2 Eutectic Composites with the Ordered Lamellar Structure, December 2010 IEEE Nuclear Science Symposium conference record 1000–1003. 10.1109/NSSMIC.2010.5874090
60. Yanagida, T., Fukuda, K., Fujimoto, Y. et al. (2012). ^{6}LiF-SrF_2 doped with Eu eutectic scintillators for neutron detection. *Opt. Mater.* 34: 868–871. 10.1016/j.optmat.2011.11.022.

61. Yanagida, T., Fujimoto, Y., Fukuda, K. et al. (2013). Ce-doped LiF-SrF_2 eutectic scintillators for thermal neutron detection produced at different solidification rates. *Opt. Mater.* 35: 1449–1454. 10.1016/j.optmat.2013.02.016.
62. Yanagida, T., Kawaguchi, N., Fujimoto, Y. et al. (2013). Scintillation properties of LiF-SrF_2 and LiF-CaF_2 eutectic. *J. Lumin.* 144: 212–216. 10.1016/j.jlumin.2013.07.016.
63. Zizong, X., Zhufang, G., Jin, C. et al. (1992). Studies on decay time of fluorescence from a series of Ce-doped BaF_2 crystals. *Nucl. Instrum. Methods A.* 322: 239–242. 10.1016/0168-9002(92)90035-3.
64. Kobayashi, M., Ishii, M., Sobolev, B.P. et al. (1999). Scintillation characteristics of nonstoichiometric phases formed in MF_2–GdF_3–CeF_3 system part II. (M=Ba), scintillation of $Ba_{1-x}Ce_xF_{2+x}$ ($0.1 \leqslant x \leqslant 0.45$) fluorite type crystals. *Nucl. Instrum. Methods A.* 421: 191–198. 10.1016/S0168-9002(98)01143-7.
65. Kobayashi, M., Ishii, M., Krivandina, E.A. et al. (1991). Cerium fluoride, a highly radiation-resistive scintillator. *Nucl. Instrum. Methods A.* 302: 443–446. 10.1016/0168-9002(91)90357-V.
66. Rodnyi, P.A. (2004). Core–valence luminescence in scintillators. *Radiat. Meas.* 38: 343–352. 10.1016/j.radmeas.2003.11.003.
67. Schotanus, P., Dorenbos, P., van Eijk, C.W.E. et al. (1989). Suppression of the slow scintillation light output of BaF_2 crystals by La^{3+} doping. *Nucl. Instrum. Methods A.* 281: 162–166. 10.1016/0168-9002(89)91229-1.
68. Morgan, D. and Shea D. (2010) The Helium-3 Shortage: Supply, Demand, and Options for Congress, Congressional Research Service. 12/2010.
69. Kawahuchi, N., Okada, G., Futami, Y. et al. (2020). Scintillation and dosimetric properties of monocrystalline and polycrystalline $Li_2B_4O_7$. *Sens. Mater.* 32: 1419–1426. 10.18494/SAM.2020.2752.
70. Saint-Gobain Crystals. (2021), NaI(Tl+Li) Neutron-Gamma Scintillator. Available at: https://www.crystals.saint-gobain.com/sites/imdf.crystals.com/files/documents/nail-material-data-sheet.pdf (Accessed: 15 April 2021).
71. Kawaguchi, N., Yanagida, T., Novoselov, T. et al. (2008) Neutron responses of Eu^{2+} activated $LiCaAlF_6$ scintillator, Conference Record of the IEEE NSS/MIC, 1174–1176, Dresden, Germany, 10.1109/NSSMIC.2008.4774611
72. Kole, M., Fukazawa, Y., Fukuda, K. et al. (2015). PoGOLino: a scintillator based balloon-borne neutron detector. *Nucl. Instrum. Methods A.* 770: 68–75. 10.1016/j.nima.2014.10.016.
73. Sugimoto, D., Watanabe, K., Hirota, K. et al. (2014). Neutron TOF experiments using transparent rubber sheet type neutron detector with dispersed small pieces of $LiCaAlF_6$ scintillator. *Phys. Proc.* 60: 349–355. 10.1016/j.phpro.2014.11.047.
74. Watanabe, K., Yamazaki, T., Sugimoto, D. et al. (2015). Wavelength-shifting fiber signal readout from Transparent RUbber SheeT (TRUST) type $LiCaAlF_6$ neutron scintillator. *Nucl. Instrum. Methods A.* 784: 260–263. 10.1016/j.nima.2014.11.109.
75. Kawaguchi, N., Fukuda, K., Yanagida, T. et al. (2011). Fabrication and characterization of large size ^{6}LiF/CaF_2:Eu eutectic composites with the ordered lamellar structure. *Nucl. Instrum. Methods A.* 652: 209–211. 10.1016/j.nima.2011.02.009.
76. Fuji Electric (2021) NSN4 scintillation based neutron survey meter. Available at: www.fujielectric.co.jp/about/company/gihou_new_2020/pdf/FEJ-2020-S05.pdf (Accessed: 15 April 2021).
77. van Eijk, C.W.E. (2012). Inorganic scintillators for thermal neutron detection. *IEEE Trans. Nucl. Sci.* 59: 2242–2247. 10.1109/TNS.2012.2186154.
78. Yen, W., Shionoya, S., and Yamamoto, H. (2007). *Practical Applications of Phosphors*. New York: CRC Press.
79. Tanaka, H., Uchida, T., Tanaka, M. et al. (2009). Cosmic-ray muon imaging of magma in a conduit: degassing process of Satsuma-Iwojima volcano, Japan. *Geophys. Res. Lett.* 36: L01304. 10.1029/2008GL036451.
80. Sears, V.F. (1992). Neutron scattering lengths and cross sections. *Neutron. News.* 3: 26–37. 10.1080/10448639208218770.
81. Yanagida, T., Kim, K.J., Kamada, K. et al. (2010). Growth, optical properties, and scintillation light yield of Ce:CaF2 crystals with different Ce concentration. Jpn. J. Appl. Phys. 49: 032601-1–5. 10.1143/JJAP.49.032601.

5

Inorganic Halide Scintillators

Yutaka Fujimoto
Graduate School of Engineering, Tohoku University, Sendai, Japan

5.1 Introduction: History of Inorganic Halide Scintillator Research and Development

Scintillation detectors consisting of a scintillator material that is optically coupled to a photosensitive detector are widely used for detection and measurement of energetic photons such as X-rays and gamma-rays. These detectors are commonly used in nuclear and particle physics research, medical imaging, diffraction, nondestructive testing, nuclear treaty verification and safeguards, nuclear nonproliferation monitoring, and geological exploration. The discovery and use of scintillator materials can be traced back to 1896 with Calcium tungstate ($CaWO_4$), which was used following the discovery of the X-ray. After nearly half a century, the discovery of the photomultiplier tube (PMT) resulted in research into new scintillator materials. One of the major discoveries in these scintillator materials was in 1948–1951 by Hofstadter and Van Sciver, who identified an exceptionally bright scintillator, thallium-doped sodium iodide (NaI:Tl) [1, 2] and thallium-doped cesium iodide (CsI:Tl) [3]. These halide crystals can be used in survey meter and X-ray computed tomography (CT) scanner applications [4, 5]. Meanwhile, the poor energy resolution of the NaI:Tl and CsI:Tl of approximately 7% at 662 keV from a ^{137}Cs source limits its performance in radionuclide identification, which has motivated research into new scintillator materials with better performance to replace them. In the years 2000 and 2001, the excellent scintillation properties of $LaCl_3$:Ce and $LaBr_3$:Ce [6, 7] were initially reported by Technical University Delft together with Bern University. In particular, the $LaBr_3$:Ce have exhibited a high light yield (~60 000 ph/MeV) with a very high energy resolution (~2.8% at 662 keV) and a fast decay time of 35 ns. After 2001, rare-earth iodide-based scintillators such as YI_3:Ce, LuI_3:Ce, and GdI_3:Ce were studied by some researchers [8–10]. Those studies reported that the iodide

Phosphors for Radiation Detectors, First Edition. Edited by Takayuki Yanagida and Masanori Koshimizu.

scintillators show a very high light yield (~100 000 ph/MeV). However, they have some drawbacks due to their hygroscopicity and difficult growth. Among other cerium halide scintillators, $CeBr_3$ is proven to be a promising intrinsic scintillator in more recent studies [11]. The $CeBr_3$ has high light yield (~68 000 photons/MeV), fast decay time (~17 ns), and excellent energy resolution of ~4% at 662 keV. Furthermore, it shows low intrinsic activity and high density (5.2 g/cm^3). Originally discovered by Hofstadter in 1968 [12], europium-doped strontium iodide (SrI_2:Eu) has been very intensely investigated in recent years. Owing to recent progress in crystal growth technology, dramatically improved scintillation properties have been discovered in SrI_2:Eu crystals, which are a high light yield of above 70 000 photons/MeV, typically 3% energy resolution at 662 keV, and typical decay time of a few microseconds, depending on the Eu concentration [13–15].

As in the study described above, some binary halide scintillators have shown the potential to replace NaI:Tl and CsI:Tl in applications where excellent energy resolution and high light yield are required. But their higher cost and lower production yield compared to NaI:Tl and CsI:Tl have prevented their widespread utilization in many applications. Thus, there is a large demand for continuous research into new scintillator material with superior performance and low production costs. In recent years, there has been a discernible shift from simple binary halide scintillators toward ternary and quaternary halide-based compounds in scintillator research and development. Below we describe the basic characteristics of halide materials and crystal growth techniques for halide scintillators and review of the recently discovered novel ternary and quaternary halide scintillators and new material design approach.

5.2 Characteristics of Halide Materials

5.2.1 Formation of Color Center and Self-Trapped Exciton

Color centers are a special type of point defect owing to trapping of electrons and holes. Studies of color center in alkali halides started in 1896 by Goldstein [16] and subsequently by Pohl [17, 18]. After these early studies, many researchers focused on the color center phenomena associated with the processes in the formation and destruction of color centers, and studied these processes by some different analysis techniques in various crystalline structural material and under different conditions. The analysis techniques for these studies include the measurements of optical absorption, optical and thermal belching, luminescence, paramagnetic resonance (PMR), electron spin resonance (ESR), electron nuclear double resonance (ENDOR), optical detection magnetic resonance (ODMR), and resonance Raman scattering. The simplest type of trapped electron center is the F-center, that of an electron trapped at a halogen vacancy surrounded by some alkali metal ions. It is well known that F centers and trapped hole centers are formed by halide crystal exposure to ionizing radiation at low temperatures. Figure 5.1 shows the configuration of color centers and self-trapped excitons (STEs) in an ionic crystal. Two intrinsic hole centers, the so-called V_k centers and H centers, have similar structure. The V_k center is a covalently bounded pair of adjacent halogen ions sharing a valence hole, and is often described in terms of an X_2^- molecule ion, where X is a halogen. Thus, the two halogen ions involved in the V_k center are displaced from their normal positions toward each other. The formation of the H center corresponds to the formation of an interstitial halogen ion. This results from removing a halogen ion from

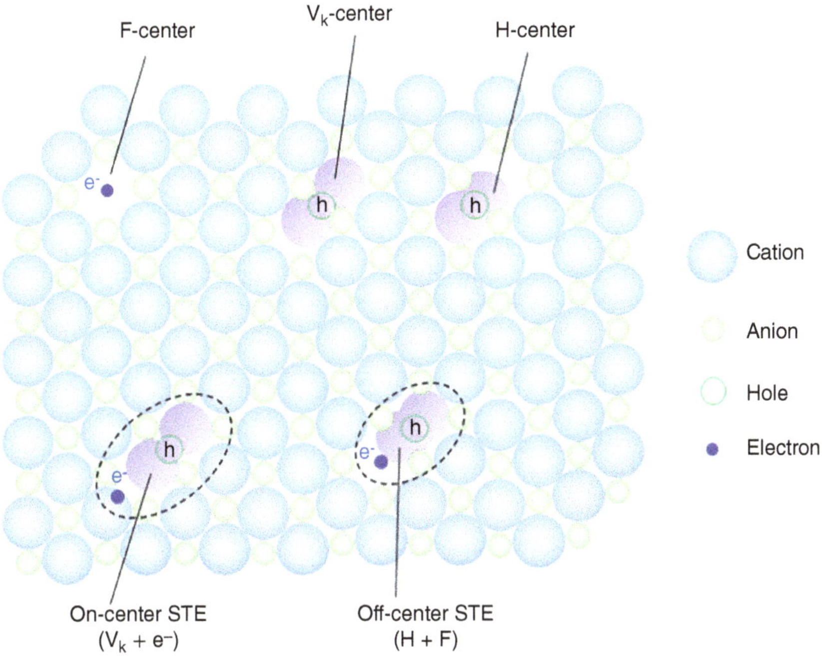

Figure 5.1 *Configuration of color centers and self-trapped excitons in an ionic crystal.*

its normal site and placing it in an interstitial position. Thus, the H center consists of a hole shared among halogen ions and is very unstable at room temperature and disappears by diffusion back to a vacant lattice site. The difference of two kinds of the self-trapped holes is that the V_k center partially occupies two anion sites and is thus effectively positively charged. Meanwhile the H centers occupy a single anion site, which results in a neutral defect. STEs are a state in which an additional electron is trapped by the V_k center, resulting in the formation of neutral Frenkel defect pairs. The structure has been determined accurately by electron paramagnetic resonance (EPR) and by ENDOR [19, 20]. Among the STEs, is an on-center STE, which an electron delocalizes in an orbit around a V_k-center ($V_k + e^-$). When the V_k center transforms into an H center and the electron is localized at the anion vacancy (F-center), this H-F pair is known as an off-center STE. The intrinsic STE emission occurs owing to a recombination of a free or self-trapped electron and a self-trapped hole. The STE emission includes the σ- and π-polarized bands. The former band generally appears in the UV wavelength range, which is a typically shorter wavelength range than that of π-emission bands. The latter can be observed only at the low temperature of liquid helium. From previous studies, the bands of both σ- and π- emissions were found in only about half of the alkali halides, while that of π-emission was observed in almost all alkali halides.

5.2.2 Hygroscopicity

Halide-based commercial scintillators, such as NaI:Tl, $LaBr_3$:Ce, $CeBr_3$, and SrI_2:Eu, are hygroscopic and must be protected from moisture under ambient atmospheric conditions by using appropriate packaging techniques, because their interaction with moisture produces degradation of physical, optical, and scintillation properties. Thus, the halide scintillators are usually enclosed in a hermetically sealed container with an optical window. The packaging design and technique, including housing, reflector, and the others components, depends upon the specific needs of a given scintillator. The hygroscopic property requires some attention, not only during the packaging process but during all processes of production from raw material selection, crystal growth, cutting, and polishing to detector assembly. Figures 5.2 and 5.3 show the moisture sorption curves of some halide scintillators [21, 22]. Various halide scintillators show characteristic trends. It is clear that the strongly hygroscopicity of CaI_2 and SrI_2:Eu, and KSr_2I_5:Eu showed a sub-linear or even polynomial relationship with exposure time. The others also exhibit a drastically increasing of the weight at the beginning, and after that the increasing rate is reduced over time. The moisture sorption curves of some

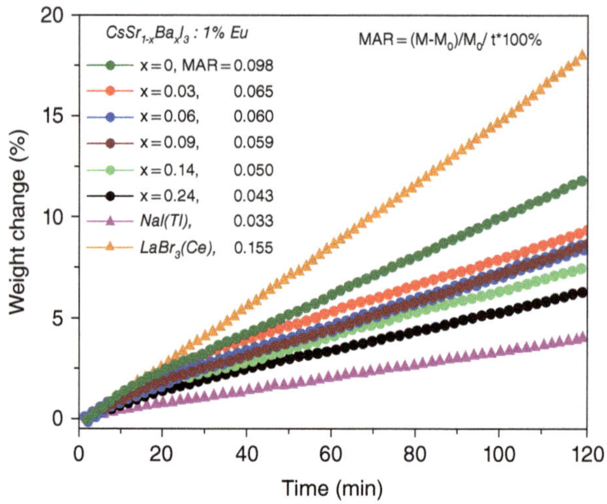

Figure 5.2 *Moisture absorption curves of a* $CsSr_{1-x}Ba_xI_3$*:Eu (1%) single crystal [21].*

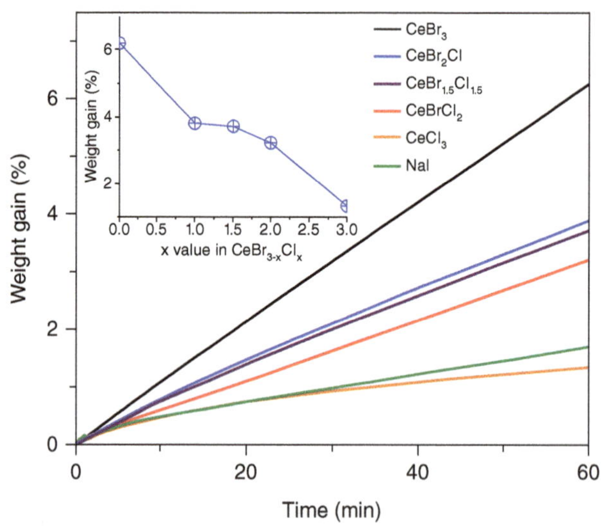

Figure 5.3 *Moisture absorption curves of* $CeBr_3$*,* $CeBr_2Cl$*,* $CeBr_{1.5}Cl_{1.5}$*,* $CeBrCl_2$*,* $CeCl_3$*, and NaI single crystal [22].*

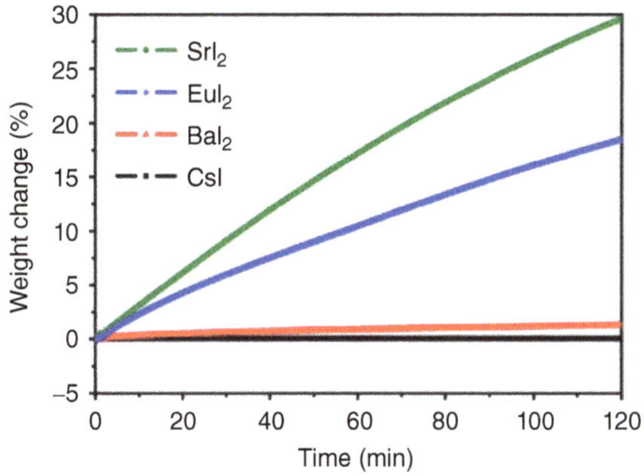

Figure 5.4 *Moisture absorption curves of SrI_2, EuI_2, BaI_2, and CaI_2 raw materials [21].*

Alkali metal iodides (A = Li, Na, K, Rb, Cs)

$2A + H_2O + 1/2O_2$	$\rightarrow$	$2AOH + I_2$
$4AI + O_2$	$\rightarrow$	$2A_2O + 2I_2$
$2AI + O_2$	$\rightarrow$	$A_2O_2 + I_2$
$AI + O_2$	$\rightarrow$	$AO_2 + I/2I_2$
$AI + 1/4O_2 + 1/2CO_2$	$\rightarrow$	$1/2A_2CO_3 + I/2I_2$

Alkali-earth metal iodides (AE = Mg, Ca, Sr, Ba)

$AE\,I_2 + H_2O$	$\rightarrow$	$AE\,O + HI_2$
$AE\,I_2 + H_2O$	$\rightarrow$	$AE\,O + I_2 + H_2$
$AE\,I_2 + 2H_2O$	$\rightarrow$	$AE\,(OH)2 + 2HI$
$2\,AE\,I_2 + O_2$	$\rightarrow$	$2\,AE\,OI + I_2$
$2\,AE\,I_2 + O_2$	$\rightarrow$	$2\,AE\,O + 2I_2$

Figure 5.5 *Basic hydration processes in alkali and alkali-earth iodides.*

iodide raw materials are shown in Figure 5.4 [21]. The CsI and BaI_2 are relatively low hygroscopic materials because of almost no change of the weight after 120 minutes of exposure to the controlled atmosphere. The high hygroscopic materials are SrI_2 and EuI_2 that showed rapid and continuous increasing of the weight.

The basic hydration processes in alkali and alkali-earth iodides are slightly different (Figure 5.5). In the case of the alkali iodides, the iodides decompose in the atmosphere of oxygen and water (H_2O, OH^-). The OH^-, CO_3^{2-}, O_2^{2-}, and O^{2-} anions can contaminate the surface and inside of the crystals and decrease optical quality and scintillation performance. The hydration reactions and processes of the alkali-earth iodides with the atmosphere are different from those of alkali halides. The important factor in relation to hygroscopicity of alkali iodides may be hydrolysis and creation of oxides, oxoiodides, and hydroxides. Previously. some researchers evaluated and determined the concentration of such anions for NaI:Tl using an IR spectroscopy technique [23, 24]. It indicates that some impurities such as NaOH (8600 cm^{-1}), $NaIO_3$ (800–770 cm^{-1}), and Na_2CO_3 (880, 1440 cm^{-1}) exist in the crystal grown in air atmosphere. They can lead to the formation of any color centers in which electrons, holes, and excitons are localized. The formation of these trapping centers results in an increase of recombination losses and, therefore, the reduction of scintillator performance, including scintillation light yield and energy resolution.

5.3 Basic Techniques for Halide Scintillation Crystal Growth

There are various techniques to prepare single crystals of halides, and among them, few techniques are suitable for production of single crystals with a large dimension for use in scintillation crystal-based ionizing radiation detectors. In solution growth techniques, chemical components are dissolved in a solvent under saturation level and then the crystals were grown around a seed crystal by slow cooling, solvent evaporation, or temperature gradient [25]. Growth rates from solution vary from 0.1 to 2 mm/day, thus the techniques are too slow for production of large volume crystals required by industry. Czochralski growth is still one of the most widely used bulk crystal growth techniques used today. The growth techniques consisting of a pulling up of a crystal from the melt contained in a heated crucible is widely used, particularly in industry for the production of metal and oxides such as Si and sapphire with a diameter of up to 10 inches. The major disadvantage of Czochralski growth is the high cost of the multiple control such as rotation and translation of seed or crucible, mass measuring of the growing crystal, and the automatic diameter control (ADC) system. Furthermore, in the case of metal halides, the growth chamber must be supplied with an inert gas atmosphere, because they are susceptible to degradation in the presence of moisture or oxygen at high temperatures. Thus, commercially available halide scintillation crystals such as NaI:Tl, CsI:Tl, and others are usually produced not by the Czochralski technique but by the Bridgman–Stockbarger growth technique. The crystal growth technique of metal halides from the melt in sealed ampoules within resistively heated furnaces was first described by Bridgman in the early twentieth century [26]. This provides an economical and efficient means of manufacturing the crystal with a large diameter at rates of a few cm per day. Some years later, this technique was modified by Stockbarger to achieve improved results for lithium fluoride crystals [27]. A schematic of the Bridgman growth station with the modifications developed by Stockbarger is shown in Figure 5.6. The growth technique commonly utilizes a two-zone furnace, either in a vertical or horizontal position. The initial (top) zone is set at a temperature above the melting point of the material, and the second (bottom) zone is set at a temperature which is lower than the melting point. Starting materials for crystal

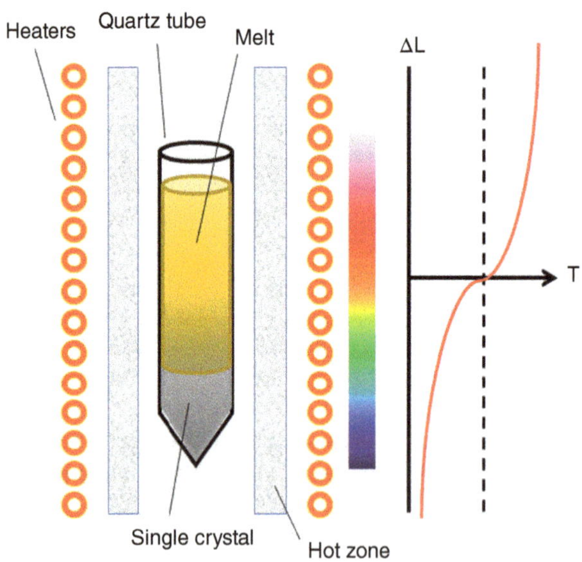

Figure 5.6 *Schematic of the Bridgman growth station with the modifications developed by Stockbarger.*

growth are typically loaded into a sealed ampoule which is evacuated or filled with an inert gas to prevent exposure to ambient moisture and oxygen. The sealed ampoule travels through the temperature gradient controlled by the furnace and enters the colder zone. A single crystal can be obtained by growing from a seed crystal located in the lowest portion. The only parameters required to control the growth process are basically the furnace temperatures and the pulling rate of the ampoule. Various commercial halide scintillation crystals are generally grown by the Bridgman technique. Among them, the NaI:Tl crystals with a diameter of 5 inches to more than 30 inches are in commercial use [28].

5.4 Novel Ternary and Quaternary Halide Scintillators

5.4.1 Alkali Halide-Rare Earth Halide (AX–REX_3)

Several Ce^{3+}-activated alkali halide-rare earth halide compounds have been found to be excellent scintillators. In general, the Ce^{3+} ion on trivalent lattice sites has the simplest electron configuration, and the Ce^{3+}emission results from electron transitions between the 4f ground state and the lowest $5d^1$ excited state. Table 5.1 shows the main results.

In 1995, Spijker and co-researchers were the first to report the scintillation in K_2LaCl_5 doped with Ce^{3+} and found good scintillation properties with an energy resolution of 5.1% at 662 keV and a high light yield of 29 000 ph/MeV [29]. The K_2LaCl_5 is isotypic with Y_2HfS_5 and the trivalent La^{3+} sites that Ce^{3+} ions occupy are surrounded by seven Cl^- anions forming a distorted mono-capped trigonal prism [41]. The density and Z_{eff} are 2.89 g/cm^3 and 44,

Table 5.1 *Discovered and developed Ce^{3+}-activated alkali-halide-rare-earth halide compounds.*

Compound	Year	LY (ph/MeV)	ΔE/E (%) @662 keV	Ref.
K_2LaCl_5:Ce	1995	29000	5.1	[29]
$RbGd_2Br_7$:Ce	1997	54000	3.8	[30]
K_2LaBr_5:Ce	2005	40000	5	[30]
K_2LaI_5:Ce	2005	55000	4.5	[31]
$Cs_3Lu_2I_9$:Ce	2007	22800	9	[31]
Cs_3LuI_6:Ce	2007	15000	–	[31]
$Cs_3Gd_2I_9$:Ce	2007	26000	–	[32]
K_2CeCl_5	2007	30000	5.8	[33]
K_2CeBr_5	2008	50000	6.3	[34]
Cs_3CeCl_6	2010	19000	8.4	[34]
$CsCe_2Cl_7$	2010	26000	7.5	[34]
$CsCe_2Br_7$	2010	40000	8	[34]
Cs_3CeBr_6	2010	28000	9	[35]
Cs_3EuI_5	2011	39000	~10	[36]
KGd_2Cl_7:Ce	2011	30000	~10	[37]
Cs_3GdCl_6:Ce	2014	24500	4.5	[37]
Cs_3GdBr_6:Ce	2014	47000	4	[38]
Cs_3LaCl_6:Ce	2014	16000	8.6	[38]
Cs_3LaBr_6:Ce	2014	32500	4.9	[39]
Rb_2CeCl_5	2019	36000	15	[40]

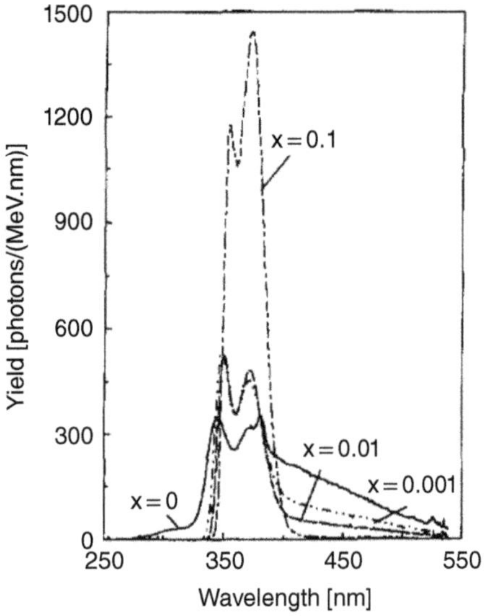

Figure 5.7 *The X-ray excited scintillation spectra of K_2LaCl_5:Ce [29].*

respectively. In the X-ray excited scintillation spectra (Figure 5.7), the characteristic $^2D_{3/2}$--$^2F_{5/2}$ and $^2D_{3/2}$-$^2F_{7/2}$ emission lines of Ce^{3+} can be observed at 380 and 340 nm, respectively. Scintillation properties of Ce^{3+}-activated K_2LaX_5 (X = Cl, Br, I), which are isostructural series, have also been studied by the same group [31, 42]. Among the compounds, 10% Ce-doped K_2LaI_5 appears to be the best in terms of scintillation light yield, decay time, and energy resolution. It has a high light yield of 55 000 ph/MeV, a decay time of 24 ns, and an energy resolution of 4.5% at 662 keV. The scintillation light yield of the K_2LaX_5:Ce (X = Cl, Br, I) increases in the sequence Cl → Br → I (Table 5.2), while the decay time becomes faster in the sequence Cl → Br → I (Figure 5.8). They are possibly related to the migration rates of STEs and V_k centers that tend to increase with larger size of the anion and with the smaller size of the cation.

In 19, the scintillation properties of $RbGd_2Br_7$ doped with Ce^{3+} were first reported by Dorenbos from Delft University [30]. The $RbGd_2Br_7$ crystal have an orthorhombic structure

Table 5.2 *Scintillation light yield and energy resolution of K_2LaX_5:Ce (X = Cl, Br, I) under ^{137}Cs 662 keV-gamma-ray [30].*

Host	(Ce^{3+}) (%)	Photon yield (10^3 ph/MeV) 0.5 μs	3 μs	10 μs	Energy resolution R (%)
K_2LaCl_5	0.23	4.5	11	14	–
K_2LaCl_5	0.7	8	17	21	–
K_2LaCl_5	10	23	27	30	5.0
K_2LaBr_5	0.7	–	–	26	7.5
K_2LaBr_5	10	–	–	40	5.0
K_2LaI_5	0.7	48	52	52	5.7
K_2LaI_5	10	53	55	55	4.5

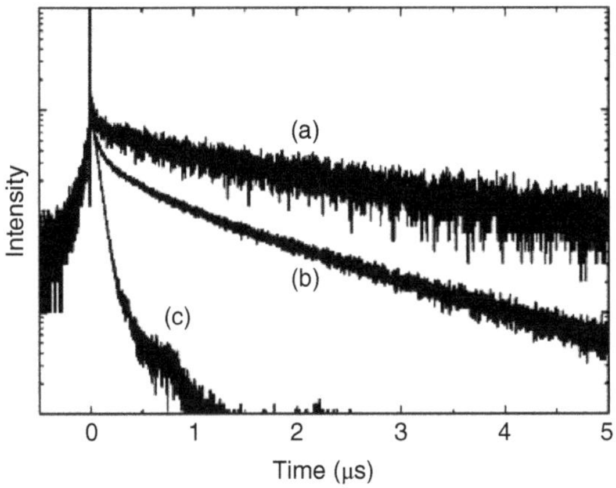

Figure 5.8 *Scintillation decay curves of (a) K_2KaCl_5:Ce (b) K_2KaBr_5:Ce and (c) K_2KaI_5:Ce at room temperature [31].*

with a density of 4.7 g/cm^3 and a Z_{eff} of 50. The compound melts congruently at 590 °C and, therefore, its crystals can be grown using the conventional Bridgman method. For a 5% Ce-doped sample, the light yield reaches 54 000 ph/MeV with a main decay time of 60 ns. It shows excellent energy resolution of 3.8% for the 662 keV full absorption peak, as shown in Figure 5.9. The characterizations of the $RbGd_2Br_7:Ce^{3+}$ for thermal neutron detection were reported by another group in 2007 [43], because two isotopes of Gd, ^{157}Gd, and ^{155}Gd have very high thermal neutron absorption cross-sections of about 2.55×10^5 b and 6.1×10^4 10 b, respectively. In the pulse height spectrum measurements with the $RbGd_2Br_7:Ce^{3+}$, two peaks at ~34 and ~74 keV were successfully detected resulting from thermal neutron reaction.

Some Ce-concentrated binary halide compounds were intensely studied because of their high light yield with a low concentration quenching, despite a high Ce concentration. In 2007, K_2CeCl_5 was first introduced as a scintillator by Roy et al. [33]. The K_2CeCl_5 crystallized in an orthorhombic structure with lattice parameters a = 8.022 Å, b = 8.815 Å, and c = 12.717 Å [44]. The polished K_2CeCl_5 crystal is highly transparent and is much less hygroscopic compared to $LaBr_3$. The scintillation spectrum of K_2CeCl_5 is shown in Figure 5.10a.

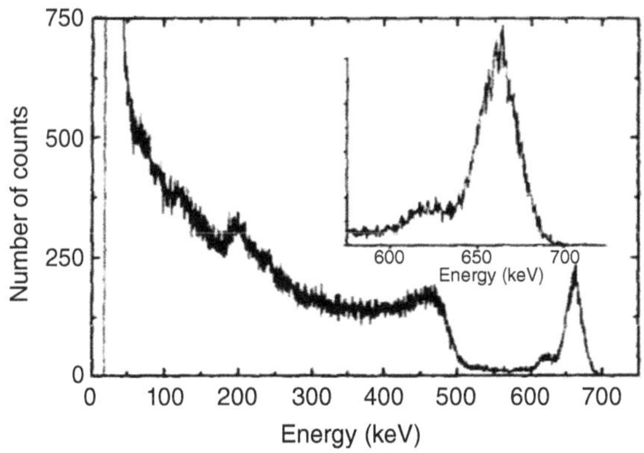

Figure 5.9 *^{137}Cs-gamma-ray irradiated pulse height spectrum of $RbGd_2Br_7$:Ce [30].*

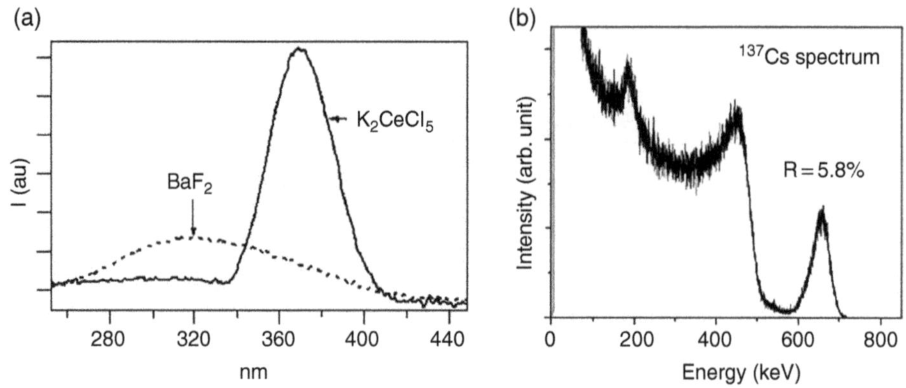

Figure 5.10 *(a) Scintillation spectrum and (b) ^{137}Cs-gamma-ray irradiated pulse height spectrum of K_2CeCl_5 [33].*

The spectrum consists of a single broad band peaking at 370 nm, owing to 5d-4f transition of Ce^{3+}. From the pulse height spectrum using the K_2CeCl_5 (Figure 5.10b), a high light yield of 30 000 ph/MeV with an energy resolution of 5.8% at 662 keV is obtained. A year later, Hawrami et al. reported the crystal growth and scintillation properties of K_2CeBr_5 that show a high light yield of 50 000 ph/MeV with an energy resolution of 6.3% at 662 keV [34]. These Ce-concentrated halide compounds might well be considered as the isostructural La-based ternary halide compounds doped with 100% Ce^{3+}, and as described above, an influence of Ce concentration quenching on the scintillation light yield is less than that of the Ce^{3+}-doped oxide and other compounds.

Except for the cerium compounds, an Eu-concentrated halide Cs_3EuI_5 showing a high light yield was discovered by Yang from the University of Tennessee in 2011 [36]. The density and Z_{eff} of Cs_3EuI_5 are 4.55 g/cm^3 and 55, respectively and the crystal structure is hexagonal. Figure 5.11a shows excitation and emission spectra of the Cs_3EuI_5. The broad excitation band from 270 to 460 nm is similar to the excitation band of Eu^{2+} for Ba_2CsI_5:Eu and BaBrI:Eu [45, 46]. Thus, the emission band peaking at 470 nm corresponds to the 5d–4f transition of Eu^{2+}. The pulse height spectrum of the Cs_3EuI_5 compared with a BGO reference is shown in Figure 5.11b. The scintillation light yield is calculated to be 39 000 ph/MeV, with ~10% energy resolution at 662 keV. Improved energy resolution can be expected resulting in an improvement of crystalline quality with better transparency and less colorization.

5.4.2 Alkali Halide-Alkalin Earth Halide (AX–AEX_2)

Eu^{2+}-activated alkali halide-alkalin earth halide compounds, including in particular KSr_2I_5, KBa_2I_5, $KCaI_3$, and $CsBa_2I_5$, often exhibit higher light yields with a more excellent energy resolution than Ce^{3+}-activated compounds. Eu^{2+} ions are widely used as emission centers for phosphor materials, where the Eu^{2+} ion is on a divalent lattice site because of its high fluorescence quantum efficiency and suitable emission wavelength for photodetectors, owing to the inter-configurational $4f^65d$-$4f^7$ ($^8S_{7/2}$) transition of Eu^{2+}. Table 5.3 compiles the main results of inspecting the Eu^{2+}-activated compounds.

The family of compounds given by the generic formula AB_2X_5 (A = Cs, Rb, K; B = Ca, Sr, Ba; X = Cl, Br, I) is the current trend in scintillator research and development. In 2009, Bourret-Courchesne et al. reported the discovery of a new bright scintillator,

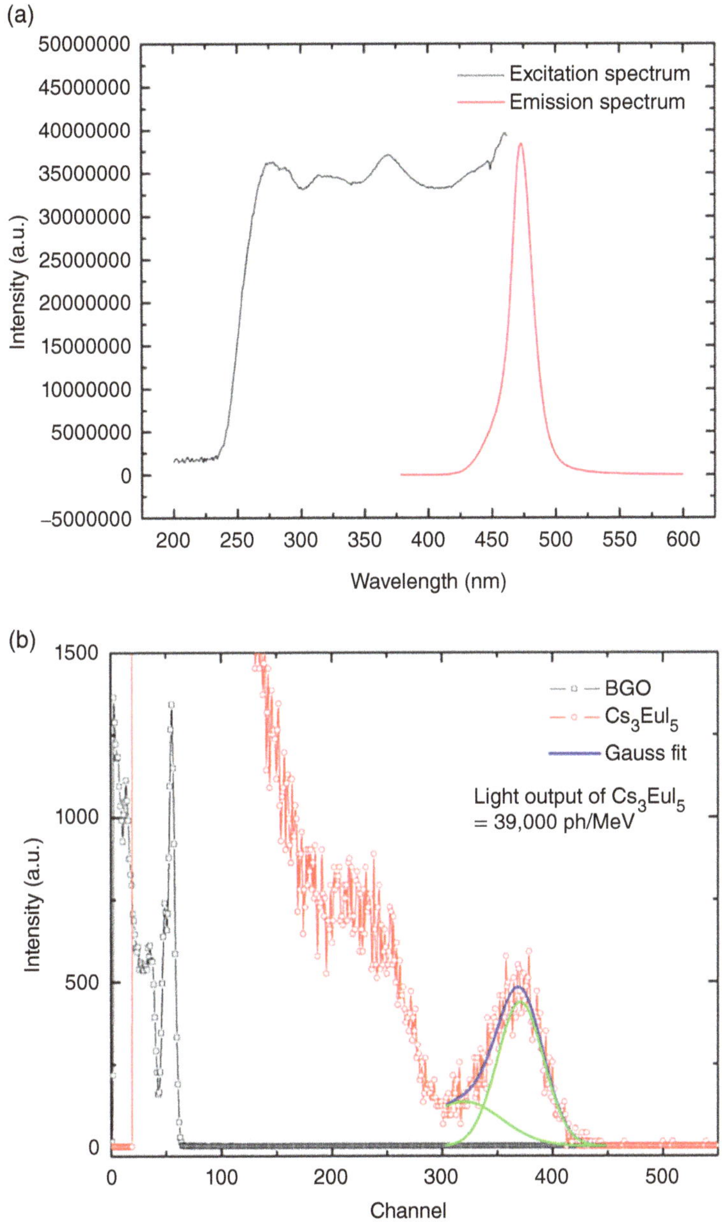

Figure 5.11 *(a) Excitation and emission spectra and (b) ^{137}Cs-gamma-ray irradiated pulse height spectrum of Cs_3EuI_5 [36].*

Eu-doped $CsBa_2I_5$ with a light yield of 97 000 ph/MeV and an energy resolution of 3.8% (Figure 5.12) [45]. The $CsBa_2I_5$ have a monoclinic structure $P2_1/c$ space group with cell parameters a = 10.541(3) Å, b = 9.256(3) Å, c = 14.637(5) Å, β = 90.194(6), which is similar to that of $CsBa_2Br_5$, and it melts congruently at 610 °C. In 2011, the same group reported further improvements in their scintillation performance with an energy resolution of 2.6% and a light yield of 100 000 ph/MeV [57]. Temperature and concentration

Table 5.3 Discovered and developed Eu^{2+}-activated alkali-halide-alkalin-earth halide compounds.

	Year	LY (ph/MeV)	ΔE/E (%) @662 keV	Ref.
$CsBa_2I_5$:Eu	2009	97 000	3.8	[45]
$CsSrI_3$:Eu	2011	73 000	3.9	[47]
$CsCaI_3$:Eu	2012	38 500	8	[48]
$CsCaCl_3$:Eu	2012	18 000	8.9	[48]
$CsSrBr_3$:Eu	2012	~31 300	8.8	[49]
$CsCaBr_3$:Eu	2013	28 000	9.3	[50]
KSr_2I_5:Eu	2015	94 000	2.4	[51]
$KCaI_3$:Eu	2015	72 000	3	[52]
KBa_2I_5:Eu	2016	90 000	2.4	[53]
K_2BaI_4:Eu	2016	63 000	2.9	[53]
$LiSr_2I_5$:Eu	2016	60 000	3.5	[54]
$LiCa_2I_5$:Eu	2016	90 000	5.6	[54]
$LiSr_2Br_5$:Eu	2016	32 000	6.1	[54]
$RbSr_2Br_5$:Eu	2017	64 700	4	[55]
$RbSr_2I_5$:Eu	2017	90 400	3	[55]
Cs_4CaI_6:Eu	2018	51 800	3.6	[56]
Cs_4SrI_6:Eu	2018	62 300	3.3	[56]

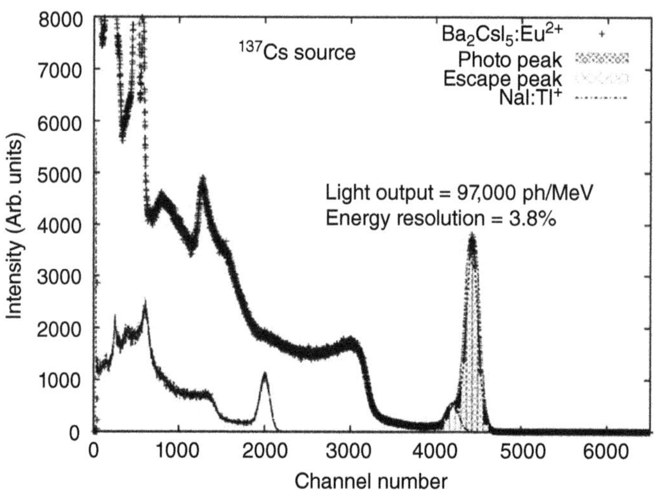

Figure 5.12 Comparison of pulse height spectra of $CsBa_2I_5$:Eu and NaI:Tl measured under ^{137}Cs gamma-ray irradiation [45].

dependency and thermoluminescence properties of the Eu-doped $CsBa_2I_5$ compound were reported by Alekhin et al. in 2014 [58]. Figure 5.13 shows the decay time constants as a function of temperature of undoped, 0.5, and 5% Eu-doped $CsBa_2I_5$. The decay time increases with increasing temperature and Eu concentration due to self-absorption of the Eu^{2+} emission. Although the scintillation decay time constants are generally consistent with those of photoluminescence, at temperatures below 295 K, 0.5% Eu-doped $CsBa_2I_5$

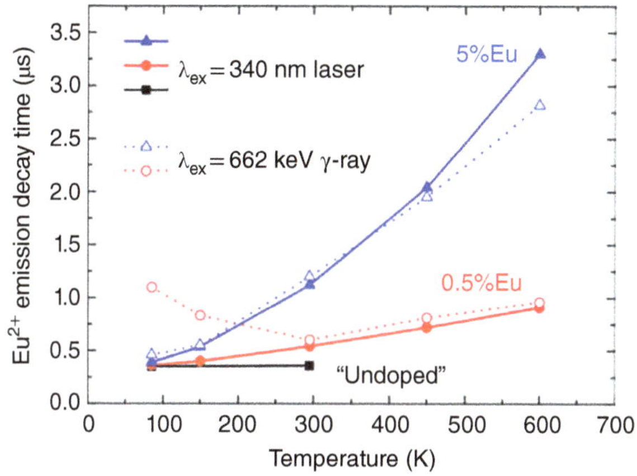

Figure 5.13 *Eu^{2+} emission decay time constants as a function of temperature of $CsBa_2I_5$ and $CsBa_2I_5$:Eu [58].*

has a significantly longer scintillation decay component as compared to the photoluminescence one. This is attributed to a delayed energy transfer from host lattice to Eu^{2+} emission centers.

In 2015, Eu-doped KSr_2I_5 was discovered as a new high light yield scintillator by Stand et al. from the University of Tennessee [51]. The crystal structure of KSr_2I_5 is a monoclinic structure with space group $P12_1/c1$ and a density of 4.39 g/cm^3 [59]. Single crystals of 13 and 22 mm in diameter were successfully grown from melt by using the Bridgman technique. The 4% Eu-doped KSr_2I_5 has a measured light yield of 94 000 ph/MeV and an excellent energy resolution of 2.4% at 662 keV, as shown in Figure 5.14a. Moreover, the compound has an excellent proportional response to a wide range of gamma-ray energies (see Figure 5.14b). At almost the same time, Lindsey from the same group reported on the crystal growth and characterizations of Eu-doped $KCaI_3$ [52]. The orthorhombic crystal belongs to the ABX_3 family of perovskite-type compounds (A = Cs, Rb, K; B = Ca, Sr, Ba; X = Cl, Br, I). A high scintillation light yield of 72 000 ph/MeV and good energy

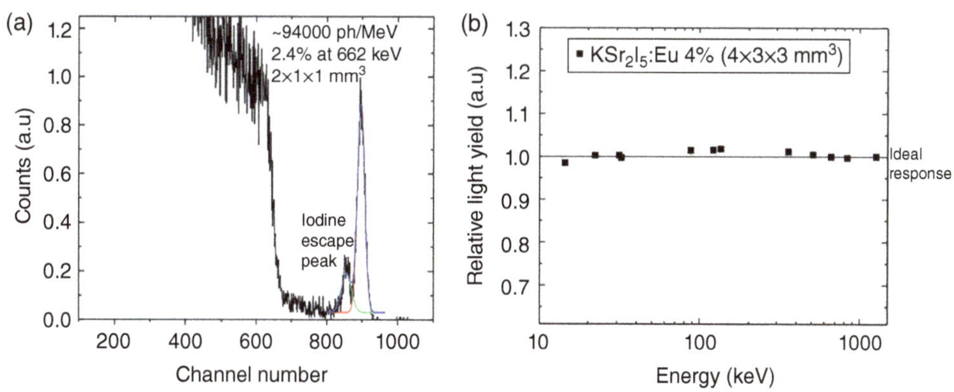

Figure 5.14 *(a) ^{137}Cs-gamma-ray irradiated pulse height spectrum and (b) non-proportionality measurements of KSr_2I_5:Eu [51].*

resolution of 3% at 662 keV and 6.1% at 122 keV was obtained from small samples of approximately 15 mm^3. In 2019, they were successful at growing the crystal with a diameter of φ38 mm without cracking [60]. The grown crystal show an energy resolution of ~4% at 662 keV. It is known that the performance of Eu^{2+}-doped compounds worsens significantly when large volume crystals are prepared, which is due to some trapping centers or self-absorption effects.

Within other alkali halide-alkalin earth halide families, two new scintillators, Eu-doped KBa_2I_5 and K_2BaI_4,were reported by the Tennessee groups. Compared to the Ca- and Sr-containing halide families, these Ba-containing compounds are expected to have higher density and less deliquescent. A sample of a 4% Eu-doped KBa_2I_5 has a light yield of 90 000 ph/MeV with an energy resolution of 2.4% and that of 7% Eu-doped K_2BaI_4 is 63 000 ph/MeV with an energy resolution of 2.9% at 662 keV. Both crystals have an excellent proportional response to a wide range of gamma-ray energies.

5.4.3 Elpasolite

Halide elpasolite (double perovskite) compounds have a general formula of A_2BREX_6, where A and B are monovalent alkali metals, *RE* is a trivalent rare earth, and X is a halogen. The compounds can be viewed as cubic or nearly cubic crystalline structures over a wide range of compositions that are expected to reduce the impact of thermally mechanical stress during crystal growth. Due to their high structural symmetry, they were originally well-known as tunable solid-state laser hosts. A new thermal neutron scintillator, Ce-doped Cs_2LiYCl_6 from the elpasolite family, was discovered by Combes from Delft University in 1999 [61], that shows a reasonable light yield under gamma-ray excitation (~18 400 ph/MeV). Moreover, the same group have shown that the Ce-doped Cs_2LiYCl_6 can be used with a pulse shape discrimination (PSD) technique to distinguish neutrons from gamma-rays due to its core-valence (CV) luminescence under gamma-ray excitation [62]. The ^6Li-included compound can convert incident neutrons into secondary ionizing particles via $^6Li(n, \alpha)t$ neutron capture reaction. The Q-value of this capture reaction amounts to 4.78 MeV and the reaction products give rise to the scintillation phenomenon. The CV luminescence shows a very fast decay rate produced by the radiative recombination of a valence electron with an outermost core hole. Except for Cs_2LiYCl_6 elpasolite, it was reported in some halide compounds such as $LiBaF_3$ [63], BaF_2 [64], Cs_2ZnCl_4 [65], and $CsCaCl_3$ [66], that the fast CV luminescence is observable under gamma-ray and X-ray excitation. After the discovery of Ce-doped Cs_2LiYCl_6, the crystalline and electronic structure and scintillation properties of several other halide elpasolite compounds are also studied by replacing other compositions. Also, recently, some researchers show that the Ce-doped Cs_2LiYCl_6 can detect fast neutrons with the $^{35}Cl(n, p)^{35}S$ reaction [67, 68]. Figure 5.15 shows the PSD scatter plot (top) and energy spectra (bottom) recorded for the Ce-doped Cs_2LiYCl_6 under excitation with a 1.6 MeV fast neutron beam. In the scatter plot and energy spectra, three distinct neutron related features can be observed. A peak in the center is due to the thermal neutrons captured by ^6Li ions, while a left peak and a right continuum are attributed to the fast neutrons captured by the ^{35}Cl ions and interacted with the ^6Li ions. The Ce-doped Cs_2LiYCl_6 therefore has a useful capability to discriminate between fast- and thermal-neutron and gamma-rays. Thus, Ce-doped Cs_2LiYCl_6 was commercialized as a neutron scintillator due to its high light yield and good neutron/gamma PSD properties.

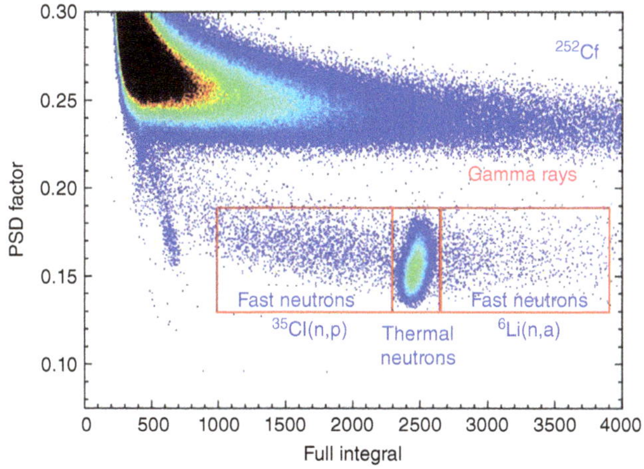

Figure 5.15 *PSD scatter plot recorded for Cs_2LiYCl_6:Ce under irradiation from a ^{252}Cf source [68].*

5.5 Mixed-Anion Halide Scintillators

Study of mixed-anion halide scintillators shows that they have potential as a new material design technique to improve scintillation properties. A recent significant example of an enhancement of the scintillation light yield in the mixed-anion halide is the study of the $LaBr_{3-x}Cl_x$:Ce scintillator by the controlling of Br-Cl composition. In the mixed-anion halide of LaBr (66%)–LaCl (34%), the scintillation light yield was found to be 138% as compared to LaBr:Ce [69]. In 2008, Birowosuto et al. reported $LaBr_{3-x}I_x$:Ce scintillators with four different compositions of x (x = 0.75, 1.5, 2, and 2.25) [70]. XRD patterns of the $LaBr_{3-x}I_x$:Ce are shown in Figure 5.16. The $LaBr_{2.25}I_{0.75}$:Ce crystallized in the UCl_3 structure with space group $P6_3/m$ as $LaBr_3$, while the $LaBr_{1.5}I_{1.5}$:Ce and $LaBr_{0.75}I_{2.25}$:Ce showed the $PuBr_3$ structure with space group *Cmcm*. Table 5.4 shows the summary of scintillation properties of the $LaBr_{3-x}I_x$:Ce. From the results, the highest light yield of 58 000 ph/MeV was found for

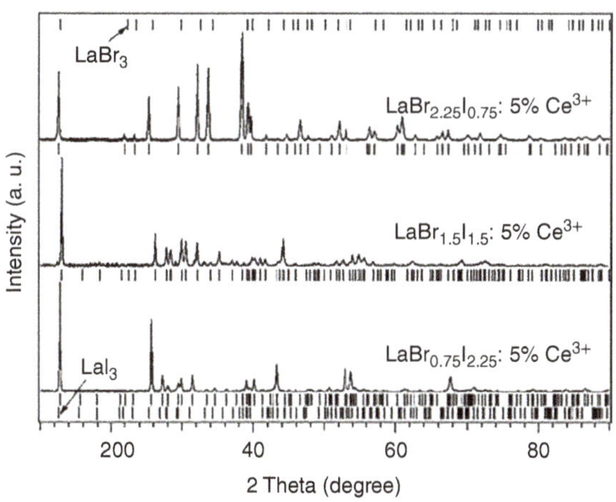

Figure 5.16 *XRD patterns of $LaBr_{2.25}I_{0.75}$:Ce, $LaBr_{1.5}I_{1.5}$:Ce, and $LaBr_{0.75}I_{2.25}$:Ce [70].*

Table 5.4 *Summary of scintillation properties for* $LaBr_{2.25}I_{0.75}$*:Ce,* $LaBr_{1}I_{2}$*:Ce,* $LaBr_{0.75}I_{2.25}$*:Ce and* $LaBr_{0.75}I_{2.25}$*:Ce [70].*

Compounds $LaBr_{3-x}I_x$	Photoelectron (10^3 photoelectrons/MeV)		Light yield (10^3 photons/MeV)		Energy resolution (%)	Decay components (relative contributions to total light yield) ns (%)
	0.5 μs	10 μs	0.5 μs	10 μs		
$LaBr_{2.25}I_{0.75}$:5%Ce^{3+}	13.8	14.9	42	45	4.1	31 (70), 244 (30)
	12.7	13.7	38	41	7.3	
$LaBr_{1.5}I_{1.5}$:5%Ce^{3+}	15.9	16.9	55	58	14.6	28 (100)
$LaBrI_2$:5%Ce^{3+}	6.6	7.5	22	25	25.9	13 (100)
$LaBr_{0.75}I_{2.25}$: 5%Ce^{3+}	6.6	6.7	22	22	35.9	12 (100)

$LaBr_{1.5}I_{1.5}$:Ce, whereas the fastest decay time of 12 ns was found for $LaBr_{0.75}I_{2.25}$:Ce. The energy resolution of the $LaBr_{3-x}I_x$:Ce was poor. which is possibly due to a crystalline inhomogeneity.

5.6 Next Generation of Halide Scintillators

5.6.1 Hf- and Tl-Based Halide Scintillators

In recent years, a new class of inorganic halide scintillators based on hafnium (Hf) and thallium (Tl) ion have been discovered and reported by some groups for the detection of X- and gamma-rays (Table 5.5). Due to its large Z-number of Hf (Z = 72) and Tl (Z = 81), most of the developed scintillator is expected to show large Z_{eff} and density. These Hf- and Tl-based scintillators are supposed to be ideal for X- and gamma-ray detectors, because the interactions of X- and gamma-rays with matter, and in particular, photoelectric absorption (sometimes called photo-absorption) probabilities depend on ρZ_{eff}^4.

In 2015, a new intrinsic scintillator Cs_2HfCl_6, showing a high light yield of ~35 000 ph/MeV with an energy resolution of 3.3% at 662 keV, was initially reported by Burger et al. [71].

Table 5.5 *Discovered and developed Hf- and Tl-based halide scintillators.*

Compound	Year	LY (ph/MeV)	ΔE/E (%) @662 keV	Z_{eff}	Ref.
Cs_2HfCl_6	2015	54 000 (silicon-CCD) ~35 000 (SBA-PMT)	3.3	58	[71]
$Tl_2LiGdCl_6$:Ce	2015	58 000	4.6	70	[72]
Rb_2HfCl_6	2016	24 100	–	55	[73]
$TlMgCl_3$	2016	46 000	5	71.6	[74]
$Tl_2LiGdBr_6$:Ce	2016	17 400	17	66	[75]
Tl_2LiYCl_6:Ce	2016	26 000	4.2	69	[76]
Tl_2LaBr_5:Ce	2017	43 000	6.3	67	[77]
Tl_2LaCl_5:Ce	2017	76 000	3.4	70	[78]
$Tl_2LiLuCl_6$:Ce	2017	27 000	5.6	72	[79]
$TlCaCl_3$	2017	30 600	5	65.5	[80]
$TlCaI_3$	2017	42 200	6.2	65	[81]
$TlSr_2Br_5$	2017	37 600	4.6	58.5	[82]
$TlCdCl_3$	2017	2200	–	68.7	[83]
Cs_2HfBr_6	2018	61 500	–	54.2	[84]
$Cs_2HfCl_4Br_2$	2018	18 600	4.4	56	[85]
Cs_2HfCl_6:Te	2018	13 100	–	58	[86]
Tl_2GdCl_5:Ce	2018	53 100	5	71	[87]
$TlSr_2I_5$:Eu	2018	70 000	4.2	61	[88]
Tl_2HfCl_6	2018	~27 000	~3.7	72.7	[89]
Tl_2ZrCl_6	2018	~50 800	~3.4	70.7	[89, 90]
$TlMgCl_3$:Br	2019	59 000	–	71.6	[91]
$TlMgCl_3$:K	2020	55 000	–	71.6	[92]
$TlSr_2Cl_5$	2020	19 000	–	63.7	[96]
$TlSrCl_3$	2020	21 300	10	69	[97]

The Cs_2HfCl_6 belongs to a K_2PtCl_6 cubic crystalline structure and shows a low hygroscopic and a good Z_{eff} of 58, and also includes no self-radioactivity. In 2020, Hawrami from the same group succeeded in crystal growth of Cs_2HfCl_6 with a large diameter of 16 mm and 1 in., as shown in Figure 5.17 [95]. At the same time, an excellent energy resolution of 2.8 at 662 keV and a light yield of 27 000 ph/MeV was achieved by an improvement of crystalline quality due to growth technique upgrade [94]. The electric structure, optical properties, and excitation and emission dynamics in the Cs_2HfCl_6 were intensely studied by some researchers using quantum chemical calculations, EPR, and temperature dependency of the scintillation properties [97–99]. These studies suggest a three excited state level model in which the intrinsic emissions are from two kinds of STEs excited levels (Figure 5.18). After the discovery of the Cs_2HfCl_6, other K_2PtCl_6 families such as Rb_2HfCl_6 [73], Cs_2HfBr_6 [84], and $Cs_2HfCl_4Br_2$ [85] showing a high light yield have been reported by some researchers. Among the compounds, the Cs_2HfBr_6 discovered by Saeki et al. exhibit the highest scintillation light yield of 61 500 ph/MeV.

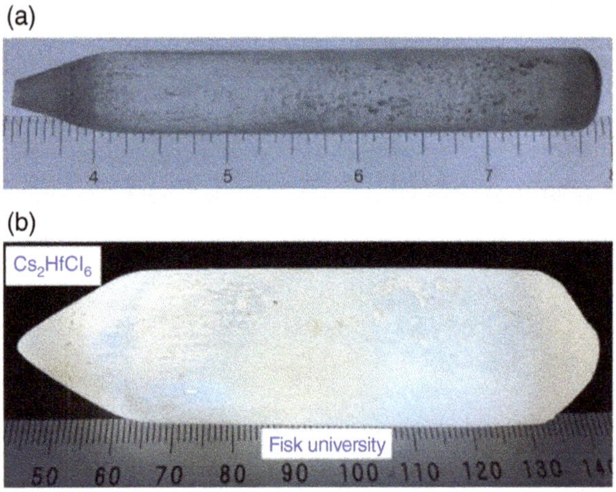

Figure 5.17 As-retrieved grown Cs_2HfCl_6 with a diameter of (a) 16 mm and (b) 1 in. [93].

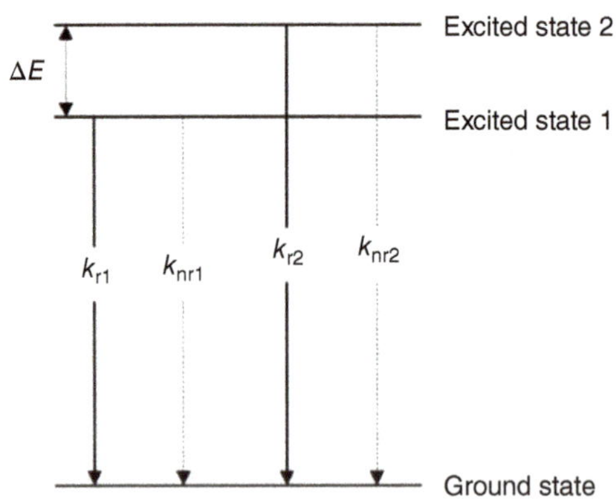

Figure 5.18 Schematic diagram of proposed three-level model for a non-luminescent center in Cs_2HfCl_6 [99].

Initial studies of Tl-concentrated elpasolite halides activated with Ce was reported in 2015 by Kim et al. [72]. The developed $Tl_2LiGdCl_6$:Ce is a very hygroscopic material like other halide scintillators but is a promising candidate material for γ-ray spectroscopy applications because of its higher light yield of 58 000 ph/MeV and higher Z_{eff} of 70 than the commercially available $Lu_{2-x}Y_xSiO_5$:Ce (LYSO), as shown in Figure 5.19a,b. A year later, Glodo from Radiation Monitoring Devices Inc. (RMD) reported the scintillation and neutron/gamma discrimination properties of Tl_2LiYCl_6:Ce [76]. The summary of properties of Tl_2LiYCl_6:Ce compared with commercially available Cs_2LiYCl_6:Ce is listed in Table 5.6. The Tl_2LiYCl_6:Ce has a light yield of 26 000 ph/MeV with an energy resolution of 4.2% at 662 keV gamma-ray, while the light yield from a single thermal neutron is approximately 47 000 ph/n. The differences of scintillation decay time between gamma-ray and neutron

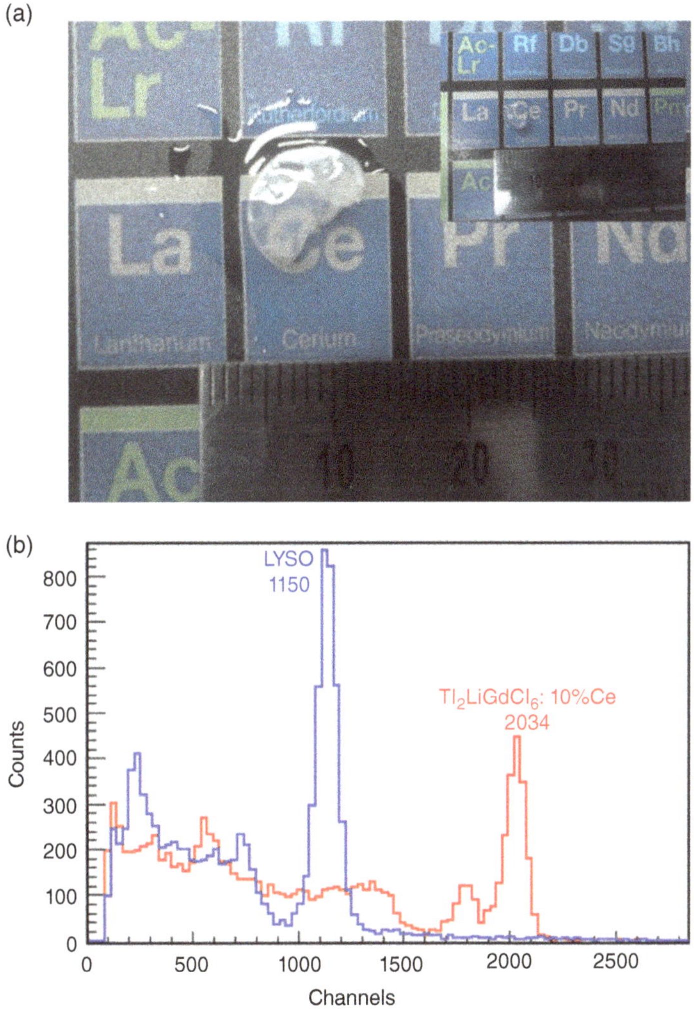

Figure 5.19 *(a) Photographic view of the $Tl_2LiGdCl_6$:Ce and (b) ^{137}Cs-gamma-ray irradiated pulse height spectra of the $Tl_2LiGdCl_6$:Ce and LYSO:Ce [72].*

Table 5.6 *Properties of Tl_2LiYCl_6:Ce compared with commercially available Cs_2LiYCl_6: Ce [76].*

Scintillator	NaI:Tl	CLYC
Crystal structure	Cubic	Cubic
melting point (°C)	661	640
Density	3.67	3.31
Z_{eff}	50	45
P photofraction at 662 keV	0.1	0.08
Light yield (γ-ray) (ph/MeV)	38,000	18000
Emission wavelength (nm)	420	370
Decay time (ns)	230	60–5500
Projected ER (at 662 keV)	6–7%	≤4%
GEE for thermal neutron (MeV)	N/A	3.2
N^0 efficiency (at 1.5 cm)	N/A	80
γ-n PSD capability	N/A	Yes
Fast neutron spectroscopy	No	Yes

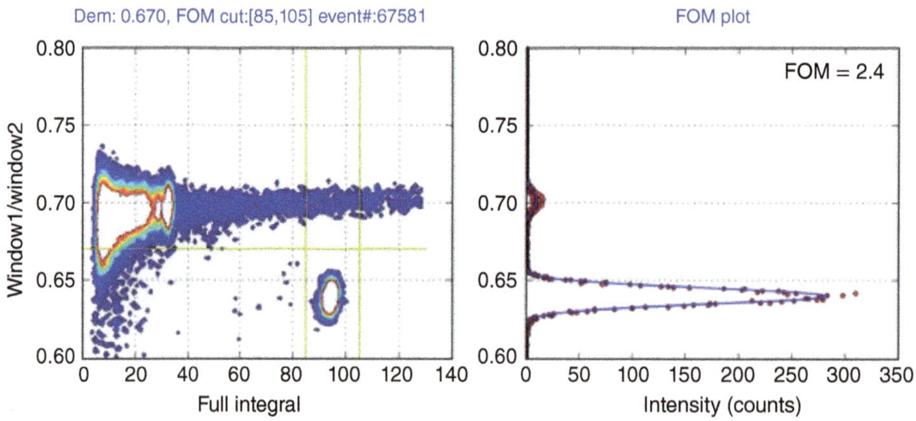

Figure 5.20 *PSD scatter plot created using traces collected with Tl_2LiYCl_6:Ce using a $^{241}Am/^{9}Be$ mixed neutron source [76].*

interactions are enough to achieve good PSD in this material, resulting in the figure-of-merit of approximately 2.4. The Tl_2LiYCl_6:Ce can compete with the Cs_2LiYCl_6:Ce due to its high density and large Z_{eff} (Figure 5.20).

In 2016, a $TlMgCl_3$ intrinsic scintillator was initially discovered by Fujimoto et al. [74]. The crystalline structure has been determined to comprise $GdFeO_3$-type perovskites (Pnma, oP20). The $TlMgCl_3$ had high density (=5.26 g/cm^3) and large Z_{eff} (~71.6), and also appeared as a very low hygroscopic. The excellent scintillation properties include high light yield of 46000 ph/MeV, good energy resolution of 5%, and fast decay time of 60–350 ns (see Figures 5.21 and 5.22). In more recent years, the same group succeeded in improving the light yield for the $TlMgCl_3$ from 46000 to 55000–59000 ph/MeV with cation and anion substitutions [91, 92]. They possibly attributed to the decreasing of the trapping center

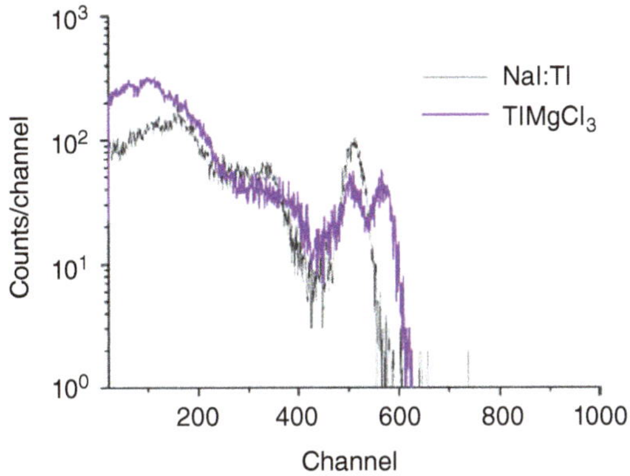

Figure 5.21 *[137]Cs-gamma-ray irradiated pulse height spectra of the $TlMgCl_3$ in comparison with a NaI:Tl [74].*

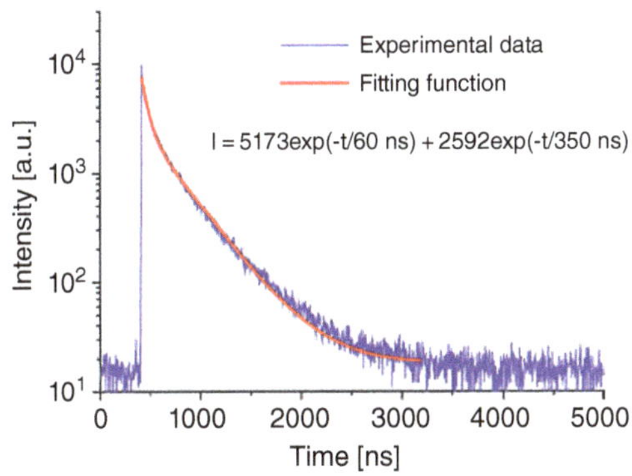

Figure 5.22 *Scintillation decay time profile under excitation at pulsed X-rays of the $TlMgCl_3$ [74].*

concentration with the substitutions. Unfortunately, the origin and mechanism of the bright intrinsic emission from the $TlMgCl_3$ is not yet fully understood, thus further studies such as temperature dependency of the intrinsic emission and band-gap measurements, and electric structure analysis are required.

References

1. Hofstadter, R. (1948). Alkali halide scintillation counters. *Phys. Rev.* 74: 100.
2. Hofstadter, R. (1949). The detection of gamma-rays with thallium-activated sodium iodide crystals. *Phys. Rev.* 75: 796.
3. Van Sciver, W. and Hofstadter, R. (1951). Scintillations in thallium-activated CaI_2 and CsI. *Phys. Rev.* 84: 1062.
4. Yamamura, S., Nakamura, T., Itou, K. et al. (2008). Development of wide-energy range X/γ-ray survey-meter. *J. Nucl. Sci. Technol.* 45: 178.
5. Totsuka, D., Yanagida, T., Fujimoto, Y. et al. (2012). Afterglow suppression by codoping with Bi in CsI:Tl crystal scintillator. *Appl. Phys. Express* 5: 052601.

6. van Loef, E.V.D., Dorenbos, P., van Eijk, C.W.E. et al. (2000). High-energy-resolution scintillator: Ce^{3+} activated $LaCl_3$. *Appl. Phys. Lett.* 77 (10): 1467.
7. van Loef, E.V.D., Dorenbos, P., van Eijk, C.W.E. et al. (2001). High-energy-resolution scintillator: Ce^{3+} activated $LaBr_3$. *Appl. Phys. Lett.* 79: 1573.
8. Shah, K.S., Glodo, J., Klugerman, M. et al. (2004). LuI_3:Ce – a new scintillator for gamma-ray spectroscopy. *IEEE Trans. Nucl. Sci.* 51: 2302.
9. Birowosuto, M.D., Dorenbos, P., van Eijk, C.W.E. et al. (2005). Scintillation properties of $LuI_3:Ce^{3+}$ – high light yield scintillators. *IEEE Trans. Nucl. Sci.* 52 (4): 1114.
10. van Loef, E.V., Higgins, W.M., Glodo, J. et al. (2008). Crystal growth and characterization of rare earth iodides for scintillation detection. *J. Cryst. Growth* 310: 2090.
11. Shah, K.S., Glodo, J., Higgins, W. et al. (2005). $CeBr_3$ scintillators for gamma-ray spectroscopy. *IEEE Trans. Nucl. Sci.* 52: 3157.
12. Hofstadter, R. (1968) US patent 3,373,279, 12 March 1968.
13. van Loef, E.V., Wilson, C.M., Cherepy, N.J. et al. (2009). Crystal growth and scintillation properties of strontium iodide scintillators. *IEEE Trans. Nucl. Sci.* 56: 869.
14. Sturm, B.W., Cherepy, N.J., Drury, O.B. et al. (2011) Characteristics of undoped and Europium-doped SrI_2 scintillator detectors, 2011 IEEE Nuclear Science Symposium Conference Record, Valencia, 7–11.
15. Yanagida, T., Koshimizu, M., Okada, G. et al. (2016). Comparative study of nondoped and Eu-doped SrI_2 scintillator. *Opt. Mater.* 61: 119.
16. Goldstein, E. (1896). *Zeits. f. Instrumentenkunde* 16: 211.
17. Pohl, R.W. (1937). Electron conductivity and photochemical processes in alkali-halide crystals. *Proc. Phys. Soc.* 49: 3.
18. Pohl, R.W. (1938). Summary report on electron conduction and photochemical processes in alkali halide crystals. *Phys. Zeits.* 39: 36.
19. Castner, T.G. and Kanzig, W. (1957). The electronic structure of V-centers. *J. Phys. Chem. Solid* 3: 178.
20. Gazzinelli, R. and Mieher, R.L. (1968). Electron-nuclear double resonance of the self-trapped hole in LiF. *Phys. Rev.* 175: 395.
21. Wei, H., Martin, V., Lindsey, A. et al. (2014) The scintillation properties of CeBr3-xClx single crystals, *J. Lumin.* 156: 175.
22. Wei, H., Zhuravleva, M., Young, K. et al. (2013) Effect of Ba substitution in CsSrI3:Eu2+, *J. Cryst. Growth.* 384: 27.
23. Kostenko, N.S., Mokhir, E.P., and Mustafina, R.K. (1965). Effect of anionig impurities developing during the growth of NaI(Tl) single crystals upon their luminescence and scintillation properties. *Zh. Prikl. Spektrosk.* 3: 573.
24. Mustafina, R.K. and Panova, A.N. (1967). Luminescence of NaI(Tl) crystals with oxygen-containing impurities. *Zh. Prikl. Spektrosk.* 6: 370.
25. Harami, T., Yamagishi, E., and Okuda, A. (1974). Growth of alkali halide crystals from aqueous solutions. *Jpn. J. Appl. Phys.* 13: 924.
26. Bridgman, P.W. (1925). Certain physical properties of single crystals of tungsten, antimony, bismuth, tellurium, cadmium, zinc, and tin. *Proc. Am. Acad. Arts Sci.* 60: 305.
27. Stockbarger, D.C. (1936). The production of large single crystals of lithium fluoride. *Rev. Sci. Instrum.* 7: 133.
28. Imagawa, K. (2005). NaI scintillation detector. *Horiba Tech. J. Readout* 9: 98.
29. van't Spijker, J.C., Dorenbos, P., and de Haas, J.T.M. (1995). Scintillation properties of K_2LaCl_5 with Ce doping. *Radiat. Meas.* 24: 379.
30. Dorenbos, P., van't Spijker, J.C., Frijns, O.W.V. et al. (1997). Scintillation properties of $RbGd_2Br_7$: Ce^{3+} crystals: fast, efficient, and high density scintillators. *Nucl. Instrum. Methods Phys. Res. B.* 132: 728.
31. van Loef, E.V.D., Dorenbos, P., van Eijk, C.W.E. et al. (2005). Scintillation properties of $K_2LaX_5:Ce^{3+}$(X=Cl, Br, I). *Nucl. Instrum. Methods Phys. Res.* A537: 232.
32. Birowosuto, M.D., Dorenbos, P., van Eijk, C.W.E. et al. (2007). Scintillation and luminescence properties of Ce^{3+} doped ternary cesium rare-earth halides. *Phys. Stat. Sol. (a)* 204: 850.
33. Roy, U.N., Groza, M., Cui, Y. et al. (2007). K_2CeCl_5: a new scintillator material. *Nucl. Instrum. Methods Phys. Res. A.* 579: 46.
34. Hawrami, R., Batra, A.K., Aggarwal, M.D. et al. (2008). New scintillator materials (K_2CeBr_5 and Cs_2CeBr_5). *J. Cryst. Growth* 310: 2099.
35. Zhuravleva, M., Yang, K. Rothfuss, H. et al. (2010) Crystal Growth and Scintillation Properties of Cs_3CeX_6 and $CsCe_2X_7$ (X = Cl, Br), IEEE Nuclear Science Symposuim & Medical Imaging Conference, Knoxville, TN, 1296–1299

36. Yang, K., Zhuravleva, M., and Melcher, C.L. (2011). Crystal growth and scintillation properties of Cs_3 EuI_5 crystals. *J. Cryst. Growth* 318: 833.
37. Zhuravleva, M., Yang, K., Green, A. et al. (2011). Crystal growth and scintillation properties of Ce^{3+}-doped KGd_2Cl_7. *J. Cryst. Growth* 318: 796.
38. Samulon, E.C., Gundiah, G., Gascón, M. et al. (2014). Luminescence and scintillation properties of Ce^{3+}-activated $Cs_2NaGdCl_6$, Cs_3GdCl_6, $Cs_2NaGdBr_6$ and Cs_3GdBr_6. *J. Lumin.* 153: 64.
39. Wei, H., Zhuravleva, M., Tyagi, M. et al. (2014). Scintillation properties of Cs_3LaCl_6:Ce^{3+} and Cs_3LaBr_6:Ce^{3+}. *IEEE Trans. Nucl. Sci.* 61: 390.
40. Fujimoto, Y., Saeki, K., Nakauchi, D. et al. (2019). Photoluminescence and scintillation properties of Rb_2CeCl_5 crystal. *Sens. Mater.* 31: 1241.
41. Schurz, C.M., Schleida, T., and Meyer, G. (2010). K_2LaCl_5. Section E Structure Reports Online 66. *Acta Crystallogr.* E66: i78.
42. van Loef, E.V.D., Dorenbos, P., van Eijk, C.W.E. et al. (2003). Influence of the anion on the spectroscopy and scintillation mechanism in pure and Ce^{3+}-doped K_2LaX_5 and LaX_3 (X=Cl, Br, I). *Phys. Rev. B.* 68: 045108.
43. Shah, K.S., Cirignano, L., Grazioso, R. et al. (2002). $RbGd_2Br_7$:Ce scintillators for gamma-ray and thermal neutron detection. *IEEE Trans. Nucl. Sci.* 49: 1655.
44. Seifert, H.J., Sandrock, J., and Thiel, G. (1986). Thermochemical studies on the systems. ACl/$CeCl_3$ (A = Na-Cs). *J. Thermal Anal.* 31: 1309.
45. Bourret-Courchesne, E.D., Bizarri, G., Borade, R. et al. (2009). Eu^{2+}-doped Ba_2CsI_5, a new high-performance scintillator. *Nucl. Instrum. Methods Phys. Res. A.* 612: 138.
46. Bourret-Courchesne, E.D., Bizarri, G., Hanrahan, S.M. et al. (2010). BaBrI:Eu^{2+}, a new bright scintillator. *Nucl. Instrum. Methods Phys. Res. A.* 613: 95.
47. Yang, K., Zhuravleva, M., Rothfuss, H. et al. (2010). Optical and scintillation properties of single crystal $CsSr_{1-x}EuxI_3$, IEEE Nuclear Science Symposium Conference Record, Tennessee, 1603–1606.
48. Zhuravleva, M., Blalock, B., Yang, K. et al. (2012). New single crystal scintillators: $CsCaCl_3$:Eu and $CsCaI_3$:Eu. *J. Cryst. Growth* 352: 115.
49. Cherginets, V.L., Grippa, A.Y., Rebrova, T.P. et al. (2012). Production and properties of $CsSrBr_3$:Eu^{2+} scintillator. *Funct. Mater.* 19: 429.
50. Grippa, A.Y., Rebrova, N.V., Gorbacheva, T.E. et al. (2013). Crystal growth and scintillation properties of $CsCaBr_3$:Eu^{2+} ($CsCa1\text{-}_xEuxBr_3$, ($0 \leq x \leq 0.08$)). *J. Cryst. Growth* 371: 112.
51. Stand, L., Zhuravleva, M., Lindsey, A. et al. (2015). Growth and characterization of potassium strontium iodide: a new high light yield scintillator with 2.4% energy resolution. *Nucl. Instrum. Methods Phys. Res. A.* 780: 40.
52. Lindsey, A.C., Zhuravleva, M., Stand, L. et al. (2015). Crystal growth and characterization of Europium doped $KCaI_3$, a high light yield scintillator. *Opt. Mater.* 48: 1.
53. Stand, L., Zhuravleva, M., Chakoumakos, B. et al. (2016). Scintillation properties of Eu^{2+}-doped KBa_2I_5 and K_2BaI_4. *J. Lumin.* 169: 301.
54. Pandian, L.S., Hawrami, R., Glodo, J. et al. (2016). Lithium alkaline halides-next generation of dual mode scintillators. *IEEE Trans. Nucl. Sci.* 63: 490.
55. Stand, L., Zhuravleva, M., Johnson, J. et al. (2017). New high performing scintillators: $RbSr_2Br_5$:Eu and $RbSr2I_5$:Eu. *Opt. Mater.* 73: 408.
56. Stand, L., Zhuravleva, M., Chakoumakos, B. et al. (2018). Crystal growth and scintillation properties of Eu^{2+} doped Cs_4CaI_6 and Cs_4SrI_6. *J. Cryst. Growth* 486: 162.
57. Bizarri, G., Bourret-Courchesne, E.D., Yan, Z. et al. (2011). Scintillation and optical properties of BaBrI:Eu^{2+} and CsBa2I5:Eu^{2+}. *IEEE Trans. Nucl. Sci.* 58: 3403.
58. Alekhin, M.S., Biner, D.A., Krämer, K.W. et al. (2014). Optical and scintillation properties of $CsBa_2I_5$:Eu^{2+}. *J. Lumin.* 145: 723.
59. Schilling, G. and Meyer, G. (1996). Ternäre Bromide und Iodide zweiwertiger Lanthanide und ihre Erdalkali-Analoga vom Typ AMX3 und AM2X5. *Zeitschrift für anorganische und allgemeine Chemie* 622 (5): 759.
60. Loyd, M., Lindsey, A., Wu, Y. et al. (2019). Growth of large size (≥38 mm diameter) $KCaI_3$:Eu scintillator crystals. *Nucl. Instrum. Methods Phys. Res. A.* 914: 8.
61. Combes, C.M., Dorenbos, P., van Eijk, C.W.E. et al. (1999). Optical and scintillation properties of pure and Ce^{3+}-doped Cs_2LiYCl_6 and Li_3YCl_6:Ce^{3+} crystals. *J. Lumin.* 82: 299.
62. Bessiere, A., Dorenbos, P., van Eijk, C.W.E. et al. (2005). Luminescence and scintillation properties of Cs_2LiYCl_6:Ce^{3+} for gamma and neutron detection. *J. Lumin.* 537: 242.
63. Knitel, M.J., Dorenbos, P., de Haas, J.T.M. et al. (1996). $LiBaF_3$, A thermal neutron scintillator with optimal neutron-gamma discrimination. *Nucl. Instrum. Methods Phys. Res. A.* 374: 197.

64. Dinca, L.E., Dorenbos, P., De Haas, J.T.M. et al. (2002). Alpha–gamma pulse shape discrimination in CsI:Tl, CsI:Na and BaF_2 scintillators. *Nucl. Instrum. Methods* 486: 141.
65. Yahaba, N., Koshimizu, M., Sun, Y. et al. (2014). X-ray detection capability of a Cs_2ZnCl_4 single crystal scintillator. *Appl. Phys. Express* 7: 062602.
66. Koshimizu, M., Yahaba, N., Haruki, R. et al. (2014). Scintillation and luminescence properties of a single $CsCaCl_3$ crystal. *Opt. Mater.* 36: 1930.
67. D'Olympia, N., Chowdhury, P., Guess, C.J. et al. Optimizing Cs_2LiYCl_6 for fast neutron spectroscopy. *Nucl. Instrum. Methods Phys. Res. A.* 694: 140.
68. Glodo, J., Hawrami, R., and Shah, K.S. (2013). Development of Cs_2LiYCl_6 scintillator. *J. Cryst. Growth* 60: 864.
69. Srivastava, A.M. Scintillator compositions, and related processes and articles of manufacture US Pat. No. 7,084,403, General Electric Company.
70. Birowosuto, M.D., Dorenbos, P., Krämer, K.W. et al. (2008). Ce^{3+} activated $LaBr_3$-xIx: high-light-yield and fast-response mixed halide scintillators. *J. Appl. Phys.* 103: 103517.
71. Burger, A., Rowe, E., Groza, M. et al. (2015). Cesium hafnium chloride: a high light yield, non-hygroscopic cubic crystal scintillator for gamma spectroscopy. *Appl. Phys. Lett.* 107: 143505.
72. Kim, H.J., Rooh, G., Park, H. et al. (2015). Luminescence and scintillation properties of the new Ce-doped $Tl_2LiGdCl_6$ single crystals. *J. Lumin.* 164: 86.
73. Saeki, K., Wakai, Y., Fujimoto, Y. et al. (2016). Luminescence and scintillation properties of Rb_2HfCl_6 crystals. *Jpn. J. Appl. Phys.* 55: 110311.
74. Fujimoto, Y., Koshimizu, M., Yanagida, T. et al. (2016). Thallium magnesium chloride: a high light yield, large effective atomic number, intrinsically activated crystalline scintillator for X-ray and gamma-ray detection. *Jpn. J. Appl. Phys.* 55: 090301.
75. Kim, H.J., Rooh, G., Park, H. et al. (2016). Investigations of scintillation characterization of Ce-activated $Tl_2LiGdBr_6$ single crystal. *Radiat. Meas.* 90: 279.
76. Hawrami, R., Ariesanti, E., Pandian, L.S. et al. (2016). Lithium alkaline halides: next generation of dual mode scintillators. *IEEE Trans. Nucl. Sci.* 63: 2838.
77. Kim, H.J., Rooh, G., Khan, A. et al. (2017). New Tl_2LaBr_5:Ce^{3+} crystal scintillator for γ-rays detection. *Nucl. Instrum. Methods Phys. Res. A.* 849: 72.
78. Hawrami, R., Ariesanti, E., Wei, H. et al. (2017). Tl_2LaCl_5:Ce, high performance scintillator for gamma-ray detectors. *Nucl. Instrum. Methods Phys. Res. A.* 869: 107.
79. Rooh, G., Kim, H.J., Jang, J. et al. (2017). Scintillation characterizations of $Tl_2LiLuCl_6$:Ce^{3+} single crystal. *J. Lumin.* 187: 347.
80. Khan, A., Rooh, G., Kim, H.J. et al. (2017). Intrinsically activated $TlCaCl_3$: A new halide scintillator for radiation detection. *Radiat. Meas.* 107: 115.
81. Hawrami, R., Ariesanti, E., Wei, H. et al. (2017). Intrinsic scintillators: $TlMgCl_3$ and $TlCaI_3$. *J. Cryst. Growth* 475: 216.
82. Rooh, G., Khan, A., Kim, H.J. et al. (2017). $TlSr_2Br_5$: New intrinsic scintillator for X-ray and G-ray detection. *Opt. Mater.* 73: 523.
83. Fujimoto, Y., Saeki, K., Yanagida, T. et al. (2017). Luminescence and scintillation properties of $TlCdCl_3$ crystal. *Radiat. Meas.* 106: 151.
84. Saeki, K., Fujimoto, Y., Koshimizu, M. et al. (2018). Luminescence and scintillation properties of Cs_2HfBr_6 and Cs_2ZrBr crystals. *Jpn. J. Appl. Phys.* 57: 030310.
85. Lam, S., Guguschev, C., Burger, A. et al. (2018). Crystal growth and scintillation performance of Cs_2HfCl_6 and $Cs_2HfCl_4Br_2$. *J. Cryst. Growth* 483: 121.
86. Fujimoto, Y., Saeki, K., Nakauchi, D. et al. (2018). Photoluminescence, photoacoustic, and scintillation properties of Te^{4+}-doped Cs_2HfCl_6 crystals. *Mater. Res. Bull.* 105: 291.
87. Khan, A., Rooh, G., Kim, H.J. et al. (2018). Ce^{3+}-activated Tl_2GdCl_5: Novel halide scintillator for X-ray and γ-ray detection. *J. Alloys Compd.* 741: 878.
88. Kim, H.J., Rooh, G., Khan, A. et al. (2018). Scintillation performance of the $TlSr2I_5$ (Eu^{2+}) single crystal. *Opt. Mater.* 82: 7.
89. Fujimoto, Y., Saeki, K., Nakauchi, D. et al. (2018). New intrinsic scintillator with large effective atomic number: Tl_2HfCl_6 and Tl_2ZrCl_6 crystals for X-ray and gamma-ray detections. *Sens. Mater.* 30: 1577.
90. Phan, Q.V., Kim, H.J., Rooh, G. et al. (2018). $Tl2ZrCl_6$ crystal: efficient scintillator for X- and γ-ray spectroscopies. *J. Alloys Compd.* 766: 326.

91. Arai, M., Fujimoto, Y., Koshimizu, M. et al. (2019). Luminescence and scintillation properties of TlMg(Cl_1-xBrx)$_3$ crystals. *Mater. Res. Bull.* 120: 110589.
92. Arai, M., Fujimoto, Y., Koshimizu, M. et al. (2020). Scintillation and photoluminescence properties of (Tl_1-xAx) $MgCl_3$ (where A = alkali metal). *J. Alloys Compd.* 823: 153871.
93. Hawrami, R., Ariesanti, E., Buliga, V. et al. (2020). Advanced high-performance large diameter Cs_2HfCl_6 (CHC) and mixed halides scintillator. *J. Cryst. Growth* 533: 125473.
94. Ariesanti, E., Hawrami, R., Burger, A. et al. (2020). Improved growth and scintillation properties of intrinsic, non-hygroscopic scintillator Cs_2HfCl_6. *J. Lumin.* 217: 116784.
95. Kang, B. and Biswas, K. (2016). Carrier self-trapping and luminescence in intrinsically activated scintillator: cesium hafnium chloride (Cs_2HfCl_6). *J. Phys. Chem. C.* 120: 12187.
96. Arai, M., Takahashi, K., Fujimoto, Y. et al. (2020). $TlS_{r2}Cl_5$: new intrinsically activated crystalline scintillator. *Radiat. Meas.* 134: 106328.
97. Yanagida, T., Fujimoto, Y., Arai, M. et al. (2020). Comparative studies of scintillation properties of Tl-based crystals. *Sens. Mater.* 32: 1351.
98. Král, R., Babin, V., Mihóková, E. et al. (2017). Luminescence and charge trapping in Cs_2HfCl_6 single crystals: optical and magnetic resonance spectroscopy study. *J. Phys. Chem. C.* 121: 12375.
99. Koshimizu, M., Saeki, K., Fujimoto, Y. et al. (2018). A three-state model for describing the temperature variation of the scintillation properties of Cs_2HfCl_6. *Jpn. J. Appl. Phys.* 57: 032401.

6

Semiconductor Scintillators

Naoki Kawano
Graduate School of Engineering Science, Akita University, Japan

6.1 Introduction

Semiconductor scintillators are phosphor materials, exploiting optical properties of semiconductors. Semiconductors include various materials such as IVB (C, Si, Ge, etc.), IVB–IVB (SiC, SiGe, etc.), IIIB–VB (GaAs, etc.), and IIB–VIB (ZnS, ZnSe, etc.). They are used in many applications such as light emitting diodes and laser diodes, as well as scintillators. The well-known original semiconductor scintillator was the ZnS-based phosphor materials reported around 1900, which emit scintillation photons where alpha and beta particles can be detected by naked eye [1]. Up to now, it has been demonstrated that various semiconductor scintillators such as Ga_2O_3 and Te-doped CdS show bright scintillation with a short decay time. Many types of emissions from semiconductors under ionizing radiation are observed, depending on the chemical composition, band structure, and defective state. In this chapter, fundamental photoluminescence and scintillation mechanisms in semiconductors are first recapped, and the examples of the previously reported semiconductor scintillators are subsequently summarized (Table 6.1).

Semiconductor scintillators exploiting the quantum size effect are also introduced. Recent advances in nanotechnology enable us to fabricate a wide variety of materials with low dimensional systems, such as a quantum well (two-dimensional structure), a quantum wire (one-dimensional structure), and a quantum dot structure (zero-dimensional structure). When electrons and holes are confined in the low dimensional systems of the order of nanometers or less, electronic states and optical properties change due to the quantum size effect. Due to this effect, efficient scintillation with a short decay time can be observed at room temperature from semiconductor nanostructures, which can be one of the potential candidates for fast response scintillators. In this chapter, we briefly introduce the theory of the quantum size effect and

Phosphors for Radiation Detectors, First Edition. Edited by Takayuki Yanagida and Masanori Koshimizu.

Table 6.1 *Scintillation characteristics of various semiconductor scintillators.*

Scintillator	Density g/cm^3	Wavelength nm	Relative Intensity	Light yield(γ) photons/MeV	Light yield(α) photons/5.5 MeV-α	Decay time ns	References
PbI_2	6.2	~495/~510–530	0.4 (12 K)/ $Bi_4Ge_3O_{12}$			0.2/0.6/1.8 (11 K)	[2]
HgI_2	6.3	535/565	1.6 (12 K)/ $Bi_4Ge_3O_{12}$			0.6/2.0/> 5 (14 K)	[2]
ZnO	5.6	400/550			500	4.3	[3]
ZnS	4.0	440/520			500	9.7/8.9 μs	[4]
Ga_2O_3	5.9	380		15 000		8/793	[5]
CdS:Te	4.8	640		17 000		18/270	[6]
ZnSe:Te	5.4	645	1.0 /$CdWO_4$			about 50 μs	[6]
ZnSe:O	<5.3	600	1.3 /CsI:Tl				[7]
ZnO:Ga	5.6	370–410	2.0 (14 K)/ $Bi_4Ge_3O_{12}$			0.3/0.6/ (14 K)	[8]
CdS:In	4.8	500–550	0.13 (13 K)/ $Bi_4Ge_3O_{12}$			0.1/1.7 2/7/21 (13 K)	[9, 10]
ZnS:Ag	4.1	450	1.3/NaI:Tl			~70	[11]
ZnS:Ag,Cl	<4.1	450	6.0/$YAlO_3$:Ce			about 100 μs	[11–13]
$(n\text{-}C_3H_7NH_3)_2PbBr_4$	2.7	438	0.065/NaI:Tl			2.8/18/130	[14]
$(n\text{-}C_6H_{13}NH_3)_2PbI_4$	2.4	558	0.11/NaI:Tl			0.4/3.8/16	[14]
$(n\text{-}C_4H_9NH_3)_2PbBr_4$	2.4			<1500		1.1/8	[15]
$(C_6H_5CH_2NH_3)_2PbBr_4$	2.2			4000		2.8/10	[15]
$(C_6H_5C_2H_4NH_3)_2PbBr_4$	2.3	437		14 000		9.6/55	[15, 16]
$(C_6H_5C_2H_4NH_3)_2Pb_{1-x}Ba_xBr_4$	2.3	440		19 000		9.2/32	[17]
$(C_6H_5C_2H_4NH_3)_2Pb_{1-x}Sr_xBr_4$	2.3	440		20000		4.5/13	[18]

describe the photoluminescence and scintillation properties of an organic–inorganic perovskite compound, which is a representative example of semiconductor scintillators using the quantum size effect. The reader can find the textbooks which cover the detailed theory of the quantum size effect [19, 20].

6.2 Photoluminescence and Scintillation Mechanisms in Semiconductors

Semiconductors consist of individual atoms bonded together in a periodic crystal structure, and the energy state of an electron in semiconductors becomes a continuous band structure where the conduction and valence band are the bands closest to the Fermi level, as shown in Figure 6.1. The conduction band is the lowest range of vacant electronic states, whereas the valence band is the highest range of electronic states, which is assumed to be fully occupied by electrons at absolute zero temperature ($T \rightarrow 0$). These bands are separated by a band-gap where no electronic state can exist. In the case where a luminescence center such as Ce^{3+} and Eu^{2+} is not doped in semiconductors, they exhibit emissions due to various radiative processes such as recombination of a free electron–hole pair and recombination of excitons.

Prior to an introduction on the optical properties of semiconductors, we will describe a band structure in an intrinsic semiconductor, which is a pure semiconductor without any dopant species. Density of state $\rho(E)$, and Fermi–Dirac distribution $f(E)$ in an intrinsic semiconductor, are shown in Figure 6.1. E_c is the minimum energy of the conduction band, E_v is the maximum energy of the valence band, and E_f is the Fermi energy. Density of state $\rho(E)$ is the density of electron or hole states between E and $E+\Delta E$ energies. For example, the following equation holds for the density state of electrons in the conduction band: $dN(E) = V \times \rho_c(E)\,dE$, where $N(E)$ is the number of states between E and $E+dE$ energies, V is the volume of the material, and $\rho_c(E)$ is the density state of electrons in the conduction band. Based on theoretical studies, the density of state for electrons in the conduction band $\rho_c(E)$ and the density of state for holes in the valence band $\rho_v(E)$ can be expressed as follows [21, 22]:

$$\rho_c(E) = \frac{1}{2\pi^2}\left(\frac{2m_e}{\hbar^2}\right)^{3/2}(E-E_c)^{1/2}, E \geq E_c \tag{6.1}$$

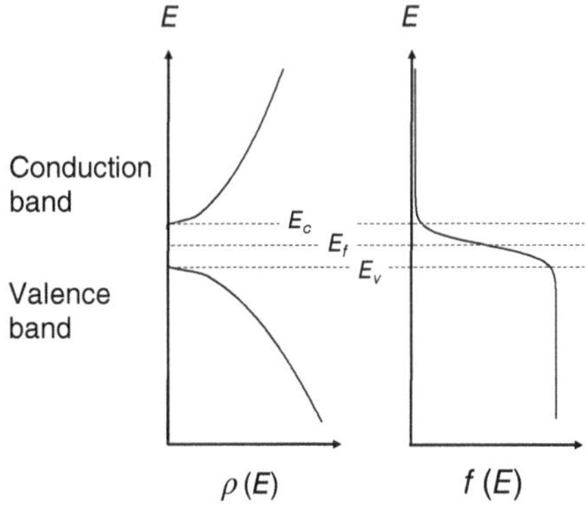

Figure 6.1 Density of state ρ(E) and Fermi–Dirac distribution f (E) in an intrinsic semiconductor.

$$\rho_v(E) = \frac{1}{2\pi^2}\left(\frac{2m_h}{\hbar^2}\right)^{3/2}(E_v - E)^{1/2}, E \le E_v \tag{6.2}$$

Here, m_e is the mass of an electron and m_h is the mass of a hole in the approximation of effective mass. From the above equation, it can be seen that the density of state for electrons in the conduction band $\rho_c(E)$ is proportional to $\sqrt{E}$. In addition, an electron concentration n and a hole concentration p are expressed as follows:

$$n = \int_{E_c}^{\infty} \rho_c(E) f(E) dE \tag{6.3}$$

$$p = \int_{-\infty}^{E_v} \rho_v(E)\left[1 - f(E)\right] dE \tag{6.4}$$

In an intrinsic semiconductor, the electron concentration n is equivalent to the hole concentration p. Here, the energy distribution of electrons $f(E)$ (probability of electrons with the energy E) can be expressed by the following Fermi–Dirac distribution function:

$$f(E) = \frac{1}{1 + e^{(E - E_f)/kT}} \tag{6.5}$$

k is the Boltzmann constant, T is the absolute temperature, and kT is the thermal kinetic energy, which is approximately 25 meV at room temperature (300 K). At absolute zero ($T \to 0$), all electrons are assumed to be occupied at the energy level below E_f. As the temperature increases, electrons can exist at energy levels higher than E_f. For an intrinsic semiconductor, E_f lies in the middle of the band-gap.

We now provide more details on the radiative recombination of a free electron–hole pair in an intrinsic semiconductor. The occurrence of emission from semiconductors depends on band structure, lattice defects, and dopant species. In an intrinsic semiconductor, we can observe emissions due to the recombination of an electron in the conduction band and a hole in the valence band. At first, the emissions from the intrinsic semiconductor can be classified into emissions from a direct band-gap semiconductor and an indirect band-gap semiconductor. As shown in Figure 6.2, the bottom of the conduction band and the top of the valence band are at the same wavenumber position in the direct band-gap semiconductor. In this case, the electron in the conduction band and the hole in the valence band recombine at the same k position, and photons with energy corresponding to the energy difference between the electron and hole are emitted. In this process, energy and momentum are conserved without the contribution of phonons, resulting in an efficient emission. On the other hand, in an indirect gap semiconductor, an electron in the conduction band and a hole in the valence band lie at a different k position, thus phonons must be involved to satisfy momentum conservation, resulting in inefficient emission. Therefore, a direct band-gap semiconductor is generally used for scintillators. Let us consider radiative recombination of an electron with E_2 energy in the conduction band and a hole with E_1 energy in the valence band at the same k_0 positon in an intrinsic semiconductor. In this case, a photon with energy $h\nu = E_2 - E_1$ is emitted by the

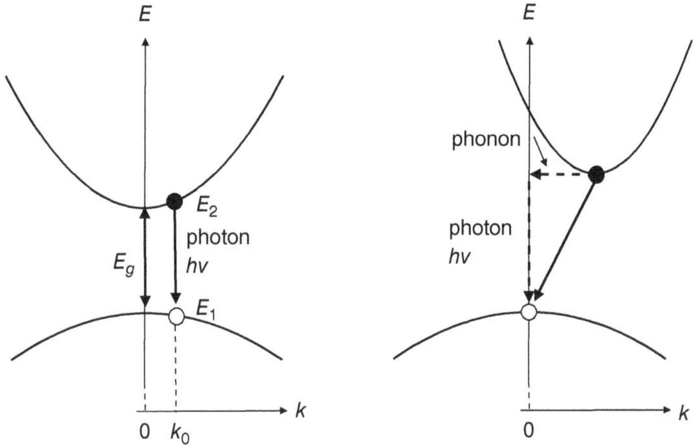

Figure 6.2 *Energy band structure of a direct band-gap semiconductor and an indirect band-gap semiconductor.*

recombination of the electron and hole, and probability density $I(h\nu)$ of the emission can be expressed by the following equation [23]:

$$I(h\nu) \approx |M|^2 f_e(\nu)\rho(\nu) \tag{6.6}$$

Here, $f_e(\nu)$ is the probability of a suitable configuration of occupancy of conduction and valence band states, $\rho(\nu)$ is the joint density of electrons and holes states, and $|M|^2$ is the quantum-mechanical probability of the transition in the form of the square of the absolute value of the matrix element of the transition. The quantum-mechanical probability is not covered in this book, but a detailed theory is described in references [23, 24].

We will now provide more details of $f_e(\nu)$ and $\rho(\nu)$. When an electron with E_2 energy is formed in the conduction band and a hole with E_1 energy is formed in the valence band, to take place the radiative recombination of a free electron–hole pair, $f_e(\nu)$ can be expressed as follows using the occupancy factors f_c and f_v in the conduction and valence bands, respectively.

$$f_e(\nu) = f_c f_v = f(E_2)\left[1 - f(E_1)\right] \tag{6.7}$$

$f(E)$ is the Fermi–Dirac distribution function described above. Here, let us assume that the recombination of a free electron–hole pair occurs at a low temperature in an intrinsic semiconductor with a large band-gap energy under a low fluence excitation. In this case, $kT \ll (E_g/2) \approx E_2 - E_f \approx E_f - E_1$ is fulfilled, thus the $f_e(\nu)$ can be approximated as follows:

$$f_e(\nu) = f(E_2) \times \left[1 - f(E_1)\right] \approx \exp\left[-\frac{E_2 - E_f}{kT}\right] \times \exp\left[-\frac{E_f - E_1}{kT}\right] \approx \exp(-h\nu / kT) \tag{6.8}$$

Furthermore, the joint density of state $\rho(\nu)$ is related to the difference energy band $E_c(k) - E_v(k)$, and can be obtained by subtracting $E_v(k)$ from $E_c(k)$ mathematically when the radiative transitions are vertical in the k space. Based on the density of state in the conduction band $\rho_c(E)$ and valence band $\rho_v(E)$, as mentioned above, the joint density of state $\rho(v)$ can be obtained as follows:

$$\rho(v) = \frac{(2m_r)^{3/2}}{\pi\hbar^2}(hv - E_g)^{1/2},\ hv \geq E_g \tag{6.9}$$

Here, m_r is the reduced mass of the electron and hole. Using the above formulas, the emission spectrum due to the recombination of a free electron–hole pairs can be expressed as follows [23]:

$$I_{sp}(v) \approx (hv - E_g)^{1/2} \exp\left[-(hv - E_g)/kT\right] \tag{6.10}$$

Therefore, the density of state is considered to be one of the most important factors for the emission spectrum. The density of state depends on the band structure, which is affected by the crystal structure and chemical composition of the semiconductors. Moreover, when an electron and a hole are confined in semiconductors with low-dimensional systems, the density of state changes according to the dimensionality of the systems due to the quantum size effect, resulting in the enhancement of emission intensity. The effect of optical properties due to the quantum size effect will be introduced in Section 6.4.

We will now describe radiative decay processes in an intrinsic semiconductor. Generally, semiconductor scintillators show emissions due to a wide variety of radiative decay processes such as: (i) recombination of a free electron–hole pair (band-to-band recombination); (ii) recombination of a free hole with a neutral donor; (iii) recombination of a free electron with a neutral acceptor; (iv) recombination of a donor–acceptor pair; and (v) recombination of excitons. The illustrations of these radiative decay processes are summarized in Figure 6.3. An emission due to recombination of a free electron–hole pair has been already been described above, we will briefly introduce in order from: recombination of a free electron (hole) with a neutral acceptor (donor); recombination of a donor–acceptor pair; and recombination of excitons.

An impurity atom in semiconductors can become an acceptor (donor) when the valence of the impurity atom is smaller (larger) by one electron than that of the atom at the substituted

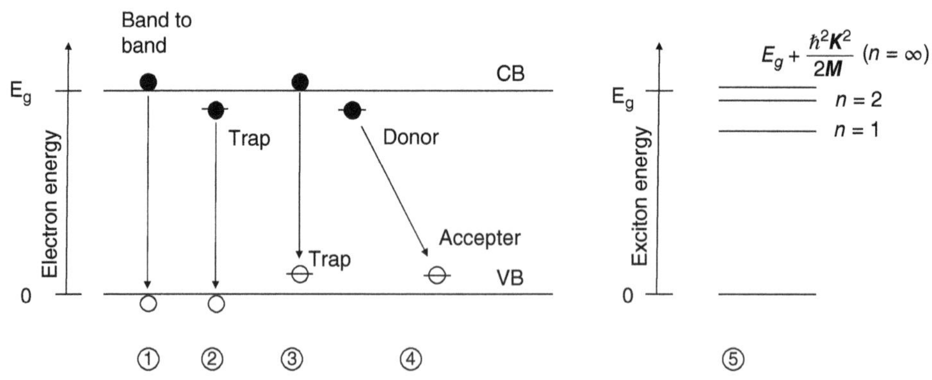

Figure 6.3 *Examples of radiative decay processes in semiconductors.*

site. In this case, emissions from the semiconductors occur by recombination of a free electron (hole) with a neutral acceptor (donor). The spectrum shape of the emission corresponds to that of the recombination of a free electron–hole pair, but the spectrum shifts to a lower energy side by ionization energy (E_A or E_D) of the acceptor or donor, respectively [23].

$$I_A(v) \approx \left(hv - E_g + E_A\right)^{1/2} \exp\left[-\left(hv - E_g + E_A\right)/kT\right] \tag{6.11}$$

$$I_D(v) \approx \left(hv - E_g + E_D\right)^{1/2} \exp\left[-\left(hv - E_g + E_D\right)/kT\right] \tag{6.12}$$

(iv) Recombination of a donor–acceptor pair

When the neutral donor and acceptor are formed simultaneously in semiconductors, radiative recombination can occur due to recombination of an electron trapped at the donor and a hole trapped at the acceptor. The donor and acceptor are electrically neutral before the recombination, but they return to the state of a positively ionized donor and a negatively ionized acceptor after the recombination. Therefore, the final state is stabilized by Coulomb interaction, so the emission energy can be expressed as follows [23]:

$$E(R) = hv(R) = E_g - \left(E_A + E_D\right) + e^2 / \left(4\pi\varepsilon_0\varepsilon_r R\right) \tag{6.13}$$

where R is the distance between the donor and acceptor. It can be seen that the energy of the photons emitted by the donor–acceptor recombination depends on the distance R between the donor and acceptor.

In addition, the transition probability of the donor–acceptor recombination is affected by the overlap of wave function of an electron trapped at the donor and a hole trapped at the acceptor. Let us consider the case where a hole is strongly bound to the acceptor and an electron is weakly bound to the donor. In this case, the overlap of the wave functions can be considered to depend on the squared value at the position of the distance r of the electron wave function, since the wave function of an electron is assumed to be more widely spread than that of a hole. Therefore, the transition probability of the donor–acceptor recombination can be expressed as follows:

$$I(r) = I_{\text{max}} \left|\exp\left(-r/r_D\right)\right|^2 = I_{\text{max}} \exp\left(-2r/r_D\right) \tag{6.14}$$

where r_D is the Bohr radius of the electron wave function. For the same reason, if a hole is weakly bound to the acceptor and an electron is strongly bound to the donor, the transition probability of the donor–acceptor recombination can be expressed as follows:

$$I(r) = I_{\text{max}} \left|\exp\left(-r/r_A\right)\right|^2 = I_{\text{max}} \exp\left(-2r/r_A\right) \tag{6.15}$$

where r_A is the Bohr radius of the hole wave function.

(v) Recombination of excitons

Excitons are a kind of excited state in which an electron and a hole are bound by Coulomb interaction. Wannier excitons with a weakly coupled electron–hole pair are often formed in semiconductors among various excitons such as Frenkel excitons and charge transfer excitons. In the case of Wannier excitons, the excitons move through the crystal lattice translationally in semiconductors, and the following quantities are related to the translational movement: the exciton wave vector $\boldsymbol{K}$ and exciton mass $\boldsymbol{M} = m_e + m_h$, where m_e and m_h are the effective mass of an electron and a hole, respectively. Then, the total energy of the Wannier exciton can be expressed as follows [25]:

$$E_{ex} = E_g + \frac{\hbar^2 \boldsymbol{K}^2}{2\boldsymbol{M}} - \frac{R^*}{n^2} \qquad R^* = R_H\left(\mu / m_0 \varepsilon_r^2\right) \tag{6.16}$$

Here, n is the principal quantum number, and R^* is the exciton binding energy for $n = 1$ in the three-dimensional structure (bulk structure), which can be expressed as above using the ionization energy of hydrogen atoms $R_H = 13.6\,\text{eV}$, the relative dielectric constant of the semiconductor ε_r, the exciton reduced mass μ, and the free electron mass m_0. The typical values of μ/m_0 and ε_r in conventional semiconductors are 0.1 and 10, thus the exciton binding energy R^* is usually about several tens meV. In addition, lattice defects and impurities such as a donor, an acceptor, and isoelectronic ones are present in the actual semiconductors, and the excitons can be trapped at such sites to emit photons. Please refer to the references [23, 25] for details on the optical properties of such trapped excitons.

Various radiative decay processes in semiconductors have been described so far. Now, the scintillation mechanism in semiconductors will be briefly described [26–28]. In scintillation processes, the final radiative decay process is the same as that in photoluminescence, but additional processes occur under ionizing radiation. When semiconductors are irradiated by a single quanta of high energy radiation, a high energy primary photoelectron and a hole in an inner shell are generated, followed by the relaxation of the atom with the hole by a cascade of radiative transitions (X-ray), nonradiative Auger transition in the time scale of 10^{-15}–10^{-13} s. Furthermore, the secondary X-rays are reabsorbed mainly by the photoelectric effect, and the generated photoelectrons and the Auger electrons interact with the shells of atoms, resulting in an avalanche of secondary carriers. When the electron energies fall below the ionization threshold, the holes and electrons are thermalized by intraband transitions via electron–phonon interactions. In the case of semiconductors, the holes and electrons retain a diffuse band state, and become excited states such as excitons and trapped carriers at defects and impurities in the time scale of 10^{-12}–10^{-11} s. Finally, these excited states return to the ground state by emitting photons or by non-radiative processes. The final radiative processes for scintillation are the previously mentioned radiative recombination processes in semiconductors, such as the recombination of excitons and a donor–acceptor pair. Therefore, the final radiative decay process for scintillation differs, depending on the crystal structure and chemical composition of semiconductors, presence of defects, and impurities.

6.3 Various Semiconductor Scintillators

This section introduces examples of previously reported semiconductor scintillators. Semiconductor scintillators can be roughly classified into undoped semiconductor scintillators and doped semiconductor scintillators, where an impurity is intentionally incorporated. In undoped semiconductor scintillators, emissions due to excitons and lattice defects such as vacancies in host materials are often observed at low temperature where they show an efficient scintillation response characterized by a high light yield and a fast decay time. However, they do not generally exhibit efficient scintillation at room temperature due to thermal disturbance.

In recent years, it has been discovered that some undoped semiconductor scintillators exhibit efficient scintillation at room temperature with various mechanisms involving self-trapped carriers and lattice defects. In the case of doped semiconductor scintillators, scintillation from the scintillators depends on the presence of lattice defect and doped impurities, which can be classified into isoelectronic, donor, and acceptor impurities. In this section, the scintillation characteristics of undoped and doped semiconductor scintillators are summarized.

6.3.1 Undoped Semiconductor Scintillator

We now describe the scintillation properties of undoped semiconductor scintillators at low temperature. Undoped PbI_2 is one of the famous undoped semiconductor scintillators that show efficient scintillations at low temperature [2]. It shows scintillation due to the recombination of free and trapped excitons with a peak approximately at ~495 nm, and due to the donor–acceptor recombination possibly associated with iodine vacancies and lead vacancies with a broad peak approximately at ~510–530 nm at low temperature. It has been demonstrated by the measurement of the wavelength-dependent decay curves that the scintillation intensity of the exciton emissions is found to be weak compared with that of the longer wavelength scintillation. Furthermore, the scintillation intensity of undoped PbI_2 at 12 K is 0.4 relative to that of $Bi_4Ge_3O_{12}$ (8200 photons/MeV) at 295 K, and scintillation decay time of the main component at 11 K is about 1.8 ns, which is extremely fast compared with that of the conventional inorganic scintillators such as Ce-doped $YAlO_3$. However, the scintillation intensity of undoped PbI_2 decreases drastically as the temperature increases from 11 to 165 K, as shown in Figure 6.4.

Another example of undoped semiconductor scintillators is undoped HgI_2 [2]. It exhibits scintillation at low temperature with a peak approximately at 535 nm arising from free and trapped excitons, and at 565 nm originating from featureless impurity/defect, which might be associated with iodine vacancies, mercury vacancies, and defects related to stoichiometric deviation. The scintillation intensity of longer wavelength scintillation is found to be more intense, as in the case of PbI_2. Furthermore, scintillation intensity of undoped HgI_2 at 12 K is 1.6 times that of $Bi_4Ge_3O_{12}$ (8200 photons/MeV) at 295 K, and the total scintillation lifetime of undoped HgI_2 is about 1.6–0.5 ns from 14 to 102 K. As shown in Figure 6.5, scintillation

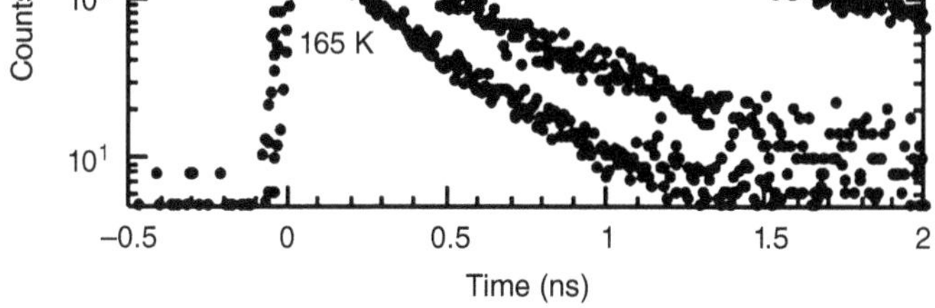

Figure 6.4 X-ray induced scintillation decay curves of undoped PbI_2 at low temperature [2].

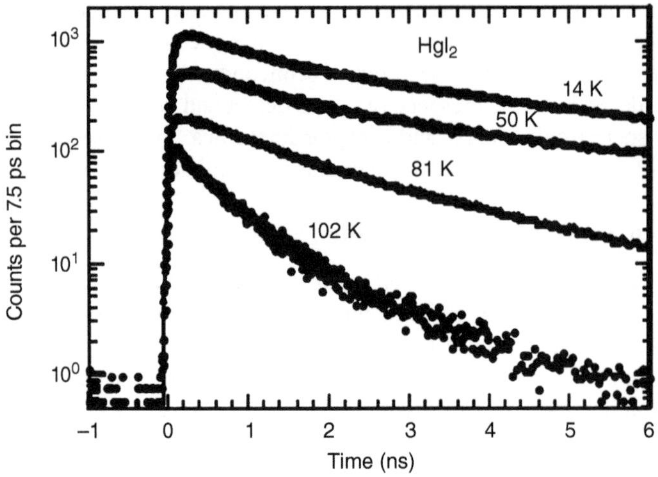

Figure 6.5 *X-ray induced scintillation decay curves of undoped HgI_2 at low temperature [2].*

intensity of undoped HgI_2 decreases drastically with increasing temperature as well as undoped PbI_2. As described above, the undoped semiconductor scintillators show efficient scintillation due to the excitons and lattice defects at a low temperature, but it is generally difficult to observe efficient scintillation from undoped semiconductor scintillators at room temperature.

One of the examples for semiconductor scintillators where scintillation can be observed at room temperature is undoped ZnO [3]. In X-ray induced scintillation spectrum, it exhibits two intense emission peaks at 400 and 550 nm. The sharp peak at 400 nm is ascribed to free and trapped excitons, and the broad peak at 550 nm is related to oxygen vacancy, respectively (Figure 6.6). The scintillation lifetime of free and trapped exciton emissions observed at 400 nm is about 4 ns, which is confirmed to be a very short lifetime. As shown in Figure 6.7, measurement of the pulse height spectrum with different shaping time is performed under 5.5 MeV α-ray irradiation from ^{241}Am. The relative scintillation intensity increases by increasing the shaping time, and it saturates at shaping times longer than 3 μs. These results suggest that the main scintillation component is not exciton recombination observed at 400 nm but oxygen vacancy observed at 550 nm. Assuming that main scintillation photons are originated from the oxygen vacancy, the overall scintillation light yield is estimated to be approximately 500 photons/5.5 MeV-α.

Undoped ZnS is another example of a semiconductor scintillator, where scintillation can be detected at room temperature [4]. It shows scintillation peaks at approximately 440 and 520 nm originating from zinc vacancies and sulfur vacancies, respectively. Furthermore, a clear peak height peak is observed in the pulse height spectrum under ^{241}Am α-ray irradiation, and the scintillation light yield is estimated to be 500 photons/MeV-α. However, the scintillation light yield is low compared to Ag-doped ZnS, which is one of the famous doped semiconductor scintillators described below.

Ga_2O_3 is an excellent semiconductor scintillator to show a high light yield under γ-ray irradiation at room temperature [5]. As shown in the X-ray induced scintillation spectrum in Figure 6.8, a broad peak is observed at 380 nm, and the main component of this emission is

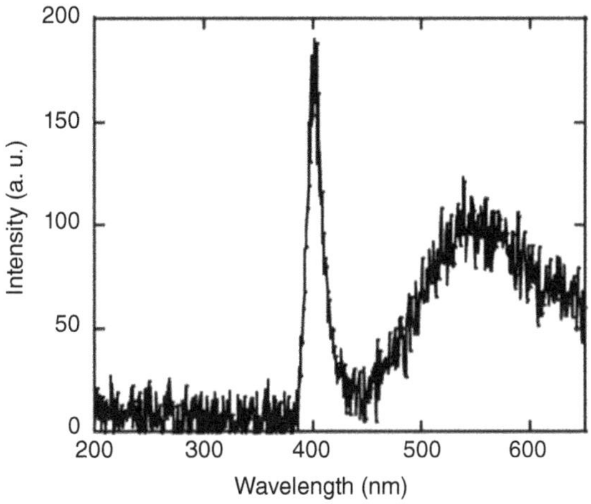

Figure 6.6 *X-ray induced scintillation spectra of undoped ZnO at room temperature [3].*

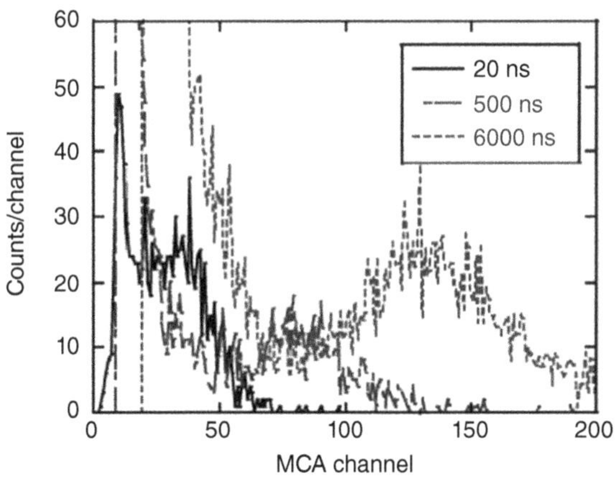

Figure 6.7 *^{241}Am α-ray induced pulse height spectra of undoped ZnO with different shaping times [3].*

considered to originate from the recombination of free electrons and self-trapped holes. In addition, two exponential components of 8 and 977 ns are observed in the X-ray induced scintillation decay time profile, and the decay times of the first component are generally faster than that of conventional inorganic scintillators such as Ce-doped Lu_2SiO_5 and $Gd_3(Al, Ga)_5O_{12}$. Moreover, a pulse height peak under γ-ray irradiation from ^{137}Cs is successfully obtained, as shown in Figure 6.9, and the scintillation light yield is estimated to be about 15 000 photons/MeV. Such a clear pulse height peak from semiconductor scintillators under γ-ray irradiation is observed in very few examples such as Te-doped CdS and Te-doped ZnSe. Compared with these materials, the scintillation light yield of undoped Ga_2O_3 is higher than that of Te-doped ZnSe and is equivalent to Te-doped CdS.

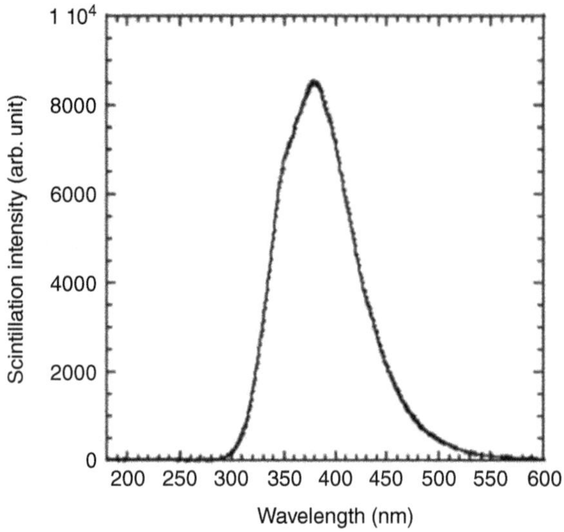

Figure 6.8 *X-ray induced scintillation spectra of undoped Ga_2O_3 at room temperature [5]. Source: Copyright (2016) The Japan Society of Applied Physics.*

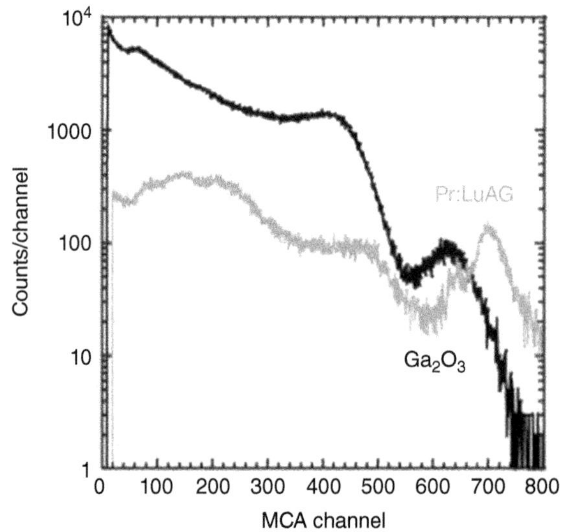

Figure 6.9 *γ-ray induced pulse height spectra of undoped Ga_2O_3 at room temperature [5]. Source: Copyright (2016) The Japan Society of Applied Physics.*

6.3.2 Doped Semiconductor Scintillator

We learned in Subsection 6.3.1 about the scintillation properties of undoped semiconductor scintillators. From now, scintillation characteristics in doped semiconductor scintillators to which impurities are intentionally added will be described. The doped semiconductor scintillators include various semiconductor materials with isoelectronic, donor, and acceptor impurities. In the previous subsection, radiative decay processes such as the recombination of a free electron (hole) with a neutral acceptor (donor) and recombination of a donor–acceptor pair in doped semiconductor scintillators have been already described. It should be noted that the final scintillation process in doped semiconductor scintillators depends on not only the species of dopant impurities but also the presence of lattice defects.

6.3.2.1 Semiconductor Scintillator with Isoelectronic Impurities

First, we introduce semiconductor scintillators with isoelectronic traps where the host element is replaced with another that has the same valence. When the replacement element is more electropositive or electronegative, it becomes hole or electron trap, respectively. Generally, the replacement element can be a luminescence center for recombination with the other carrier to show scintillation.

Te-doped CdS is one of the examples of semiconductor scintillators with an isoelectric hole trap, and Te can be considered as an isoelectronic hole trap [6]. Te-doped CdS shows a broad peak at approximately 640 nm under X-ray irradiation. The broad scintillation band consists of two band centers at 600 and 730 nm, and these bands originate from excitons trapped at Cd interstitials and S vacancies, respectively. In this case, the isoelectronic Te ions promote the formation of lattice defects. From a pulse height spectrum measurement using the silicon photodiode, a clear peak height is observed from Te-doped CdS under ^{137}Cs γ-ray irradiation, and the scintillation light yield at 4 μs shaping time is estimated to be 17 000 photons/MeV, which is almost equivalent to that of undoped Ga_2O_3, as described above. Based on the peak position in the scintillation spectra, Ga_2O_3 (380 nm) and Te-doped CdS (640 nm) are suitable for use with a photomultiplier tube and silicon photodiode, respectively.

Te-doped ZnSe is another semiconductor scintillator with an isoelectronic hole trap [6]. Te-doped ZnSe exhibits a broad peak at approximately 645 nm under X-ray excitation. The origin of the scintillation is complicated, and is considered to be based on excitons localized at isoelectronic traps. A clear peak height can be detected in the pulse height spectrum using a silicon photodiode under ^{137}Cs γ-ray irradiation, and the scintillation light yield of Te-doped ZnSe at 12 μs shaping time is found to be of same order of magnitude as for that of $CdWO_4$.

O-doped ZnSe is an example of semiconductor scintillators with an isoelectronic electron trap [7]. O-doped ZnSe shows a broad scintillation peak at around 600 nm, and the scintillation intensity of O-doped ZnSe is about 1.28 times higher than that of Tl-doped CsI from the X-ray induced scintillation spectra measured with a silicon photodiode. Furthermore, the afterglow level of O-doped ZnSe is found to be lower than that of Tl-doped CsI. Moreover, clear pulse height peaks are obtained under γ-ray irradiation using ^{241}Am (59.5 keV) and ^{137}Cs (661 keV), as shown in Figure 6.10, and the energy resolution of a pulse height peak under γ-ray irradiation from ^{137}Cs is approximately 7.4%. In addition, O-doped ZnTe is another semiconductor scintillator with an isoelectronic electron trap, and further details are given in reference [29].

6.3.2.2 Semiconductor Scintillators with Donor Impurities

Ga-doped ZnO is one of the semiconductor scintillators with a donor impurity [8]. The replacement of Zn with Ga leads to the formation of a degenerate donor band overlapping the bottom of the conduction band. The Ga-doped ZnO shows a broad near-band-edge emission across the wavelengths of 370–410 nm due to the recombination of the donor band electrons and ionizing holes. The X-ray induced scintillation decay time profiles of Ga-doped ZnO are measured in the temperature range of 14 to 365 K, as shown in Figure 6.11,

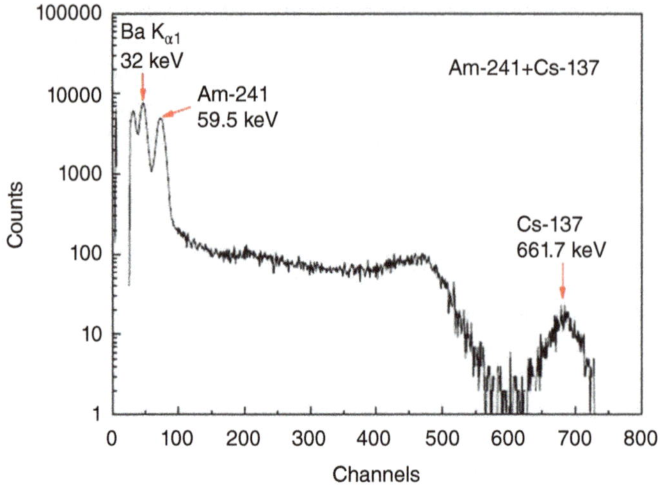

Figure 6.10 Pulse height spectrum of ZnSe:O under ^{137}Cs and ^{241}Am γ-ray irradiation [7].

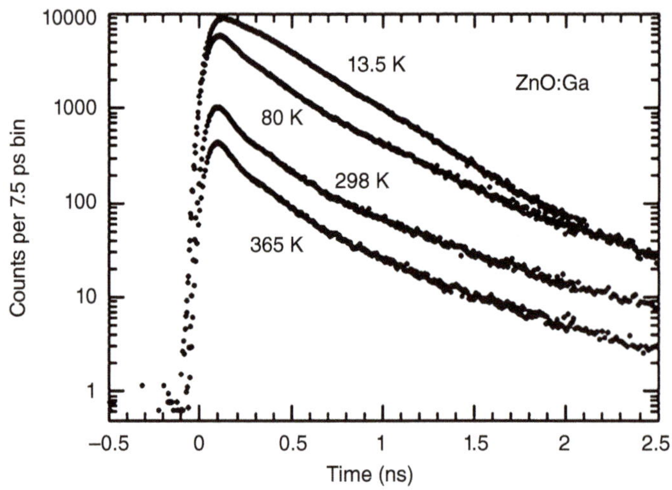

Figure 6.11 X-ray induced scintillation decay time profiles of ZnO:Ga [8].

and the scintillation decay time of the main component is 0.3 ns at 14 K and 0.1 ns at 365 K. The relative scintillation intensity of Ga-doped ZnO for $Bi_4Ge_3O_{12}$ (8200 photons/MeV) at 295 K is 2.0 at 14 K, and 0.06 at 365 K. The other ZnO-based scintillators with a donor impurity include Sc-doped ZnO, and the detail properties are introduced in reference [30].

In-doped CdS is also a semiconductor scintillator with a donor impurity [8–10] where the conduction band is degenerated and partly filled with donor electrons, and shows a featureless near-band-edge emission at around 500–550 nm. The scintillation decay time is very fast (~0.2 ns) and is almost constant over 15–300 K. However, the scintillation intensity is not so high, and the relative scintillation intensity of In-doped CdS compared with $Bi_4Ge_3O_{12}$ (8200 photons/MeV) at 295 K is 0.13 at 13 K and 0.001 at 295 K.

6.3.2.3 Semiconductor Scintillators with Acceptor Impurities

Ag-doped ZnS is one of the well-known semiconductor scintillators with an acceptor impurity for a long time [11–13]. It shows photoluminescence and scintillation peaking at approximately 450 nm due to the donor–acceptor recombination involving S vacancies as the donor and Ag substituting for Zn as the acceptor. Ag-doped ZnS is obtained as a polycrystalline powder, thus it is used as thin films for the detection of alpha particles and heavy charged particles. The scintillation intensity of Ag-doped ZnS is 1.3 times higher than that of Tl-doped NaI under α-ray irradiation, and the decay time of the fast component obtained from the decay time profile under α-ray irradiation is moderate (~72 ns).

6.3.2.4 Semiconductor Scintillators with a Donor–Acceptor Pair

Ag and Cl-doped ZnS is one of the semiconductor scintillators with a donor–acceptor pair where Cl and Ag are regarded as a donor and an acceptor impurity, respectively [26]. It exhibits an emission at 450 nm due to the donor–acceptor recombination. It is reported that the luminosity of Ag and Cl-doped ZnS is 6 times higher than that of Ce-doped $YAlO_3$, but the decay time is long (~100 μs) due to low spatial overlap of the wave functions of a donor and an acceptor. Another candidate material of the semiconductor scintillators includes Ag and Cl-doped CdS, but there are few reports on the scintillation characteristics of semiconductor scintillators with an intentional donor–acceptor pair.

6.4 Quantum Size Effect

In Section 6.3, scintillation properties of various semiconductor scintillators have been introduced. At room temperature, a main origin of scintillation from the undoped semiconductor scintillators such as ZnO and ZnS is often lattice defects, and it is difficult to observe efficient exciton emission at room temperature due to thermal disturbance. With the outstanding improvement of semiconductor technology, especially the fabrication technology of quantum structures in an order of nanometers, it is possible to fabricate quantum nanostructures with low dimensional systems such as a quantum well, a quantum wire, and a quantum dot. When excitons are confined in a quantum nanostructure, the exciton binding energy and oscillator strength of excitons can be enhanced due to the quantum size effect. The enhancement in physical properties of excitons enables us to fabricate fast response scintillators, which is characterized by high light yield and fast decay time. In this section, the electronic energy structures and optical properties of quantum nanostructures with low dimensional systems are briefly introduced. Further details on the quantum size effect can be found in references [20, 23].

Illustrations of three-dimensional (3D), two-dimensional (2D), one-dimensional (1D), and zero-dimensional (0D) quantum nanostructures are shown in Figure 6.12. In these illustrations, a bulk crystal, a quantum well, a quantum wire, and a quantum dot are shown, and the illustrations explain that electrons are confined in each quantum nanostructure by the potential barriers. In the actual quantum nanostructures, a semiconductor with a certain band-gap is usually surrounded by semiconductors with a larger band-gap to form these quantum nanostructures. It should be noted that the basic crystal structure (unit cell) remains the same on going from three-dimensional to low dimensional systems. In addition, there are some cases of quantum nanostructures, depending on the band alignments of the two semiconductors where electrons and

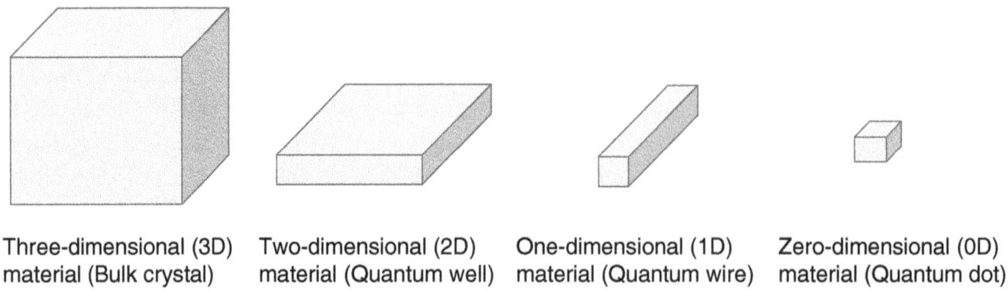

Figure 6.12 *Fundamental types of low dimensional quantum nanostructures.*

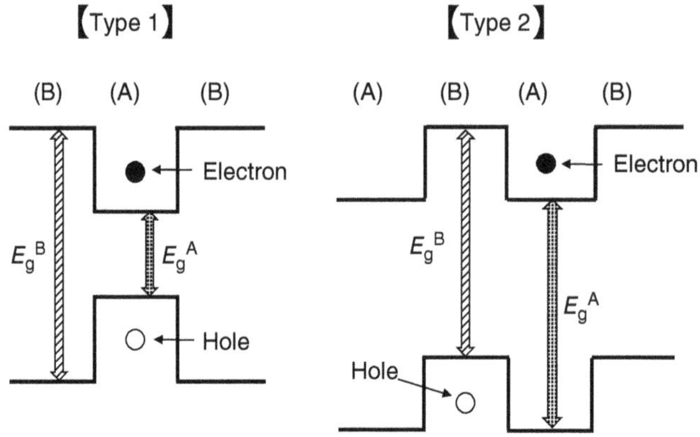

Figure 6.13 *Basic types of quantum wells and their illustration.*

holes are confined in the same semiconductor layer (Type I), and where electrons and holes are confined in different semiconductor layers (Type II), as shown in Figure 6.13. In this book, electronic structure and optical properties are explained based on the Type I. There are various types of the band alignment, so please refer to [31].

Figure 6.14 shows the density of states for an electron in the conduction band of 3-, 2-, 1-, and 0-dimensional quantum nanostructures. The density of states is characterized by a square-root form in the three-dimensional system and a step function in the two-dimensional system. Furthermore, the density of states in the one-dimensional system decrease as $1/\sqrt{E}$ with increasing energy, and the density of the states in the zero-dimensional system have the form of discrete delta functions [20, 23]. In this way, the density of states is greatly affected by the dimensionality of quantum nanostructures. In addition, band-gap energy tends to increase with decrease of the dimensionality due to the change in the energy level at the bottom of the conduction band and at the top of the valence band. The band-gap energy has an influenced on the peak wavelength of absorption and emission, and the density of states affects the peak width and temperature dependence of the absorption and emission peaks. Therefore, the quantum size effect enables us to obtain functional materials with desired optical properties tailored to each application such as scintillators.

Now we will describe the effect of dimensionality on the physical properties of Wannier excitons. In a three-dimensional bulk structure, the Schrödinger equation in the relative motion of Wannier excitons can be expressed as follows in an effective mass approximation [32]:

$$\left[-\left(\frac{\hbar^2}{2\mu}\right)\nabla^2 - \frac{e^2}{4\pi\varepsilon_0\varepsilon_r r}\right]\psi(r) = (E - E_g)\psi(r) \tag{6.17}$$

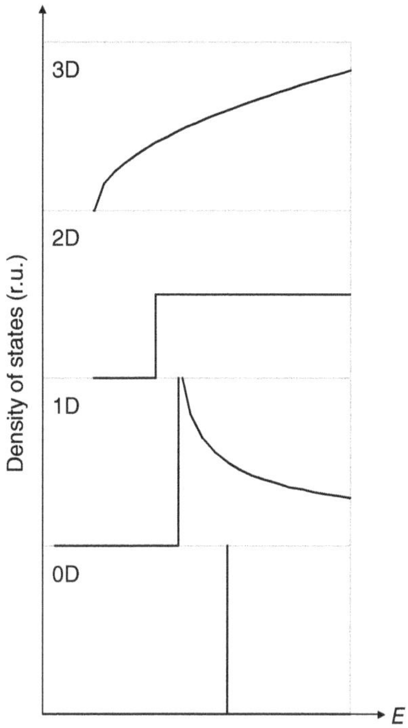

Figure 6.14 Density of states in the conduction band of each semiconductor quantum structures.

In this formula, E is the exciton energy, μ is the reduced mass of the exciton, E_g is the band-gap energy, and r is the relative distance vector. The Schrödinger equation is the simple hydrogen-atom problem in the case of an isotropic structure, but an additional anisotropic potential should be considered in an anisotropic structure such as a quantum well, and the effective mass changes depending on the structural orientation. Through the fractional dimensional model [33, 34], anisotropic excitation dynamics can be studied by solving the relevant Schrödinger equation in a non-integer dimensional space, where the excitation is assumed to be in an isotropic environment. Using this model, the Schrödinger equation can be expressed using the Laplacian proposed by Stillinger [35]:

$$\left[-\frac{\hbar^2}{2\mu r^{\alpha-1}}\frac{\partial}{\partial r}\left[r^{\alpha-1}\frac{\partial}{\partial r}\right]+\frac{l^2}{2\mu r^2}-\frac{e^2}{4\pi\varepsilon_0\varepsilon_r r}\right]\psi(r,\theta)=\left(E-E_g\right)\psi(r,\theta) \tag{6.18}$$

Here, α is the dimensionality $(1<\alpha<3)$, and r is the radial distance $(0\leq r\leq\infty)$. Related angle θ $(0\leq\theta\leq\pi)$ measured relative to an axis passing through the origin has two coordinates describing r in the α-dimensional (αD) space. In addition, $\boldsymbol{l}^2$ is the angular-momentum operator as follows [32]:

$$\boldsymbol{l}^2=-\frac{\hbar^2}{\sin^{\alpha-2}\theta}\frac{\partial}{\partial\theta}\left[\sin^{\alpha-2}\theta\frac{\partial}{\partial\theta}\right] \tag{6.19}$$

By solving the equation, the physical properties of the excitons can be obtained as a function of dimensionality.

Now, the resulting eigenenergies of the bound states (exciton binding energy) are shown as follows:

$$E_n(3\mathrm{D}) = R^*/n^2 \; (n = 1,\ 2,\ 3 \cdots\cdot) \tag{6.20}$$

$$E_n(2\mathrm{D}) = R^*/(n-1/2)^2 \; (n = 1,\ 2,\ 3 \cdots\cdot) \tag{6.21}$$

$$E_n(1\mathrm{D}) = R^*/(n-1)^2 \; (n = 1,\ 2,\ 3 \cdots\cdot) \tag{6.22}$$

where n is the principal quantum number, and R^* is the Wannier exciton binding energy in the bulk crystal (three-dimensional system) for $n = 1$. It can be seen that the exciton binding energy of the two-dimensional system for all quantum numbers is higher than that of the three-dimensional system, indicating that exciton binding energy increases due to the quantum size effect. For the ground state of an exciton $n = 1$, the exciton binding energy in the two-dimensional system is four times higher than that in the three-dimensional system.

Furthermore, the corresponding values of the orbital radii are as follows:

$$a_n(3\mathrm{D}) = n^2 a_\mathrm{e} \; (n = 1,\ 2,\ 3 \cdots\cdot) \tag{6.23}$$

$$a_n(2\mathrm{D}) = (n-1/2)^2 a_\mathrm{e} \; (n = 1,\ 2,\ 3 \cdots\cdot) \tag{6.24}$$

$$a_n(1\mathrm{D}) = (n-1)^2 a_\mathrm{e} \; (n = 1,\ 2,\ 3 \cdots\cdot) \tag{6.25}$$

$$a_\mathrm{e} = (m_0 \varepsilon_r / \mu) a_\mathrm{H} \tag{6.26}$$

Here, m_0 is a free electron mass and a_H is the Bohr radius of hydrogen atom about 0.53Å. It can be seen that the exciton Bohr radius decreases as the dimensionality decreases. In addition, the oscillator strength f of the excitons scales reciprocally with the third power of the quantum number n.

$$f(3\mathrm{D}) \approx n^{-3} \; (n = 1,\ 2,\ 3 \cdots\cdot) \tag{6.27}$$

$$f(2\mathrm{D}) \approx (n-1/2)^{-3} \; (n = 1,\ 2,\ 3 \cdots\cdot) \tag{6.28}$$

$$f(2\mathrm{D})/f(3\mathrm{D}) = n^3/(n-1/2)^3 > 1 \; (n = 1,\ 2,\ 3 \cdots\cdot) \tag{6.29}$$

For $n = 1$, the oscillator strength in the two-dimensional system is eight times higher than that in the three-dimensional system. As described above, the quantum size effect leads to an increase in the exciton binding energy, an increase in the oscillator strength, and a decrease in the exciton Bohr radius.

We now describe the effect of the quantum size effect on the absorption spectra of quantum nanostructures in the low dimensional systems. The shape of the absorption spectrum based on the square root form of joint density of states in the three-dimensional system is affected by an exciton effect, resulting in the observation of several sharp exciton peaks. The intensity of exciton absorption for $n = 1$ is usually the highest, and the intensity tends to decrease with increasing the quantum number n. In the two-dimensional system, the joint density of state is characterized by a step function and several excitons peaks affected by the joint density of state are observed in Figure 6.15. The intensity of discrete absorption peaks

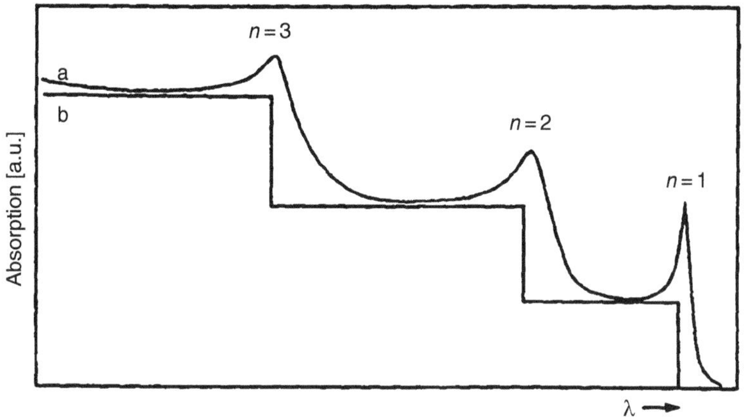

Figure 6.15 *Illustration of absorption spectra of two-dimensional systems (a) with exciton effect and (b) without exciton effect. [31].*

and exciton binding energies are usually higher than those in the three-dimensional system due to the quantum size effect. For the one-dimensional system, the shape of the joint density of states scales as inverse square root, and the absorption intensity for the $n = 1$ exciton is higher than that for the other excitons. In this way, the enhancement effect of exciton binding energy and oscillator strength due to the quantum size effect can be observed in the absorption spectra of quantum nanostructures in low dimensional systems.

The quantum size effect can also be observed in the luminescence spectra. Generally, excitonic emission from the materials with the three-dimensional system cannot usually be detected at room temperature. Thanks to the quantum size effect, some quantum nanostructures with low dimensional systems can show efficient exciton emissions due to the increase in exciton binding energy and oscillator strength. By exploiting the unique optical properties due to the quantum size effect, it is expected to fabricate fast response scintillators with a high light yield and a fast decay time. In addition, the exciton absorption and emission energies tend to shift to a higher energy as the quantum size decreases, although the detailed mechanism is not covered in this book. As shown in Figure 6.16, the emission wavelength in a GaAs/AlAs quantum well structure is shifted to the shorter wavelength side by decreasing the layer thickness [36]. Like this example, the exciton absorption and emission wavelength can be controlled by tuning the layer thickness of the active layer. This wavelength tunability can be applied to match the optimum sensitivity wavelength of the photodetectors, such as a photomultiplier tube. As described above, it is possible to develop a fast response scintillator by exploiting the unique optical properties of quantum nanostructures due to the quantum size effect.

6.5 Organic–Inorganic Perovskite-Type Compounds

6.5.1 Introduction

Organic–inorganic perovskite type compounds are a kind of semiconductor scintillator that were proposed by Keisuke Asai around 2000. Organic–inorganic perovskite type compounds form various structures in a self-organized manner, depending on the species and chemical

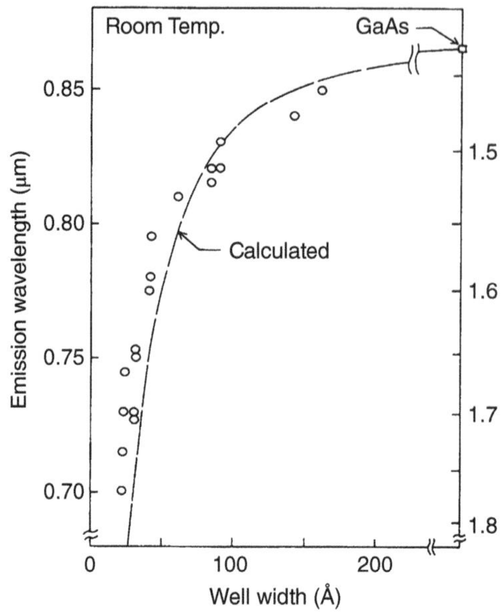

Figure 6.16 Photoluminescence emission wavelength versus quantum well size [36]. Source: Copyright (1981) The Physical Society of Japan and The Japan Society of Applied Physics

compositions of precursors used. The self-organized structures include a bulk structure and quantum nanostructures such as a quantum well, a quantum wire, and a quantum dot structure, where inorganic layers are sandwiched between organic layers as potential barriers. In the quantum nanostructures, excitons confined in the inorganic layer are known to show unique optical properties due to the quantum size effect. In the self-organized manner, creation of lattice defects, and size distribution at the interface can be avoided as compared with quantum nanomaterials fabricated by epitaxial methods such as a molecular beam epitaxy. For this reason, organic–inorganic perovskite type compounds were first used with an aim to investigate exciton properties of quantum nanostructures after the study on alkylammonium leadiodide compounds was reported in 1986 [37, 38]. Recently, organic–inorganic perovskite type compounds are expected to be applied to various applications such as solar cells, electroluminescence devices, and scintillators. In this section, fundamental optical and scintillation properties of organic–inorganic perovskite type compounds will be described. Please refer to the references [39–43] for details about the properties of excitons in the low dimensional systems and other applications.

6.5.2 Materials and Structures

The organic–inorganic layered perovskite-type compounds have a perovskite structure with a chemical composition AMX_3. As shown in Figure 6.17, the unit cell of the perovskite is a cubic structure where the cation M (a divalent metal ion) is located at the body center, and the anion X (a halogen ion) is located at the face center. In addition, the cation A (an ammonium ion) is located at the eight corner of the cubic structure, being shared by eight unit cells.

Organic–inorganic perovskite type compounds can form several structures based on $[MX_6]^{4-}$ octahedra, and their octahedra can share 0, 2, 4, and 6 corners to form 0-, 1-, 2-, and 3-dimensional

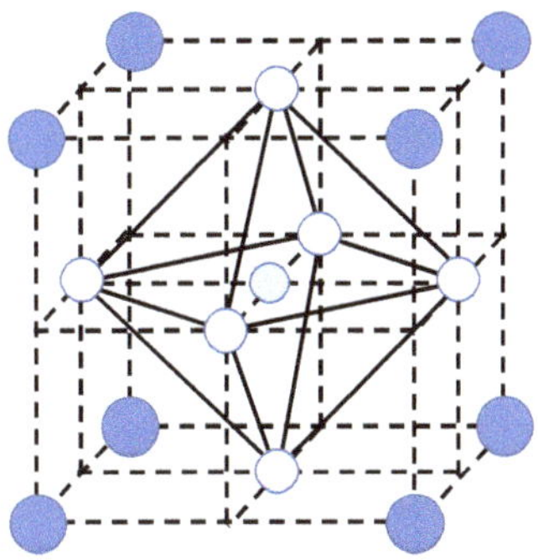

Figure 6.17 *Unit cell of perovskite structure where the closed circle is divalent cation A, shaded circle is cation M, and the open circle is anion X.*

quantum nanostructure, respectively. In these compounds, there are three different bonds to form the structure: (i) van der Waals bond between the organic amines: (ii) hydrogen bond between the halogens and the ammoniums; (iii) ionic bond between metal ions and halogen ions. When a methylamine is present at the A site, a bulk structure where $[MX_6]^{4-}$ octahedra are arranged three-dimensionally is usually formed, as shown in Figure 6.18. The compounds with the three-dimensional system show excellent performances in optoelectronic devices such as light emitting devices, sensitive photodetectors, and field effect transistors. Furthermore, the compounds with the two-dimensional system have a quantum well structure where a two-dimensional inorganic layer is sandwiched between organic amines such as alkylamines at the A site, and the alkylammonium chain is oriented perpendicular to the inorganic layers. In the compounds, the inorganic layer is regarded as a well layer and the organic layer is regarded as a barrier layer, since the band-gap energy of the inorganic layer is usually lower than that of the organic layer. The compounds with the two-dimensional system show unique optical properties owing to the quantum size effect, which can be exploited for scintillators and nonlinear optical materials. In addition, the compounds with the one-dimensional and zero-dimensional systems (Figure 6.18) can be formed by selecting particular organic amines and chemical compositions. Up to now, scintillation properties of the compounds with the two-dimensional and three-dimensional systems have been mostly investigated. In Subsection 6.5.4 and 6.5.5, photoluminescence and scintillation properties of these compounds are mainly introduced. In particular, Pb-based compounds are focused upon because they tend to show efficient photoluminescence and scintillation although there are various elements such as Pb, Ge, Sn, and Cu as candidates for inorganic elements of the organic–inorganic perovskite type compounds [44].

6.5.3 Sample Preparation

Organic–inorganic perovskite type compounds have self-organized structures, which are formed by hydrogen, van der Waals, and ionic bonding, as described above. Generally, the compounds can be fabricated by selecting appropriate precursors and chemical compositions in a chemical-based procedure. In the case of the compounds in a form of a polycrystalline powder, the compounds usually can be obtained by mixing precursors in a solvent at a stoichiometric ratio for several hours and then evaporating off the solvent.

On the other hand, various methods have been used for growing single crystals of the compounds. At the beginning of the basic investigation into the exciton properties of the

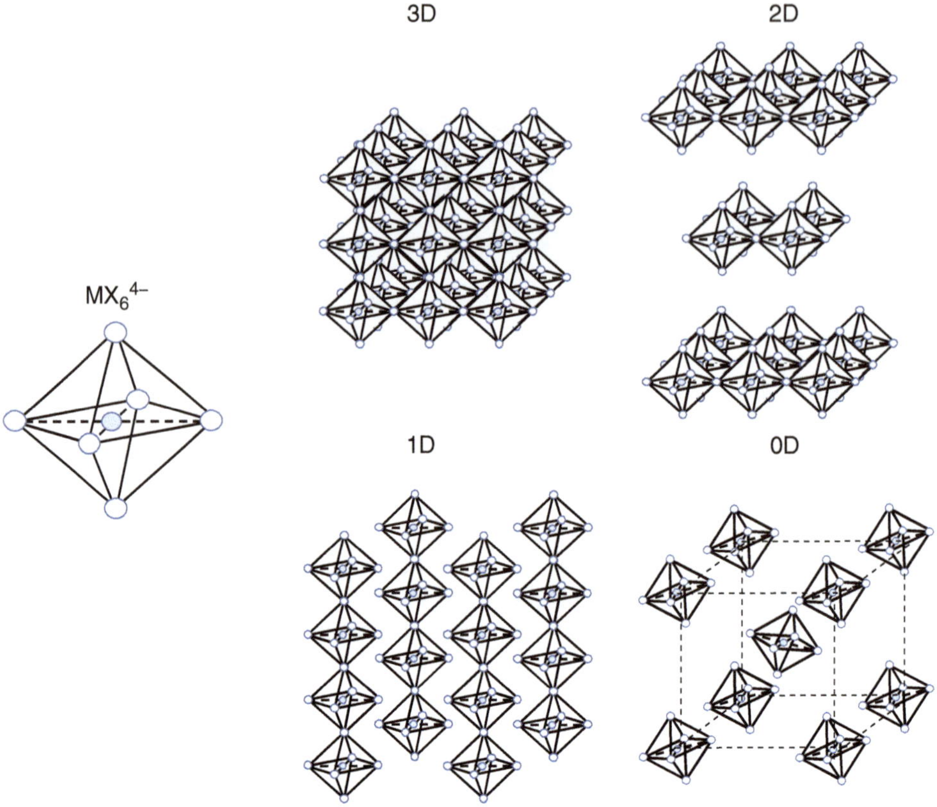

Figure 6.18 Structures based on $[PbX_6]^{4-}$ octahedra.

compounds with low dimensional systems, single crystals of the compounds were fabricated by the silica-gel method. However, a gel remains on the surface of the single crystal, and impurities may become mixed inside the single crystal by this method [45]. After the KI acetone method and gel-free new solution method were considered [45–47], the temperature-gradient crystallization method and poor-solvent diffusion method are now mainly performed for the crystal growth [48, 49].

A single crystal is grown by changing temperature gradually in the temperature-gradient crystallization method, which is widely used for the fabrication of compounds with the three-dimensional system. For example, the fabrication procedure of $(CH_3NH_3)PbBr_3$ single crystals is as follows. A 1 M solution of $(CH_3NH_3)PbBr_3$ is prepared by mixing $PbBr_2$ and (CH_3NH_3Br) at a stoichiometric ratio in dimethylforamide. Subsequently, a single crystal can be obtained by increasing the temperature slowly with a hot plate, as shown in Figure 6.19, since the solubility of $(CH_3NH_3Br)PbBr_3$ in dimethyforamide is about 0.8 g/ml at room temperature and about 0.3 g/ml at 80 °C [48].

The poor solvent diffusion method is mainly used for the crystal growth of the compounds with the two-dimensional system, and their single crystals are fabricated by decreasing the solubility of the solution slowly due to the vapor diffusion of the poor solvent. The illustration of a poor solvent diffusion method is shown in Figure 6.20. At first, a polycrystalline

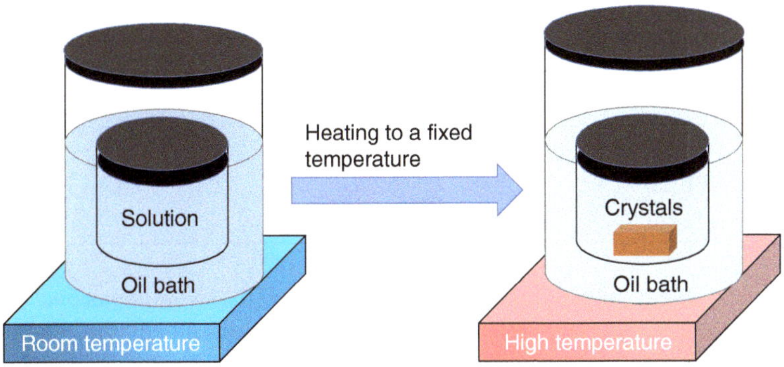

Figure 6.19 *Illustration of the temperature-gradient crystallization method.*

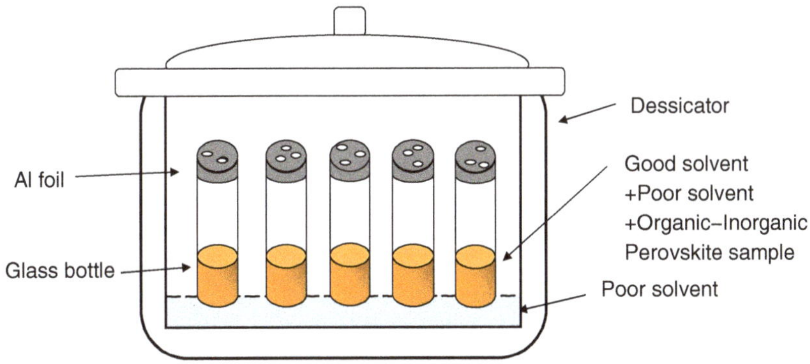

Figure 6.20 *Poor solvent diffusion method.*

powder of the compound with the two-dimensional system is dissolved into a good solvent, which is often dimethylforamide or acetone. Then, a poor solvent such as nitromethane and acetone is dropped into the solution just before precipitation. The solution is stored for one month in a bottle in the desiccator where a poor solvent is poured at the bottom, as shown in Figure 6.20. The vapor of the poor solvent is gradually diffused into the solution to decrease the solubility. Finally, single crystals can be obtained from the bottom of the bottle.

6.5.4 Fundamental Optical Property

In this subsection, fundamental photoluminescence properties of organic–inorganic layered perovskite-type compounds are introduced. As described above, organic–inorganic layered perovskite-type compounds can have a variety of structures such as bulk, quantum well, quantum wire, and quantum dot structures. Table 6.2 summarizes the exciton properties of $(CH_3NH_3)PbI_3$, $(n\text{-}C_{10}H_{21}NH_3)_2PbI_4$, and $(CH_3NH_3)_4PbI_6{\cdot}2H_2O$, which correspond to 3-, 2-, and 0-dimensional quantum nanostructures, respectively [50]. The energy of the lowest exciton state blue-shifts markedly as the dimension decreases. Moreover, $(n\text{-}C_{10}H_{21}NH_3)_2PbI_4$ and $(CH_3NH_3)_4PbI_6{\cdot}2H_2O$ with the low dimensional systems show higher exciton binding energy and oscillator strength than $(CH_3NH_3)PbI_3$ with the three-dimensional system due to

Table 6.2 *Exciton energy (E_{ex}), bandgap energy (E_g), oscillator strength (f_{ex}), Bohr radii (Å), and the binding energy (E_B) of $(CH_3NH_3)PbI_3$, $(n\text{-}C_{10}H_{21}NH_3)_2PbI_4$, and $(CH_3NH_3)_4PbI_6{\cdot}2H_2O$.*

	E_{ex} (eV)	E_g (eV)	f_{ex}	a_B(Å)	E_B (eV)	References
$(CH_3NH_3)PbI_3$	1.6	1.7	0.02 ± 0.004	28	0.04	[50]
$(n\text{-}C_{10}H_{21}NH_3)_2PbI_4$	2.6	2.9	0.7	<7	0.32	[50]
$(CH_3NH_3)_4PbI_6{\cdot}2H_2O$	3.4	–	0.4 ± 0.1	–	2.3	[50]

the quantum size effect. In particular, the compounds with the two-dimensional system tend to show the highest exciton absorption and emission intensities among the compounds with the 3-, 2-, 1-, and 0-dimensional systems. For this reason, the compounds with the two-dimensional system have been mainly focused on for scintillator applications.

The exciton properties of the organic–inorganic layered perovskite-type compounds are affected by the crystal structure of the inorganic layer [44]. The compounds form a stable structure self-assembly according to the chemical composition and species of the precursors. As shown in Figure 6.21, the inorganic structure often has structural distortions between the PbX_6^{4-} octahedra and inside the PbX_6^{4-} octahedron [49, 51]. Crystal structure and exciton absorption peak of $(RNH_3)_2SnI_4$ (R: organic amine) are summarized in Table 6.3. The structural distortions in the inorganic layer and exciton absorption peak are found to be affected by organic species in the organic layer.

Such a change in the structure has an effect on the band structure, absorption, and photoluminescence properties. The band structure of the PbI_2-based organic–inorganic perovskite compound is shown in Figure 6.22. The bottom of the conduction band consists of Pb(6p)–I(5p) and Pb(6p)–I(5s) orbitals, and the top of the valence band consists of Pb(6s)–I(5p) orbitals. The transfer integral between the adjoining atoms depends on the overlap integral of those orbitals. Band calculation has demonstrated that one of the causes for the bandwidth narrowing in the

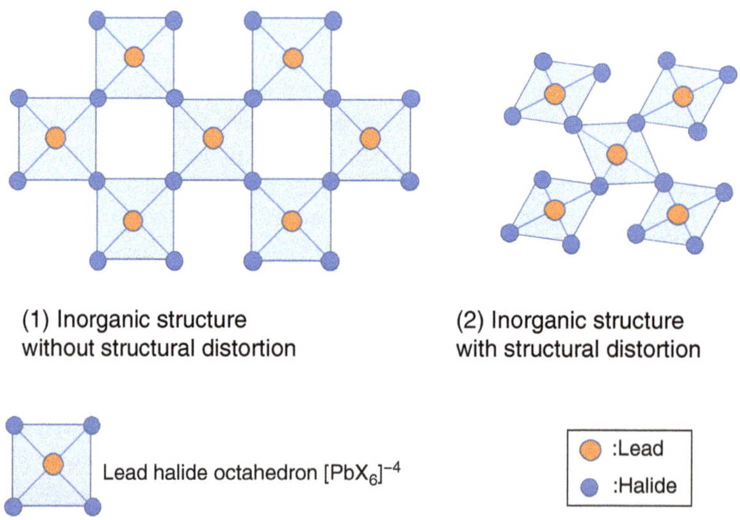

Figure 6.21 *Illustration of the inorganic structure without structural distortion (1) and with structural distortion (2).*

Table 6.3 *Exciton peak and crystal structure in the inorganic layer. Distortion of SnI_6^{4-} octahedron was defined as $\Sigma abs(di - \langle d \rangle)/6$, where di is the summation over the six Sn–I distances and $\langle d \rangle$ represents the average bond length.*

Organic amine	Exciton peak (nm)	Sn–Sn bond angle (deg)	Distortion of SnI_6^{4-} octahedron	References
4-Chlorophenethylamine	615	157.0	0.017	[44]
Phenethylamine	609	156.5	0.012	[44]
4-Fluorophenethylamine	609	156.4	0.020	[44]
3-Fluorophenethylamine	599	154.2	0.033	[44]
2-Fluorophenethylamine	588	153.3	0.013	[44]
2-Chlorophenethylamine	586	154.8	0.006	[44]
2-Bromophenethylamine	557	148.7	0.005	[44]
n-Butylamine	607	159.6	0.011	[44]
n-Dodecylamine	580	155.7	0.007	[44]

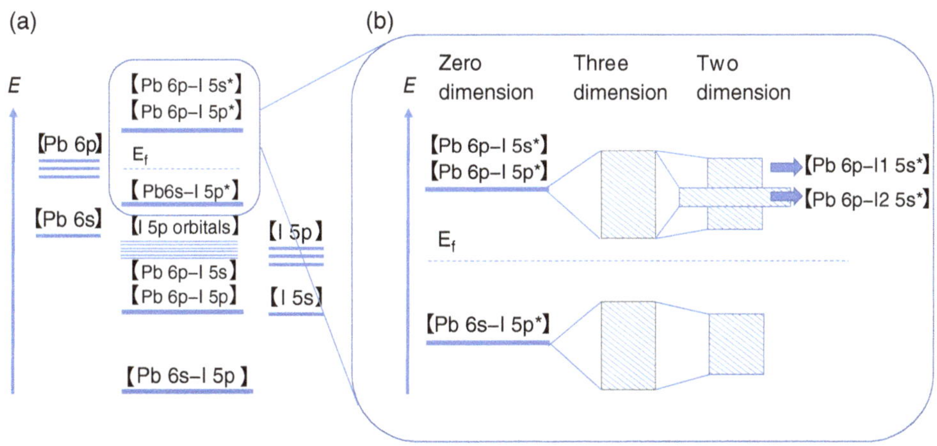

Figure 6.22 *Bonding diagram of $[PbI_6]^{4-}$ cluster (a), and 3D and 2D crystal at the top of the valence band and the bottom of the conduction band (b). [51].*

structure with the two-dimensional system is the structural distortion in the inorganic layer, as shown in Figure 6.21. Figure 6.23 shows the calculated band structure of $(n\text{-}C_4H_9NH_3)_2PbI_4$ without structural distortion (a) and with structural distortion (b). The valence band and conduction bond of the $(n\text{-}C_4H_9NH_3)_2PbI_4$ with structural distortion are narrower than those without structural distortion, indicating that the structural distortion in the inorganic layer leads to the bandwidth narrowing. The band narrowing is remarkable at the top of conduction band and the bottom of the valence band. The smaller band dispersion induced by the structural distortion results in a larger effective mass of electrons, holes, and excitons.

The effect of the structure in the inorganic layer on the photoluminescence properties of organic–inorganic layered perovskite-type compounds has been investigated by photoluminescence spectroscopy and single crystal structure analysis [49]. Table 6.4 summarizes the

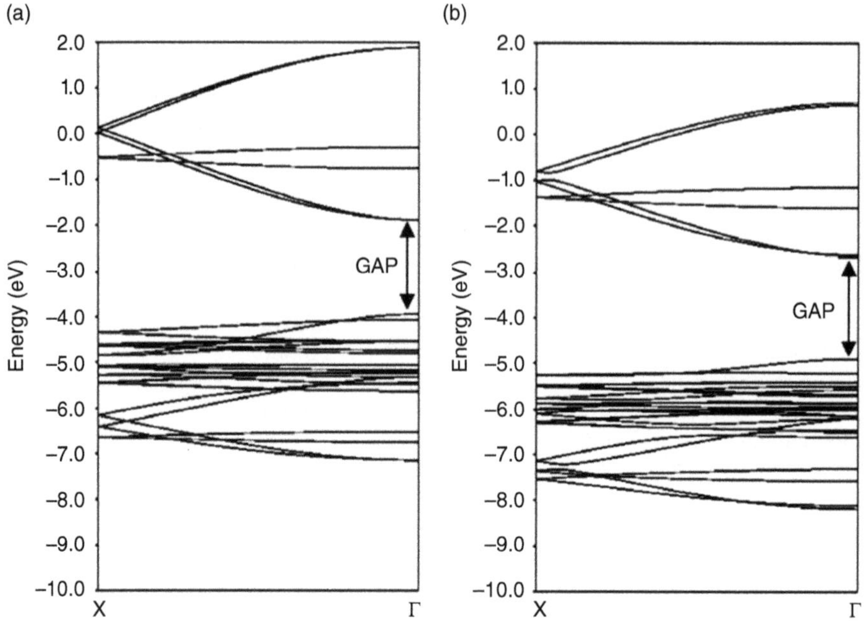

Figure 6.23 *Calculation of band structure of* $(n\text{-}C_4H_9NH_3)_2PbI_4$ *(a) without structural distortion and (b) with structural distortion. [51].*

Table 6.4 *Physical properties of excitons and crystal structure in the inorganic layer.*

	$C4PbBr_4$	$BenPbBr_4$	$PhePbBr_4$
Quantum efficiency (—)	0.004	0.04	0.22
PL decay time (ns)	1.7	3.9	5.1
Radiative decay rate (k_f) ($10^6\,s^{-1}$)	2.2	9.9	43
Non-radiative decay rate (k_{nr}) ($10^6\,s^{-1}$)	6.0	2.5	1.6
Pb–Br–Pb bond angle between $PbBr_6^{4-}$ octahedra (degree)	155	150	152
Br–Pb–Br bond angle inside $PbBr_6^{4-}$ octahedron (degree)	180	180	171
References	[49]	[49]	[49]

※$C4PbBr_4$ ($(n\text{-}C_4H_9NH_3)_2PbBr_4$), $BenPbBr_4$ ($(C_6H_5CH_2NH_3)_2PbBr_4$), $PhePbBr_4$ ($(C_6H_5C_2H_4NH_3)_2PbBr_4$).

physical properties of excitons and structures in the inorganic layer of $(n\text{-}C_4H_9NH_3)_2PbBr_4$, $(C_6H_5CH_2NH_3)_2PbBr_4$, and $(C_6H_5C_2H_4NH_3)_2PbBr_4$. These compounds with the two-dimensional system show exciton emissions from the inorganic layer, and it can be seen that the quantum efficiency (Φ) and exciton emission lifetime (τ) change by the replacement of the organic moiety.

The radiative decay rate (k_f) and non-radiative decay rate (k_{nr}) are calculated using the following formulas:

$$\Phi = k_f / (k_f + k_{nr}) \tag{6.30}$$

$$\tau = 1 / (k_f + k_{nr}) \tag{6.31}$$

The quantum efficiencies are found to be well correlated with the radiative decay rate, suggesting that photoluminescence properties mainly depend on the exciton properties of the compounds. Focusing on the structure in the inorganic layer, the Pb–Br–Pb bond angle for $(n\text{-}C_4H_9NH_3)_2PbBr_4$ is 155 degrees, which is higher than that of 150 degrees for $(C_6H_5CH_2NH_3)_2PbBr_4$ and 152 degrees for $(C_6H_5C_2H_4NH_3)_2PbBr_4$. In addition, the bond angle of Br–Pb–Br for $(C_6H_5C_2H_4NH_3)_2PbBr_4$ is 171 degrees, which is lower than that of 180 degrees for $(n\text{-}C_4H_9NH_3)_2PbBr_4$ and $(C_6H_5CH_2NH_3)_2PbBr_4$. These bond angles indicate that $(C_6H_5C_2H_4NH_3)_2PbBr_4$ has both structural distortions between the adjoining $PbBr_6^{4-}$ octahedra and inside the $PbBr_6^{4-}$ octahedron. As shown in Table 6.3, $(C_6H_5C_2H_4NH_3)_2PbBr_4$ shows the highest quantum efficiency and radiative decay rate. This is because an increase in the effective mass of excitons induced by the structural distortions leads to a decrease in the Bohr radius of excitons [49]. Therefore, photoluminescence properties of organic–inorganic layered perovskite-type compounds are governed by the exciton properties, which depend on the structure in the inorganic layer.

6.5.5 Scintillation

Now, we introduce the scintillation properties of organic–inorganic layered perovskite-type compounds. As well as photoluminescence, organic–inorganic layered perovskite-type compounds show scintillation due to the exciton recombination in the inorganic layer. In particular, the compounds with the two-dimensional system exhibit efficient scintillation whose decay time constants are usually several hundred picoseconds to several nanoseconds due to the quantum size effect. In recent years, some compounds with the two-dimensional system exhibit a higher scintillation intensity than plastic scintillators, which are characterized by a moderate scintillation light yield (~10 000 photons/MeV) and a short decay time (several nanoseconds). Moreover, they show an excellent afterglow level comparable to $Bi_4Ge_3O_{12}$, which is well known for low afterglow levels. In Subsection 6.5.5, scintillation properties of organic–inorganic layered perovskite-type compounds with the two- and three-dimensional systems will be introduced, focusing on their scintillation light yield, scintillation decay time, and afterglow level.

6.5.5.1 Scintillation Light Yield

Scintillation light yield under ionizing radiation has been evaluated in various organic–inorganic layered perovskite-type compounds with the two- and three-dimensional systems. Efficient scintillation due to the exciton recombination from the compounds with the two-dimensional system can be observed at room temperature owing to the quantum size effect. On the other hand, exciton binding energy in the compounds with the three-dimensional system is usually so low that efficient exciton emission cannot be detected at room temperature, although the compounds have a possibility to show high scintillation light yield at low temperature such as $(CH_3NH_3)PbBr_3$ (~ 90 000 photons/MeV at 77 K) due to low band-gap energy [52]. In fact, scintillation light yield of $(CH_3NH_3)PbI_3$ and $(CH_3NH_3)PbBr_3$ are estimated to be less than 1000 photons/MeV at room temperature. From the viewpoint of practical use as scintillators, the compounds with the two-dimensional system are currently more promising compounds than those with the three-dimensinal system [14, 53].

Previous studies have shown that the scintillation light yield under ionizing radiation is affected by the organic moiety in the compounds with the two-dimensional system, and their scintillation light yield mainly depends on the quantum yield of exciton emission from the inorganic layer. Relative quantum efficiency in photoluminescence and relative scintillation light yield are summarized in Figure 6.24 [14, 15]. The relative scintillation light yield is

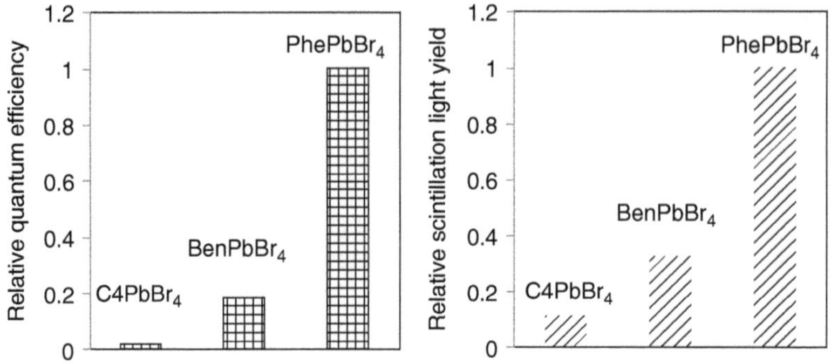

Figure 6.24 *Relative quantum efficiency and relative scintillation light yield. ※C4PbBr$_4$ ((n-$C_4H_9NH_3)_2PbBr_4$), BenPbBr$_4$ (($C_6H_5CH_2NH_3)_2PbBr_4$), PhePbBr$_4$ (($C_6H_5C_2H_4NH_3)_2PbBr_4$).*

Table 6.5 *Scintillation properties of organic–inorganic layered perovskite-type compounds.*

	NaI(Tl)	C3PbBr$_4$	C6PbI$_4$	C4PbBr$_4$	BenPbBr$_4$	PhePbBr$_4$
Emission wavelength (nm)	410	438	558			437
Light yield (photons/MeV)	41 000			<1500	4200	14 000
Relative output (%)	100	6.5	11			
Decay time (ns)	230	2.8, 18, 130	0.4, 3.8, 16	1.1, 8	2.8, 10	9.6, 55
Density (g/cm^3)	3.67	2.65	2.40	2.37	2.23	2.26
References	[14]	[14]	[14]	[15]	[15]	[15, 16]

※C3PbBr$_4$ ((n-$C_3H_7NH_3)_2PbBr_4$), C6PbI$_4$ ((n-$C_6H_{13}NH_3)_2PbI_4$), C4PbBr$_4$ ((n-$C_4H_9NH_3)_2PbBr_4$), BenPbBr$_4$(($C_6H_5CH_2NH_3)_2PbBr_4$), PhePbBr$_4$ (($C_6H_5C_2H_4NH_3)_2PbBr_4$).

well correlated with the relative quantum efficiency, thus the quantum efficiency is considered to be one of the important factors for the enhancement of scintillation light yield. Among the compounds shown in Table 6.5, $(C_6H_5C_2H_4NH_3)_2PbBr_4$ with the highest quantum efficiency shows the highest scintillation light yield [16].

Pulse height spectrum of $(C_6H_5C_2H_4NH_3)_2PbBr_4$ under γ-ray irradiation from ^{137}Cs is measured, as shown in Figure 6.25. A broad pulse height peak is clearly detected, and the scintillation light yield of $(C_6H_5C_2H_4NH_3)_2PbBr_4$ is estimated to be 14 000 photons/MeV, which is higher than common plastic scintillators BC452 (~10 000 photons/MeV). However, energy resolution of $(C_6H_5C_2H_4NH_3)_2PbBr_4$ is approximately 24%, which is poorer than that of conventional inorganic scintillators. One of the reasons is self-absorption of exciton in the inorganic layer due to the low Stokes shift.

Recently, a different divalent metal ion (Ca^{2+}, Cd^{2+}, Mn^{2+}, Fe^{2+}, etc.) in the inorganic layer is incorporated into the inorganic layer to enhance the scintillation intensity of $(C_6H_5C_2H_4NH_3)_2PbBr_4$. Past studies show that the photoluminescence intensity of $(C_6H_9C_2H_4NH_3)_2PbI_4$ increases by doping the divalent metal ions, although the detailed mechanism has not been demonstrated. We found that Sr-doped or Ba-doped $(C_6H_5C_2H_4NH_3)_2PbBr_4$ show high quantum yield (~30%) and high scintillation light yield (~20 000 photons/MeV) with the short decay time of less than 10 ns [17, 18]. The doped compounds should be promising candidates for fast response scintillators.

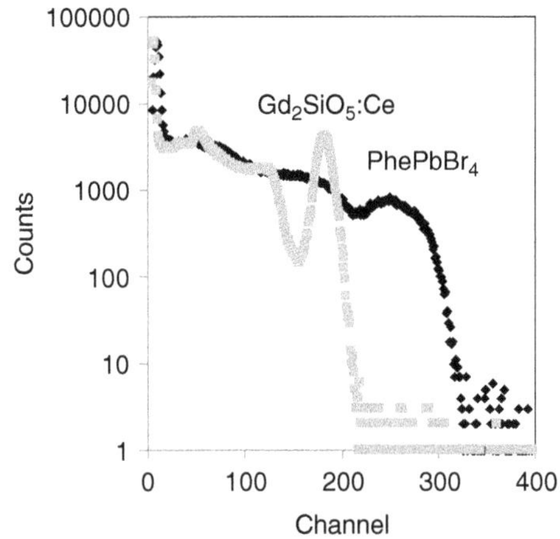

Figure 6.25 *^{137}Cs induced pulse height spectra of $PhePbBr_4$ and Gd_2SiO_5:Ce. ※$PhePbBr_4$ $((C_6H_5C_2H_4NH_3)_2PbBr_4)$.*

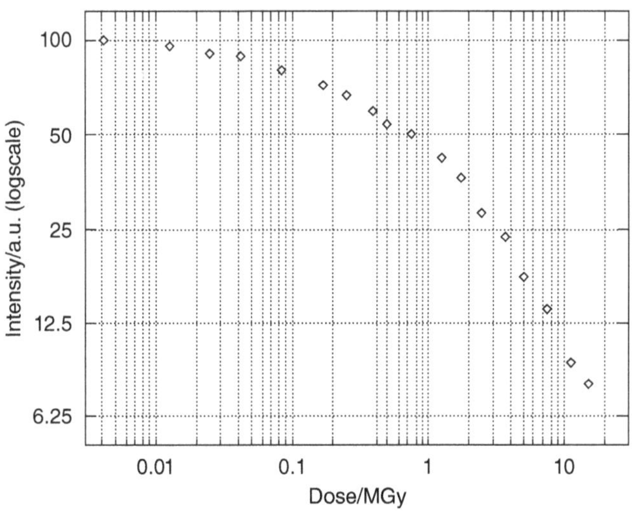

Figure 6.26 *Correlation between the peak intensity of excitonic emission from $(C_6H_{13}NH_3)_2PbI_4$ and irradiation dose. The initial intensity is normalized to be 100 [54].*

In addition to $PbBr_2$-based compounds, research on the scintillation properties of $PbCl_2$-based compounds with the two-dimensional system has recently began. (EDBE)$PbCl_4$ (EDBE:2,2′-(ethylenedioxy)bis(ethylammonium)) is reported to exhibit high exciton binding energy and a moderate scintillation light yield of 9000 photons/MeV [53]. There are few reports on the scintillation properties of $PbCl_2$-based compounds, and they can become a promising scintillator due to high scintillation intensity and large Storks shift.

Furthermore, irradiation dose dependence of scintillation intensity of $(n\text{-}C_6H_{13}NH_3)_2PbI_4$ is evaluated under a 2.0 MeV proton beam, as shown in Figure 6.26 [54]. Scintillation light yield attenuates due to radiation damage as the irradiation dose increases, and it decreases by half after 8×10^5 Gy irradiation. The radiation hardness of the compound is acceptable for practical use such as a radiation detector for high energy colliding beam facilities, which is required to endure 10^2–10^4 Gy/year [55].

6.5.5.2 Scintillation Decay Time

Scintillation decay curves of $(n\text{-}C_3H_7NH_3)_2PbBr_4$ and $(n\text{-}C_6H_{13}NH_3)_2PbI_4$ with the two-dimensional system are measured under a pulsed electron beam from a linear accelerator (LINAC) [14]. $(n\text{-}C_3H_7NH_3)_2PbBr_4$ have a scintillation lifetime of 2.8 ns (38%), 18 ns (39%), and 130 ns (23%), as shown in Figure 6.27. The first component is due to radiative recombination of the free excitons. The second and the third components are likely due to excitons trapped at lattice defects. Therefore, $(n\text{-}C_3H_7NH_3)_2PbBr_4$ with the two-dimensional system is found to show scintillation due to exciton emission with very short decay times. The scintillation decay time profile is also recorded for $(n\text{-}C_6H_{13}NH_3)_2PbI_4$ with the two-dimensional system. The obtained decay time constants are 390 ps (30%), 3.8 ns (30%), and 16 ns (40%). The first component comes from free excitons, and the second and third components are possibly due to some lattice imperfections. It can be seen that the organic–inorganic perovskite type compounds with the two-dimensional system exhibit an excellent fast response.

Scintillation properties of organic–inorganic layered perovskite-type compounds with the two-dimensional system are shown in Table 6.5. The compounds have a scintillation decay time in an order of several nanosecond, which is usually faster than conventional inorganic scintillators with a luminescence center such as Gd_2SiO_5:Ce, $YAlO_3$:Ce and NaI:Tl. The fast decay time originating from the exciton emissions is due to the quantum size effect, since electrons and holes generated by ionizing radiation are confined to the inorganic layer. Generally, it is difficult for scintillators with a luminescence center to achieve a scintillation decay time of several nanoseconds, since photoluminescence and scintillation decay time are mainly determined by the radiative transition probability of a luminescence center. Exceptionally, BaF_2 is one of the inorganic material scintillators with an extremely fast scintillation lifetime. The decay time constant is about 800 ps, but its emission wavelength's range is 190–220 nm, which is not a favorable wavelength for a photomultiplier tube. On the

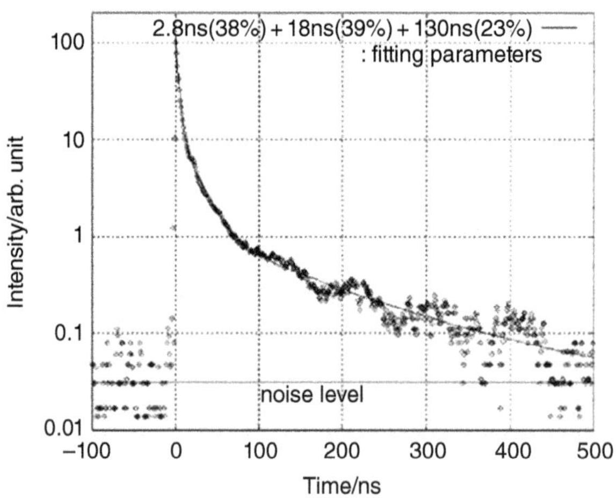

Figure 6.27 Scintillation decay time profile of $(n\text{-}C_3H_7NH_3)_2PbBr_4$ [14]. Source: Copyright (2004) The Japan Society of Applied Physics.

other hand, the emission wavelength of organic–inorganic perovskite compounds is usually across 400–550 nm, which is suitable for the detection of the photomultiplier tube. In addition, other fast response scintillators are plastic scintillators, which exhibit scintillation with moderate light yield (~10 000 photons/MeV) and fast decay time (several ns). The plastic scintillators show an excellent fast response, but have a low density of approximately 1.0 g/cm^3. The density of the compounds with the two-dimensional system (~2.0 g/cm^3) is higher than that of plastic scintillators, and the $(C_6H_5C_2H_4NH_3)_2PbBr_4$ (14 000 photons/MeV) shows higher scintillation light yield than conventional plastic scintillators.

The scintillation decay time of organic–inorganic layered perovskite-compounds with the two-dimensional system has been described so far. However, there are a few reports on the scintillation decay time profile of organic–inorganic layered perovskite-type compounds with the three-dimensional system. It is reported that $(CH_3NH_3)PbBr_3$ shows a scintillation lifetime of 2 and 50 ns at 8 K under X-ray irradiation [52]. In the future, it is expected that scintillation decay time of various organic–inorganic perovskite-type compounds with the three-dimensional system will be evaluated.

6.5.5.3 Afterglow

Another advantage of organic–inorganic perovskite-type compounds is low afterglow levels. The dynamics of scintillation process are generally complicated by slower components due to charge carrier trapping and detrapping processes, which result in an afterglow with a lifetime of typically a few ms under ionizing radiation. Low afterglow level is essential for applications such as X-ray computed tomography, because high afterglow reduces the image quality.

The afterglow profile of $(C_6H_5C_2H_4NH_3)_2PbBr_4$ is shown in Figure 6.28. In this case, afterglow levels are evaluated using the following formula [56]:

$$I_{AG}[\%] = 100 \times (I_2 - I_0)/(I_1 - I_0) \tag{6.32}$$

I_1 is the averaged intensity during X-ray irradiation, I_2 is the averaged intensity approximately at 20 ms after the X-ray irradiation, and I_0 is the averaged intensity before X-ray irradiation.

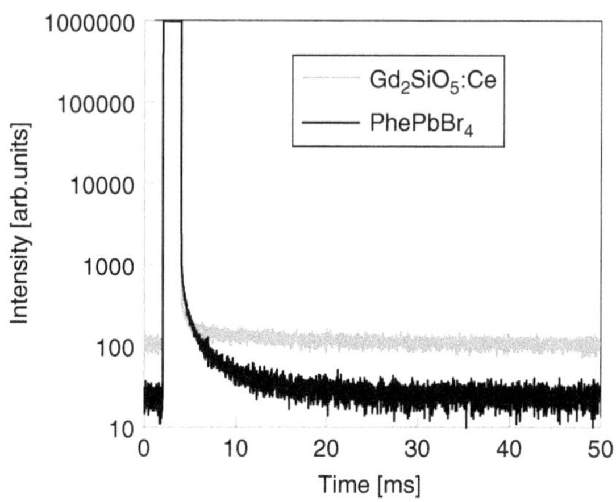

Figure 6.28 *Afterglow profiles of* $PhePbBr_4$ *and* Gd_2SiO_5*:Ce. ※*$PhePbBr_4$ *(*$(C_6H_5C_2H_4NH_3)_2PbBr_4$*).*

The afterglow level for $(C_6H_5C_2H_4NH_3)_2PbBr_4$ is found to be 5 ppm, which is comparable to conventional inorganic scintillators such as $Bi_4Ge_3O_{12}$, Gd_2SiO_5:Ce [54]. One of the reasons for low afterglow levels is to suppress the formation of lattice defects, which affect afterglow levels due to their self-organized manner. In addition, such low afterglow levels are reported in the compounds with the three-dimensional system such as $(CH_3NH_3)PbI_3$ and $(CH_3NH_3)PbBr_3$, and the obtained afterglow levels are lower than that of lanthanide aluminum perovskite such as $LuAlO_3$:Ce and $YAlO_3$:Ce [53]. Organic–inorganic layered perovskite-type compounds should become fast response scintillators with low afterglow levels.

References

1. Kolar, Z.I. and den Hollander, W. (2004). 2003: a centennial of spinthariscope and scintillation counting. *Appl. Radiat. Isot.* 61: 261–266.
2. Klintenberg, M.K., Weber, M.J., and Derenzo, D.E. (2003). Luminescence and scintillation of PbI_2 and HgI_2. *J. Lumin.* 102–103: 287–290.
3. Yanagida, T., Fujimoto, Y., Yamanoi, K. et al. (2012). Optical and scintillation properties of bulk ZnO crystal. *Phys. Status. Solidi C.* 9: 2284–2287.
4. Yanagida, T. and Fujimoto, Y. (2014). Evaluation of pure zinc sulfide crystal scintillator. *Jpn. J. Appl. Phys.* 53: 032601.
5. Yanagida, T., Okada, G., Kato, T. et al. (2016). Fast and high light yield scintillation in the Ga_2O_3 semiconductor material. *Appl. Phys Express* 9: 042601.
6. Shotanus, P., Dorenbos, P., and Ryzhikov, V.D. (1992). Detection of CdS(Te) and ZnSe(Te) scintillation light with silicon photodiodes. *IEEE Trans. Nucl. Sci.* 39: 546–550.
7. Kim, Y.K., Kim, J.K., Lee, W.G. et al. (2007). Properties of semiconductor scintillator ZnSe:O. *Nucl. Instrum. Phys. Res. A.* 580: 258–261.
8. Derenzo, S.E., Weber, M.J., and Klintenberg, M.K. (2002). Temperature dependence of the fast near-band-edge scintillation from CuI, HgI_2, PbI_2, ZnO:Ga and CdS:Ln. *Nucl. Instrum. Phys. Res. A.* 486: 214–219.
9. Derenzo, S.E., Bourret-Courchesne, E., Weber, M.J. et al. (2005). Scintillation studies of CdS(ln): effect of various semiconductor doping strategies. *Nucl. Instrum. Phys. Res. A.* 537: 261–265.
10. Derenzo, S.E., Weber, M.J., Bourret-Courchesne, E. et al. (2003). The quest for the ideal inorganic scintillator. *Nucl. Instrum. Phys. Res. A.* 505: 111–117.
11. Kobayashi, M. (2014) Radiation Measurement with Scintillators, V2-Solution.
12. Knoll, G.F. (2000). *Radiation Detection and Measurement.* New York: Wiley.
13. Wu, Y., Darge, A.W., Trofimov, A.A. et al. (2019). Fabrication and characterization of ZnS:Ag-based ultrafiltration membrane scintillator. *Opt. Mater.* 88: 424–428.
14. Shibuya, K., Koshimizu, M., Murakami, H. et al. (2004). Development of ultra-fast semiconducting scintillators using quantum confinement effect. *Jpn. J. Appl. Phys.* 43: L1333–L1336.
15. Kawano, N., Koshimizu, M., Horiai, A. et al. (2016). Effects of organic moieties on the scintillation properties of organic–inorganic layered perovskite-type compounds. *Jpn. J. Appl. Phys.* 55: 110309.
16. Kawano, N., Koshimizu, M., Okada, G. et al. (2017). Scintillating organic–inorganic layered perovskite-type compounds and the gamma-ray detection capabilities. *Sci. Rep.* 7: 14754.
17. Nakauchi, D., Kawano, N., Kawaguhi, N. et al. (2020). Luminescence and scintillation properties of $(C_6H_5C_2H_4NH_3)_2(Pb,Ba)Br_4$ with self-organized bi-dimensional quantum-well structures. *Jpn. J. Appl. Phys* 59: SCCB04.
18. Akatsuka, M., Kawano, N., Kato, T. et al. (2020). Development of scintillating 2D quantum confinement materials – $(C_6H_5C_2H_4NH_3)2Pb_{1-x}Sr_xBr_4$. *Nucl. Instrum. Phys. Res. A.* 95: 161372.
19. Davies, J.H. (1998). *The Physics of Low-Dimensional Semiconductors: An Introduction.* Cambridge: Cambridge University Press.
20. Haug, H. and Koch, S.W. (2008). *Quantum Theory of the Optical and Electronic Properties of Semiconductors.* World Scientific.
21. Kittel, C. (1976). *Introduction to Solid State Physics*, 5e. New York: Wiley.
22. Saleh, B.E.A. and Teich, M.C. (1991). *Fundamentals of Photonics.* New York: Wiley.

23. Pelant, I. and Valenta, J. (1998). *Luminscence Spectroscopy of Semiconductors*. Oxford: Oxford University Press.
24. Bebb, H.B. and Williams, E.W. (1976). *Photoluminescence I. Theory. In Semiconductors and Semimetals*, vol. 8, 181. New York: Academic Press.
25. Toyozawa, Y. (2003). *Optical Processes in Solids*. Cambridge: Cambridge University Press.
26. Derenzo, S.E., Bourret-Courshesne, E., Bizarri, G. et al. (2016). Bright and ultra-fast scintillation from a semiconductor? *Nucl. Instrum. Phys. Res. A.* 805: 36–40.
27. Rodnyi, P.A., Dorenbos, P., and Eijk, C.W.E. (1995). Energy loss in inorganic scintillators. *Phys. Stat. Sol. B.* 187: 15–29.
28. Lempicki, A., Wojtowicz, A.J., and Berman, E. (1993). Fundamental limits of scintillator performance. *Nucl. Instrum. Phys. Res. A.* 333: 304–311.
29. Nagarkar, V.V., Gaysinskiy, V., Ovechkina, O.E. et al. (2010). Bright semiconductor scintillator for high resolution X-ray imaging. *IEEE Trans. Nucl. Sci.* 57: 923–930.
30. Zhang, K., Ouyang, X., Song, Z. et al. (2005). An ideal scintilaltor – ZnO:Sc for sub-nanosecond pulsed radiation detection. *Nucl. Instrum. Phys. Res. A.* 756: 14–18.
31. Papavasilliou, G.C. (1997). Three- and low-dimensional inorganic semiconductors. *Prog. Solids St. Chem.* 25: 125–270.
32. He, X.F. (1990). Excitons in anistropic solids: the model of fractional-dimensional space. *Phys. Rev. B.* 43: 2063–2069.
33. He, X.F. (1990). Anisotropy and isotropy: a model of fraction-dimensional space. *Solid State Commun.* 75: 111–114.
34. He, X.F. (1990). Fractional dimensionality and fractional derivative spectra of Interband optical transitions. *Phys. Rev. B.* 42: 11751–11756.
35. Stillinger, F.H. (1977). Axiomatic basis for spaces with noninteger dimension. *J. Math. Phys.* 18: 1224–1234.
36. Ishibashi, T., Suzuki, Y., and Okamoto, H. (1981). Photoluminescence of an AlAs/GaAs superlattice grown by MBE in the 0.7–0.8 μm wavelength region. *Jpn. J. Appl. Phys.* 20: L623–L626.
37. Dolzhenko, Y.I., Inabe, T., and Maruyama, Y. (1986). *In situ* X-ray observation on the intercalation of weak interaction molecules into perovskite-type layered crystals $(C_9H_{19}NH_3)_2PbI_4$ and $(C_{10}H_{21}NH_3)_2CdCl_4$. *Bull. Chem. Soc. Jpn.* 59: 563.
38. Ishihara, T., Takahashi, J., and Goto, T. (1989). Exciton state in two-dimensional perovskite semiconductor $(C_{10}H_{21}NH_3)_2PbI_4$. *Solid State Commun.* 69: 933.
39. Ogawa, T. and Kanemitsu, Y. (1995). *Optical Properties of Low-Dimensional Materials*. World Scientific.
40. Tan, Z.K., Moghaddam, R.S., Lai, M.L. et al. (2014). Bright light-emitting diodes based on organometal halide perovskite. *Nat. Nanotechnol.* 9: 687–692.
41. Chondroudis, K. and Mitzi, D.B. (1999). Electroluminescence from an organic-inorganic perovskite incorporating a quaterthiophene dye within lead halide perovskite layers. *Chem. Mater.* 11: 3028–3030.
42. Dou, L., Yang, Y.M., You, J. et al. (2014). Solution-processes hybrid perovskite photodetectors with high detectivity. *Nat. Commun.* 5: 5404-1–5404-6.
43. Jin, X.Y., Cortecchia, D., Yin, J. et al. (2015). Lead iodine perovskite light-emitting field-effect transistor. *Nat. Photon.* 6: 7383-1–7383-9.
44. Saparov, B. and Mitzi, D.B. (2016). Organic-inorganic perovskites: structural versatility for functional materials design. *Chem. Rev.* 116: 4558–4596.
45. Ishihara, T., Takahashi, J., and Goto, T. (1990). Optical properties due to electronic transitions in two-dimensional semiconductors $(C_nH_{2n+1}NH_3)_2PbI_4$. *Phys. Rev. B.* 42: 11099–11107.
46. Ito, R., Xu, C., and Kondo, T. (1991). *Solid State Materials* (eds. S. Radhakrishna and A. Daud). New Delhi: Narosa Publishing House.
47. Hirasawa, M. University (1994) PhD Thesis, Tohoku.
48. Saidaminov, M., Abdelhady, A.L., Murali, B. et al. (2015). High-quality bulk hybrid perovskite single crystals within minutes by inverse temperature crystallization. *Nat. Commun.* 6: 7586.
49. Kawano, N., Koshimizu, M., Sun, Y. et al. (2014). Effects of organic moieties on luminescence properties of organic–inorganic layered perovskite-type compounds. *J. Phys. Chem. C* 118: 9101–9106.
50. Hirasawa, M., Ishihara, T., and Goto, T. (1994). Exciton features in 0-, 2-, and 3-dimensional networks of $[PbI_6]^{4-}$ octahedra. *J. Phys. Soc. Jpn.* 63: 3870–3879.
51. Umebayashi, T., Asai, K., Kondo, T. et al. (2003). Electronic structures of lead iodide based low-dimensional crystals. *Phys. Rev. B.* 67: 155405.
52. Mykhaylk, V.B., Kraus, H., and Saliba, M. (2019). Bright and fast scintillation of Organolead perovskite $MAPbBr_3$ at low temperature. *Mater. Horiz.* 6: 1740–1747.

53. Birowosuto, M.D., Cortecchia, D., Drozdowski, W. et al. (2016). X-ray scintillation in Lead halide perovskite crystals. *Sci. Rep.* 6: 37254.
54. Shibuya, K., Koshimizu, M., Takeoka, Y. et al. (2002). Scintillation properties of $(C_6H_{13}NH_3)_2PbI_4$: exciton luminescence of an organic/inorganic multiple quantum well structure compound induced by 2.0 MeV protons. *Nucl. Instrum. Methods Phys. Res. Sect. B.* 194: 207–212.
55. Caffrey, A.J., Heath, R.L., Ritter, P.D. et al. (1986). Radiation damage studies on BAF_2 and BGO scintillator materials. *IEEE Trans. Nucl. Sci.* 33: 230–234.
56. Yanagida, T., Fujimoto, Y., Ito, T. et al. (2014). Development of X-ray-induced afterglow characterization system. *Appl. Phys. Lett.* 7: 062401.

7

Thermally Stimulated Luminescent (TSL) Materials

Kiyomitsu Shinsho
Department of Radiological Sciences, Tokyo Metropolitan University, Tokyo, Japan

7.1 Introduction

In 1664, Sir Robert Boyle, who is well known for Boyle's Law, noticed that when natural diamonds are heated by contact with human skin, they momentarily emit a tiny amount of light. Boyle recorded this as "glimmering light" [1]. In 1895, W.C. Röntgen discovered X-rays, which turned out to be linked to the phenomenon that Boyle had observed over 200 years earlier [2]. Although the true nature of this light was that the natural diamonds had accumulated natural radiation energy and emitted it as light in response the heat, Boyle died without knowing the reason. Because the existence of radiation was not known at that time and there were no devices that could detect this "glimmering light," it was difficult to gain a deeper understanding of the true nature of this light.

However, when X-rays were discovered, this situation changed. The phenomenon of fluorescence was used as a method for detecting radiation such as cathode rays and X-rays, and consequently research into fluorescence rapidly advanced along with the study of radiation. In the same year as the discovery of X-rays, the "glimmering light" discovered by Boyle was named "thermoluminescence" (TL) by Wiedemann and Schmidt [3]. Furthermore, in 1941, the RCA Corporation of America invented the photomultiplier tube (PMT) [4], which can detect even weak light, and an abundance of research into luminescence began to be conducted, not only for TL but also scintillation and other phenomena. Starting from the 1960s, research into TL combined with storage-based luminescence phenomena such as photostimulated luminescence (PSL) expanded, and the term thermally stimulated luminescence (TSL) began to also be used for TL. In this chapter, "TL" is unified as "TSL."

Phosphors for Radiation Detectors, First Edition. Edited by Takayuki Yanagida and Masanori Koshimizu.

Other TSL materials, besides natural diamonds, were discovered in 1913 through research into ionization and spark potentials by Harlow [5]. Lime and aluminum phosphate were found to exhibit yellow and blue TSL, respectively. In 1927, Laird demonstrated that mixed crystals of $CaSO_4$ with addition of 2% $MnSO_4$ exhibited TSL, and that the square of the sustained duration is proportional to the total energy of absorbed soft X-rays [6]. This result formed the basis for dosimetry using TSL. The development of TSL materials truly started from the 1950s. The trigger for this was the investigation of the TSL properties of over 3000 rocks and minerals by Daniels and his group, who observed TSL in 75% of them [7]. This showed that TSL is a general property often observed in rocks and minerals. Furthermore, Daniels and his group posited that they could investigate the correlation between radiation dose and amount of TSL starting from low doses because weak TSL can be detected by a PMT. They then successfully created practical dosimetry based on TSL for the first time by using LiF, which is the main component of griceite. This was the beginning of the thermoluminescence dosimeter (TLD). In this chapter, "TLD" is unified as "TSLD." This development came 289 years after Boyle discovered TSL.

From the 1950s to the 1980s, many TSLDs were developed that were suitable for dosimetry. In 1968, Yamashita et al. reported that $CaSO_4$ doped with Tm and $CaSO_4$ doped with Dy were around 30 times more sensitive than LiF [8]. $CaSO_4$ doped with Tm is widely used for environmental monitoring personal exposure dosimetry, and other applications; one example is the Matsushita UD-110S. Many other TSL materials have been developed, including CaF [9, 10], Al_2O_3 [11, 12], BeO [13], and $Li_2B_4O_7$ [14], and research into trapping levels and emission levels also advanced at the same time.

A theoretical interpretation of TSL was first proposed by Randall et al. [15] in 1945. This was eight years before Daniels et al. [7] successfully performed dosimetry. This theory was based on the simplest assumption that light is emitted by recombination of all the electrons and holes trapped in electron and hole trapping centers, respectively, when the TSL element is heated at a fixed heating rate after exposure to radiation. The relationship between the temperature of the TSL element and TSL intensity is called the glow curve. The glow curve is expressed by the following equation according to the Randall–Wilkins first-order model:

$$I(T) = sn_0 \exp\left(-\frac{E}{kT}\right) \exp\left[-\frac{s}{\beta}\int_{T_0}^{T} \exp\left(-\frac{E}{kT'}\right) dT'\right] \tag{7.1}$$

Here, $I(T)$ is the TSL intensity at temperature T (K), n_0 is the total number of trapped electrons, s is the frequency factor (s^{-1}), E is the trap depth (eV), k is the Boltzmann constant, and β is the heating rate ($K\,s^{-1}$), where the formula is identical to Equation (1.35) (Figure 7.1).

$$\left(E = 1.3\,\text{eV}, s = 10^{13}\,\text{s}^{-1}, \beta = 5\,\text{K}\,\text{s}^{-1}\ \text{and}\ n_0 = 10^3\right) \tag{7.2}$$

Subsequently, in 1948, Garlick et al. reported a second-order model that takes re-trapping into account [16]. In 1964, May et al. [17] used experimental methods to study intermediate models that cannot be explained by first- or second-order kinetics models.

Furthermore, research into two-dimensional (2-D) TSL, which has made great advances in recent years, was performed immediately after Daniels et al. [18] successfully performed dosimetry by TSLD in 1953. In 1967, Nakajima et al. [19] developed a thin-film luminescent phosphor of diameter 10 mm and thickness 80 μm using $CaSO_4$:Mn made by Matsushita

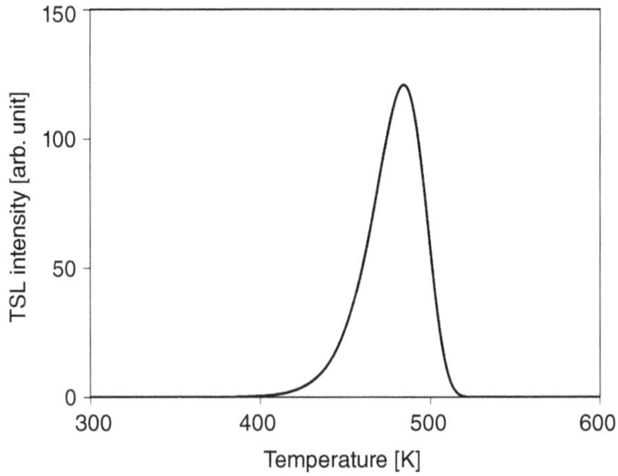

Figure 7.1 *Glow curve from the Randall–Wilkins first-order model equation.*

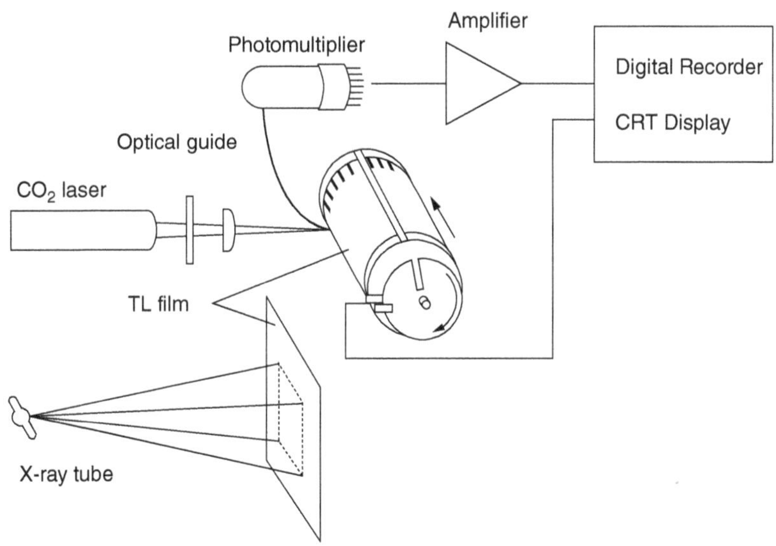

Figure 7.2 *Scanning heating system with CO_2 laser and PMT [21].*

Electric Industrial Co. Ltd. (now Panasonic Corporation). However, this thin-film luminescent phosphor was for the purpose of measuring skin dose and was not used for radiation imaging. Cutting-edge research into radiation imaging by TSLD was the development of the image storage panel (ISP) using TSL by Broadhead and Newman [20] in 1972. Their ISP completely fused a luminescent layer of $CdSO_4$-Mn-Sm-TI-NaF with a heating layer, and images could be obtained by using a camera or film. In 1983, Yasuno et al. [21] enlarged the aforementioned $CaSO_4$:Mn thin film fluorescent phosphor as a TSL film, up to 100 mm × 100 mm, and successfully performed radiation imaging by using a readout method (Figure 7.2) that employs a scanning CO_2 laser beam and PMT. However, this readout method was not developed further because it was extremely expensive and also difficult to perform.

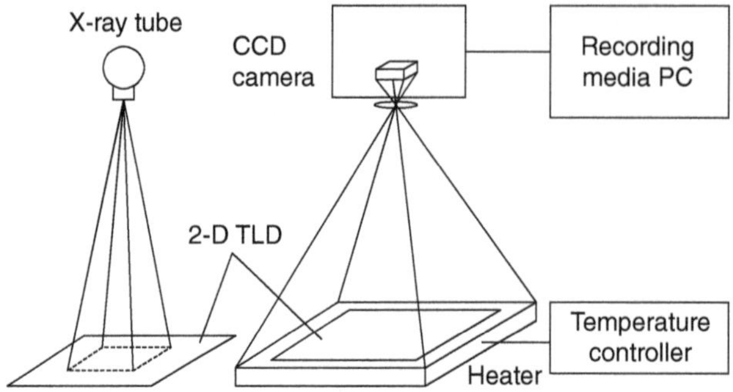

Figure 7.3 2-D TSL readout system with CCD camera and heater [22].

Over 10 years after laser heating systems were first developed, charge-coupled device (CCD) image sensors (hereinafter, CCD cameras) could be purchased relatively cheaply, and measurement systems using CCD cameras began to be developed. In 1997, Duller et al. [22] successfully obtained TSL distributions at a maximum spatial resolution of 17 μm/pixel by using a CCD camera. Figure 7.3 shows an overview of the CCD system. This CCD system consists of a heater and CCD camera, and has become the main 2-D TSL measurement method. Recently, several research groups have reported 2-D TSL readout systems using a CCD camera, and the performance and convenience are improving through advances such as larger areas.

7.2 TSL Phenomenon

7.2.1 Basic Principles of TSL

TSL can be divided into the process of storing radiation energy and the process of releasing stored radiation energy, which can be represented as shown by the energy bands in Figure 7.4.

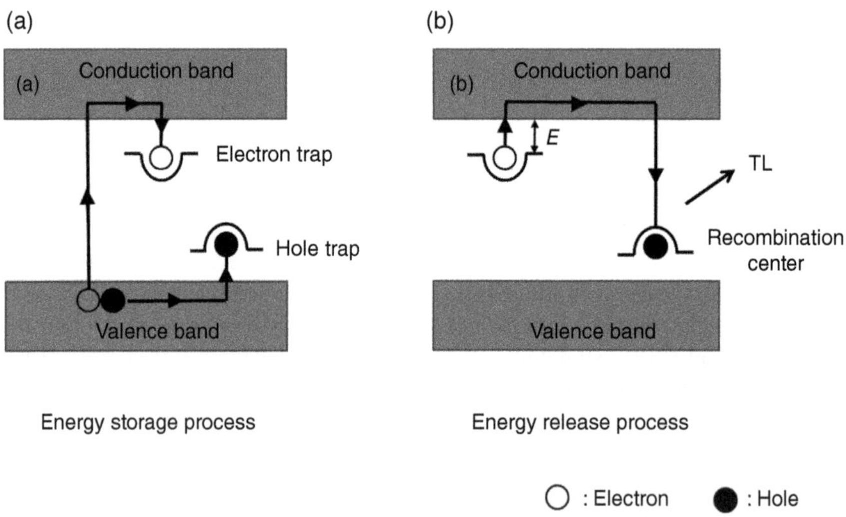

Figure 7.4 The processes of storing radiation energy and releasing stored radiation energy in TSL.

Energy storage process: Exposing a TSL element to radiation such as X-rays generates free electrons (e^-) and holes (h^+) within its crystal structure. These electrons and holes are captured by the trapping centers associated with impurities and lattice defects (electron traps and hole traps) and enter a meta-stable state. This state is maintained at room temperature. Furthermore, since the number of electrons and holes generated increases with the amount of absorbed radiation energy, the number of trapped electrons and holes also naturally increases.

Energy release process: Upon thermal excitation of a TSL element in a state with stored radiation energy (i.e., one with electron trapping centers and hole trapping centers), the trapped electrons become free electrons again and move to the conduction band. These free electrons recombine with the holes, and when this occurs, the excess energy is emitted as light (TSL).

TSL occurs only when the thermal energy applied is greater than or equal to the activation energy (E) needed in order to release the trapped electrons. Furthermore, TSL becomes stronger in proportion to the number of electrons and holes that recombine. Thus, TSL is also proportional to the absorbed radiation dose.

7.2.2 Theory and Measurement of Glow Curves

7.2.2.1 Theory of Glow Curves

7.2.2.1.1 First-Order Kinetics

TSL is detected by measuring TSL element temperature and TSL intensity when the TSL element is heated, and the relationship between them is called a glow curve. Figure 7.5 shows the glow curve for $CaSO_4$:Tm. A theoretical interpretation of the glow curve based on the basic principles of TSL was first proposed by Randall et al. [15] in 1945. This theory suggests that when the TSL element is heated at a constant heating rate, light is emitted by the recombination of all of the electrons and holes that were captured by one species of the respective electron and hole trapping centers. The glow curves are given theoretically by this model in terms of activation energy E (eV), frequency factor s (s^{-1}), and heating rate

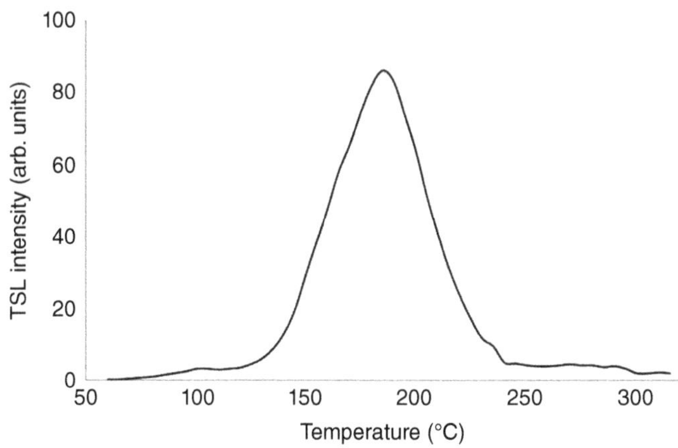

Figure 7.5 *Glow curve of $CaSO_4$:Tm (heating rate: 0.2 °C·s^{-1}).*

β (K·s^{-1}). Because E is a parameter that arises from the thermal stability of the electrons trapped in the electron trapping centers, the glow peak temperature increases (decreases) with increasing (decreasing) E. The parameter s is related to the lattice vibration frequency of the TSL element, and arises from the probability for the number of trapped electrons and holes being emitted per unit time. As the value of s increases, the probability of a trapped electron being released at a certain temperature increases, which reduces the glow peak temperature and the full width at half-maximum (FWHM) of the glow curve. The peak temperature, FWHM, and other parameters of the glow curve change depending on the parameters E, s, and β that define the glow curve, and therefore can be deduced from a glow curve obtained by actual measurement. The following equation shows the Randall–Wilkins first-order model [15]:

$$I(T) = sn_0 \exp\left(-\frac{E}{kT}\right) exp\left[-\frac{s}{\beta}\int_{T_0}^{T} exp\left(-\frac{E}{kT}\right) dT\right] \tag{7.3}$$

Here, I (T) is the TSL intensity at temperature T (K), n_0 is the total number of trapped electrons, s is the frequency factor (s^{-1}), E is the activation energy (eV), k is the Boltzmann constant, and β is the heating rate (K·s^{-1}), where the equation is identical to Equation (1.35).

7.2.2.1.2 Second-Order Kinetics and General-Order Kinetics

First-order kinetics assumes that light is emitted via the recombination of all electrons and holes trapped in the electron and hole trapping centers, but in reality, re-trapping also occurs via route (2) in Figure 7.6b in addition to charge recombination via route (1).

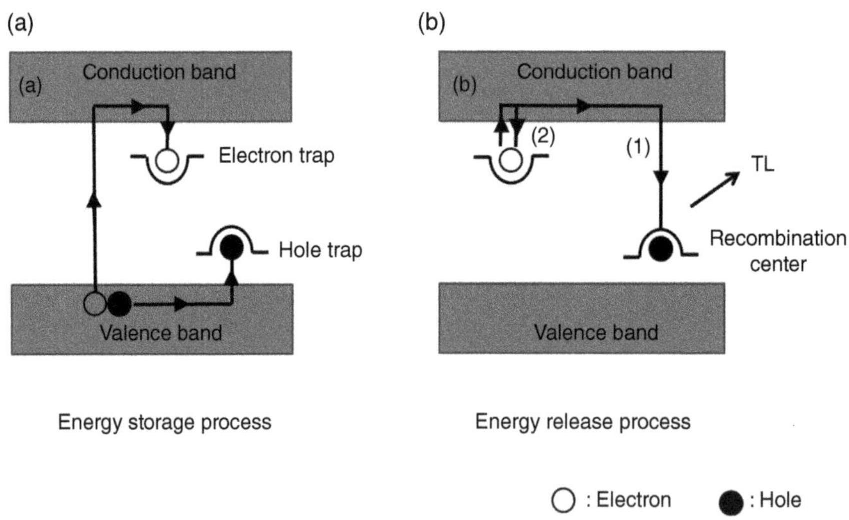

Figure 7.6 The processes of storing radiation energy and releasing stored radiation energy include re-trapping process in TSL.

A model that takes this re-capturing into account is called second-order kinetics and was reported by Garlick et al. in 1948 [16]. This model is given as follows:

$$I(T) = \frac{s' n_0^2 \exp\left(-\frac{E}{kT}\right)}{\left[1 + \frac{s' n_0}{\beta} \int_{T_0}^{T} \exp\left(-\frac{E}{kT'}\right) dT'\right]^2} \tag{7.4}$$

with

$$s' = \frac{s}{N} \tag{7.5}$$

where N is the trap density (cm^{-3}).

The width at half-maximum of the glow curve is broader in second-order kinetics than in first-order kinetics because re-trapping is taken into account. When actual glow curves were analyzed, however, they could not be explained by either first-order or second-order kinetics models, so an intermediate model was needed in which the degree of re-trapping can be set arbitrarily. Therefore, general-order kinetics was proposed by May et al. in 1964 as an empirical method [17]. Assuming there is one trap energy level, general-order kinetics can be described by

$$I(T) = s n_0 \exp\left(-\frac{E}{\mathrm{k}T}\right)\left[1 + (b-1)\frac{s}{\beta}\int_{T_0}^{T} \exp\left(-\frac{E}{\mathrm{k}T'}\right) \mathrm{d}T'\right]^{-\frac{1}{b-1}} \tag{7.6}$$

where b is the order number ($1 < b \leq 2$). This is equivalent to second-order kinetics when $b = 2$, and reduces to first-order kinetics as b approaches 1, where the formula is identical to Equation (1.36).

Figure 7.7 shows the general-order kinetics for $b = 1.5$. The parameters in the figure were set to $E = 1.3$, $s = 10^{12}$, and $n_0 = 1 \times 10^{10}$. Given that general-order kinetics is an intermediate

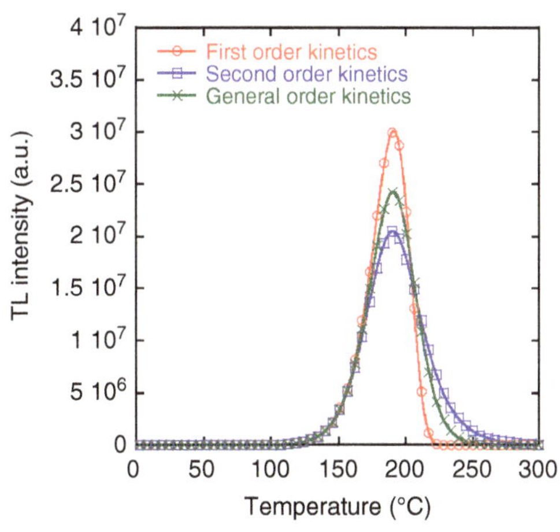

Figure 7.7 *Glow curve for each model.*

model between first- and second-order kinetics, the width at half-maximum of the glow curve increases in the order of first-order kinetics < general-order kinetics ≤ second-order kinetics. Furthermore, even if the re-trapping probability changes, the glow peak temperature does not change since it is determined by *E* and *s*.

7.2.2.1.3 Relationship Between Glow Curve and Parameters

Glow curves as each of the parameters in general-order kinetics are changed, as shown below. Figure 7.8 shows glow curves when the number of traps n_0 is varied, and it can be seen that the overall intensity is increased by increasing only n_0. Figure 7.9 shows glow

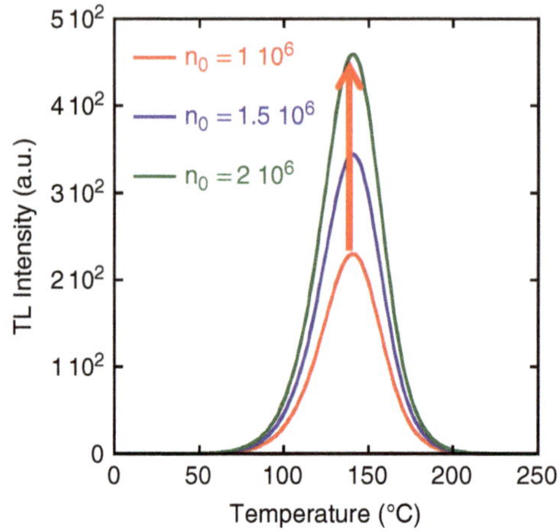

Figure 7.8 Change in glow curve according to n_0.

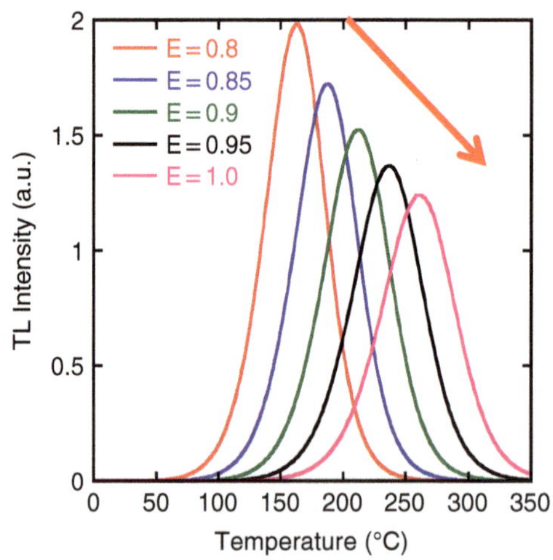

Figure 7.9 Change in glow curve according to *E*.

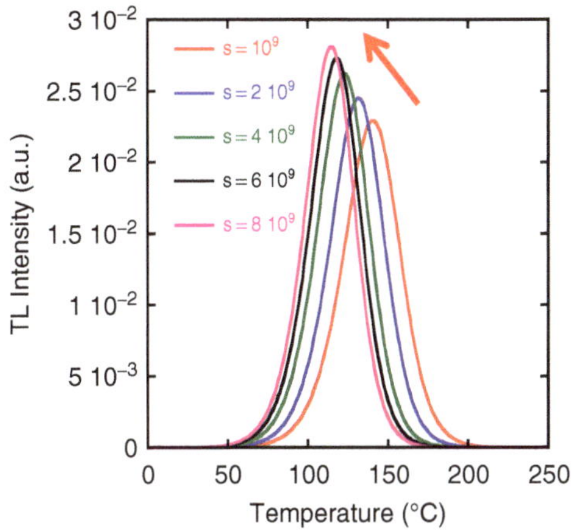

Figure 7.10 *Change in glow curve according to s.*

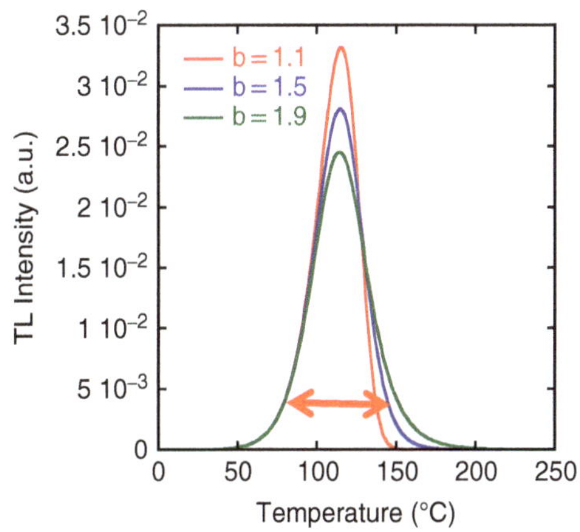

Figure 7.11 *Change in glow curve according to b.*

curves when the trap depth E is varied. Here the peak temperature shifts toward higher temperature and the width at half-maximum increases as E is increased. Although the maximum intensity decreases with increasing E, the area under the curve (integrated amount of TSL) does not change. Figure 7.10 shows glow curves for various values of the frequency factor s, which represents lattice vibrations. As this value increases, the peak temperature shifts toward lower temperature and the width at half-maximum decreases. Figure 7.11 shows glow curves when varying, b, which represents the order number, that is, the probability of re-trapping. As b increases, the re-trapping probability increases and the width at half maximum broadens, but the peak temperature does not change.

The trends in the behavior of the glow curves for each of the parameters above are the same for first-order kinetics and second-order kinetics.

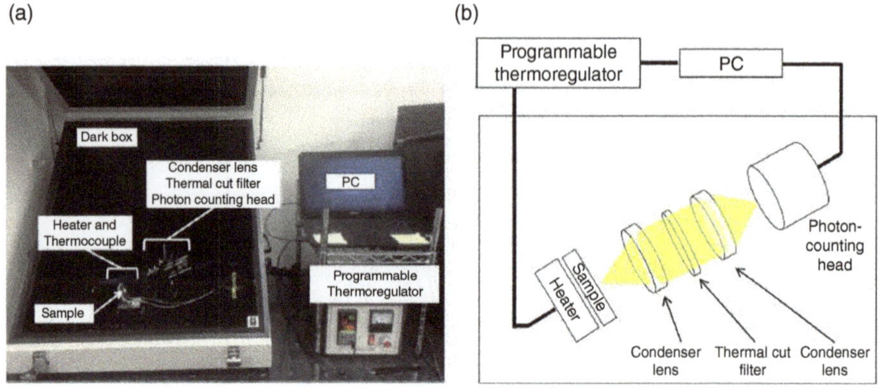

Figure 7.12 *Glow curve measurement system and schematic diagram [23].*

7.2.2.2 Measurement of Glow Curves

Glow curves are acquired using a dedicated measurement device consisting primarily of a dark box, a heater for heating the TSL element, and a PMT for detecting the TSL and converting it into an electrical signal. When the glow curve has been acquired, the accumulated energy is released. Thus, thermoluminescence dosimeter (TSLD) can be reused after heat treatment. The read time for a TSLD is short, requiring around 10 seconds of heating from room temperature to 300 °C. Although a hot air system is often used for the heater, there are also cases where other heaters are used, such as tungsten heaters, ceramic heaters, and infrared heaters. A short read time is useful for dose measurement, but produces a discrepancy between the heater temperature and TSL element temperature, so accurate glow curve measurements cannot be performed for investigating TSL properties in detail. For that purpose, measurement should be performed with a very slow heating rate. Furthermore, the glow curve shape differs according to various factors, such as the crystallinity of the TSL material, its density, differences in the synthesis method, and the species used as additive materials, and consideration also needs to be given to the various differences in the properties of the radiation. Figure 7.12 shows an example of a glow curve measurement system [23].

In actual dose measurement with a TSLD, it is necessary to construct a dose conversion table in advance showing the relationships of glow curve peak intensity or the amount of TSL in a certain temperature range with known radiation doses from ^{137}Cs and ^{60}Co γ sources, high-energy X-rays, and other types of radiation. The dose is then calculated from this dose conversion table and the peak intensity of the glow curve and the area under the glow curve (integrated amount of TSL) is obtained by actual measurement.

7.3 TSL Materials: Fluoride, Oxides, Sulfates, and Borate

7.3.1 Fluorides

7.3.1.1 Lithium Fluoride (LiF)

LiF is the main TSL material used for human exposure dosimetry. Although its glow curve is complicated and its sensitivity is low, LiF has been used for a long time and continues to be used due to its suitability for a wide variety of dose measurements, including human exposure dosimetry. This is because the effective atomic number of LiF is almost equivalent to that of human tissue (water) and LiF also has excellent TSL dose responsivity. TLD-100, which is widely used commercially, consists of LiF co-doped with Mg and Ti.

Figure 7.13 shows the crystal structure of LiF [24]. It has a face-centered cubic lattice in which Li^+ ions and F^- ions are arranged in alternate rows. The Mg^{2+} ions substituted for Li^+ ions generate vacancies at Li^+ ion positions due to charge compensation. Furthermore, Farbe centers (F centers) are formed from F^- vacancies that have captured an electron.

Figure 7.14 shows the glow curve of LiF:Mg,Ti (TLD-100) after irradiated with 100 Gy of ^{60}Co γ-rays [25]. Although the glow component with the lowest glow peak temperature has been shown to disappear due to fading, a total of 10 glow components have been reported. Furthermore, it is known that the shape of the glow curve changes readily, with the proportions of these components (peak intensity ratios) strongly depending on the irradiation conditions, including dose and linear energy transfer (LET).

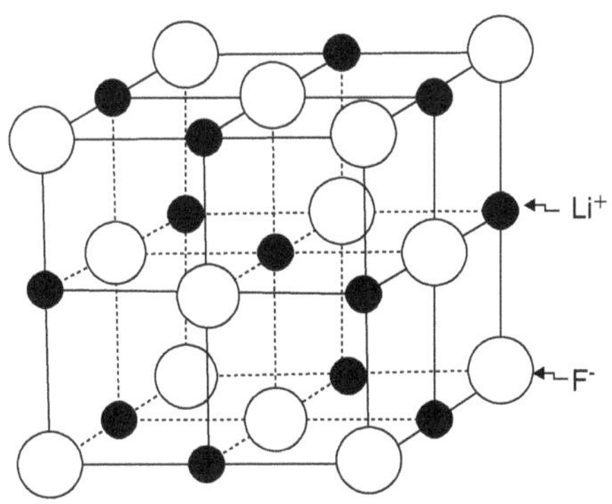

Figure 7.13 *Crystal structure of LiF [24].*

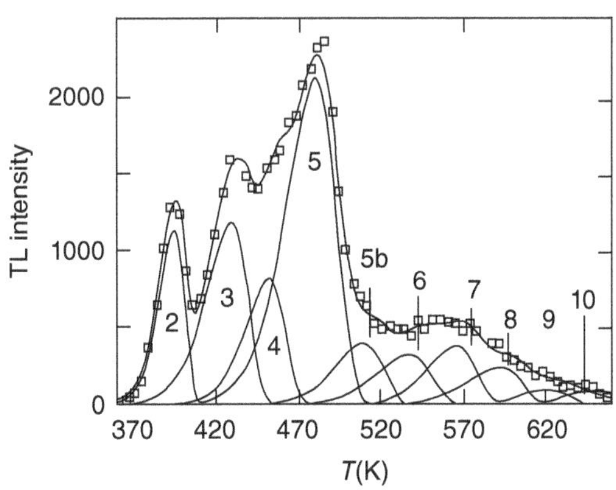

Figure 7.14 *Typical glow curve of LiF:Mg,Ti (TLD-100) irradiated with 100 Gy of ^{60}Co γ-rays [25].*

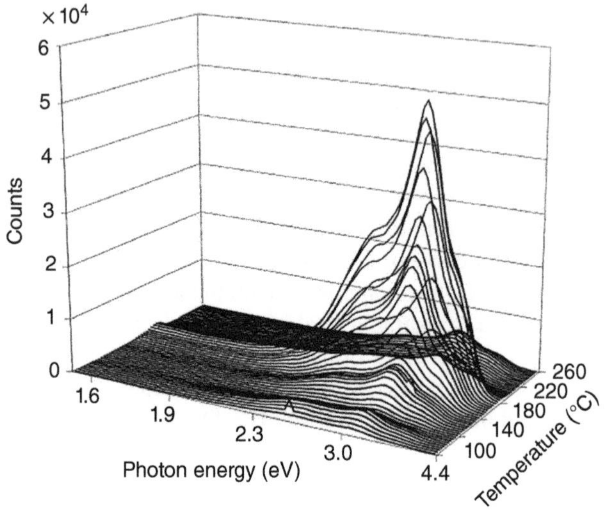

Figure 7.15 TSL emission spectrum of LiF:Mg,Ti (TLD-100) during heating after irradiation with 200 Gy of $^{90}Sr/^{90}Y$ β-rays [26].

Figure 7.15 shows TSL emission spectrum of LiF:Mg,Ti (TLD-100) following irradiation with 200 Gy of $^{90}Sr/^{90}Y$ β-rays [26]. Five of the observed emission bands, at 2.49 eV (498 nm), 2.64 eV (470 nm), 2.84 eV (437 nm), 3.1 eV (400 nm), and 3.59 eV (345 nm), do not depend on the sample temperature or dose, indicating that multiple glow peaks have the same luminescence center.

One TSL element based on LiF other than LiF:Mg,Ti (TLD-100) is LiF:Mg,Cu,P. Mg is an important dopant that has the strongest effect on TSL sensitivity in both LiF:Mg,Ti and LiF:Mg,Cu,P [24]. As described above, the Mg^{2+} ions are substituted into the LiF lattice for Li^{+} and form a dipole with the Li^{+} vacancies generated by charge compensation. This dipole is thought to form a cluster and act as a trapping center. Although there have been various discussions regarding other dopants, the details are not clear. Figure 7.16 shows the glow curves of LiF:Mg,Ti and LiF:Mg,Cu,P [27]. Despite differences in peaks 3 and 4 of LiF:Mg,Cu,P and the high temperature components, both form complicated glow curves and the peak temperatures are also highly similar. LiF:Mg,Cu,P has a TSL sensitivity 30 times higher than that of LiF:Mg,Ti, but the characteristics of LiF:Mg,Cu,P also differ in other ways, for example, in terms of dose and LET dependence. Although LiF:Mg,Cu,P has high sensitivity, it exhibits high variability and its characteristics are also readily changed by the annealing temperature, so care is required when it is used.

7.3.1.2 Calcium Fluoride (CaF)

Not only synthetic materials but also natural fluorite can be used for dosimetry. Calcium fluoride (CaF_2), the main constituent of natural fluorite, has long been used as a TSL material [9, 10]. Because it has higher sensitivity than LiF:Mg,Ti and also has a simple glow curve, CaF_2 is used as a TSL element that is particularly suitable for measuring environmental radiation. The main CaF_2 TSL element is CaF_2:Mn, which was developed by Ginther et al. [9] in 1957. It was later released commercially by Harshaw as TLD-400. In addition, CaF_2:Dy (TLD-200) and CaF_2:Tm (TLD-300) are also widely used.

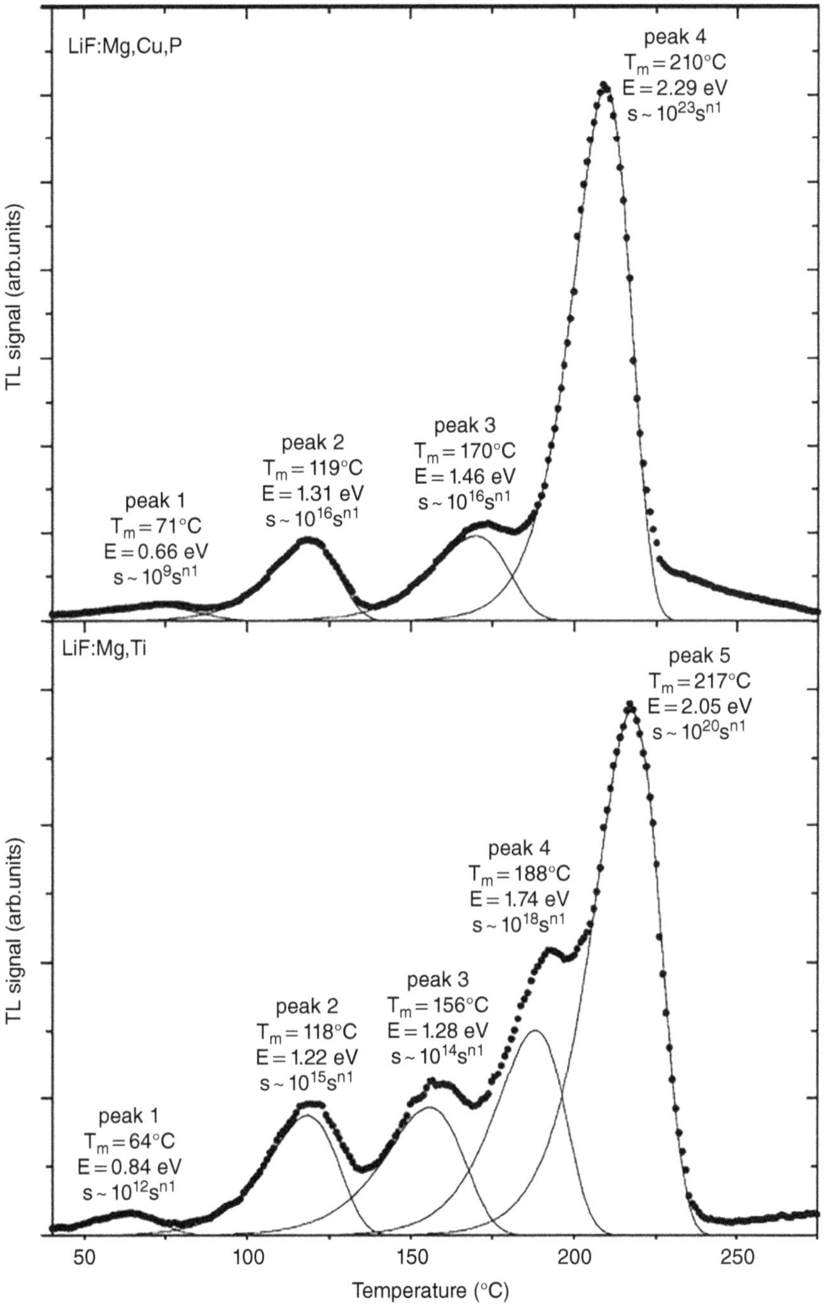

Figure 7.16 *Glow curves of LiF:Mg,Ti and LiF:Mg,Cu,P (heating rate 2 °C s^{-1}, D = 1.5 mGy) [27].*

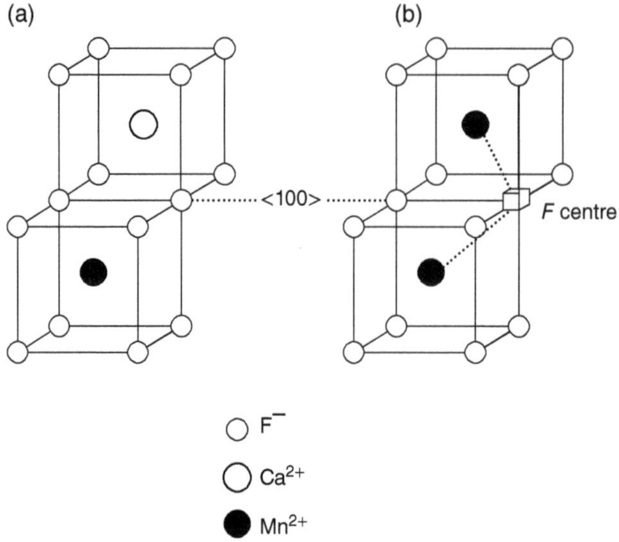

Figure 7.17 *Crystal lattice of CaF_2:Mn (a) before and (b) after irradiation [28].*

Figure 7.17 shows the crystal structure of CaF_2:Mn and the defect structure after irradiation [28]: (a) The Mn^{2+} ions are substituted for Ca^{2+}, there are 8 coordinates, and the symmetry is octahedral (O_h); (b) After irradiation, the F center is thought to be stabilized by the two Mn^{2+} ions. In the cases of CaF_2:Dy and CaF_2:Tm, it is thought that Dy^{3+} and Tm^{3+} are substituted into Ca^{2+} sites, the same as the Mn^{2+} ions. However, defects different from those in CaF_2:Mn are generated by compensation for the different charges. As a result, the TSL characteristics differ greatly among the different dopants.

Figure 7.18 shows the glow curve of CaF_2:Mn (TLD-400) [29]. The white circles show measured values, the dotted lines show the individual glow components analyzed using the computerized glow curve deconvolution (CGCD) method, and the solid line shows the sum of the glow components analyzed by the CGCD method. Although the glow curve has a simple shape, the existence of multiple glow components has been confirmed, not only by the CGCD method, but also by the initial rise method and glow peak shape analysis. However, theoretical interpretation is challenging because the glow component analysis results depend strongly on the annealing temperature, time, and irradiation history. Figure 7.19 shows the TSL emission spectrum [30]. It has a peak at 495 nm, and luminescence is observed based on the ${}^4T_1({}^4G) \rightarrow {}^6A_1({}^6S)$ transition of the dopant Mn^{2+} ions [31].

Figure 7.20 shows the glow curve of CaF_2:Dy (TLD-200) [32]. The white circles show measured values, the solid line shows the analyzed glow components, and the bottom part shows the difference between the measured values and the sum of each of the analyzed glow components. Although the glow curve of CaF_2:Mn (TLD-400) consists of around five components with simple shapes, the glow curve of CaF_2:Dy (TLD-200) consists of six components, and the glow curve shape and glow peak temperature also differ greatly from those of CaF_2:Mn (TLD-400). Figure 7.21 shows the TSL emission spectrum of CaF_2:Dy (TLD-200) near the glow peak (150 °C) [30]. Table 7.1 shows the luminescence originating from the dopants, the same as for CaF_2:Mn (TLD-400).

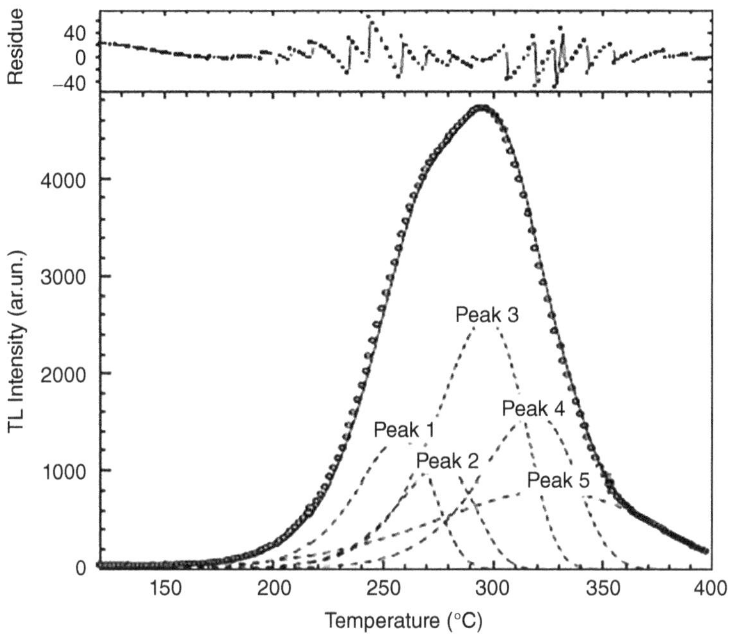

Figure 7.18 *Typical glow curve of CaF_2:Mn (TLD-400) at a dose of 3.85 Gy and glow curves obtained by the CGCD method ($\beta = 1\,°C\,s^{-1}$) [29].*

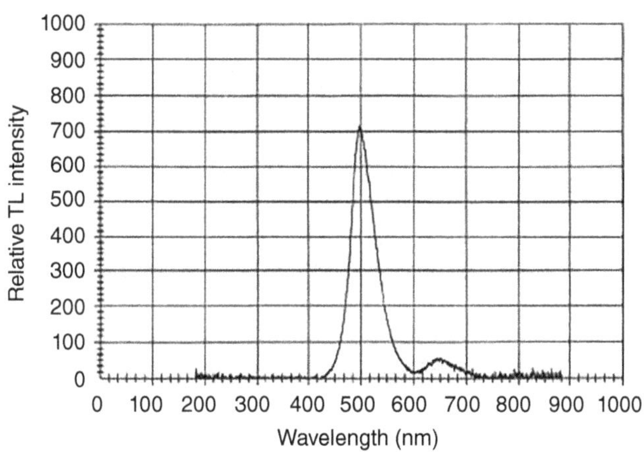

Figure 7.19 *TSL emission spectrum of CaF_2:Mn (TLD-400) at 320 °C [30].*

Figure 7.22 shows the glow curve of CaF_2:Tm (TLD-300). The inset is a semi-logarithmic graph of the TSL intensity to clearly show the glow components in the low TSL intensity region. The black circles show the experimental data, the thin lines show the separated glow peaks, and the thick line shows the sum of all of the separated glow peaks. Massillon et al. [34] reported decomposition into seven glow components. The shape is similar to the

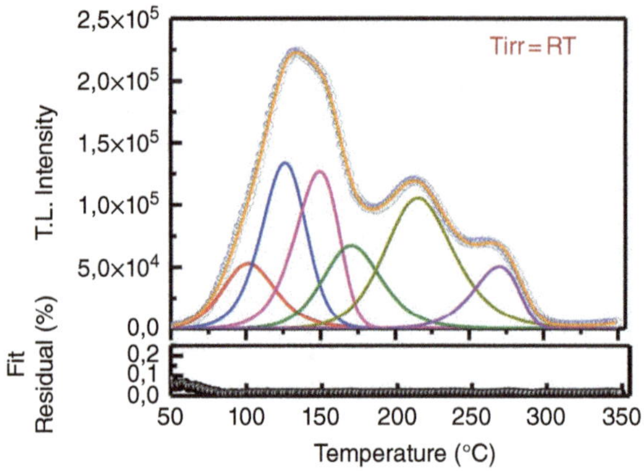

Figure 7.20 A typical TL glow-curve of CaF_2:Dy (TLD-200) [32].

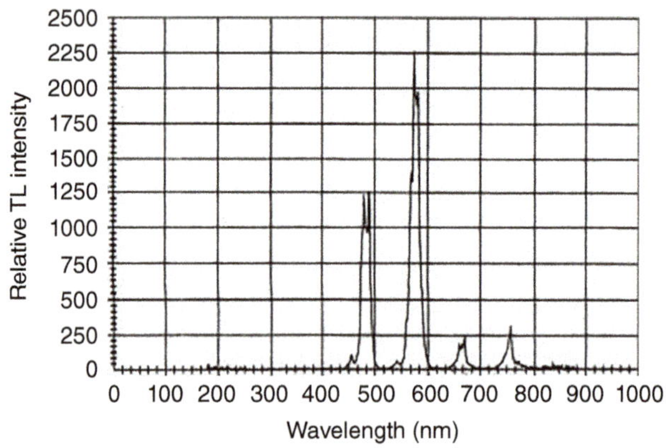

Figure 7.21 TSL emission spectrum of CaF_2:Dy (TLD-200) at 150°C [30].

Table 7.1 Measured wavelengths and line widths of the main spectral lines for CaF_2:Dy (TLD-200) [30].

Wavelength (nm)	Width (nm)	Dy^{3+} transition
455	6	$^4I_{15/2} \rightarrow {}^6H_{15/2}$
479/488	20	$^4F_{9/2} \rightarrow {}^6H_{15/2}$
541	6	$^4I_{15/2} \rightarrow {}^6H_{13/2}$
573/580	20	$^4F_{9/2} \rightarrow {}^6H_{13/2}$
658/670	20	$^4F_{9/2} \rightarrow {}^6H_{11/2}$
758	6	$^4G_{11/2} \rightarrow {}^6H_{5/2}$

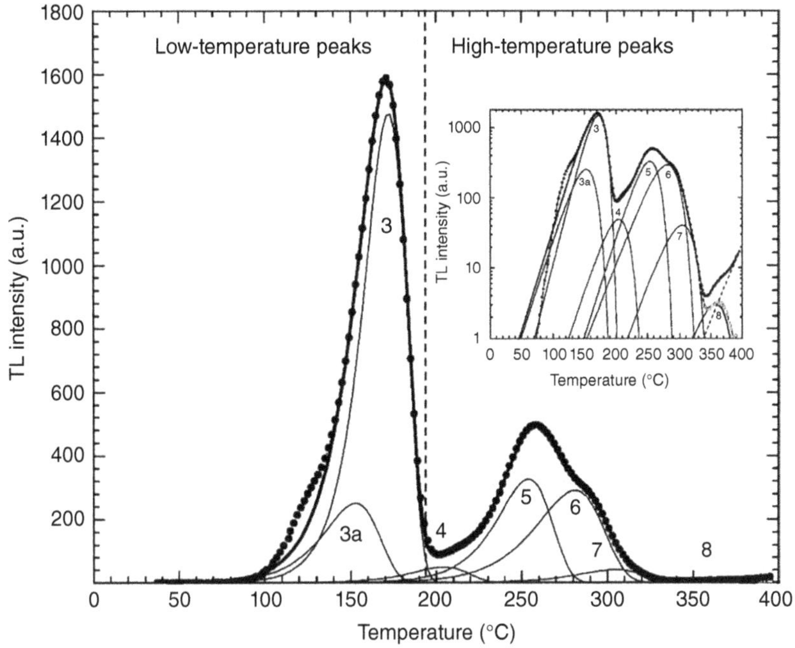

Figure 7.22 Typical CaF_2:Tm (TLD-300) glow curve [33].

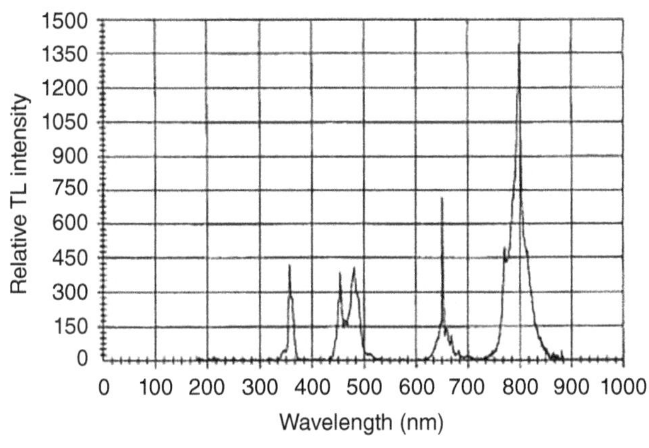

Figure 7.23 TSL emission spectrum of CaF_2:Tm (TLD-300) at 140 °C [30].

glow curve for CaF_2:Dy (TLD-200), the glow peak temperatures were also almost the same, and the TSL characteristics have many points in common. Given that the charges of Tm^{3+} and Dy^{3+} are the same and their ionic radii are also almost identical, there is a high likelihood that similar defects contribute to the trapping levels. Furthermore, the TSL emissions exhibit luminescence originating from the dopant (Tm^{3+}), the same as for CaF_2:Mn (TLD-400) and CaF_2:Dy (TLD-200) (Figure 7.23 and Table 7.2) [30].

Table 7.2 *Measured wavelengths and line widths of the main spectral lines for CaF_2: Tm (TLD-300) [30].*

Wavelength (nm)	Width (nm)	Tm^{3+} transition
290	8	$^3P_0 \rightarrow {}^3H_6$
356	8	$^1D_2 \rightarrow {}^3H_6$
455	9.5	$^3P_0 \rightarrow {}^3F_4$
482	17	$^1G_4 \rightarrow {}^3H_6$
650	2.5	$^1G_4 \rightarrow {}^3H_4$
799	18	$^3F_4 \rightarrow {}^3H_6$

7.3.2 Oxides

7.3.2.1 Aluminum Oxide (Al_2O_3)

Aluminum oxide (Al_2O_3) is the main component of natural minerals such as sapphires and rubies. It is synthesized as γ-alumina or α-alumina (also known as corundum). This material has been extensively studied since the start of the 1950s. Although various Al_2O_3 materials have been reported, including undoped and Ti-doped (sapphire) [34, 35], Si and Ti co-doped [36], Mg and Y co-doped [37], Cr-doped (ruby) [38], and Cr and Ni co-doped [39], C-doped Al_2O_3 [40, 41] has the highest sensitivity and is suitable for dose measurement. Here, we describe α-Al_2O_3:C and the basic properties of Al_2O_3:Cr that can be used for large area radiation imaging as developed by Shinsho et al. [23, 42, 43].

Carbon-doped aluminum oxide (α-Al_2O_3:C) was first developed by Akselrod et al. [40, 41] in Russia. Currently, it is sold commercially as TLD-500 K, and has sensitivity more than 80 times that of LiF:Mg,Ti (TLD-100). Sun et al. [44] also reported α-Al_2O_3:C with an even higher sensitivity that was formed by the vertical gradient freezing (VGF) method. Currently, α-Al_2O_3:C is widely used for human exposure dosimetry because it also has excellent optical stimulated luminescence (OSL). Although the crystal structure is actually rhombohedral, it approximately belongs to the hexagonal system. If two electrons are captured by an O^{2-} vacancy, a neutral F center is formed, and if a single electron is captured, a positively charged F^+ center is formed. It is thought that C^{2+} substitutes for Al^{3+} to form F^+ centers that increase the TSL sensitivity, but the directly contributing electron trapping sites have not yet been determined experimentally, and the detailed mechanisms remain unclear. The TSL emission peak is 420 nm, and this luminescence is thought to be attributable to a process in which F centers that have been excited by recombination of electrons with F^+ centers (F^*) relax to the ground state through the emission of photons in the range of 410–420 nm [45].

$$F^+ + e^- \Rightarrow F^* \Rightarrow F + h\nu_{420\,\mathrm{nm}} \tag{7.7}$$

Figure 7.24 shows the relationship between TSL intensity and the temperature and wavelength of α-Al_2O_3:C (TLD-500 K) [46]. The peak emission wavelength is 420 nm, the glow peak temperature is around 190 °C at a temperature rise rate of 2.5 °C s^{-1}, and the glow curve has a simple shape.

Al_2O_3:Cr, which was developed by Shinsho et al. [23, 42, 43], can be used for wide-area radiation imaging and has been confirmed to be useful for applications such as verifying the geometric precision of exposure position in high-precision radiotherapy (Section 7.5).

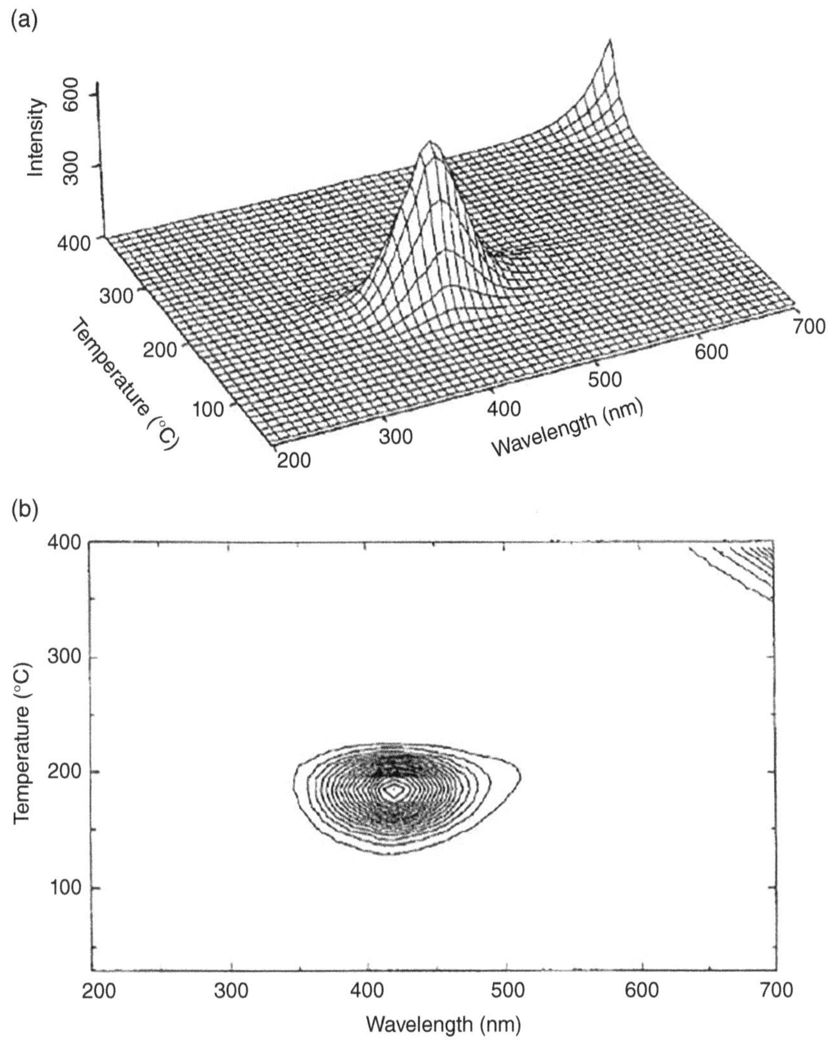

Figure 7.24 *(a) Isometric plot of emission intensity versus temperature and wavelength for TSL from α-Al_2O_3:C (TLD-500 K) irradiated with 50 mGy and heating at 2.5 °C s^{-1}. (b) Contour plot of the same data shown in (a) [46].*

This section explains only the basic properties of this material. Al_2O_3 doped with 0.5 wt% Cr is used as a ruby scintillator for monitoring accelerator beams and other purposes. Shinsho et al. reported that the TSL characteristics vary greatly depending on the amount of Cr doping, and that the sensitivity is dramatically increased by doping with a very small amount of Cr (0.05 wt%) [23]. Figure 7.25 shows the glow curve for Al_2O_3:Cr as the amount of Cr dopant is varied. Al_2O_3 doped with 0.05 wt% Cr has the highest TSL sensitivity, and the glow peak temperature is hardly affected by fading at 300 °C. Furthermore, the TSL emission at 693 nm arises from the $^2E \rightarrow {}^4A_{2g}$ transition in Cr^{3+} ions (Figure 7.26) [23, 45]. Although doped Cr^{3+} substitutes for Al^{3+}, it is difficult for defects to form through charge compensation because the charges are equal. Furthermore, Cr^{3+} acts as an electron trapping center, so Cr^{2+}

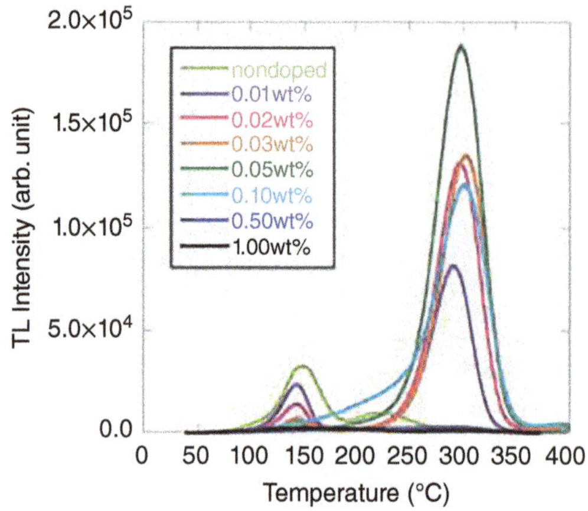

Figure 7.25 *Glow curves of undoped and Cr-doped* Al_2O_3 *[23].*

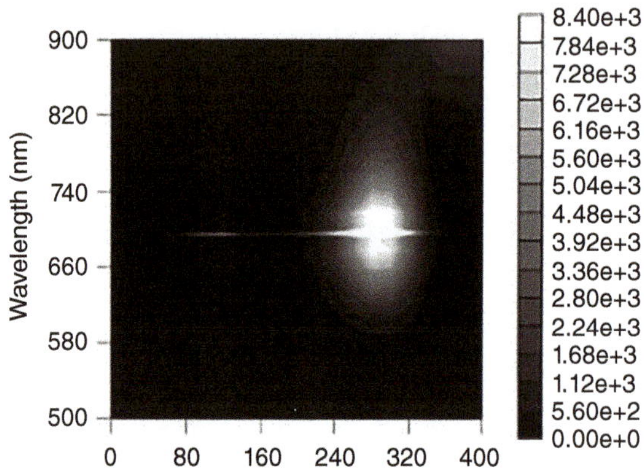

Figure 7.26 *TSL emission spectra of 0.05 w% Cr-doped* Al_2O_3 *during heating [23].*

is generated by irradiation. The Cr^{2+} ions that have captured an electron recombine with a hole (h^+) under heating and transition to Cr^{3+*} in an excited state (^2E), and this transitions to the ground state ($^4A_{2g}$) of Cr^{3+} by emission of a 693 nm photon.

$$Cr^{2+} + h^+ \Rightarrow Cr^{3+}* \Rightarrow Cr^{3+} + h\nu_{693\,nm} \tag{7.8}$$

7.3.2.2 Beryllium Oxide (BeO)

Beryllium oxide (BeO) is an important oxide TSL material. BeO not only has an effective atomic number ($Z_{eff} = 7.21$) similar to that of human soft tissue (water), but also has TSL sensitivity similar to that of LiF:Mg,Ti (TLD-100). BeO has long attracted the interest of researchers because it is a rigid, easy-to-handle TSL material that can also be used as an OSL element. Although manufacturing costs have increased and manufacturing facilities have also become limited since the strong toxicity of BeO was discovered in recent years, this

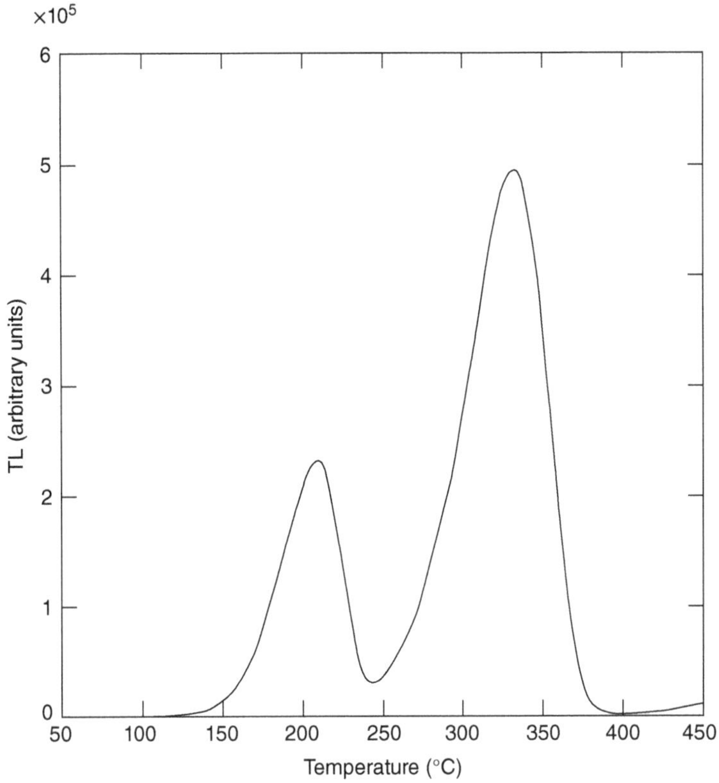

Figure 7.27 *Glow curve of BeO (Thelmalox™ 995) irradiated with 1 Gy from a ^{137}Cs source and heated at 1°C s^{-1} [47].*

toxicity is accompanied by advantages that are difficult to replace. Because BeO has a significantly smaller neutron reaction cross-section than that of other materials, it has again begun to attract attention as the only TSL element able to measure the γ-rays contained in a neutron field. In next-generation radiotherapy such as boron neutron capture therapy, accurate measurement of γ-rays within a mixed field is essential, so advances are expected in areas such as two-dimensionalization in solids where dispersal of powder does not occur.

Although the main BeO elements include Na-doped BeO and Na and Li co-doped BeO, these are all powders, so the risk of dispersal is high and sales have been halted. We therefore describe the characteristics of Thermalox™ 995 (Brush Wellman Inc.), a BeO element that can be sintered into disks of a size suitable for dose measurement (6.0–9.0 mm in diameter and 0.9–1.5 mm thick).

Figure 7.27 shows the glow curve of BeO (Thermalox 995) [47]. It has two broad glow peaks at around 200 and 340 °C. Each of these glow curve components has been reported to be a superposition of multiple components. Although the glow peak intensity at 340 °C is high at a dose of 1 Gy, the glow curve is dose dependent not only in terms of the peak intensity ratio, but also in terms of the full width at half-maximum of the glow curve [48,49,50]. The TSL emission spectrum has a peak at 335 nm (Figure 7.28). Although Becker et al. [49] showed by spectral analysis that the main impurity contained in Thermalox 995 is Si, details

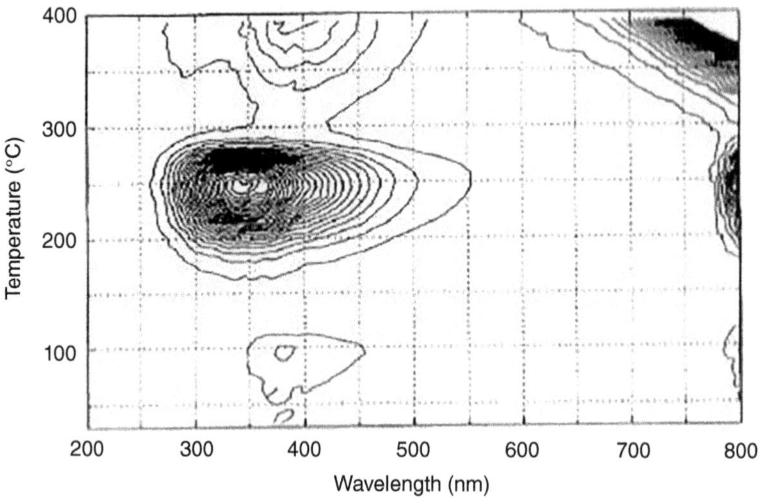

Figure 7.28 TSL emission spectra of BeO (Thelmalox 995) irradiated with 10 Gy from a ^{137}Cs source and heated at 2.5 °C s^{-1} [24].

of the emission mechanisms, such as the contribution of Si, have not yet been clarified. One possibility is a powerful emission mechanism proposed by Kortov et al., in which Al^{3+} ions present as an impurity capture electron to form Al^{2+}, followed by emission of 335 nm photons via recombination with holes and return to the Al^{3+} ground state by heating [51]

$$Al^{2+} + h^{+} \Rightarrow Al^{3+} + h\nu_{335\,nm} \tag{7.9}$$

Magnesium oxide (MgO) is another oxide used for dose measurement not only of X-rays, but also ultraviolet rays [52,53]. It has a lower glow peak temperature and is more readily affected by fading compared with other TSL elements.

7.3.3 Sulfates

7.3.3.1 Calcium Sulfate ($CaSO_4$)

The main sulfate TSL elements are $CaSO_4$:Tm and $CaSO_4$:Dy, which have a main ingredient of calcium sulfate ($CaSO_4$). Although Watanabe [54] reported the TSL characteristics of $CaSO_4$:Mn in 1951, the fading was large and it did not lead to practical application. In 1968, Yamashita et al. [8] successfully synthesized $CaSO_4$:Tm and $CaSO_4$:Dy TSLD elements with a sensitivity 30 times higher than that of LiF:Mg,Ti (TLD-100), and $CaSO_4$:Tm has come to be widely used for environmental monitoring and human exposure dosimetry as Matsushita UD-200. $CaSO_4$:Tm and $CaSO_4$:Dy have both high sensitivity in TSL dosimeters and a simple glow curve compared with other elements, and therefore are considered suitable for clarifying TSL mechanisms, a topic that has been pursued by many researchers. Huzimara et al. [55] and Nambi et al. [56] suggested that SO_4^- and SO_3^- formed by radiation exposure act as trapping sites. Furthermore, Morgan et al. [57,58] reported detailed analysis of electron spin resonance (ESR) of monocrystals and found that four types of SO_4^- acted as trapping sites. However, Gundu Rao et al. [59] reported from analysis of the ESR and glow curves of $CaSO_4$:Dy,Na that the behavior of SO_4^- reported by Morgan et al. was not correlated with TSL. There are many unknowns regarding the radicals that directly contribute to TSL.

The rare earth (RE) dopants Dy^{3+} and Tm^{3+} play multiple roles in $CaSO_4$. They substitute for Ca^{2+} and generate Ca^{2+} vacancies by charge compensation, but it has been shown experimentally that they capture electrons to form Dy^{2+} and Tm^{2+}, which act as electron trapping centers. Experiments have also shown that the glow curves of RE-doped $CaSO_4$ have similar shapes and glow peak temperatures regardless of the RE element, that the glow curves are a superposition of multiple glow components [60], and that the TSL emission wavelength is dominated by luminescence based on energy state transitions in the trivalent RE dopants [56]. Figure 7.29 shows glow curves of RE-doped $CaSO_4$. Although there are differences in TSL sensitivity, the glow curve shapes and main glow peak temperatures (170–180 °C) are similar regardless of the species of RE dopant (Tm, Dy, Sm, Tb). Furthermore, these main glow peaks appear to have a single component, but various analysis results have shown that there are 8 or more glow components (Figure 7.30) [56,60]. Figure 7.31 shows the TSL emission

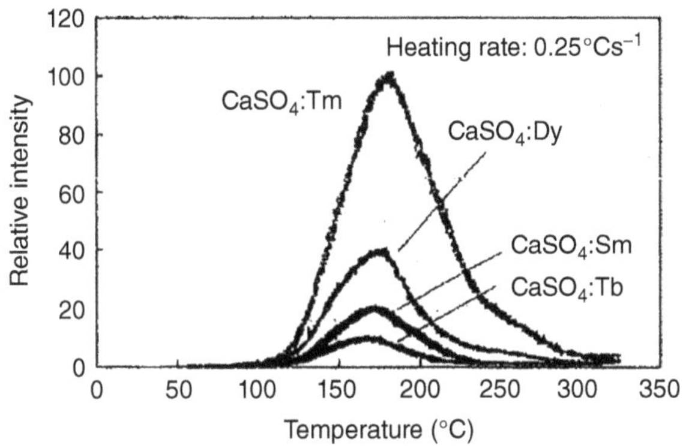

Figure 7.29 Glow curves of $CaSO_4$:RE (Tm, Dy, Sm and Tb) for a heating rate of 0.25 °C s^{-1} [60].

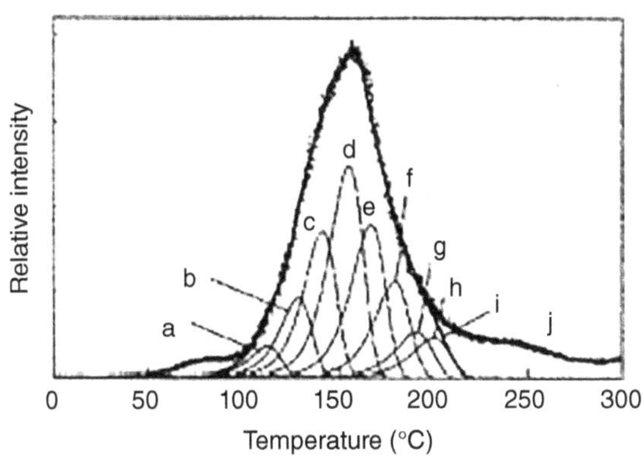

Figure 7.30 Typical examples of theoretical glow curve fitting for the positions of eight components (determined by the post-annealing method for $CaSO_4$:Dy) [60]. s set in all 10^{19} s^{-1}: (a) $E = 1.66$ eV; (b) $E = 1.70$ eV; (c) $E = 1.76$ eV; (d) $E = 1.82$ eV; (e) $E = 1.86$ eV; (f) $E = 1.92$ eV; (g) $E = 1.97$ eV; (h) $E = 2.01$ eV; (i) total; and (j) measured value.

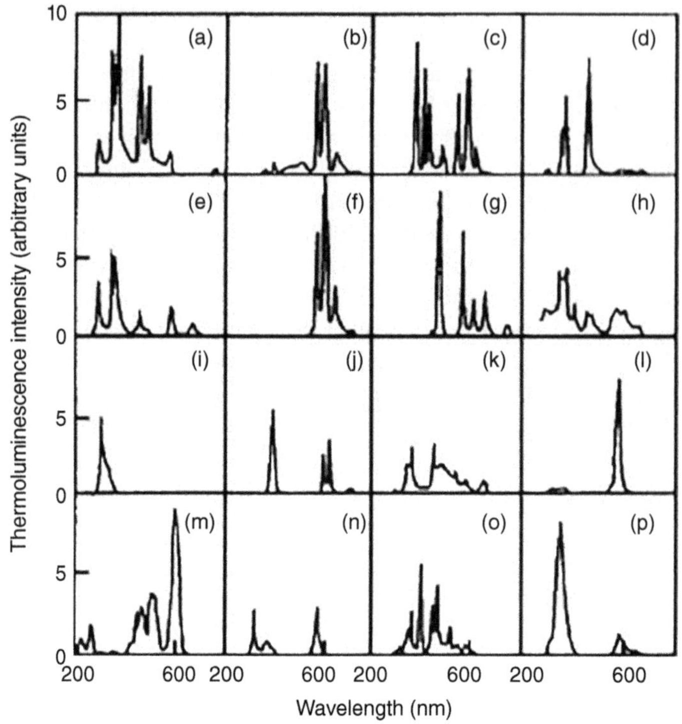

Figure 7.31 *TSL emission spectra of $CaSO_4$ with and without RE dopants: (a) undoped; (b) Nd; (c) Tb; (d) Tm; (e) La; (f) Sm; (g) Dy; (h) Yb; (i) Ce; (j) Eu; (k) Ho; (l) Lu; (m) Pr; (n) Gd; (o) Er; and (p) Y [56].*

spectrum of $CaSO_4$ TSL doped with RE. In all cases, luminescence is based on the emission level of the trivalent RE element used as a dopant [56].

7.3.4 Borates

Borates such as $Li_2B_4O_7$ and MgB_4O_7 are TSL materials that are particularly attractive for human exposure dosimetry because they are nearly equivalent to human soft tissue (water) and also have high sensitivity, the same as LiF. Therefore, these materials have long been subjects of research.

7.3.4.1 Lithium Borate ($Li_2B_4O_7$)

In 1967, Schulman et al. [14] first reported the TSL properties of $Li_2B_4O_7$ doped with 0.1 wt% Mn. Although the sensitivity of this material is somewhat lower than that of LiF, its energy dependence is low because it is nearly equivalent to human soft tissue (water), and its manufacturing costs are also low. Figure 7.32 shows the glow curve of $Li_2B_4O_7$:Mn [61]. There is one main peak, and the glow peak temperature is 185 °C when the temperature rise rate is 10 °C s^{-1}. However, this glow peak temperature is strongly dependent on Mn concentration [61] and the dopant Mn acts as a recombination center, so the luminescence wavelength is orange and is not compatible with the sensitive region of PMTs. In 1980, Takenaga

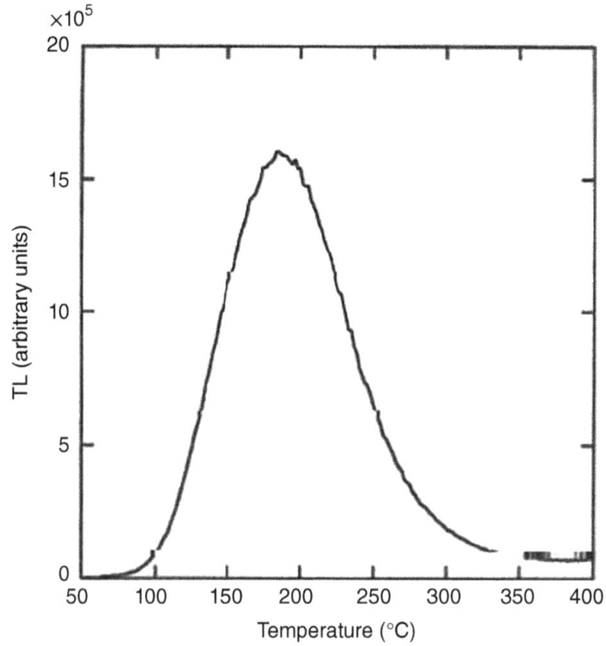

Figure 7.32 *Glow curve of $Li_2B_4O_7$:Mn irradiated with 1 Gy of ^{137}Cs γ-rays and heated at 10°C s^{-1} [61].*

et al. [62] reported a Cu-doped $Li_2B_4O_7$ TSL element that exhibits luminescence at 368 nm, which is in the sensitive region of PMTs, and successfully increased the sensitivity 20-fold. Subsequent attempts were made to improve the hygroscopicity by addition of Si [61], and it was shown that the luminescence of RE-doped $Li_2B_4O_7$ was based on the energy levels of the dopant trivalent RE, the same as in $CaSO_4$, and that the effective dopants are Dy, Tm, and Tb (Figure 7.33) [63].

7.3.4.2 Magnesium Borate (MgB_4O_7)

Magnesium borate (MgB_4O_7), which is another widely used borate, was first reported by Kazanskaya et al. [64] in 1974. It exhibits favorable properties with respect to not only X-rays, but also γ-rays, β-rays, and neutron radiation. Shortly after this, MgB_4O_7:Dy and MgB_4O_7:Tm were reported as high-sensitivity TSL elements by Prokić [65] and Barbina et al. [66], and came into use for dose measurement. Figure 7.34 shows the glow curves of MgB_4O_7:Dy, MgB_4O_7:Tm, and MgB_4O_7:Mn [67]. The main glow peaks of MgB_4O_7:Dy and MgB_4O_7:Tm were both observed around 200 °C. Furthermore, it has been reported that component analysis of the glow curves has shown that the main peak is a superposition of at least 11 glow components [68]. For MgB_4O_7:Mn, the main glow peak temperature is 383 °C and the glow curve characteristics vary greatly, and it has been reported that the glow peak temperature depends on the Mn concentration.

Figure 7.35 shows the TSL emission spectrum of MgB_4O_7:Dy. The TSL emission spectrum of MgB_4O_7 is based on the energy levels of dopants, the same as in $CaSO_4$ and $Li_2B_4O_7$. In the case of Dy^{3+}, 480 nm and 575 nm emission bands are attributable to the $^4F_{9/2} \rightarrow {}^6H_{15/2}$ transition.

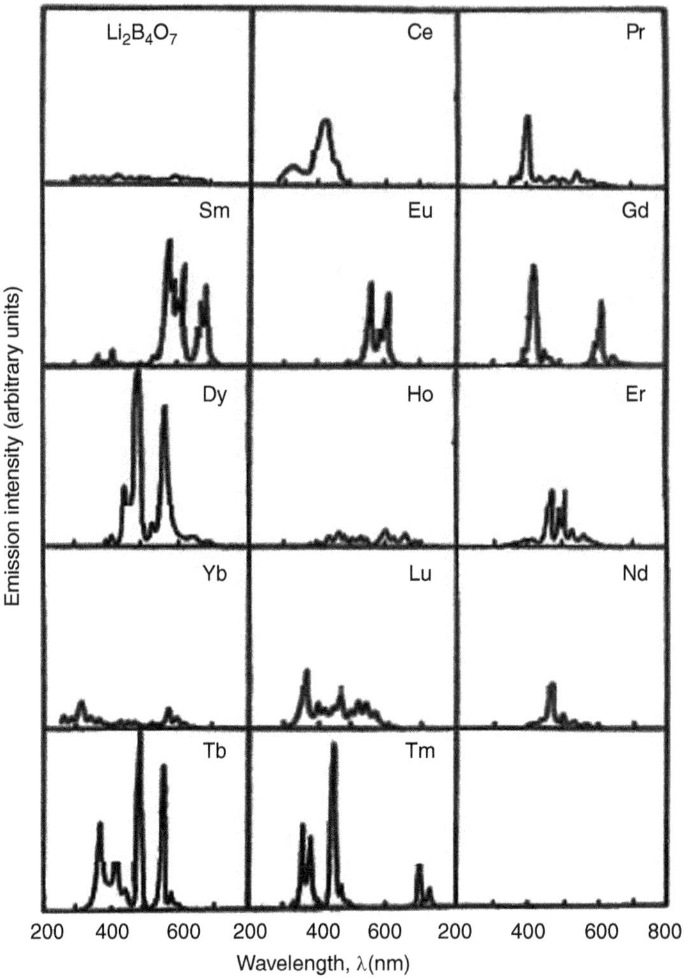

Figure 7.33 TSL emission spectra of $Li_2B_4O_7$ with and without RE dopants after irradiation with 10^4 Gy of ^{60}Co γ-rays [63].

7.4 TSL Dosimetric Properties for Photons, Charged Particles, and Neutrons

Materials that exhibit TSL are mainly inorganic compounds, with many inorganic materials exhibiting this physical phenomenon. Because of the versatility of these materials, TSL is used in detectors for various types of radiation. TSL materials are often selected that have a large cross-section for interaction with the radiation to be measured.

7.4.1 TSL Dosimetric Properties for Photons

Table 7.3 shows the properties of the main TSL dosimeter (TSLD) elements most commonly used for photons [69]. About nine types of TSL materials are used commercially as TSLDs,

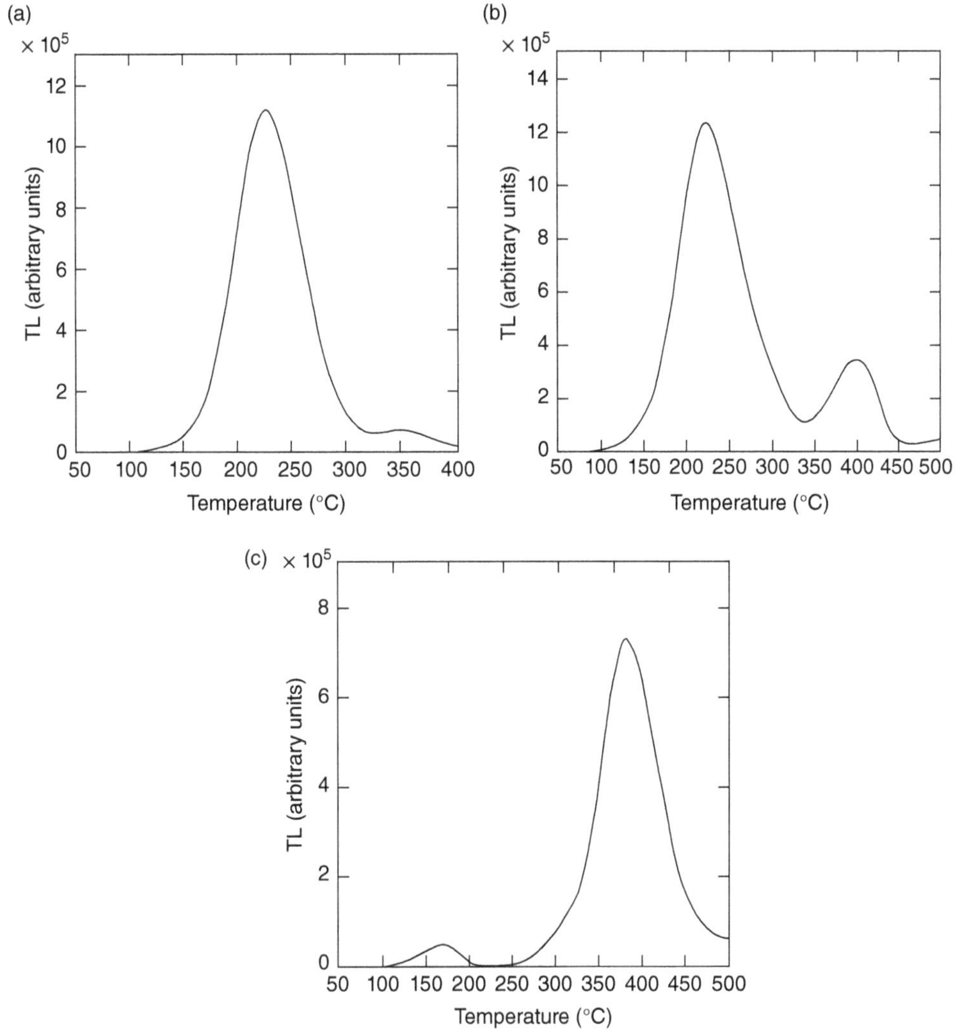

Figure 7.34 *Glow curves of MgB_4O_7:Dy, MgB_4O_7:Tm and MgB_4O_7:Mn after 1 Gy of ^{137}Cs γ-rays and heating at 1 °C s^{-1} [67].*

including those in the forms of powder, powder-filled glass tubes, rods, and thin films. Figure 7.36 shows some of the various forms of TSLDs.

The common advantages of TSLDs include the following:

1. TSL is emitted in proportion to the absorbed radiation energy.
2. Electrons and trapping centers are eliminated by TSL reading and annealing to restore the initial state of TSLDs, so they can be used repeatedly.
3. A wide range of doses can be measured.
4. Their compactness allows them to be used for localized dose measurement.
5. There is no dependence on dose rate up to 10^9 Gy s^{-1}.

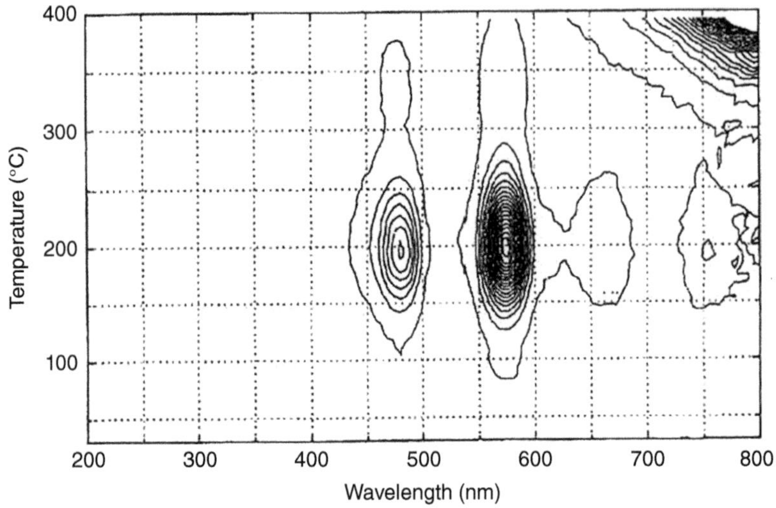

Figure 7.35 *TSL emission spectra of MgB_4O_7:Dy irradiated with 5 Gy of ^{137}Cs γ-rays and heated at 2.5 °C s^{-1} [67].*

There are many more types of TSLDs than optically stimulated luminescence (OSL) dosimeters and the characteristics vary between elements. Accordingly, the specific characteristics of TSL elements need to be determined before use. In particular, because the TSL response depends on the amount of energy absorbed through the interaction between photons and the TSL material, it is desirable to select a TSL element with an effective atomic number (Z_{eff}) close to that of water, the main constituent of the human body, when measuring the absorbed dose in the human body.

The TSL response to a fixed dose can be expressed as a function of incident photon energy. In the case of photon beam irradiation, the variation in TSL response to absorbed radiation energy is defined by the mass energy absorption function of the material, that is, μ_{en}/ρ. The photon energy response $S_E(E)$ to a reference material (normally air) is defined as follows:

$$S_E(E) = \frac{(\mu_{en}/\rho)_m}{(\mu_{en}/\rho)_{ref}} \tag{7.10}$$

For mixed materials, taking the mass proportion of the *i*th constituent element as W_i gives

$$(\mu_{en}/\rho) = \Sigma(\mu_{en}/\rho)_i W_i \tag{7.11}$$

The specific breakdown of energy losses includes the photoelectric effect, Compton scattering, and electron-pair production. These loss processes depend on the incident photon energy and the effective atomic number (Z_{eff}) of the TSL. The photoelectric effect component of the mass energy absorption function is approximately proportional to the cube of the Z_{eff}, the Compton scattering component is proportional to Z_{eff}/M, and the electron-pair production component is proportional to the square of Z_{eff}. In the low-energy regime of incident

Table 7.3 *Properties of the main TSL dosimeter (TSLD) elements most commonly used for photons [69].*

	Commercial names[a]	ρ (g/cm^3)	Z_{eff}[b]	Main glow peak temp[c] (°C)	Emission[d] (nm)	TL sensitivity vs. LiF	k_Q 30 keV/^{60}Co	Fading of dosimetry peak at normal temp
LiF:Mg·Ti (TL)	TLD-100	2.6	8.31	~235	~410	Referent	1.3	5% in 3–12 months
^{6}LiF:Mg,Ti (TL)	TLD-600	2.6	8.31	~235	~410	1.0	1.3	5% in 3–12 months
^{7}LiF:Mg,Ti (TL)	TLD-700	2.6	8.31	~235	~410	1.0	1.3	5% in 3–12 months
LiF:Mg,Cu,P (TL)	TLD-100H	2.5	8.31	~200	~370	30	1.25	2% in 3 months
$Li_2B_4O_7$:Mn (TL)	TLD-800	2.3	7.32	~185	~600	0.3	0.9	5–10% in 3 months
CaF_2:Dy (TL)	TLD-200	3.18	16.90	~160, 185, 245, 290	480, 575, 660, 750	30	14	25% in 4 week
CaF_2:Mn (TL)	TLD-400	3.18	16.90	~300	~495	10	14	15% in two to 4 week
$CaSO_4$:Dy (TL)	TLD-900	2.61	15.62	~220	480, 575, 660, 750	15	12	6% in 6 months
Al_2O_3:C(OSL)	nanoDot™	3.95[e]	11.28	~200	~410	N/A	3.7	4% in 3 months
BeO (OSL)	Thermalox 995	2.85	7.21	~210.330	~335 (TL), ~390(OSL)	N/A	0.82	5–10% in 3 months

[a] Commercial name used by Thermo Fisher Scientific, Inc. (formerly Harshaw, Inc.), Materion Ceramics, and Landauer Inc.
[b] Values from Bos. (Bos, A.J.J. (2001) High sensitivity thermoluminescence dosimetry. Nucl Instrum Methods Phys Res B. 184 3–28.)
[c] These are approximate values, as the TL peak temperature varies with heating rate and may also vary considerably with impurity content and source of material.
[d] The symbol "~" indicates broad emission bands. Emission from Dy3 + in Dy-doped materials are characterized by sharp emission lines.
[e] The density of Al_2O_3 is 3.95; however, the active nanoDot crystal is powder embedded in a polyester tape and includes many air pockets. (Asena, A., Crowe, S.B., Kairn, T. et al. (2014) Response variation of optically stimulated luminescence dosimeters. Radiat Meas. 61 21–24). The density of the active volume has been found to be in the range of 1.41–2.45 g/cm.

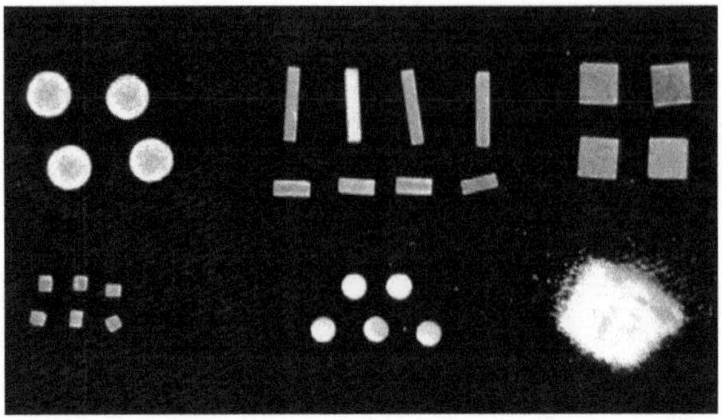

Figure 7.36 TSL dosimeter forms, including disks, rods, chips (top row, from left), micro-cubes, disks, and powder (bottom row, from left) [69].

photons, the photoelectric effect is dominant, and as the incident photon energy increases, Compton scattering becomes dominant. Electron-pair production occurs at high energies (E) of 1.02 MeV or higher. The actual TSL response curve to incident photon energy is defined by the relative energy response (RER) using the TSL response to 1.25 MeV photons from ^{60}Co radiation source as the reference. This gives the following:

$$(RER)_E = \frac{S_E(E)}{S_E(1.25\,MeV\ ^{60}Co)} \tag{7.12}$$

Figure 7.37 shows RER versus incident photon energy, as found by Spurny et al. [70]. The values of Z_{eff} for LiF and CaF are 8.3 and 16.9, respectively. As Z_{eff} increases, RER increases up to 100 keV. This is because the photoelectric effect becomes more dominant as Z_{eff} increases in the low-energy regime. As a result, the TSL elements that are suitable for measuring the absorbed dose in the human body mainly consist of LiF, BeO, and $Li_2B_4O_7$, which have an effective atomic number close to that of water, making them nearly equivalent to human soft tissue (water). If a TSL element with a large Z_{eff} is used, some mechanism is required, such as combination with a filter, in order to appropriately attenuate low-energy photons and correct the energy dependence.

Other factors also affect the measurement precision of TSL dosimeters, as described in the following:

1. Fading
   In this phenomenon, the amount of TSL is degraded as a result of electrons trapped at trapping centers escaping during the period from irradiation to read out. If the activation energy (E) of the electron trapping centers is small (i.e., if the glow peak temperature is low), the fading effect is large because the force holding the trapped electrons is weak.
2. Optical fading
   Electrons trapped in trapping centers can be excited to the conduction band not only by heat, but also by light. TSL elements are affected by exposure to light after irradiation. Although the degree of this effect varies between elements, the dosimeter should ideally be shielded from light in the period from irradiation to measurement, regardless of the element used.

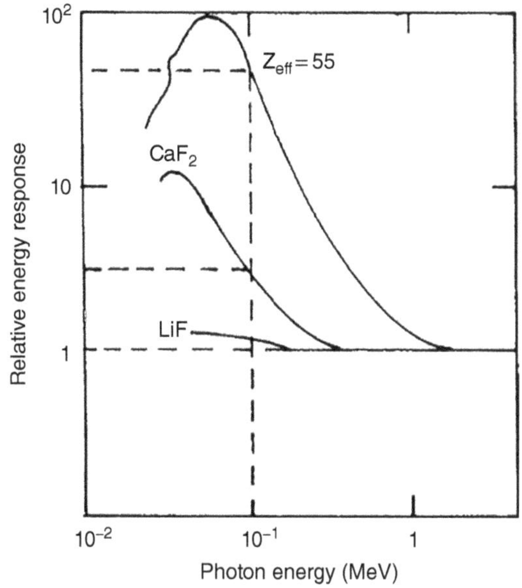

Figure 7.37 *Relative photon energy response of LiF (Z_{eff} = 8.14), CaF_2 (Z_{eff} = 16.3), and a theoretical material (Z_{eff} = 55). The reference material is air [70].*

3. Triboluminescence
   Friction due to mechanical stimuli (e.g., when something is dropped) forms electrons and holes, and luminescence phenomena such as scintillation and TSL may be observed, the same as after radiation. Although the strength of this effect differs depending on the type of TSL element, care is required during handling since physical impacts give rise to large errors in dose measurement.
4. Supralinearity
   Although the amount of TSL is proportional to the absorbed radiation energy, above a certain dose, the conversion efficiency to TSL increases and results in an excess response. The region where supralinearity is observed depends on the type of TSL element. In some TSL elements, the TSL conversion efficiency decreases above a certain dose and the element exhibits sublinearity.
5. Annealing
   Annealing refers to heat treatment performed at a temperature determined for each type of Thermally stimulated luminescent dosimeter (TSLD), and returns a TSLD to its initial state by releasing electrons trapped at trapping centers. Annealing is performed before use in order to avoid the effects of natural radiation and residual readings. Annealing conditions affect the sensitivity of TSLDs, so it is important to perform annealing under the conditions determined for the TSLD to be used.

7.4.2 TSL Dosimetric Properties for Charged Particles

TSL response is generally dependent on LET. In the case of heavy charged particles with high LET, the conversion of light energy to the TSL energy is focused and localized around the particle track. This trend becomes stronger as LET increases. The energy transfer

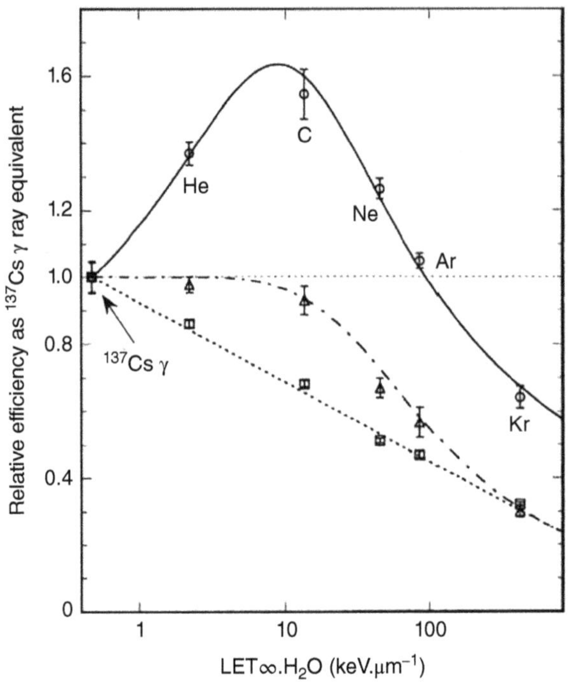

Figure 7.38 *Relative efficiencies of TSLD-BeO:Na (UD-170A) (○), TSLD-Mg_2SiO_4:Tb (TLMS) (△), and RPL glass (□) referenced to ^{137}Cs γ-rays as a function of $LET_\infty H_2O$ [71].*

becomes more localized with increasing ionization density, and thus the energy accumulation per unit dose becomes saturated above some LET. As a result, the TSL response per irradiated unit dose becomes smaller than that in the case of low-LET radiation, although the degree differs between TSL elements. Yasuda et al. [71] reported the LET dependence of TSLD-BeO:Na, TSLD-Mg_2SiO_4:Tb, and radiophotoluminescence (RPL) glass (Figure 7.38). The TSL response of all three dosimeters degraded under high-LET radiation, when the TSL response to ^{137}Cs γ rays was taken as the reference. However, the behavior under low-LET radiation differs for each type of dosimeter, indicating the complexity of this phenomenon. Because these kinds of TSLDs exhibit large LET dependence, they are difficult to use for dose measurement in heavy particle beam therapy. However, the use of elements with differences in LET dependence may be useful for various purposes, such as inferring LET. This is because the difference in LET dependence between these two TSLDs is uniquely determined by the value of LET within a limited LET range. Hirata et al. [72,73] reported a dose inference method for heavy charged particle beams using two different types of OSL with different LET dependences. Although the application of TSLDs in the particle beam regime is thought to be useful for skillfully using the LET dependence shown here, Koba et al. [74] reported that the LET dependence of Al_2O_3:Cr was weaker than that of other TSLDs, and also expected the development of new TSLD materials for particle beams (Table 7.4).

Furthermore, the LET dependence of the glow curve has long been known, and Jähnert et al. investigated the properties of ^{7}LiF:Mg, Ti (TLD-700) for proton beams and α beams and reported that the glow curves differed depending on the radiation quality [75]. This

Table 7.4 *LET dependence of alumina-based ceramic [74].*

	LET (keV/ μm)	nL/D				
		0.5 Gy	1 Gy	2 Gy	5 Gy	Average
X-ray	0.2	0.99	1.00	0.97	0.99	0.99
H	0.5	0.98	0.91	0.90	0.90	0.92
He	2.0	0.99	0.98	0.90	0.89	0.94
C	13.2	0.93	0.97	0.92	0.87	0.92
Ne	29	0.94	0.91	0.89	0.85	0.90
Ar	86	0.99	0.94	0.83	0.74	0.88

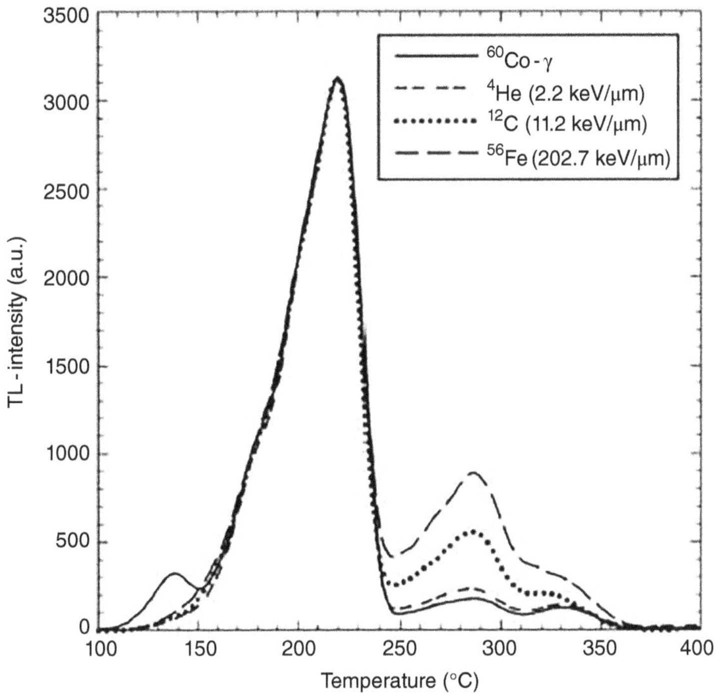

Figure 7.39 *Glow curves LET dependence of* ^{7}*LiF:Mg, Ti (TLD-700) with 1 mGy of* ^{60}*Co gamma rays and 1 mGy of different Heavy ions. Glow curves have been normalized to main peak height [83].*

result attracted the interest of many researchers, and TSL properties related to LET were reported one after another [25, 75, 76–82]. Figure 7.39 shows the glow curve LET dependence of ^{7}LiF:Mg,Ti as reported by Berger, Hajek, and co-workers [83]. They reported that the intensity ratios of the glow peaks differed depending on differences in LET. The high-temperature glow components increase with increasing LET of the particle. Even today, advances are being made on new uses of this property, for example, in methods for inferring LET.

7.4.3 TSL Dosimetric Properties for Neutrons

There are two methods of generating neutrons. One is by nuclear fission via neutron capture, as shown below for ^{235}U:

$$^{235}\mathrm{U} + \mathrm{n} \rightarrow {}^{236}\mathrm{U} \rightarrow \mathrm{A} + \mathrm{B} + (2 \sim 3)\mathrm{n} + 200\,\mathrm{MeV} \tag{7.13}$$

where A and B are the fission fragments. The other is by accelerating electrons, protons, and deuterons for nuclear reaction with a target.

$$^{7}\mathrm{Li} + \mathrm{p} \rightarrow {}^{7}\mathrm{Be} + \mathrm{n} - 1.65\,\mathrm{MeV} \tag{7.14}$$

$$^{9}\mathrm{Be} + \mathrm{p} \rightarrow {}^{9}\mathrm{Be} + \mathrm{n} - 1.85\,\mathrm{MeV} \tag{7.15}$$

In addition, the radiation generated by a high-energy X-ray therapy device of 10 MV or higher also includes secondary neutrons in addition to X-rays. In either case, it is difficult to generate only neutrons, and the neutron field always contains a mixture of radiation, including α rays, β rays, γ rays, and X-rays. Furthermore, since materials that exhibit TSL do not have sensitivity to neutrons alone, two types of TSLDs are required in order to detect neutrons in a neutron field with a mixture of multiple types of radiation. If one type sensitive to both photons and neutrons is used together with another type sensitive to only photons, the neutron component can be determined from the difference. The TSLDs that are used for these types of measurement are ^{6}LiF (TLD-600) and ^{7}LiF (TLD-700). The thermal neutron cross-section of TLD-600 is over 1000 times that of TLD-700, making TLD-600 a high sensitivity thermal neutron detector, while the response of TLD-700 is basically limited to only the photon beam component. Neither type of TSLD is suitable for the measurement of neutrons that have higher energy than thermal neutrons. Because of this, measuring neutrons that have energies higher than thermal neutrons requires indirect measurement of the neutron energy by combining the TLD-600 and TLD-700 pair with a neutron moderator, such as a Bonner sphere to decelerate the neutrons to thermal neutron energies.

7.5 Two-Dimensional (2-D) TSL Dosimetry

7.5.1 Introduction

With the development of radiation therapy technologies in recent years, high-precision radiotherapy such as intensity modulated radiotherapy (IMRT) and stereotactic radiotherapy (SRT) that can locally irradiate a tumor have come into clinical use. Although film dosimetry has conventionally been used for dose distribution verification during radiotherapy, ionization chamber and semiconductor array detectors are becoming widespread due to their simplicity of measurement. However, the array spacing of array detectors is from 5 to 10 mm, and the obtained dose distribution is interpolated by calculations using the measured values from the detectors. Furthermore, in ionization chamber array detectors, since the sensitive volume is large, it has been reported that over- and underestimation occur in areas where there is a steep dose gradient [84]. This kind of dose distribution verification requires the use of film dosimetry with uniform sensitivity and high spatial resolution, and the demand for 2-D passive dosimetry as exemplified by film dosimetry remains high.

Against this background, we conducted research into 2-D TSLDs as a new tool for 2-D passive dosimetry. Although Nakajima et al. [19] developed a 2-D TSL phosphor in 1967, it did not lead to further research due to the background of the time and various technical difficulties. Although the non-uniformity in the sensitivity of 2-D TSLDs is currently much larger than that of film dosimeters, 2-D TSLDs can obtain dose distributions at a similar spatial resolution of 75–300 dpi (0.08–0.34 mm/pixel). According to previous reports, 2-D TSLDs have been used in photon beam radiotherapy [43, 84–87], brachytherapy [88, 89], and particle beam radiotherapy [90–92], as well as in 2-D TSL measurement of synchrotron radiation [93, 94] and thermal neutrons [95. In sum, 2-D TSLDs are expected to have a wide range of applications as a new 2-D passive dosimetry.

7.5.2 Types of 2-D TSLDs

Given that a wide range of materials exhibit TSL, many types of 2-D TSLDs have been developed in the past. 2-D TSLDs have been reported that uses the tissue-equivalent materials lithium fluoride [84–86, 93, 94, 96] and lithium borate [87, 88], which have effective atomic numbers of around 7.0–8.5 – close to the 7.42 of water. 2-D TSLDs have also been reported that use non-tissue equivalent materials such as calcium sulfate [19, 21, 86, 92, 96], barium sulfate [88, 89], and aluminum oxide [23, 42, 43], which have higher effective atomic numbers.

Although there are also pellet-type 2-D TSLDs in which a fixed pressure is applied to the TSL element, thin-film TSLDs are mainly used. The thin films are made from small particles of TSL elements spread out in 2-D, and similar to film dosimeters, 2-D TSLDs are flexible and can be freely processed by cutting. Thin-film TSLDs have been developed by multiple research groups, including TSL foils [84–86, 89–91] and TSL sheets [88, 89, 93–95]. The typical methods for fabricating thin-film TSLDs are to mix TSL elements with resin and form a film shape. The TSL elements are then spread on a plastic sheet and hot-pressed. Furthermore, there are few reports on pellet TSLDs because the maximum size is up to 2 cm in diameter, it is difficult to increase the volume, and the shapes that can be manufactured are limited.

Another type of TSLD is called TSL slab [87, 88], which is thicker than thin-film TSLDs. Some TSL slabs are cast using only TSL elements, and some are cast using a mixture of TSL elements and resin. Furthermore, recently reported ceramic plates of aluminum oxide that exhibit TSL [23, 42, 43] are also categorized as TSL slabs.

Table 7.5 shows a comparison of the characteristics of the types of TSLDs shown in the Figures 7.40–7.42.

Table 7.5 *Comparisons of characteristics of each type of 2-D TSLD.*

Characteristic	TSL foil or TSL sheet	TSL slab	Ceramic TSL slab
Tissue equivalent	○	◎	△
Thickness	◎ 0.2–0.3 mm	× 2.5 mm	○ 0.7 mm
Reproducibility	×	△	◎
Dose calibration	△	○	◎
Sensitivity uniformity	△	△	◎
Other	Waterproof, flexible	3-D measurement	Waterproof, stiff

The characteristics are evaluated as follows: ◎, excellent; ○, good; △, moderate; ×, poor.

Figure 7.40 TSL foils made from LiF:Mg,Cu,P and $CaSO_4$:Dy [96].

(a) (b)

Figure 7.41 Tissue-equivalent TSL slab dosimeter made from: (a) $Li_2B_4O_7$:Mn,Al [88]; and (b) $Li_3B_7O_{12}$:Cu [87].

7.5.3 Measurement Systems

Figure 7.43 shows the 2-D TSL measurement system for TSL slab by Shinsho et al. [42]. It consists of a charge coupled device (CCD) camera (Atik383L+; resolution: 3362×2504 pixels; pixel size: 5.40 μm×5.40 μm; ADC: 16-bit); a custom-made 100 mm×100 mm hot plate (Sakaguchi E.H. Voc Corp); and a dark box (Seisei Manufactory Co. Ltd.). A control panel is installed on the front of the measuring device and is used for configuring settings such as the heater temperature and number of sheets to measure. A sliding door for inserting and removing Al_2O_3:Cr ceramic plates is installed in the side, and light is completely blocked by a solenoid lock. The interior of the dark box is fitted with three pins and a heater.

Figure 7.42 Low melting point Cr-doped Al_2O_3 ceramic plates (Ceramic TSL slab) [23].

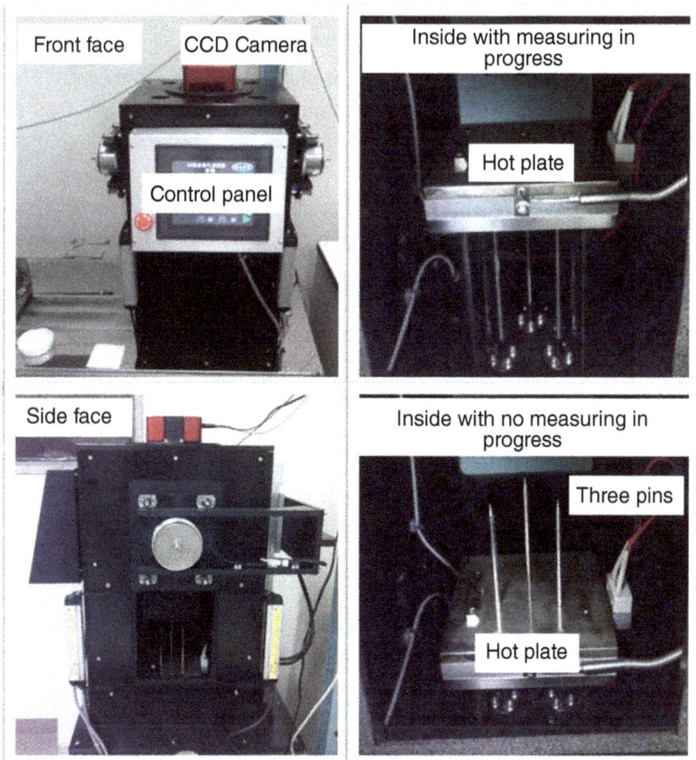

Figure 7.43 TSL imaging measurement system [42].

The ceramic plate is placed on the three pins, and then after the door is closed, the heater rises up and begins heating the TSL slab while simultaneously beginning measurement. The TSL distribution data captured by the CCD camera is saved to a PC for analysis. The maximum temperature of the heater is 450 °C. The spatial resolution depends on the CCD camera used and the particle size of the TSL slab. The spatial resolution when Cr-doped Al_2O_3 is used with this system is around 50 μm.

Although there have also been various reports of measuring systems for TSL foils [84–86, 89–92, 96] and TSL sheets [88, 89, 93–95], there are no major differences in the basic configuration, which consists of a CCD camera that is used for reading the TSL distribution, a device for heating the 2-D TSLDs, a dark box, and a PC for analysis.

7.5.4 Application of 2-D TSLDs in Photon Beam Radiotherapy

7.5.4.1 TSL Foils

In recent years, research into applications of 2-D TSLDs in photon beam radiotherapy has been reported by one research group in each of Poland and Italy, as well as one research group in Japan. In 2008, Kobat et al. [84] reported verification of 6 MV X-rays in a small radiation field by using TSL foil. The TSL foil was a waterproof 2-D TSLD with dimensions of 50 mm × 50 mm and thickness of 0.2 mm composed of LiF:Mg,Cu,P and Teflon. In addition to the 2-D dose distribution, they also verified the dose profile in the depth direction in water (percentage depth dose) and the dose profile in the lateral direction at the maximum depth (off-axis ratio) together with a treatment planning system. Although the measurement results in that report had large errors and highly precise verification was not performed, it demonstrated that it is possible to apply a 2-D TSLD in photon beam radiotherapy. Furthermore, in 2010 Kisielewicz et al. [85] reported dose distribution verification of IMRT by using TSL foil of LiF:Mg,Cu,P enlarged to 200 mm × 200 mm. This report used gamma analysis to evaluate both the dose error and the position error as a method for analyzing the dose distribution. The results of gamma analysis were an acceptance rate of 92.7% on acceptance criteria of 3%/3 mm. Although a good result was obtained, there was considerable noise in the dose profile, and the signal-to-noise ratio decreased with increasing dose. As noted in that report, the sensitivity uniformity currently needs improvement, and there is insufficient precision for use in clinical settings. Furthermore, in the same year Klosowski et al. [89] developed a TSL foil of thickness 0.3 mm composed of $CaSO_4$:Dy and perfluoroalkoxy alkane resin. In 2013, Marrazzo et al. [86] reported verification of the dose profile and dose distribution of 6 MV X-rays using two types of TSL foils. However, a variation of 5–10% was observed in the dose profile, and the effect of non-uniformity in the sensitivity of the TSL foil was also considered to be large. Although TSL foil has the advantages of being able to be used in water and used repeatedly, Kłosowski et al. [96] reported a decrease in sensitivity due to repeated measurement using two types of TSL foils. According to this report, the TSL intensity decreased by around 8% after 10 uses, and it was thought that degradation of the resin might be the cause. Although heat-resistant resin is used in 2-D TSLD, it is difficult to completely remove the effect of resin degradation on TSL properties by repeated measurement and annealing.

7.5.4.2 TSL Slab

In 2011, the Japanese research group of Shinsho et al. [88] cast $Li_2B_4O_7$:Mn,Al onto a plate, and developed a tissue-equivalent phantom TSLD that had an effective atomic number and density equivalent to those of soft tissue. This used not only a 2-D TSLD with dimensions of

80 mm × 80 mm and thickness of 2.5 mm, but also a new form of dosimetry with the aim of obtaining the 3-D dose profile by stacking layers. This report performed verification of the off-axis ratio and 3-D dose distribution measurement, and although there was considerable noise in the dose profiles, it was a relatively good match to the measurement results obtained by ionization chamber dosimetry. Furthermore, this work was extremely important in terms of showing that 2-D TSLD has the untapped potential for 3-D dosimetry. In 2014, Shinsho et al. [87] developed a TSL slab that was a similarly stacked 3-D TSLD consisting of $Li_3B_7O_{12}$:Cu and synthetic resin, and verification was performed for the dose profile of 6 MV X-rays. In this report, dose calibration of the 2-D TSLD was performed for this first time, and it was shown that non-uniformity in the sensitivity of the TSL images after calibration was 1.2% and the off-axis ratio and percentage depth dose matched well with measurement results obtained by ionization chamber dosimetry. Although it has become easy to increase the volume by mixing with resin, degradation of the sensitivity is observed for reasons such as resin deterioration, which is also a problem for TSL foils.

7.5.4.3 Ceramic TSL Slab

For repeated usage, which is the most notable feature of 2-D TSLDs, it is desirable to fabricate 2-D TSLDs that do not contain resin. In 2016, Shinsho et al. [42] reported using a ceramic plate with commercial aluminum oxide (alumina) as the main component in a 2-D TSLD. In that report, an alumina ceramic plate with dimensions of 80 mm × 80 mm and thickness of 1 mm was used to successfully obtain a TSL distribution with better resolution than that of an IP without calibration of each pixel. Furthermore, since an alumina ceramic plate does not contain resin, it is thought that sensitivity degradation will not occur due to repeated use. In 2017, Yanagisawa et al. [43] improved the alumina ceramic plate and reported applied research into SRT for robotic radiosurgery using a ceramic TSL slab of thickness 0.7 mm composed of Al_2O_3:Cr. In that report, the repeatability of the ceramic TSL slab was around 1%, and verification of the dose profile showed an almost perfect match to the treatment planning system (Figure 7.44). Furthermore, the dose profile after calibration of each pixel agreed well with the treatment planning system, and the results of gamma

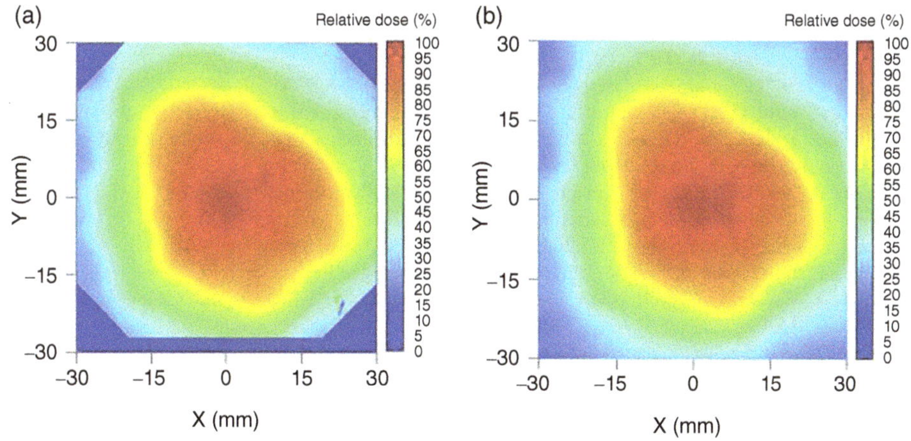

Figure 7.44 *Dose distribution of DQA test: (a) measured by the Al_2O_3:Cr TSL slab dosimeter; and (b) calculated by treatment planning system [43].*

analysis with acceptance criteria 2% / 2 mm showed a good result with an acceptance rate of 100%. Furthermore, compared to radiochromic film that was used for comparison, the dose profile was relatively free of noise, and a high-precision dose distribution was successfully verified for the first time.

7.5.5 Outlook for 2-D TSLDs

TSLDs have mainly been used for individual exposure dosimetry and environmental radiation monitoring, but they can be used repeatedly and offer the advantage of being able to perform absolute dose measurement by dose calibration, unlike other 2-D passive dosimeters. Because the film dosimeters used as 2-D passive dosimeters can be used only once, correction for non-uniformity in the sensitivity of individual dosimeters and dose calibration cannot be performed. In the future, it will probably become possible to perform dose distribution measurement with much higher precision by 2-D TSLDs by utilizing these features.

Furthermore, because high-precision radiotherapy also requires the measurement of 3-D dose distribution, it is also expected to lead to the development of 3-D TSLDs. The 3-D dosimeters currently used in clinical settings calculate the 3-D distribution from the measurement results of array semiconductor detectors with a particular face, such as the Delta4 (ScandiDos) and ArcCHECK (Sun Nuclear). Although polymer gel dosimetry has been proposed for 3-D dosimetry that can perform actual measurements, this type of dosimeter also cannot be used repeatedly, the same as film dosimetry, and consequently obtaining the absolute dose distribution is difficult. Because of this, 3-D TSLDs might be the type of dosimetry that comes closest to successfully measuring 3-D absolute dose distributions for the first time.

References

1. Boyle, R. (1664). Experiments and considerations touching colours. *Royal Soc.*: 413.
2. Röntgen, W.C. (1895). Über eine neue Art von Strahlen. *Sitzungs-Berichte Phy. med. Gesellsch., Wurzburg* 28: 132–141.
3. Wiedeman, E. and Schmidt, G.C. (1895). Ueber luminescenz. *Ann. Phys. Chem. Neue Folge* 54: 604.
4. Ramberg, E.G. (1941) Electronic Multiplier. Patent CA 400871.
5. Harlow, F.J. (1913). A lecture experiment to illustrate ionisation by collision and to show thermoluminescence. *Proc. Phys. Soc. Lond.* 26 (1): 55–60.
6. Laird, E.R. (1927). Absorption in the region of soft X-rays. *Phys. Rev.* 29 (1): 41–49.
7. Daniels, F., Boyd, C.A., and Saunders, D.F. (1953). Thermoluminescence as a research tool. *Science* 117: 343.
8. Yamashita, T., Nada, N., Onishi, H. et al. (1971). Calcium sulfate activated by thulium or dysprosium for thermoluminescence dosimetry. *Health Phys* 21: 295–300.
9. Ginther, R.J. and Kirk, R.D. The thermoluminescence of CaF_2:Mn. *Electrochem. Soc.* 104: 1957, 365–1369.
10. Binder, W., Disterhoft, S.M., and Cameron, J.R. (1968) Dosimetric properties of CaF_2:Dy. In: Proc. 2nd Int. Conf. on Luminescence Dosimetry, Gatlinburg. CONF680920 (NTIS, Springfield, VA), 43.
11. Daniels, F. and Rieman, W.P. (1954) University of Wisconsin, Final Report No. 7
12. Kossel, W., Mayer, U., and Wolf, H.C. (1954). Simultan-Dosimetrie von Strahlungsfeldern im lebenden Objekt. *Naturwiss* 41: 209.
13. Moore, L.E. (1957). Thermoluminescence of sodium sulfate and lead sulfate, and miscellaneous sulfates, carbonates, and oxides. *J. Phys. Chem.* 61: 636–639.
14. Schulman, J.H.R.D. and West, E.J. (1965) In: Proc.1st Int. Conf. on Luminescence Dosimetry, Stanford, 113.
15. Randall, J.T. and Wilkins, M.H.F. (1945). Phosphorescence and electron traps. *Proc. Roy. Soc. A* 184: –366.
16. Garlick, G.F.J. and Gibson, A.F. (1948). The electron trap mechanism of luminescence in sulphide and silicate phosphors. *Proc. Phys. Soc.* 60 (6): 308, 574–590.

17. May, C.E. and Partridge, J.A. (1964). Thermoluminescent kinetics of alpha-irradiated alkali halides. *J. Chem. Phys.* 40 (5): 1401–1409.
18. Daniels, F., Boyd, C.A., Saunders, D.F. et al. (1953). Thermoluminescence as a research tool. *Science* 117 (3040): 343–349.
19. Nakajima, T., Habu, T., Hashizume, T. et al. (1967). Thin film phosphors of $CaSO_4$(Mn) as thermoluminescent dosimeters. *Nucl. Technol.* 4 (8): 446–447.
20. Broadhead, P. and Newman, G.A. (1972). A thermoluminescent image storage panel. *J. Phys. E Sci. Instrum.* 5: 69–72.
21. Yasuno, Y., Tsutsui, O., Yamamoto, T. et al. (1983). Feasibility of two-dimensional dosimetry or an X-ray imaging method using $CaSO_4$ TL film. *Radiat. Prot. Dos.* 6 (1–4): 341–343.
22. Duller, G.A.T., Bøtter-Jensen, L., Markey, B.G. et al. (1997). A luminescence imaging system based on a CCD camera. *Radiat. Meas.* 27 (2): 91–99.
23. Shinsho, K., Maruyama, D., Yanagisawa, S. et al. (2018). Thermoluminescence properties for X-ray of Cr-doped Al_2O_3 ceramics. *Sens. Mater.* 30 (7): 1591–1598.
24. McKeever, S.W.S., Moscovitch, M., and Townsend, P.D. (1995) Thermoluminescence, Dosimetry Materials: Properties and Uses, 49.
25. Horowitz, Y.S., Oster, L., and Datz, H. (2007). The thermoluminescence dose-response and other characteristics of the high-temperature TL in LiF:Mg,Ti (TLD-100). *Radiat. Prot. Dosimetry* 124: 191–205.
26. Biderman, S., Horowitz, Y., and Oster, L. (2002). Investigation of the emission spectra of LiF:Mg,Ti (TLD-100) during thermoluminescence. *Radiat. Prot. Dos.* 100: 369–372.
27. Bilski, P. (2002). Lithium fluoride: from LiF:Mg,Ti to LiF:Mg,Cu,P. *Radiat. Prot. Dos.* 100: 199–206.
28. McKeever, S.W.S., Moscovitch, M., and Townsend, P.D. (1995) Thermoluminescence, Dosimetry Materials: Properties and Uses, 66.
29. Necmeddin Yazici, A., Metin Bedir, A., and Sökücü, S. (2007). The analysis of dosimetric thermoluminescent glow peak of CaF_2:Mn after β-irradiation. *Nucl. Instrum. Methods Phys. Res. B.* 259: 955–965.
30. Vilaithong, T., Wanwilairat, S., Rhodes, M. et al. (2002). High resolution emission spectra of TL materials. *Radiat. Prot. Dos.* 100: 211–216.
31. Chakrabarti, K., Sharma, J., Mathur, V.K. et al. (1995). Optical properties of radiation defect centers involving single and paired Mn^{2+} centers in CaF_2:Mn. *Phys. Rev.* 51: 16541–16548.
32. Karsu Asal, E.C., Polymeris, G.S., and Kitis, G. (2018). Thermoluminescence response of CaF_2:Dy as a function of irradiation temperature between room temperature and 270 °C. *Radiat. Meas.* 118: 14–19.
33. Massillon-JL, G., Gamboa-deBuen, I., Buenfil, A.E. et al. (2008). CaF_2:Tm (TLD-300) thermoluminescent response and glow curve induced by γ-rays and ions. *Nucl. Instrum. Methods Phys. Res. B.* 266: 772–780.
34. McDougall, R.S. (1970). Thermoluminescent dosimetry of aluminum oxide. *Health Phys.* 19: 281–283.
35. Rike, J.K. and Daniels, F. (1957). Thermoluminescence studies of aluminum oxide. *J. Phys. Chem.* 61: 629–633.
36. Methta, S.K. and Sengupta, S. (1976)). Gamma dosimetry with Al_2O_3 thermoluminescent phosphor. *Phys. Phys. Med. Biol.* 21: 955.
37. Osvay, M. and Biro, T. (1980). Aluminium oxide in TL dosimetry. *Nucl. Instrum.Methods* 175: 60–61.
38. Lapraz, D., Iacconi, P., Daviller, D. et al. (1991). Thermostimulated luminescence and fluorescence of α-Al_2O_3:Cr^{3+} samples (ruby). *Phys. Status Solidi A.* 126: 521–531.
39. Pokorny, P. and Ibarra, A. (1993). On the origin of the thermoluminescence of Al_2O_3:Cr,Ni. *J. Phys. Condens. Matter* 5: 7387.
40. Akselrod, M.S., Kortov, V.S., Kravetsky, D.J. et al. (1990). Highly sensitive thermoluminescent anion-defective alpha-Al_2O_3:C single crystal detectors. *Radiat. Prot. Dosimetry* 32: 15–20.
41. Akselrod, M.S., Kortov, V.S., and Gorelova, E.A. (1993). Preparation and properties of alpha-Al_2O_3:C. *Radiat. Prot. Dosimetry* 47: 159–164.
42. Shinsho, K., Kawaji, Y., Koba, Y. et al. (2016). X-ray imaging using the thermoluminescent properties of commercial Al_2O_3 ceramic plates. *Appl. Radiat. Isot.* 111: 117–123.
43. Yanagisawa, S., Shinsho, K., Inoue, M. et al. (2017). Applicability of two-dimensional thermoluminescence slab dosimeter based on Al_2O_3:Cr for the quality assurance of robotic radiosurgery. *Radiat. Meas.* 106: 326–330.
44. Sun, T., Tang, K., Cui, H. et al. (2019). Thermoluminescence of newly developed highly sensitive α-Al_2O_3:C by the vertical gradient freezing method. *Radiat. Prot. Dosimetry* 184: 174–178.
45. Summers, G.P. (1984). Thermoluminescence in single crystal alpha-Al_2O_3. *Radiat. Prot. Dosimetry* 8: 69–80.
46. McKeever, S.W.S., Moscovitch, M., and Townsend, P.D. (1995) Thermoluminescence, Dosimetry Materials:Properties and Uses, 121.

47. G. Busuoli, L. Lembo, R. Nanni, I. Sermenghi (1983). Use of BeO in Routine Personnel Dosimetry.Radiat. Prot. Dosimetry 6: 317–320.
48. Crase, K.W. and Gammage, R. (1975). Improvements in the use of ceramic BeO for TLD. *Health Phys.* 29: 739.
49. Becker, K., Cheka, J.S., and Oberhofer, M. (1970). Thermally stimulated exoelectron emission, thermoluminescence, and impurities in LiF and BeO. *Health Phys.* 19: 391–403.
50. Tochilin, E., Goldstein, N., and Miller, W.G. (1969). Beryllium oxide as a thermoluminescent dosimeter. *Health Phys.* 16: 1–7.
51. Kortov, V.S., Milman, I.I., Slesarcv, A.I. et al. (1993). New BeO ceramics for TL ESR dosimetry. *Radiat. Prot. Dosimetry* 47: 267–270.
52. Takeuchi, N., Inabe, K., Yamashita, J. et al. (1976). Thermoluminescence of MgO single crystals for UV dosimetry. *Health Phys.* 31: 519–521.
53. Kortov, V.S., Milman, I.I., Monakhov et al. (1993). Combined TSL-ESR MgO detectors for ionising and UV radiations. *Radiat. Prot. Dos.* 47: 273–276.
54. Watanabe, S. (1951). Properties of $CaSO_4$: Mn phosphor under vacuum ultraviolet excitation. *Phys. Rev.* 83: 785.
55. Huzimura, R. and Atarashi, K. (1982). The role of sulphur-oxy radicals in thermoluminescence and exoelectron emission of $CaSO_4$ phosphors. *Phys. Status Solidi A* 70: 649.
56. Nambi, K.S.V. and Bapat, V.N. (1974). Thermoluminescence of $CaSO_4$ doped with rare earths. *J. Phys. C* 7: 4403.
57. Morgan, M.D. and Stobe, T.G. (1989). Optical absorption and luminescent processes in thermoluminescent $CaSO_4$:Dy. *J. Phys. C* 1: 5773.
58. Morgan, M.D. and Stobe, T.G. (1990). An electron spin resonance study of thermoluminescence mechanisms in $CaSO_4$:Dy. *J. Phys. C* 2: 1619.
59. Gundu Rao, T.K., Bhatt, B.C., Srivastava, J.K. et al. (1993). On the sulphoxy radicals in $CaSO_4$:Dy,Na thermoluminescent phosphor: electron paramagnetic resonance studies. *J. Phys.: Condens. Matter* 5: 1791.
60. Shinsho, K., Suzuki, Y., Harada, K. et al. (2006). Multilevel based analysis of the thermoluminescence of $CaSO_4$: RE (RE=Tm, Dy, Tb, and Sm). *J. Appl. Phys.* 99: 043506.
61. Christensen, P., Bøtter-Jensen, L., and Majborn, B. (1973) Proc. Reg. Conf. on Radiation Protion, Jerusalem (Israel AEC), 194.
62. Takenaga, M., Yamamoto, O., and Yaashita, T. (1980). Preparation and characteristics of $Li_2B_4O_7$: Cu phosphor. *Nucl. Instrum. Methods* 175: 77–78.
63. Rzyski, B.M. and Morato, P. (1980). Luminescence studies of rare earth doped lithium tetraborate. *Nucl. Instrum. Methods* 175: 62–64.
64. Kazanskaya, V.A., Kuzmin, V.V., Minaeva, E.E. et al. (1974) Magnesium borate radiothermoluminescent detectors. Proc. 4th Int. Conf. Luminescence Dosimetry, Krakow, Poland, 581–592
65. Prokic, M. (1980). Development of highly sensitive $CaSO_4$: Dy/Tm and MgB_4O_7 : Dy/Tm sintered thermoluminescent dosimeters. *Nucl. Instrum. Methods* 175: 83–86.
66. Barbina, V., Contento, G., Furetta, C. et al. (1981). Preliminary results on dosimetric properties op MgB_4O_7:Dy. *Radiat. Eff. Lett.* 67: 55–62.
67. McKeever, S.W.S., Moscovitch, M., and Townsend, P.D. (1995) Thermoluminescence, Dosimetry Materials: Properties and Uses, 187–192
68. Souza, J.H., Ferrari, V.A., and de Freitas, L.C. (1993). TL peak structure of MgB_4O_7:Dy. *Radiat. Prot. Dosimetry* 47 (1–4): 239–242.
69. AAPM TG 191: Clinical use of luminescent dosimeters: TLDs and OSLDs.
70. Spurny, Z. (1980). Some new materials for TLD. *Nucl. Instrum. Methods* 175: 71–73.
71. Yasuda, H. and Fujitaka, K. (2001). Responses to TLD-BeO:Na (UD-170A) to heavy ions and space radiation. *Radiat. Prot. Dosimetry* 94: 275–280.
72. Hirata, K., Watanabe, Y., Yoshihashi, S. et al. (2017). Particle dependence of quenching effect in an optical-fiber-type optically stimulated luminescence dosimeter. *Sens. Mater.* 29: 1455–1464.
73. Hirata, Y., Watanabe, K., Uritani, A. et al. (2017). Correction of quenching effect of a small size OSL dosimeter using Eu:BaFBr and Ce:CaF_2. *Radiat. Meas.* 106: 246–251.
74. Koba, Y., Chang, W., Shinsho, K. et al. (2016). Response of alumina-based ceramic thermoluminescence detector to charged particles. *Sens. Mater.* 28: 881–886.
75. Jähnert, B. (1972). The response of TLD-700 thermoluminescent dosemeters to protons and alpha particles. *Health Phys.* 23: 112–114.

76. Waligo'rski, M.P.R. and Katz, R. (1980). Supralinearity of peak 5 and peak 6 in TLD-700. *Nucl. Instrum. Methods* 175: 48–50.
77. Waligo'rski, M.P.R. and Katz, R. (1980). Supralinearity of peak 5 and peak 6 in TLD-700. *Nucl. Instrum. Methods* 172: 463–470.
78. Pradhan, A.S., Rassow, J., and Meissner, P. (1985). Dosimetry of d(14)+ Be neutrons with the two-peak method of LiF TLD-700. *Phys. Med. Biol.* 30: 1349–1354.
79. Horowitz, Y.S., Satinger, D., Puks, E. et al. (2003). On the use of LiF:Mg,Ti TLDs in space: a critical review. *Radiat. Prot. Dos.* 106: 7–24.
80. Massillon-Jl, G., Gamboa-de Buen, I., and Brandan, M.E. (2006). Onset of supralinear response in TLD-100 exposed to ^{60}Co gamma rays. *J. Phys. D: Appl. Phys.* 39: 262–268.
81. Bilski, P. (2006). Response of various LiF thermoluminescent detectors to high energy ions: results of the ICCHIBAN experiment. *Nucl. Instrum. Methods Phys. Res., Sect. B.* 251: 121–126.
82. Berger, T., Reitz, G., Hajek, M. et al. (2006). Comparison of various techniques for the exact determination of absorbed dose in heavy ion fields using passive detectors. *Adv. Space Res.* 37: 1716–1721.
83. Berger, T. and Hajek, M. (2008). On the linearity of the high-temperature emission from ^{7}LiF:Mg,Ti (TLD-700). *Radiat. Meas.* 43: 1467–1473.
84. Kobat, D., Nahajowski, D., Góra, E. et al. (2008). On the clinical applicability of large-area 2-D TL dosimetry for verifying small photon radiotherapy beams. *Radiat. Meas.* 43: 1004–1007.
85. Kisielewicz, K., Świebocka, J., Czopyk, Ł. et al. (2010). Dosimetric properties of TL foils based on LiF:Mg,Cu,P (MCP-N) phosphors for clinical applications. *Radiat. Meas.* 45: 716–718.
86. Marrazzo, L., Pallotta, S., Kłosowski, M. et al. (2013). Clinical tests of large area thermoluminescent detectors under radiotherapy beams. *Radiat. Meas.* 51–52: 25–30.
87. Shinsho, K., Koba, Y., Wakabayashi, G. et al. (2014). Basic characteristics of tissue-equivalent phantom thermoluminescence slab dosimeter using new TL phosphor $Li_3B_70_{12}$:Cu. *Radiat. Meas.* 62: 15–21.
88. Iwata, K., Yue, N.J., Nath, R. et al. (2004). Near-field dosimetry of ^{125}I sources for interstitial brachytherapy implants measured using thermoluminescent sheets. *Med. Phys.* 31 (12): 3406–3416.
89. Klosowski, M., Czopyk, Ł., Olko, P. et al. (2008). TL efficiency of LiF:Mg,Cu,P (MCP-N) 2-D thermoluminescence detectors to raster-scanned carbon ion beams. *Radiat. Meas.* 43: 994–997.
90. Czopyk, Ł., Klosowski, M., Olko, P. et al. (2007). Two-dimensional dosimetry of radiotherapeutical proton beams using thermoluminescence foils. *Radiat. Prot. Dos.* 126 (1–4): 185–189.
91. Czopyk, Ł., Cirrone, G.A.P., Cuttone, G. et al. (2008). 2-D dosimetry of a proton radiotherapy beam using large-area LiF:Mg,Cu,P TL detectors. *Radiat. Meas.* 43: 977–980.
92. Kłosowski, M., Czopyk, Ł., Kisielewicz, K. et al. (2010). Novel thermoluminescence foils for 2-D clinical dosimetry, based on $CaSO_4$:Dy. *Radiat. Meas.* 45: 719–721.
93. Konnai, A., Nariyama, N., Ohnishi, S. et al. (2006). Energy response of a two-dimensional sheet-type LiF:Mg,Cu,P TL dosemeter to photons. *Radiat. Prot. Dosimetry* 120 (1–4): 125–128.
94. Nariyama, N., Konnai, A., Ohnishi, S. et al. (2006). Tissue-equivalent TL sheet dosimetry system for X- and gamma-ray dose mapping. *Radiat. Prot. Dosimetry* 120: 136–139.
95. Konnai, A., Odana, N., Nariyama, N. et al. (2006). Neutron dosimetry with ^{6}LiF-rich TL sheet. *Radiat. Prot. Dosimetry* 120 (1–4): 133–135.
96. Kłosowski, M., Liszka, M., Kopeć, R. et al. (2014). Dosimetric properties and stability of thermoluminescent foils made from LiF:Mg,Cu,P or $CaSO_4$:Dy during long-term use. *Radiat. Phys. Chem.* 104: 212–215.

8

Optically-Stimulated Luminescent Dosimeters

Hidehito Nanto and Go Okada
Co-creative Research Center of Industrial Science and Technology, Kanazawa Institute of Technology, Hakusan, Japan

8.1 Introduction

Interaction of ionizing radiation with matter is one of the most important and basic phenomena in radiation measurements and engineering. As incident radiation energy is absorbed by a material (or solid), a large number of free electrons and holes are generated via ionization and excitation processes. Usually, the generated electrons and holes collide with the host lattice and then reduce energy while generating secondary electrons. Eventually, the energy is completely lost by recombining to emit light or phonons (heat). During these processes, if localized trapping centers exist in the host, some of the radiation-generated charges such as free electrons and holes are captured by some of these trapping centers. The trapped charges are metastable or sometimes completely stable. In the metastable state, when external stimulation by heat or light is given to the material, the metastable charges relax to the ground state radiatively (meaning to emit light) and return to the original state. The luminescence phenomenon due to the thermal stimulation is called thermally-stimulated luminescence (TSL), while the one due to the optical stimulation is called optically-stimulated luminescence (OSL). The luminescence intensities of TSL and OSL are proportional to the dose of ionizing radiation delivered to the material, so these phenomena are utilized as a passive-type dosimeter to be used in applications such as environmental and personnel dose monitoring, as well as imaging plates (IP) for computed radiography (CR) in medical diagnosis.

In this chapter, we focus on the principles and applications of OSL, which are utilized today in personnel dose monitoring, environmental dose monitoring, and geographical dating.

Phosphors for Radiation Detectors, First Edition. Edited by Takayuki Yanagida and Masanori Koshimizu.

8.2 Principles of OSL Phenomenon

OSL phenomenon has been used in personnel dose monitoring and IP for CR since the 1980s. However, such OSL phenomenon was even observed in the distant past in group II–VI compound semiconductors after irradiation with UV or visible light and then stimulation by red or infrared light. It is understood that this phenomenon was first reported by Edmond Becquerel [1] and his son, Henri Becquerel [2] in the 1880s, when they were studying phosphors and ionizing radiations.

They indicated this phenomenon as the fact that phosphorescence of zinc sulfide (ZnS) and calcium sulfide (CaS) phosphors, after irradiating with ionizing radiation, can be suppressed by exposing to infrared light [3]. In addition, a German physicist, Philipp E.A. Lenard, reported the OSL phenomenon even earlier, in 1898 [4], so it is not clearly recognized when exactly the OSL phenomenon was first reported. In the field of phosphors (or luminescent materials), OSL is often referred to as photo-stimulated luminescence (PSL), and this particularly focused on detection of infrared laser light. However, the PSL and related research did not continue for a while. In the 1990s, Fuji Photo Film Co., Ltd. paid attention to BaFX:Eu (X = Br, I) [5, 6], as a new formula of PSL material for IP in CR as an alternative to the conventional photographic film in medical diagnosis, and they finally succeeded in developing and commercializing an IP by using Eu-doped BaFBr. As a result, the PSL phenomenon again received attention. Furthermore, together with the commercialization of IP, from the late 1990s to the early 2000s, a personnel dosimeter based on OSL was developed for the purpose of radiation dose monitoring and management, and even today, it is widely used in facilities dealing with ionizing radiation.

The OSL phenomenon is transitional luminescence observed by stimulation of visible or infrared light in materials such as insulators and semiconductors which are initially irradiated by ionizing radiation. Figure 8.1 illustrates an energy band diagram, which demonstrates the generation of electrons and holes (conversion process), the carrier (energy) transportation process, and the luminescence process. Furthermore, Figure 8.2 shows the

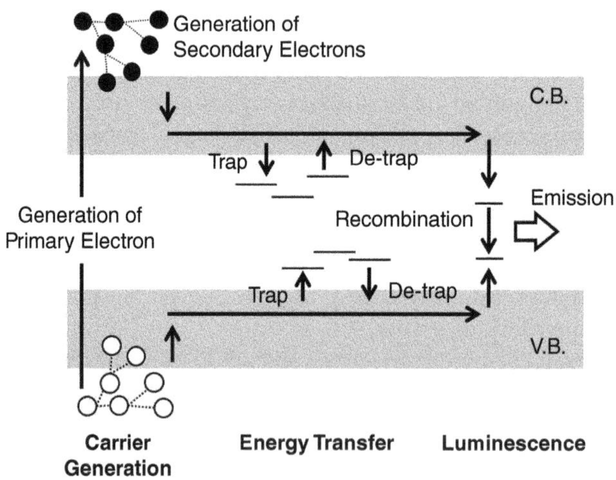

Figure 8.1 Processes of charge generation (conversion), energy transportation, and luminescence in a phosphor material upon irradiation with ionizing radiation.

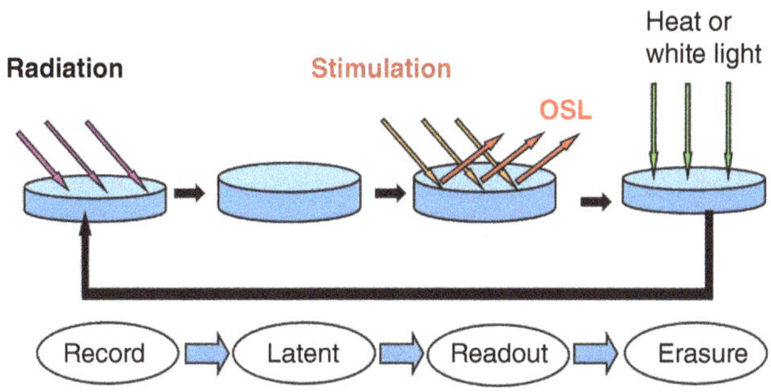

Figure 8.2 *Illustration for TSL and OSL phenomena. Upon irradiation, electrons and holes are generated (conversion), and they are trapped at local sites (latency). Next, the material is stimulated optically (stimulation), which results in photo emission of wavelengths shorter than the stimulation (readout). The remaining information can be completely removed by heat-treatment or exposing to white light, and it is reset to the initial state.*

processes of information storage, latency, readout, and erasure. Upon an incidence of radiation to phosphor material and deposition of radiation energy, excitation and ionization take place in the material. Then electrons in the valence band (V.B.) are excited to the conduction band (C.B.), while free holes are generated in the valence band. This process corresponds to the "storage" shown in Figure 8.2. The electrons excited to the conduction band freely move in the host matrix and generate many secondary electrons, which collide with the lattice and lose energy (thermalization) to relax to the bottom of the conduction band. The same processes apply to the holes in the valence band. Furthermore, these free electrons and holes commonly recombine via the radiative process and then emit light (scintillation). However, when lattice defects and impurities are present in the matrix, they generate energy levels in the forbidden band and so are able to capture free electrons and holes. As a result, the electrons generated by ionizing radiation are captured by these energy levels, and then localized trapped centers are generated in the forbidden band. Here, the energy level of the trapped center is sufficiently deep with respect to the bottom of the conduction band so that the trapped electrons remain stable at room temperature and the storage process is shown in Figure 8.2). In general, the number of trapped centers is proportional to the amount of incident radiation dose. We call this stage "latency," where information (radiation dose) is memorized stably. The information accumulated in the phosphor material, i.e., trapped centers, is metastable as mentioned above, therefore it returns back to the original state when stimulation by heat or light, known as the "readout process," shown in Figure 8.2. In this process, upon stimulation, the electrons captured by trapping centers are excited to the excitation state of localized trapped centers, travel along the conduction band, and then recombine with the trapped holes at luminescent centers. Eventually, light is emitted with the photon energy corresponding to the energy gap of excitation and ground levels of the luminescent center. The observed luminescence is OSL or TSL when the stimulation is light or heat, respectively. The OSL and TSL intensities are proportional to the incident radiation energy, and can be used as a passive-type radiation detector. A fraction of information may remain in the material after the readout process, but it can be completely removed and the detector can be initialized by heat-treatment or exposure to white light; the erasure process shown in Figure 8.2. Generally, the materials can be reused multiple times with this initialization process. In designing OSL phosphor materials, it is important to consider how the trapping center is introduced into the phosphor material. In perfect crystals, all of the electrons and holes generated by ionizing radiation are lost by recombination and trapped centers cannot be produced.

There may be two ways to introduce a trapping center into a phosphor material. One is to intentionally introduce lattice defects. For example, in ionic crystals such as alkaline halides (LiF, NaCl, KCl, CsBr, RbBr, etc.) and alkaline fluorohalides (BaF, BaFBr, BaFIBr, etc.), introducing anion vacancies can generate electron trapping centers. In Na^+Cl^- as a typical ionic crystal, Cl^- vacancies are positively charged and can capture negatively charged electrons. Then the anion vacancies are generated in the forbidden band. In general, ionic crystals having large band-gap energy do not absorb visible light and are optically transparent. However, when an electron trapped anion vacancy is generated, it absorbs visible light and takes on some color [7]. This works as a color center and is often referred to as the F-center, where F stands for Farbe (color in German). As above, alkaline halides, alkaline fluorohalides, and some oxide materials have been long recognized as OSL phosphor materials, as they show F-centers after exposure to ionizing radiation. Figure 8.3 shows typical OSL stimulation and emission spectra of X-ray irradiated Eu^{2+}-doped KBr (KBr:Eu), which is well-known as an alkaline halide-type phosphor. Also, Figure 8.4 illustrates the mechanism of OSL phenomenon in KBr:Eu by using an energy band diagram. After X-ray irradiation, on the one hand, F-centers (an electron trapped by a Cl^- ion vacancy) are generated, and the material turns purple to the naked eye. On the other hand, X-ray generated holes are trapped by Eu^{2+} ions, and so become Eu^{3+} centers. When exposing (stimulating) the latter material to red light (~ 620 nm), the electrons trapped by F-centers are excited to the conduction band and then become free electrons. Furthermore, they recombine with holes, and then the Eu^{3+} reverts back to Eu^{2+}. In this process, the recombination energy excites the Eu^{2+} ion and then radiatively relaxes to the ground state, which results in light emission of energy corresponding to the energy gap of the excited and ground state of Eu^{2+} (i.e., 420 nm). The fact that this luminescence intensity is proportional to the absorbed X-ray dose was confirmed in an earlier study [7]. Therefore, KBr:Eu has promising OSL properties for IP, as a 2-D radiation imaging medium and in personnel dose monitoring applications.

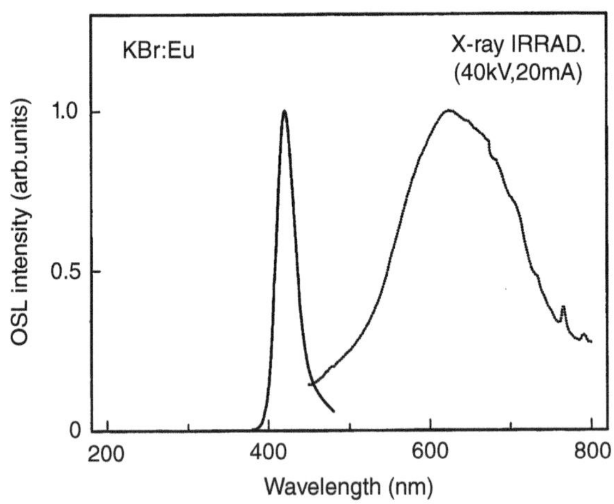

Figure 8.3 *Typical OSL emission spectrum (left) and stimulation spectrum (right) of X-ray irradiated KBr:Eu storage phosphor.*

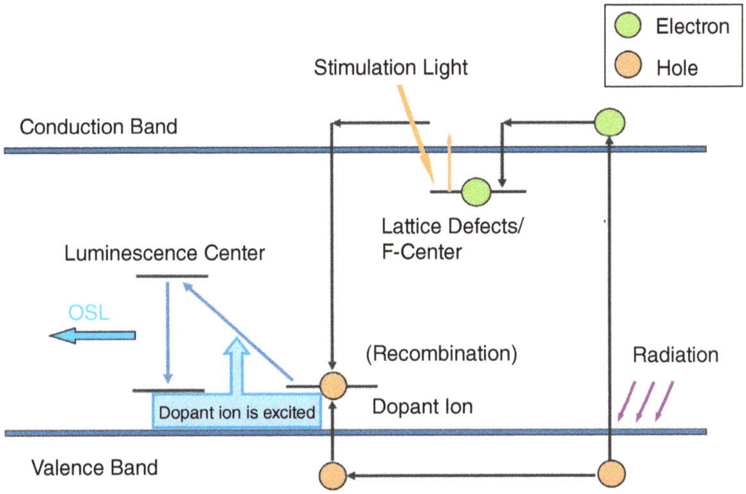

Figure 8.4 *Energy band diagram illustrating OSL phenomenon of KBr:Eu storage phosphor.*

In alkaline halide and alkaline fluorohalide phosphors, other than electron trapping centers by anion vacancies, a fraction of rare earth elements (e.g., Eu, Ce, Sm, Tb, Dy, Tm, etc.) can be introduced as luminescent centers as well as hole trapping centers. Therefore, in OSL phosphor materials, a correct amount of monovalent or divalent metal elements (other than the above rare earth elements) are added and replace with the lattice element. Also, Al_2O_3:C [8, 9] and BeO [10, 11] are known as OSL materials used in personnel dose monitoring today.

Alkaline fluorohalides such as BaFBr can be used for IP, i.e., 2-D imaging medium in medicine, and IPs using BaFBr:Eu and CsBr:Eu have been developed and are commercially utilized today. Regardless of the material, it is important to design and introduce trapping centers to stably and effectively capture electrons and holes generated by ionizing radiation.

Figure 8.5 shows typical OSL emission and stimulation spectra of each CsBr storage phosphor film doped with In_2O_3, Eu_2O_3, $SmCl_2$, $EuCl_3$, $GdCl_3$, $TbCl_3$, and $NdCl_3$. It can be seen that the $EuCl_3$-doped CsBr storage phosphor film exhibits the highest OSL emission. The peak wavelengths of OSL emission and its stimulation spectra are about 450 nm and 690 nm, respectively, as shown in Figure 8.5. Figure 8.6 shows optical absorption spectra of CsBr:Eu storage phosphor film before and after X-ray irradiation. The broad absorption band at about 690 nm in the spectrum after X-ray irradiation, coinciding with the OSL stimulation peak in Figure 8.6, is due to F centers which have been formed by Br vacancies. Figure 8.7 shows typical photoluminescence (PL) emission and its excitation spectra of CsBr:Eu storage phosphor film. It should be noted that the peak wavelength of the PL emission spectrum coincides with that of the OSL emission spectrum, implying that the origin of the OSL emission is the same as that of PL. From the excitation spectrum, it can be seen that the OSL and PL emission peaks are due to the Eu^{2+} ions.

Figure 8.8 shows a photograph of developed IP using CsBr:Eu storage phosphor film which is deposited on a heat-resistive substrate using conventional evaporation techniques. Figure 8.9 shows an SEM image of the CsBr:Eu storage phosphor film deposited on a heat resistive substrate, by the e-beam evaporation method, and indicates that the CsBr:Eu storage

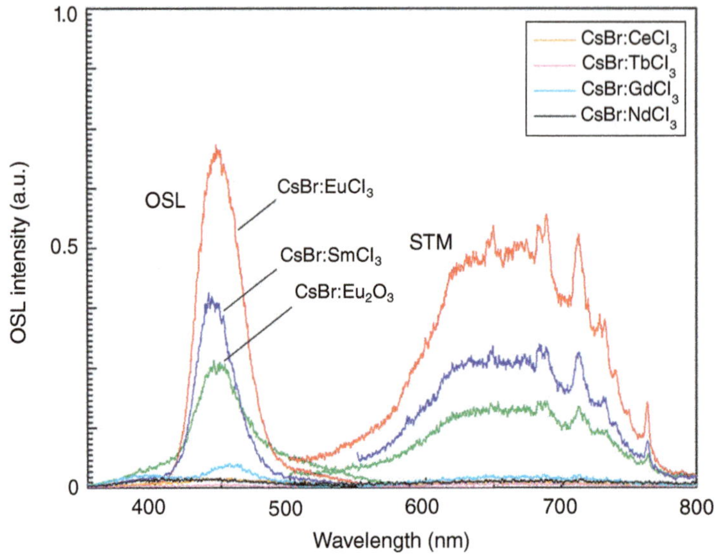

Figure 8.5 *Typical OSL emission and stimulation spectra of each CsBr storage phosphor film doped with* In_2O_3, Eu_2O_3, $SmCl_2$, $EuCl_3$, $GdCl_3$, $TbCl_3$, *and* $NdCl_3$.

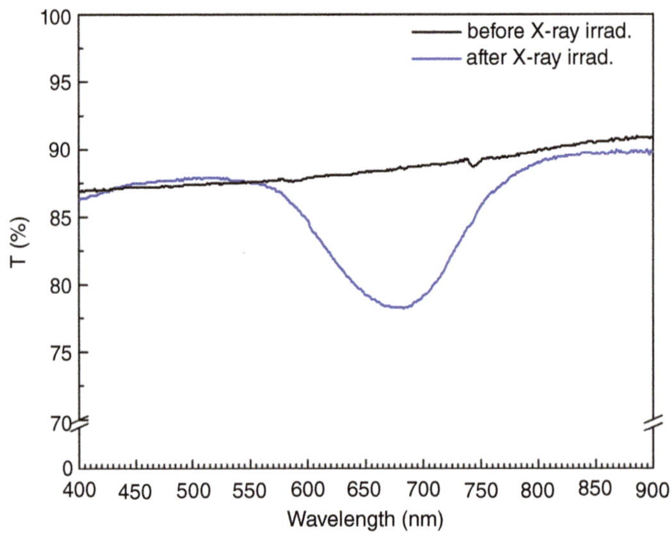

Figure 8.6 *Optical absorption spectra of CsBr:Eu storage phosphor before and after X-ray irradiation.*

phosphor film has a homogeneous columnar crystal structure. Figure 8.10 is a schematic diagram showing the light guide effects of the pillar structure in CsBr:Eu storage phosphor film. The light guide effects of homogeneous columnar structure perpendicular to the substrate produce significant improvement on OSL light collection efficiency due to reduced scattering of OSL light in the storage phosphor layer. Figure 8.11 shows transmittance

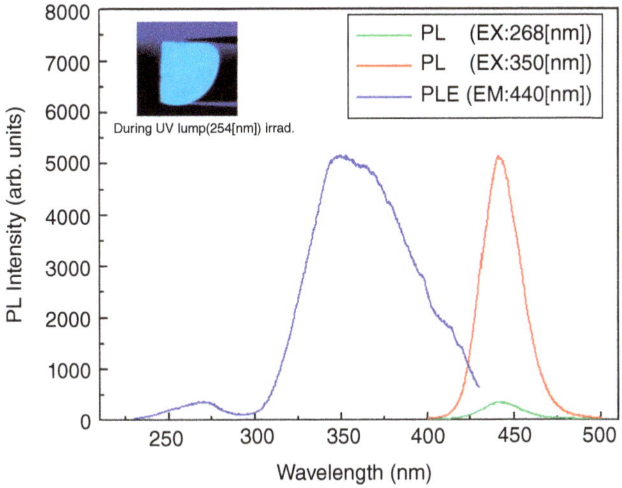

Figure 8.7 *Typical PL emission and its excitation spectra of CsBr:Eu storage phosphor. In the inset, photograph of blue PL emission is shown for UV-light (254 nm)-excited CsBr:Eu storage phosphor.*

Figure 8.8 *Photograph of developed IP sheet using CsBr:Eu phosphor storage film.*

images of IP using vapor deposited storage phosphor film and IP using coated common storage phosphor powder. It can be seen that the image of vapor deposited IP is clearer than that of coated IP.

Noise equivalent quanta (NEQ), which indicates the overall quality of image, is given by the following equation:

$$\mathrm{NEQ}(f) = \mathrm{MTF}^2(f) / \mathrm{WS} \tag{8.1}$$

where MTF indicates the modulation transfer function, WS is the Wiener spectrum, and f is the space frequency. To advance the quality of the image, it is necessary to obtain high MTF

Figure 8.9 SEM image of the CsBr:Eu storage phosphor film deposited on a glass substrate, by the e-beam evaporation method.

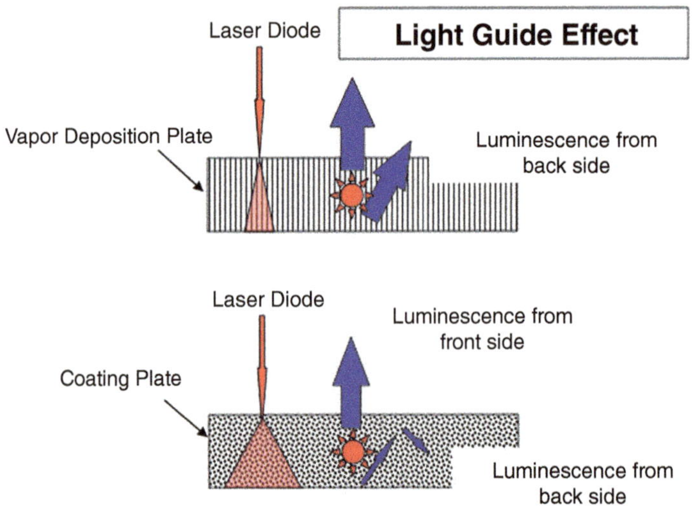

Figure 8.10 Light guide effects of a pillar structure.

and low WS. Figure 8.12 shows the Wiener spectra of the IP using CsBr:Eu storage phosphor film, compared with that of commercially-available IP using BaFI:Eu powder phosphor, thus indicating that the WS of the IP using CsBr:Eu storage phosphor film is lower than that of the IP using BaFI:Br powder phosphor. Figure 8.13 shows the MTF of the IP using CsBr:Eu storage phosphor film and that of the IP using BaFI:Eu powder phosphor. It can also be seen that the MTF of the IP using CsBr:Eu storage phosphor film is higher than that of the IP using commercially-available BaFI:Eu powder phosphor. Consequently, the NEQ (f) of the IP using CsBr:Eu storage phosphor film is also higher than that of the IP using commercially-available BaFI:Eu powder phosphor.

Figure 8.14 shows the eraser energy dependence of erasability of the OSL images for vapor deposited IP using CsBr:Eu storage phosphor film and coated IP using BaFI:Eu

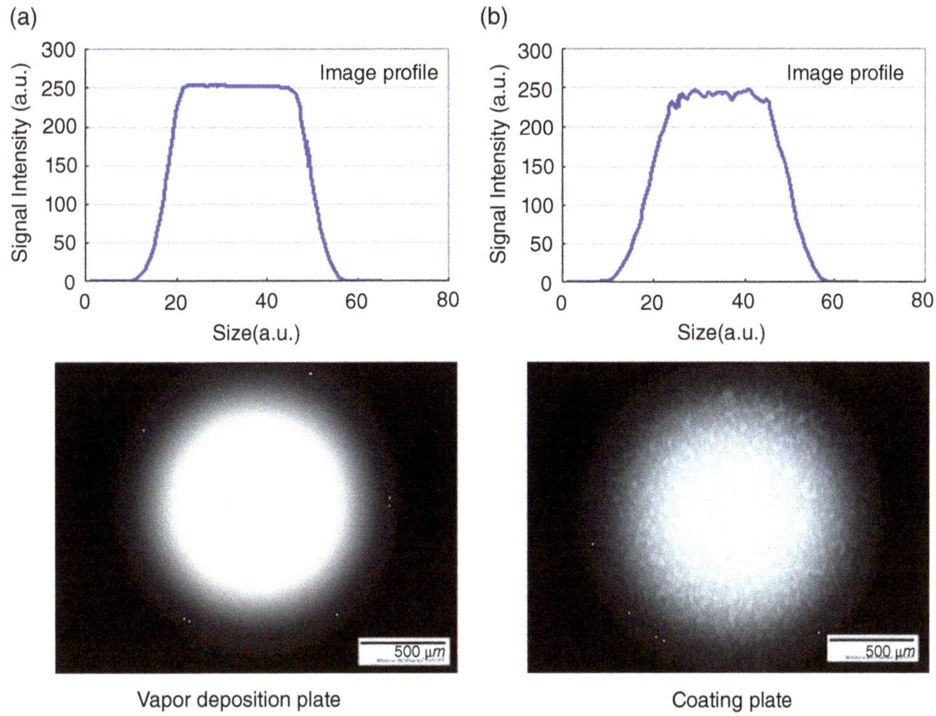

Figure 8.11 *Transmittance images of IP using vapor deposited storage phosphor film (a) and IP using coated storage phosphor powder (b). The images were observed for storage phosphor layers separated from the substrate, when the spotlight was illuminated from the back surface of the storage phosphor layer.*

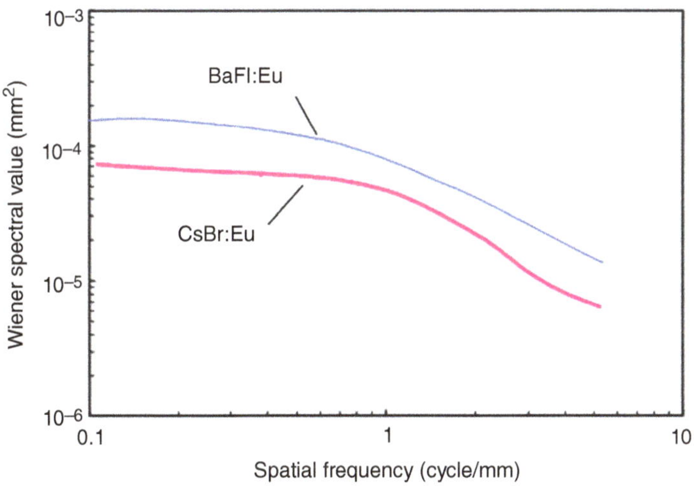

Figure 8.12 *Wiener spectra of the IP using CsBr:Eu storage phosphor film compared with that of commercially-available IP using BaFI:Eu storage phosphor.*

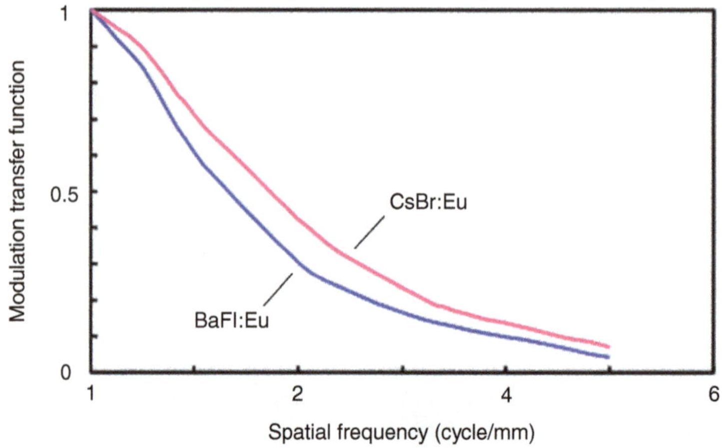

Figure 8.13 *MTF of the IP using CsBr:Eu storage phosphor film, compared with that of commercially-available IP using BaFI:Eu powder phosphor.*

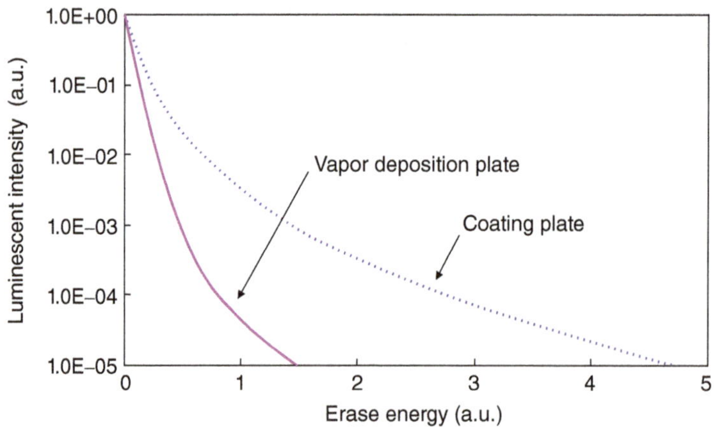

Figure 8.14 *Erasability of the OSL images for vapor deposited IP using CsBr:Eu storage phosphor film and coated IP using BaFI:Eu powder phosphor.*

powder phosphor. The erasing was done using a white light source. It can be seen that excellent erasing characteristics are obtained in the IP using CsBr:Eu storage phosphor film.

Figure 8.15 shows an X-ray image of a human breast (chest) using CsBr:Eu IP. The image is very clear. Figure 8.16 shows the chest phantom X-ray images of the IP using CsBr:Eu storage phosphor film (left) and that of the IP using BaFI:Eu powder phosphor (right), indicating that the image of the IP using CsBr:Eu storage phosphor film has better image sharpness than the image graininess of that using IP using BaFI:Eu powder phosphor. It was confirmed that the detective quantum efficiency (DEQ) value of the IP using CsBr:Eu storage phosphor is twice as high as that using BaFI:Eu powder phosphor. The IP using CsBr:Eu storage phosphor film for mammography in medical diagnosis is now commercially available from KONICA MINOLTA Co.

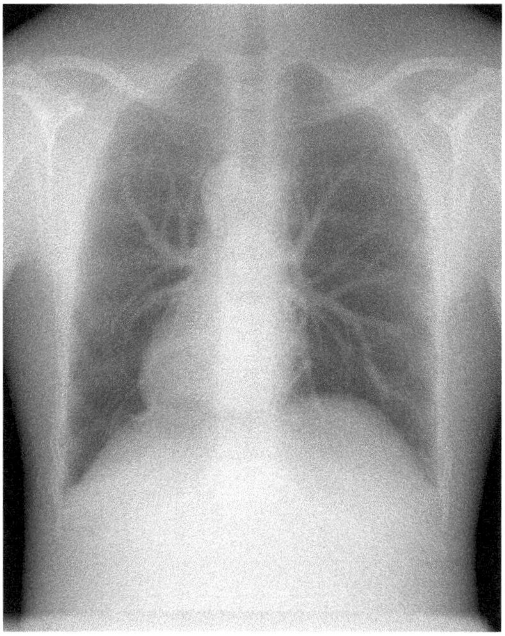

Figure 8.15 *X-ray image of human breast obtained using IP with the CsBr:Eu storage phosphor.*

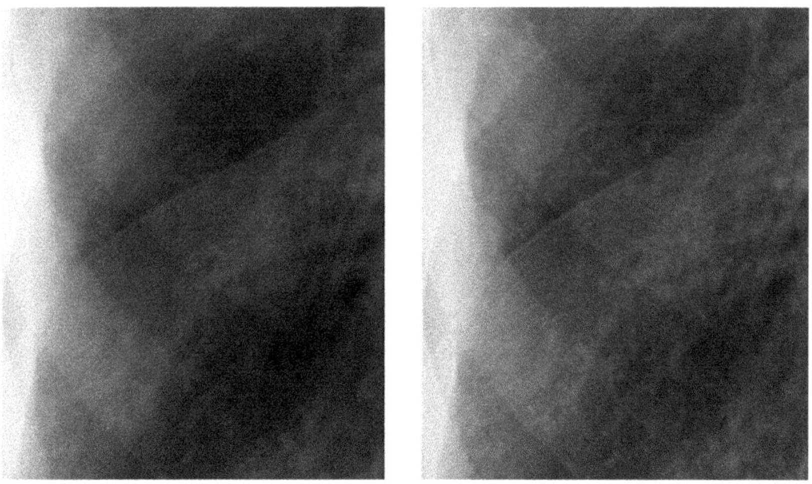

Figure 8.16 *Chest phantom X-ray images measured using the IP with the CsB r: Eu storage film (left picture) and the IP using coated BaFI:Eu phosphor powder (right picture).*

8.3 OSL Materials and Dosimeters

As discussed in Section 8.2, phosphors of alkaline halides, II–VI group compounds, oxides, diamonds, and organic compounds are known to show effective OSL phenomenon. In Table 8.1, OSL materials studied and commercialized are listed. The principles of OSL

Table 8.1 *List of OSL materials.*

Type	OSL materials	Applications	Stimulation wavelength [nm]	Emission wavelength [nm] (life time)	Effective atomic number [Z_{eff}]	References
Alkali halides	BaFX:Eu (X = Br, Cl, I)	X-ray imaging	495, 590	490 (750 ns)	~50	[10, 12–14]
	NaCl:Cu	X-ray imaging	470	353 420	~15	[7]
	KCl:Eu	X-ray UV-ray Imaging	560	420 (~1 μs)	18.1	[15]
	KBr:Eu	X-ray imaging	620	420 (~1 μs)	31.5	[16]
	RbBr:Eu	CR(X-ray)	700	360		[17]
	CsBr:Eu	CR(X-ray)	690	470	30	[18]
	$BaBr_2$:Ce	X-ray	500	360		[19]
Sulfides	CaS:Eu,Sm	X-ray, UV-ray	1180	640		[20]
	CaS:Ce,Sm	X-ray, UV-ray	1180	510, 560	18.5	[21]
	SrS:Eu,Sm	X-ray, UV-ray	1020	600	34.6	[22]
	MgS, Ce,Sm	X-ray	1050	490 (36 165 ns)	14.6	[23]
	MgS:Eu,Sm	X-ray	1180	590	14.6	[24]
Oxides	Al_2O_3:C	Individual Monitoring, Medical Space Dosimetry	530	335 (<7 ns), 420 (35 ms)	11.3	[9]
	BeO	Individual Monitoring	435	380	7.2	[10, 11, 25]
	MgO:Tb	Passive-type Dosimeters	500–570	350–700 (Tb^{3+})	10.8	[26]
	MgO:Fe	Individual Monitoring	1640	400	10.8	[27]

Quartz Etc.	Cu-doped Quartz	Dating	790	540	11.8	[28]
	Quartz	Dating	Blue	365(~30-40 μs)	11.8	[29]
	Feldspar	Dating	IR	Visible	11.5 13.7	[30]
	NaCl, Detergent Softner	Retrospective Dosimetry	Blue	Visible		[31, 32]
	Electronic-part Substrates	Retrospective Dosimetry	IR, Blue	Visible		[33–37]
	Tooth Enamel	Retrospective Dosimetry	IR	Green	15.6	[38–41]
Other materials recently reported	CaF_2:Ce	Passive-type Dosimeters	550	330	16.3	[42]
	CaF_2:Tm	Passive-type Dosimeters	580	320	16.3	[43]
	AlN	Passive-type Dosimeters	630	360		[44]
	$LiCaAlF_6$:Eu	Passive-type Dosimeters	470	370		[45]
	$LiCaAlF_6$:Ce	Passive-type Dosimeters	405	270–350		[45]
	$SrAl_2O_4$:Eu	Passive-type Dosimeters	800	510		[46]
	25SrO-$75B_2O_3$:Sn	Passive-type Dosimeters	580	400		[47]

Where CR, computed radiography.

phenomena, OSL materials, and their applications are introduced in the literature [48–57], so readers may refer to previous publications.

A dosimeter using C-doped Al_2O_3 (Al_2O_3:C) developed by Landauer, Inc. is one of those passive-type dosimeters using OSL phenomenon for personnel dose monitoring applications. Al_2O_3:C has attracted much attention, since it was first reported in the 1990s, as a material to show an intense TSL [58]. Basically, the Al_2O_3 matrix is doped with C in order to generate electron trapping centers as well as F-centers as a luminescent center, so that the material functions as a passive-type dosimeter [59]. Today, this type of dosimeter is spread worldwide, and is commercialized for personnel dose monitoring services. When OSL dosimeters are used for personnel monitoring services, important but elemental technologies to be considered are which stimulation light source to select to readout accumulated dose information from the OSL dosimeter; and how to detect the OSL signal. For the stimulation light source, in general, gas laser, solid-state laser, semiconductor laser, and LED are considered, but the most important point is that the stimulation wavelength matches that of the stimulation band. The stimulation band of many OSL materials ranges from about 450–800 nm (in the visible to near infrared range). From the viewpoint of miniaturization of the system, the semiconductor or intense LED are preferred. In this case, the stimulation light source should be selected from those available as commercial products. In addition, a photomultiplier tube (PMT) is generally used to detect the OSL signal. The PMT should be sensitive in the spectral range of OSL emission. Also, it is possible to reduce the noise signal when the stimulation light is frequency-modulated.

Another OSL material for personnel dose monitoring, which was commercialized in recent years, is recognized as a competitor to Al_2O_3:C, is BeO. As provided in Table 8.1, the effective atomic number Z_{eff} of BeO is about 7.2, and the value is close to that of biological tissue (~7.42) or so-called tissue equivalent. Therefore, from the viewpoint of dosimetry, it is advantageous compared with that of Al_2O_3 ($Z_{eff} = 11.3$), and BeO is considered to be more appropriate. The research on OSL of BeO has been intensively carried out by a group at the Technical University of Dresden [60], who have revealed that the luminescence origin is the F-center where an electron generated by a presence of an anion vacancy is captured [11]. Such BeO ceramics are currently manufactured by Mirion Technologies, Inc., USA, and RadPro International GmbH, Germany.

Despite the fact that both Al_2O_3 and BeO have been commercialized and personnel dose monitoring services are provided today, these detectors are not so sensitive. For this reason, many researchers have been intensively studying and searching for new OSL materials with higher sensitivity as alternatives. Some example materials are, as also listed in Table 8.1, oxides such as MgO:Tb, Mg_2SiO_4:Tb, and $LiAlO_2$:Tb, and sulfides such as MgS:Ce,Sm, SrS:Ce,Sm, and CaS:Ce,Sm. In contrast, OSL materials such as KCl:Eu, KBr:Eu, BaFBr:Eu, BaFBrI:Eu, and CsBr:Eu have been long studied for IP in CR, and some have been commercialized to replace conventional photographic film. In fact, these are also interesting materials for personnel dosimetric applications as they are very sensitive, while these ionic crystals may be hygroscopic.

As an alternative competitor to the OSL passive-type dosimeter, the radiophotoluminescence (RPL) dosimeter was recently developed by the Chiyoda Technol Corporation, and they have provided personnel dose monitoring services in Japan and some countries in Europe. In this type of dosimeter, Ag-doped phosphate glass is used. For more details, the reader is referred to Chapter 9, but briefly the phenomenon is understood as Ag^+

ions doped in glass matrix trap electrons and holes generated by ionizing radiation, then Ag^0 and Ag^{2+} centers are formed, which act as luminescent centers. As a result, the generated Ag^0 and Ag^{2+} centers can be excited by UV light and then emit PL peaking around 660 nm (orange emission) and 460 nm (blue emission). These luminescence intensities are proportional to the radiation dose and do not diminish, even after readout, unlike OSL and TSL. Although the Ag^0 and Ag^{2+} centers are stable under ambient conditions, these can be reversed back to the Ag^+ state by heating to around 350 °C. In recent studies, some materials were reported to show both OSL and RPL; therefore, we cannot deny a possibility that promising OSL materials will be found from those based on RPL materials. However, OSL materials, in general, are doped with rare-earth elements. Therefore, the development of these materials may be strongly affected by the depletion of resources and export control from the country of origin.

In recent years, Yanagida et al. [61, 62] have proposed an inverse relationship between scintillation phenomenon and storage luminescence phenomenon (TSL and OSL), where materials with high scintillation intensity show low storage luminescence and vice versa. This is pioneering work in the field, and we believe that this will be an important aspect to consider for designing OSL materials in the future.

8.4 Applications of OSL

Table 8.2 shows application fields of passive-type dosimeters using OSL phenomenon [47, 57]. As shown in the table, OSL dosimeters are not only used for personnel dose monitoring but also environmental radiation monitoring [63, 64], space dosimetry [65], CR for medical diagnosis and radiation therapy [56], autoradiography for drug discovery, dating, etc. It is predicted that the importance of OSL technologies will keep growing, especially in the physics and engineering fields.

Up to now, several types of OSL-based passive type dosimeters for personal dose monitoring have been commercialized by three different companies such as the Luxel detecter using C doped Al_2O_3 (Al_2O_3:C) phosphor (Landauer Inc., USA); BeO phosphor (Mirion Technologies, USA); and BeO phosphor (RadPro International GmbH, Germany). Figure 8.17 shows that the the Luxel + dosimeter consists of a Luxel detector enclosed in a multi-element filter pack, packaged in a light-tight paper wrapper, and sealed in a polycarbonate package [66].

These dosimeters utilizing OSL in Al_2O_3:C and BeO phosphors are capable of detecting low doses on an order of micrograys, whereas Al_2O_3:C and BeO phosphor are not as sensitive as KBr:Eu, phosphor reported by the authors, so it is considered that further improvements of OSL properties or development of new OSL materials are urgently needed.

In addition, extending applications of IPs (using ionic crystals such as BaFBr:Eu and CsBr:Eu) for CR systems in the field of medical diagnosis is also recognized as one of the most important fields. Today, X-ray photographic film is still used in, for example, chest radiography; however, an alternative to replace the photographic film as a 2-D imaging medium is IP. For IPs, the above-mentioned OSL materials are used, and are not only used in medical diagnosis but also in dentistry as a 2-D X-ray imaging medium. Such IPs are particularly characterized as having a wide dynamic range (the dose range linearly proportional to dose) having about 5–6 orders of magnitude, whereas that of photographic film is

Table 8.2 *Application fields of passive-type dosimeters using OPL and RPL phenomenon.*

Fields	Examples of Application	
Exposure Management	Individual Radiation Monitoring	OSLD (Al_2O_3:C)-Landaure Inc. (USA) OSLD (BeO)-Mirion Technology (USA) OSLD (BeO)-RadPro Inter. GmbH (Germany) PRLD (Ag-PG)-CHIYODA Technol. (Japan)
	Environmental Radiation Monitoring	OSLD (Al_2O_3:C), PRLD (Ag-PG) Environmental radiation dosimetry accompanying nuclear decommissioning
Space Radiation	Space Dosimetry	OSLD (Al_2O_3:C) 、TSLD (LiF:Mg, Ti) Response for Heavy particles (galaxy cosmic ray, solar particle event, earth radiation belt)
	Cosmic-ray Imaging	IP (BaFBr:Eu) Cosmic ray tracks
Medical and Biology	CR	IP (BaFBr:Eu) ① X-ray diagnosis, ② abdominal aortic angiography, ③ dental panoramic photo, ④ X-ray diffraction for muscle of bullfrog ans protein crystals, ⑤ gene analysis IP (CsBr:Eu) Mammography
	Medical treatment	OSLD (Al_2O_3:C) Radiation therapy OSLD (Al_2O_3:C) Heavy ion therapy
	Radiosurgery	OSLD (Al_2O_3:C) Gamma knife
	Real-time dosimetry (patient skin and body dose)	OSLD (Al_2O_3:C, KBr:Eu) Fiber-OSL system
	Dose mapping	OSLD (Al_2O_3:C, SrS:Ce,Sm) X-ray CT
Pharmacy	Autoradiography	IP (BaFBr:Eu) ① Electrophoretic separation of cytochrome c, ② Drug metabolism research, ③ Phamacokinetic research
Physics and Engineering	Electron beam image	IP (BaFBr:Eu) ① Detector for transmission electron microscope, ② Electron difraction
	Neutron beam image	IP (BaFBr:Eu) ① Neutron radiography
	2-D UV sensors	IP (BaFBr:Eu) ① Beam profile monitor of UV laser, ② Environmental UV monitoring
	Dating	OSL from minerals and quartz
Retrospective Accident Dose Monitoring	Accidental dosimetry	OSL from electronic parts, mobile phone, ID cards

Where OSLD, optically stimulated luminescence dosimeter; TSLD, thermally stimulated luminescence dosimeter); RPLD, radio photo luminescence dosimeter; IP imaging plate; ID, identification card).

about 1 order of magnitude. Therefore, an advantage of IPs is that they are able to detect the 2-D distribution of radiation dose from low to large doses.

Table 8.3 summarizes the properties to consider for OSL dosimeters for personnel dose monitoring and IP for medical diagnosis. Among these, the most important property is the sensitivity (①), which allows a lower detection dose limit, and also decreases radiation

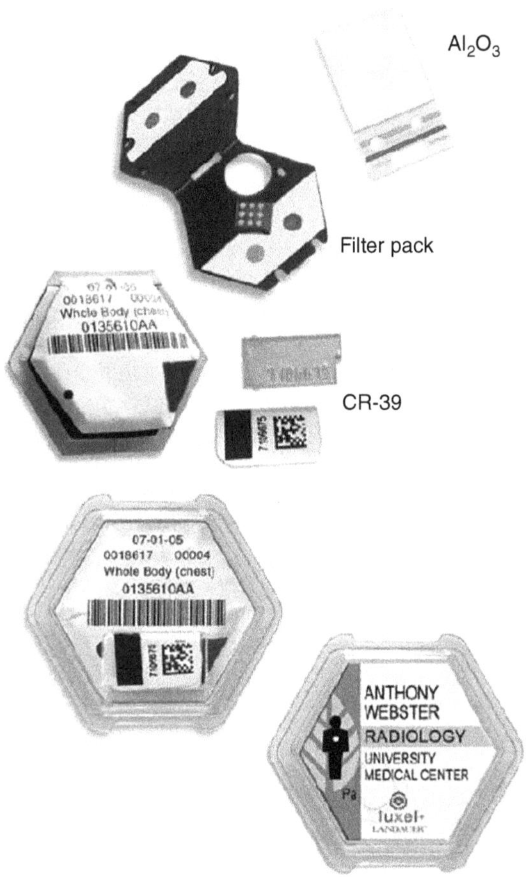

Figure 8.17 *Luxel+dosimeter consists of a Luxel detector enclosed in a multi-element filter pack, packaged in a light-tight paper wrapper and sealed in a polycarbonate package [66].*

dose to the patient in the case of medical diagnosis. The OSL emission wavelength (②) should be matched to the detection range of photodetector in the reader system. The wavelength of stimulation light should match with the most OSL effective wavelength, which is specific to each OSL material (③). Considering (②) and (③) ensures the highest possible sensitivity as an entire detection system. Although it is not listed in the table, ensuring a separation between the emission and stimulation wavelengths would help reduce the noise. Furthermore, having stable latency is a very important aspect. It results in no fading during signal storage, which is particularly important for dose evaluation and reliability. It is also important for personnel dose monitoring and evaluation of absorbed dose in biological tissue that the effective atomic number of OSL material is close and equivalent to that of biological tissue (⑨).

Table 8.4 compares some important properties: OSL sensitivity, fading property, cost, and biological equivalency or effective atomic number of commercial OSL materials, as well as those still being developed.

Table 8.3 *OSL properties to be considered for dosimetric application.*

Required performance	Characteristics as a dosimeter
① High OSL sensitivity	• Lower detection limit • Low noise
② OSL of appropriate wavelength	• Matching with sensitivity of PMP used for OSL measurement (Matching with detector)
③ Stimulation spectra of appropriate wavelength	• Matching with oscillation wavelength of laser light source used for OSL stimulation (Matching with stimulation band)
④ Life time of OSL emission	• Separation from noise • Save scan time of stimulating laser beam in read-out of OSL (in the case of IP)
⑤ Excellent fading characteristics	• Stable latent image (Stable accumulation of memorized radiation information)
⑥ Low afterglow	• Lower noise
⑦ High moisture resistance	• Improved record retention characteristics of deliquescent phosphor
⑧ Large stimulus light capture cross section	• Lower power of stimulating laser light source
⑨ Effective atomic number	• Effective for evaluating individual exposure dose
⑩ Appropriate band gap energy of OSL phosphor materials	• Latent image stabilization (Deep trap centers)
⑪ Low cost	• System cost reduction

8.5 Future Perspective

The development of dosimeters for personnel dose monitoring services is one of the most important application fields in radiation measurements. Al_2O_3:C and BeO are the currently used OSL materials, whereas further improvements of sensitivity and reliability should be considered. Also, improvements of some properties of RPL dosimeters using Ag^+-doped phosphate glass as a competitor to OSL dosimeters are also expected. Therefore, in the near future, the R&D of novel OSL and RPL materials having higher sensitivity to radiation dose will be accelerated.

Furthermore, it is considered that the market of the OSL dosimeter will continuously grow, as the use of dosimeters and radiation exposure control are regulated by law in facilities dealing with ionizing radiations such as hospitals and nuclear plants. After the accident at the Fukushima Daiichi nuclear plant in Japan, in March 2011, the plant site and neighboring region were affected by radioactive contamination. While it is extremely important to control the radiation exposure for decontamination workers, it is also important to consider development of specialized technologies, such as radiation monitoring of inside and outside of the nuclear plant, especially measuring dose distribution inside the nuclear reactor. For this, the researchers and engineers need to consider what kind of dosimeters are required, how the dose can be monitored, how the system is configured, etc. This will open up new directions of R&D for OSL technologies.

Table 8.4 *Comparison of OSL properties.*

Phosphor materials	Emission intensity (phenomenon)	Fading characteristics (Eg)	Cost	Biocompatibility (Zeff)
LiF: Mg,Ti (TLD-100:commercialized)	◎ (TSL)	○ (13.6 eV)	◎	◎ (8.2)
PG: Ag (commercialized)	◎ (RPL)	◎ (8~10 eV)	×	○ (~10)
Al_2O_3: C (TLD-500:commercialized)	△ (OSL)	◎ (8.7 eV)	○	○ (11.3)
BeO	△ (OSL)	○ (10.6 eV)	◎	◎ (8.7)
BaFBr:Eu (IP)	◎ (OSL)	△ (7.5–8.3 eV)	△	× (~50)
CsBr:Eu (IP)	◎ (OSL)	△ (~8)	△	× (~30)
KCl:Eu KBr:Eu (deliquescence)	◎ (OSL)	△ (~10 eV)	◎	× (18.1) (31.5)
Sm doped FPG	○ (RPL)	◎ (8~10 eV)	○	○ (~10)
CaF_2:Tm (TLD-300)	◎ (OSL)	○ (12.6 eV)	◎	△ (16.3)
SrO-B_2O_3:SnO	○ (OSL)	○ (8.5 eV)	○	○ (~10)
ZnO-P_2O_3:SnO	○ (OSL)	○ (~10 eV)	○	○ (~15)

E.g, band-gap energy; FPG, Fluoro phosphate glass; and Zeff, Effective Atomic Number.

References

1. Becquerel, E. (1843). Des effets produits sur les corps par les rayons solaires. *Ann. de Chem. et de Phys.* 9: 257–322.
2. Becquerel, H. (1883). Maxima et minima d'extinction de la phosphorescence sous l'influence des radiations infra-rouges. *Comptes. Rend.* 96: 1853–1856.
3. Yukihara., E.G. and Mckeever, S.W.S. (2010). *Optically Stimulated Luminescence: Fundamentals and Applications*, 2. Wiley, Incorporated.
4. Qiu, J. and Hirao, K. (1996). Research Prospective of photostimulable phosphor glasses and its application to optoelectronics. *O plus E* 203118: 125. (in Japanese).
5. Amemiya, Y. and Miyahara, J. (1988). Imaging plate illuminates many fields. *Nature* 336: 89–90.
6. Sonoda, M., Takano, M., Miyahara, J. et al. (1983). Computed radiography utilizing scanning laser stimulated luminescence. *Radiol.* 148: 833–838.
7. Nanto, H., Usuda, T., Murayama, K. et al. (1991). Photostimulable transparent NaCl:Eu single crystals for two dimensional X-ray imaging sensors. *Appl. Phys. Lett.* 59: 1838–1840.
8. Akselrod, M.S., Lucas, A.C., Polf, J.C. et al. (1998). Optically stimulated luminescence of Al_2O_3. *Radiat. Meas.* 29: 391–399.
9. Akselrod, M.S. and Mckeever, S.W.S. (1999). A radiation dosimetry method using pulsed optically stimulated luminescence. *Radiat. Prot. Dosim.* 81: 167–176.
10. Sommer, M. and Heniniger, J. (2006). Investigation of a BeO-based optically stimulated luminescence dosimeter. *Radiat. Prot. Dosim.* 119: 394–397.

11. Sommer, M., Freudenberg, R., and Henninger, J. (2007). New aspects of a BeO-based optically stimulated luminescence dosimeter. *Radiat. Meas.* 42: 617–620.
12. von Seggern, H. (1992). X-ray imaging with photostimulable phosphors, nuclear instrument and method. *Phys. Res.* A322: 467–471.
13. von Seggern, H. (1999). Photostimulable X-ray storage phosphors: a review of present understanding. *Braz. J. Phys.* 29: 254–268.
14. Rowlands, J.A. (2002). The physics of computed radiography. *Phys. Med. Biol.* 47 (R123): 254–R166.
15. Nanto, H., Endo, F., Hirai, Y. et al. (1994). Eu-doped phosphor crystals as a storage material for two-dimensional ultraviolet-ray and X-ray imaging sensors. *J. Appl. Phys.* 75: 7493–7497.
16. Douguchi, Y., Nanto, H., Sato, T. et al. (1999). Optically stimulated luminescence in Eu-doped KBr phosphor ceramics. *Radiat. Prot. Dosim.* 84: 143–148.
17. Amitani, A., Kano, K., Tsuchiy, H. et al. (1980) Adv. Printing Paper Summaries of the 26th Fall Symp. of SPSE Conf. on Electronic Imaging, San Diego, CA, p. 180.
18. Kurata, N., Kubota, N., Takei, Y. et al. (2006). Eu-doped CsBr phosphor materials for medical X-ray imaging sensors. *Radiat. Prot. Dosim.* 119: 398–401.
19. Corradi, G., Secu, M., Schweizer, S. et al. (2004). Luminescence properties of the X-ray storage phosphor $BaBr_2$: Ce^{3+}. *J. Phys. Condens. Matter* 16: 1489–1500.
20. Nanto, H., Ikeda, M., Kadota, M. et al. (1996). Photostimulated luminescence in CaS:Eu,Sm phosphor ceramics induced by excitation with ionizing radiation. *Nucl. Instr. & Meth.* 116: 262–264.
21. Sun, Y.R., Chen, Z.Y., Fan, Y.W. et al. (2011). A CaS:Ce,Sm-based dosimeter for online dosimetry measurement. *Nucl. Sci. Tech.* 22: 84–88.
22. Nanto, H., Murayama, K., Usuda, T. et al. (1993). A UV dosimeter utilizing photostimulated luminescence in SrS:Eu, Sm phosphor ceramics. *Radiat. Prot. Dosim.* 29: 254–268.
23. Dusseau, L., Plattard, D., Vaille, J.R. et al. (2000). An integrated sensor using optically stimulated luminescence for in-flight dosimetry. *IEEE Trans. Nucl. Sci.* 47: 2412–2416.
24. Rao, R.P. (1986). The preparation and thermoluminescence of alkaline earth sulfide phosphors. *J. Mater. Sci.* 21: 3357–3386.
25. Bulur, E. and Goksu, H.Y. (1998). OSL from BeO ceramics: new observations from an old material. *Radiat. Meas.* 29: 639–650.
26. Bos, A.J.J., Prokic, M., and Brouwer, J.C. (2006). Optically and thermally stimulated luminescence characteristics of MgO:Tb^{3+}. *Radia. Port. Dosimt.* 199: 130–133.
27. Kortov, V.S., Milman, I.I., Monakhov, A.V. et al. (1993). Combined TSL-ESR detector for ionizing and UV radiations. *Radiat. Prot. Dosim.* 47: 273–276.
28. Justus, B.L., Merritt, C.D., Pawlovich, K.J. et al. (1999). Optically stimulated luminescence dosimetry using doped fused wquartz. *Radiat. Prot. Dosim.* 84: 189–192.
29. Denby, P.M., Botter-Jensen, L., Murray, A.S. et al. (2006). Application of pulsed OSL to the separation of the luminescence components from a mixed quartz/feldspar sample. *Radiat. Meas.* 41: 774–779.
30. Clark, R.J. and Bailiff, I.K. (1998). Fast time-resolved luminescence emission spectroscopy in some feldspars. *Radiat. Meas.* 29: 553–560.
31. Thomsen, K.J., Botter-Jensen, L., and Murray, A.S. (2002). Household and workplace chemicals as retrospective luminescence dosimeters. *Radiat. Prot. Dosim.* 101: 515–518.
32. Larsson, C., Koslowsky, V., Gao, H. et al. (2005). Optically stimulated luminescence in forensics. *Appl. Radia. Isot.* 63: 689–695.
33. Godfrey-Smith, D.I. (2006). Applicability of moissanite, a monocrystalline form of silicon carbide, to retrospective and forensic dosimetry. *Radiat. Meas.* 41: 976–981.
34. Goksu, H.Y. (2003). Telephone chip-crds as individual dosimeters. *Radiat. Meas.* 37: 617–620.
35. Inrig, E.L., Godfrey-Smith, D.I., and Khamana, S. (2008). Optically stimulated luminescence of electronic components for forensic, retrossspective dosimetry. *Radiat. Meas.* 43: 726–730.
36. Woda, C. and Spottl, T. (2009). On the use of OSL of wire-bondchip card modules for retrospective and accident dosimetry. *Radiat. Meas.* 44: 548–553.
37. Woda, C., Greilich, S., and Beerten, K. (2010). On the OSL curve and preheat treatment of electronic components from portable electronic devices. *Radiat. Meas.* 45: 746–748.
38. DeWitt, R., Klein, D.M., Yukihara, E.G. et al. (2010). Optically stimulated luminescence (OSL) of tooth enamel and its potential use in post-exposure triage. *Health Phys.* 98: 432–439.
39. Godfrey, D.I. and Pass, B. (1997). A new method of retrospective radiation dosimetry: optically stimulated luminescence in dental enamel. *Health Phys.* 72: 744–749.

40. Godfrey, D.I. (2008). Towards *in vivo* OSL dosimetry of human tooth enamel. *Radiat. Meas.* 43: 854–858.
41. Yukihara, E.G., Mittani, J., Mckeever, S.W.S. et al. (2007). Optically stimulated luminescence (OSL) of dental enamel for retrospective assessment of radiation exposure. *Radiat. Meas.* 42: 1256–1260.
42. Yanagida, T., Fujimoto, Y., Watanabe, K. et al. (2014). Scintillation and optically stimulated luminescence of Ce-doped CaF_2. *Radiat. Meas.* 71: 162–165.
43. Nanto, H., Nakagawa, R., Yanagida, T. et al. (2015). Optically stimulated luminescence in tm-doped calcium fluoride phosphor crystal for application to a novel passive type dosimeter. *Sens. Mater.* 27: 277–282.
44. Okada, G., Fukuda, K., Kasap, S. et al. (2016). Aluminum nitride ceramic as an optically stimulable uminescence dosimeter plate. *Photonics* 3: 23–29.
45. Yanagida, Y. (2016). Ionizing radiation induced emission: scintillation and storage-type luminescence. *J. Lumin.* 169: 544–548.
46. Richter, D., Richter, A., and Dornich, K. (2015). A luminescence detection system for dosimetry, material research and dating application. *Geochronometria* 42: 202–209.
47. Hirata, Y., Watanabe, K., Yoshihashi, S. et al. (2017). Particle dependence of quenching effect in an ptical-fiber-type optically stimulated luminescence dosimeter. *Sens. Mater.* 29: 1455–1464.
48. Botter-Jensen, L., Mckeever, S.W.S., and Wintle, A.G. (2003). *Optically Stimulated Luminescence Dosimetry*. Elsevier.
49. Nanto, H. (1993). Photostimulated luminescence - application to imaging sensors for ionizing radiation and optical memories. *Solid. State. Phys.* 28: 49–58.
50. Nanto, H. (1998). Photostimulated luminescence in insulators and semiconductors. *Radiat. Eff. Defects Solids* 146: 311–321. (in Japanese).
51. Nanto, H. (1998). Advanced ionizing radiation monitoring utilizing photostimulated luminescence. *Ioniz. Rad.* 24: 97–107. (in Japanese).
52. Nanto, H. (2000). Physics of imaging plate for Radioluminography and application to advanced ionizing radiation monitoring. *Radioisot.* 49: 87–98. (in Japanese).
53. Nanto, H. (2001). Physics of imaging plate utilizing Photostimulated luminescence and application to advanced ionizing radiation monitoring. *Ioniz. Rad.* 27: 5–14. (in Japanese).
54. Nanto, H. (2011). Principle of passive -type dosimeters and its application. *Ioniz. Rad.* 37: 3–9. (in Japanese).
55. Jursinic, P.A. (2007). Characterization of optically stimulated luminescence dosimeters, OSLDs, for clinical dosimetric measurements. *Med. Phys.* 34: 4594–4604.
56. Pradhan, A.S., Lee, J.I., and Kim, J.L. (2008). Recent developments of optically stimulated luminescence materials and techniques for radiation dosimetry and clinical applications. *J. Med. Phys.* 33: 85–99.
57. Nanto, H. (2018). Photostimulable storage phosphor materials and their application to radiation monitoring. *Sens. Mater.* 30: 327–337.
58. Akselrod, M.S. and Kortov, V.S. (1990). Thermoluminescence and exoemission properties of new high-sensitive TLDα-Al_2O_3:C crystals. *Radiat. Prot. Dosim.* 33: 123–126.
59. Botter-Jensen, L., Mckeever, S.W.S., and Wintle, A.G. (2003). *Optically Stimulated Luminescence Dosimetry*, 85. Elsevier.
60. Jahn, A., Sommer, M., and Henninger, J. (2009). 2D-OSL-dosimetry using beryllium oxide. *Radiat. Mes.* 45: 674–676.
61. Yanagida, T. (2017). Inorganic scintillating materials and scintillation detectors. *Proc. of Japan Academy Series B*: 1–39.
62. Yanagida, T., Okada, G., and Kawaguchi, N. (2019). Ionizing-radiation –induced storage-luminescence for dosimetric applications. *J. Lumin.* 207: 14–21.
63. Miyamoto, Y., Kinoshita, K., Kobayashi, H. et al. (2010). Ionizing radiation sensor utilizing radiophotoluminescence in Ag-activated phosphate glass and its application to environmental radiation monitoring. *Sens. Mater.* 22: 235–245.
64. Nanto, H., Takei, Y., and Miyamoto, Y. (2011). Environmental background radiation monitoring utilizing passive solid state dosimeters. In: *Environmental Monitoring* (ed. E.O. Ekundayo), 121–136. INTEC.
65. Botter-Jensen, L., Mckeever, S.W.S., and Wintle, A.G. (2003). *Optically Stimulated Luminescence Dosimetry*, 163–217. Elsevier.
66. Yukihara, E.G. and Mckeever, S.W.S. (2010). *Optically Stimulated Luminescence: Fundamentals and Applications*, 144. Wiley, Incorporated.

9

Radiophotoluminescence (RPL)

Go Okada[1], Takayuki Yanagida[2], Hidehito Nanto[1], and Safa Kasap[3]
[1] *Co-creative Research Center of Industrial Science and Technology, Kanazawa Institute of Technology, Hakusan, Japan*
[2] *Nara Institute of Science and Technology, Nara, Japan*
[3] *Department of Electrical and Computer Engineering, University of Saskatchewan, Saskatoon, Canada*

9.1 Introduction

Radio-photoluminescence (RPL) is a luminescence phenomenon which is used for detection and measurement of radiation. Especially in the field of radiation measurement, it is often categorized as storage luminescence, as in thermally-stimulated luminescence (TSL) and optically-stimulated luminescence (OSL), as discussed in detail in earlier chapters of this book. The notable difference of RPL from the other phenomena is that the recorded signal is not lost by the readout process, and it is advantageous for a range of applications which require an extremely high accuracy of measurement. By using RPL, the readout process can be repeated multiple times to minimize statistical uncertainty. In addition, it is a very powerful tool when the measurement environment requires exposure to light or heat.

Despite the rareness of RPL studies and growing innovative works today, compared with TSL and OSL, the RPL nonetheless has a relatively long history. In 1921, the phenomenon of RPL was first reported by Przibram [1, 2] as a phosphorescence stimulated by light in activated inorganic crystals induced by γ-rays, and the term RPL started to be used at that time. Later, Schulman et al. [3] used the term RPL to refer to the creation of a new luminescent center which is metastable and cannot be destroyed, or only partly destroyed by optical excitation. This definition is more relevant to the way we use the term RPL today. Weyl et al. [4] discovered

Phosphors for Radiation Detectors, First Edition. Edited by Takayuki Yanagida and Masanori Koshimizu.

an RPL phenomenon in an Ag-doped glass. The latter material type found its application in accidental dosimetry by the US Navy. The application was limited to accidental dosimetry because the detector covered the measurement range of relatively large doses (between 0.1 and 10 Gy; or 10–1000 R) [5]. The chemical formula is 44% $Al(PO_3)_3$, 23% KPO_3, 23% $Ba(PO_3)_2$, and 8% $AgPO_3$ by weight. In the late 1950s, a similar dosimetric system based on RPL was developed in the UK and Sweden at the Atomic Energy Research Establishment (AERE). In 1961, Yokota et al. at the Tokyo Shibaura Electric Co., Ltd. (currently Toshiba Corp.) succeeded in making an RPL dosimeter material with higher sensitivity, lower energy dependence, and higher stability, by modifying the material's composition [6]. This breakthrough in RPL dosimetry advanced it into routine practices in personnel dose monitoring. In 2000, the Chiyoda Technol Corp. started personnel dose monitoring services by using RPL of Ag-doped sodium-aluminophosphate glass (manufactured by the AGC Techno Glass). The business is still currently active in Japan, and the market is increasing worldwide. In the mid-2000s, Akselrod et al. [7] proposed to utilize a fluorescent microscopic technique in order to read out the RPL signal of Al_2O_3:C,Mg, which allowed nuclear track detection by using this method.

This chapter covers a wide range of RPL technologies, from fundamentals to applications. Before starting detailed discussions, the next section will clearly define the term of RPL and then distinguish it from TSL and OSL as a different type of dosimetric technique using a phosphor medium. Section 9.3 then introduces a list of RPL materials and their fundamental properties, as reported in the literature. In addition, the principal properties as well as applications of four selected RPL materials are introduced with many illustrative materials. Section 9.4 concludes with a summary of RPL as a luminescent phenomenon used in radiation measurement, and provides future prospects.

9.2 RPL Phenomenon and the Definition

RPL is a phenomenon which is defined as the generation of a luminescent center by interaction with ionizing radiation. The luminescent center is, in general, an imperfection in the matrix which is responsible for luminescence; for example, a presence of transitional metals, rare earths, and lattice defects. At the luminescent center, an electron in the ground state can be excited by many types of energy, for example, light, mechanical energy, chemical reaction, and even ionizing radiation. The excitation process is then followed by relaxation back to the ground state, which results in an emission of light (radiative relaxation). The luminescence features such as excitation wavelength, emission wavelengths, luminescence lifetime, quantum efficiency, etc. strongly depends on the type of luminescent center; therefore, the luminescent center plays an important role in luminescence properties. A luminescent center generated as a result of RPL is called an RPL center.

Since an RPL center is a type of luminescent center, it has photoluminescence (PL) properties and the luminescence can be excited by light. Therefore, RPL can be observed as PL. The number of RPL centers increases with the radiation dose absorbed, and the PL intensity, in turn, increases. Therefore, the PL intensity as a result of RPL is utilized as an indirect probe of radiation dose. Figure 9.1 illustrates a commercial RPL material (Ag-doped sodium-aluminophosphate glass, AGC Techno Glass) irradiated with different radiation doses of γ-radiation [8]. It is clearly demonstrated that, under indoor lighting, the material pieces do not indicate notable differences, whereas, under UV light (in the dark), distinct PL in orange color is seen. In addition, the PL intensity increases with radiation dose.

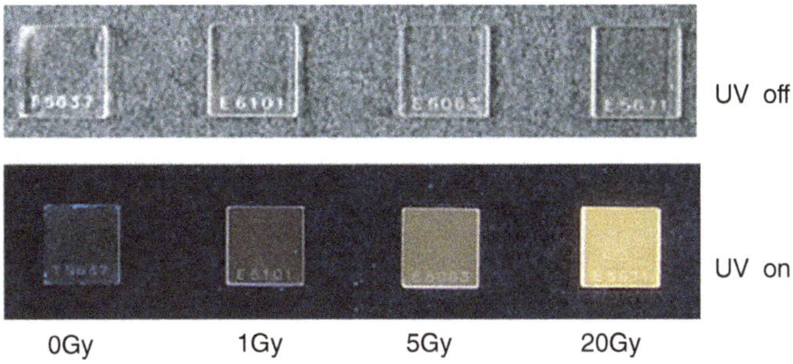

Figure 9.1 *Demonstration of RPL. A series of RPL glass plates (Ag-doped sodium-aluminophosphate glass, AGC Techno Glass) are irradiated with different gamma-ray doses, and they are illustrated (top) under room light and (bottom) UV light. Source: Reproduced from Y. Miyamoto et al. [8] by permission of Elsevier.*

It should be stressed that in scintillators, TSL materials and OSL materials, the luminescent center is intentionally introduced during synthesis, regardless of any interaction with ionizing radiation. In contrast, RPL materials, in principle, do not include luminescent centers and do not show PL, whereas, after irradiation, the RPL materials contain luminescent centers (RPL centers) and hence show PL. In practice, some RPL materials include some luminescent centers prior to irradiation; however, a different kind of luminescent center is generated as RPL.

Since the RPL phenomenon may be recognized as similar to OSL, some researchers do not correctly identify RPL, because, in both cases, luminescence phenomena can be observed by exposure to light and are used in dosimetry. However, they are completely different. The detail phenomenon of OSL is discussed in Chapter 8 but briefly they are distinguished here. In OSL, radiation-generated electronic charges are trapped at localized centers, and then the trapped charges are de-trapped by optical stimulation followed by recombination at a luminescent center. Therefore, the OSL intensity is proportional to the number of recombined charges per unit time, and it exponentially decreases with elapsed time during readout. It should be noted that the TSL phenomenon is different from OSL in the stimulation method, which is heat for TSL. The wavelength of stimulation is typically longer than that of emission in order to distinguish from PL, but it is not always so. In contrast, the RPL center is typically very stable and the resulting luminescence is due to electronic transitions within the RPL center. Therefore, the readout process is non-destructive, and the signal is consistent during readout and is not lost by the readout process itself. In addition, the luminescent center of OSL can be observed by PL regardless of pre-irradiation history, while RPL can be only observed by PL after irradiation.

9.3 RPL Materials and Applications

9.3.1 Introduction

Among the RPL materials reported today, Ag-doped sodium-aluminophosphate glasses, Al_2O_3:C,Mg, LiF, and Sm-doped materials have been worked quite actively. Including these materials, Table 9.1 summarizes a number of RPL materials reported in the literature. It should be noted that the RPL materials may not be limited to those listed in Table 9.1.

Table 9.1 *List of example RPL materials.*

	Material	Z_{eff} [a,b]	Material Form [c]	RPL Center	λ_{em} (nm)	λ_{ex} (nm)	Sensitivity (Gy)	Build-up	Fading	Reuse	References
1.	Na-Al-P-O:Ag	12.4	Glass	Ag^0, Ag^{2+}, Ag_2^+	450 630	340 310	10 μ-10	Yes	<5%/y	Yes	[9–12]
2.	Li_3PO_4-$Al(PO_3)_3$:Ag	(11.9)	Glass	Ag^{2+}	630	340					[13]
3.	KPO_3-$Al(PO_3)_3$:Ag	(13.4)	Glass	Ag^{2+}	550–700	300–350	0.1–10		Yes		[14]
4.	$NaPO_3$-$Al(PO_3)_3$:Ag	(11.6)	Glass	Ag^0, Ag^{2+}	400–500 600–700	325/339		Yes [d]			[15]
5.	$NaPO_3$-$(KPO_3)_n$:Ag		Glass	Ag^{2+}	400–600	325/339		Yes [d]			[15]
6.	$Al(PO_3)_3$-$(KPO_3)_n$:Ag		Glass	Ag^0, Ag^{2+}	400–500 600–750	325/339		Yes [d]			[15]
7.	NaCl:Ag	(15.3)	S.C [f]	Ag^0	550–650 500–600	289 339					[3, 16]
8.	KCl:Ag	(18.1)	S.C [f]	Ag^0	400–550 500–700	289 339					[16]
9.	Al_2O_3:C,Mg	11.3	S.C.	$F^+(Mg)$, F_2, $F_2^+(2\,Mg)$, $F_2^{2+}(2\,Mg)$	325 500 750 510	255 300 355 435	2 m-10	Yes	Yes [e]		[17–19]
10.	LiF:Mg	(8.3)	S.C., P.C.	F_3^+,F_2	530	450	~0.1-14 k	Yes	Yes [e]		[20, 21]
11.	LiF:Mg,Ti	(8.3)	P.C.	F_3^+,F_2			500-10 k				[22]
12.	LiF:Mg,Cu,P	(8.3)	P.C.				500-50 k				[22]
13.	KBr:Sm	(31.5)	S.C.	Sm^{2+}	730	350–450	1 m–100 m	No		Yes	[23]
14.	CaF_2:Sm	(16.9)	G.C.	Sm^{2+}	730	630	1-10 k	Yes		Yes	[24]
15.	CsBr:Sm	(49.5)	S.C.	Sm^{2+}	690	300–400	1 m-10	Yes			[25]
16.	MgF_2:Sm	(10.5)	P.C.	Sm^{2+}	700–800	340	1-1 k	No	Yes	No	[26]
17.	SiO_2-Al_2O_3:Sm	(11.7)	Glass	Sm^{2+}	683	400					[27]
18.	B_2O_3-Al_2O_3-Na_2O:Sm	11.4	Glass	Sm^{2+}	670–800	515					[28]

19.	MgF_2-AlF_3-CaF_2-SrF_2-$Sr(PO_3)_2$:Sm	(24.6)	Glass	Sm^{2+}	670–800	460	1-1 k	No		Yes	[29, 30]
20.	MgF_2-AlF_3-CaF_2-SrF_2-YF_3-BaF_2:Sm	(34.1)	Glass	Sm^{2+}	670–800	460	1-1 k	Yes		Yes	[30, 31]
21.	SrB_4O_7:Sm	(27.8)	P.C.	Sm^{2+}	670–830	408	~0.2-5 k				[32]
22.	$BaBPO_5$:Sm	(45.5)	P.C.	Sm^{2+}	670–830	408	~0.2–50	No	Yes		[32]
23.	$SrBPO_5$:Sm	(29.2)	P.C.	Sm^{2+}	670–830	408	~0.2–200	No	Yes		[32]
24.	$BaAlBO_3F_2$:Sm	(45.5)	G.C.	Sm^{2+}	670–850	300–500	10 m-10	No	No	No[f]	[33]
25.	$LiCaAlF_6$:Sm	(14.1)	P.C.	Sm^{2+}		340					[34]
26.	$CaSO_4$:Sm	(15.6)	P.C.	Sm^{2+}	600–900	300	~15 m-5	No			[35, 36]
27.	$SrSO_4$:Sm	(30.3)	P.C.	Sm^{2+}					Yes[e]		[36]
28.	$BaSO_4$:Sm	(47.0)	P.C.	Sm^{2+}					Yes[e]		[36]
29.	Ca_2SiO_4:Eu	(15.2)	P.C.	Eu^{2+}	500–800	250–500	10 m-10				[37]
30.	$BaAlBO_2F_3$:Eu	(47.1)	G.C.	Eu^{2+}	510	300–450	20 m-10	Yes		No[f]	[38]
31.	$CaSO_4$:Eu	(15.6)	P.C.	Eu^{2+}	382	300	~5 m-5			Yes	[35]
32.	NaCl:Yb	(15.3)	S.C.	Yb^{2+}	425–430	200–400	0.1–100		No		[39]
33.	C:N	(6.0)	S.C. Film	NV	689	546		Yes			[40]
34.	SiO_2-B_2O_3-Al_2O_3-Na_2O:Cu	(11.0)	Glass	Cu^{+}	500–800	240	–500	Yes			[41]
35.	MgF_2	10.5	P.C.		415 700	340	100-1 k	Yes[e]	Yes[e]		[42]
36.	Li_2CO_3	7.3	P.C.		470	340	100 μ-100 m		Yes[e]		[43]
37.	Na_2CO_3	9.5	P.C.		430	340	100 μ-10 m				[44]
38.	K_2CO_3	16.2	P.C.		440	340	1 m-10				[45]
39.	Mg_2SiO_4	11.2	S.C.		630	270	0.2–20	Yes			[46]
40.	$CaSO_4$	15.6	P.C.		690	590	400 μ-5	No	No	Yes	[47]

[a] Calculated with $m = 3.5$.
[b] The value in brackets indicates that the inclusion of dopant element is not considered in the calculation.
[c] S.C., P.C., and G.C. represent single crystal, poly-crystal, and glass–ceramic, respectively.
[d] Confirmed at low temperatures (10–300 K).
[e] Under excitation light (during readout).
[f] Information by personnel communication.

Particularly, the term of "RPL" is not limited to but often used in the field of radiation detection and measurements, and a similar phenomenon can be reported in the literature by using different terminology, which makes it more difficult to find. The effective atomic (Z_{eff}) numbers indicated in the table were calculated as

$$Z_{\text{eff}} = \left(\sum_i a_i Z_i^m\right)^{1/m} \tag{9.1}$$

where a_i is the fraction of the total number of electrons associated with element i of atomic number Z_i, which is defined by

$$a_i = \frac{f_i Z_i / M_i}{\sum_j f_j Z_j / M_j} \tag{9.2}$$

Here, f_j and M_j are the mass fraction and atomic mass of the j-th element, and m is a constant, whose value is between 3 and 4. A typical value of m for dosimetric materials is 3.5, and it is the value used in the table. λ_{em} and λ_{ex} indicate the main emission and excitation wavelengths of the RPL center. The sensitivity indicated is the dynamic range confirmed in the literature. The build-up means a post-irradiation increase of RPL signal in the absence of ionizing radiation. In contrast, the fading represents a post-irradiation decrease of RPL signal while the detector is stored. The listed materials are either doped or undoped inorganic compounds. Some dopant ions (e.g., Ag^+, Sm^{3+}, Eu^{2+}, etc.) are added in order to provide luminescence properties to the compound. The basic luminescence properties such as emission wavelength, excitation wavelength, and luminescence lifetime are predominantly controlled by the presence of dopant ions. However, other properties such as sensitivity, presence of build-up, presence of signal fading (stability), reusability etc., are difficult to predict, because no principles of material design have been established to date, which makes it more difficult to find RPL materials. The origin of RPL in undoped materials is often not well understood, while they claim that it is either by lattice defect or by residual impurity element. The following sections will provide fundamental RPL properties and applications of selected RPL materials.

9.3.2 Ag-Doped Sodium-Aluminophosphate Glasses

9.3.2.1 Introduction

RPL based on Ag-doped materials constitute the mainstream in the field of radiation measurement and dosimetry. In the 1950s to 1970s, it was extensively and competitively studied worldwide by academia and industries toward the goal of optimizing the material system for high sensitivity, stability, etc. to be used in personnel dosimetric applications. There are a large number of papers available in the literature from this era, which may be referred to by the reader who is interested in tracing the history and technologies undertaken until the development of the current technology. In 2000, the Chiyoda Technology Corporation took over the role of RPL dosimetry and led it into a business of personnel dosimetric monitoring and routine services. In the present, the RPL dosimetry is still one of the major dosimetric technologies. Despite the history and a large number of works performed, this section focuses on the principal properties of the presently commercialized RPL material, which is Ag-doped sodium-aluminophosphate glass (known as FD-7, composed of 11.00% Na, 6.12% Al, 31.55% P, 51.16% O, and 0.17% Ag by weight), and the applications of this material.

9.3.2.2 Principal Properties

Figure 9.2 illustrates the typical PL excitation and emission spectra of Ag-doped sodium-aluminophosphate glass. The emission spectrum is asymmetric and can be decomposed into two bands peaking at 450 and 630 nm. As for the emission, the excitation spectrum can also be decomposed into two bands, and the peak wavelengths are 310 and 340 nm. These two bands on excitation and emission are correlated by the fact that the 450 nm emission is predominantly excited by the 340 nm band, whereas the 630 nm emission is predominantly excited by the 310 nm excitation band. Miyamoto et al. [49] assigned the origins of these

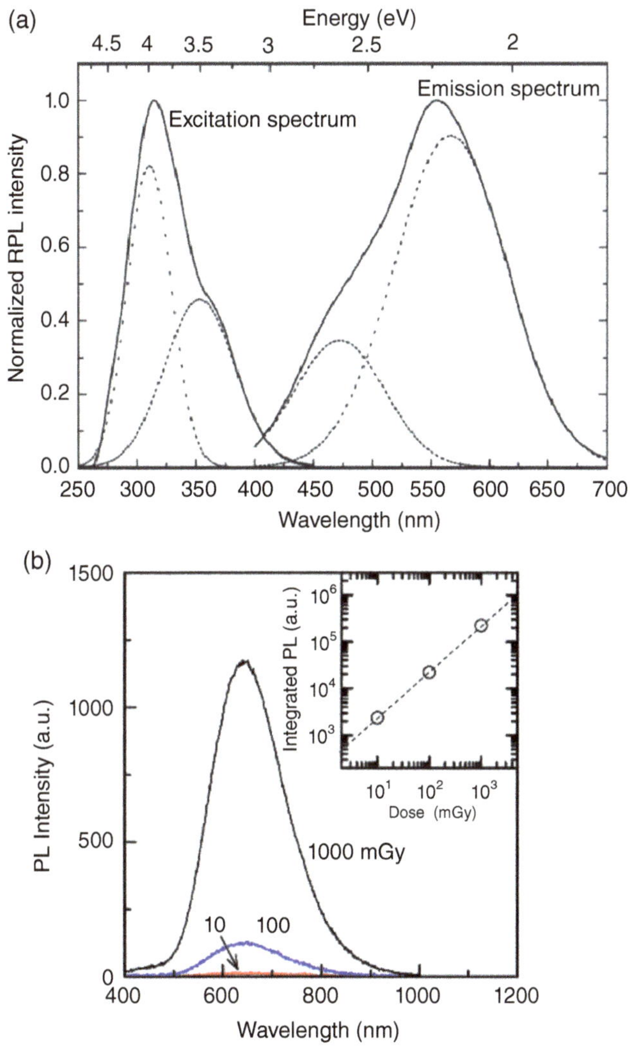

Figure 9.2 *PL spectra of Ag-doped sodium-aluminophosphate glass. (a) PL excitation and emission spectra (before a correction of the spectral response of the aperture.) Source: Reproduced from Y. Miyamoto et al. [48] by permission of Elsevier. (b) PL as a function of irradiation dose (X-rays). The inset is the dose–response curve.*

RPL centers to the Ag^0 ion (450 nm emission/340 nm excitation) and Ag^{2+} ion (630 nm emission/310 nm excitation). Although the assignment of the origin of the RPL centers is still being discussed and not completed yet, we use the above assignment to discuss the corresponding RPL materials and phenomena in this section.

The mechanism of RPL in Ag-doped sodium aluminophosphate glasses is still controversial today and has been discussed in a number of papers. Yokota et al. [50] interpreted the RPL mechanism as a charge trapping process by the doped Ag ions as:

$$2Ag^+ \xrightarrow{\text{irrad.}} 2Ag^+ + \left(e^- + h^+\right) \tag{9.3}$$

$$\xrightarrow{\text{migration}} \left(Ag^+ + e^-\right) + \left(Ag^+ + h^+\right) \tag{9.4}$$

$$\xrightarrow{\text{trapping}} Ag^0 + Ag^{2+} \tag{9.5}$$

In the initial state of the material, the Ag ion is present in the monovalent state (Ag^+). By interacting with ionizing radiation, as in many inorganic compounds and phosphor materials, a large number of electrons (e^-) and holes (h^+) are generated (Equation (9.3)). Next, these charges migrate in the host matrix and meet the Ag^+ ions (Equation (9.4)) where the electron is captured by the Ag^+ ion and then becomes the Ag^0 state while the hole is also captured by the Ag^+ ion, which then becomes the Ag^{2+} state (Equation (9.5)). These formed Ag^+ and Ag^{2+} act as luminescent centers and show emission in the visible range under UV excitation, as shown in Figure 9.2. In fact, Ag^+ also acts as a luminescent center. However, the emission and excitation wavelength is much shorter and it may not be easily measured with a readily available ordinary spectrometer; therefore, it has not been well discussed. One of a few studies [51] reported that Ag^+ has an emission peak around 300 nm under X-ray irradiation.

In fact, the charge trapping process may not be as simple as mentioned above. Miyamoto et al. [9] proposed the migration and charge trapping model, as illustrated in Figure 9.3, and the model is supported by a separate electron spin resonance (ESR) study by Yokota et al. [50]. In this model, during the migration of radiation-generated electrons and holes, the electron is directly trapped by the Ag^+ ion and then forms the Ag^0 state. In contrast, the hole is first captured by a phosphorus-oxygen (P—O) bond of the host matrix to form a

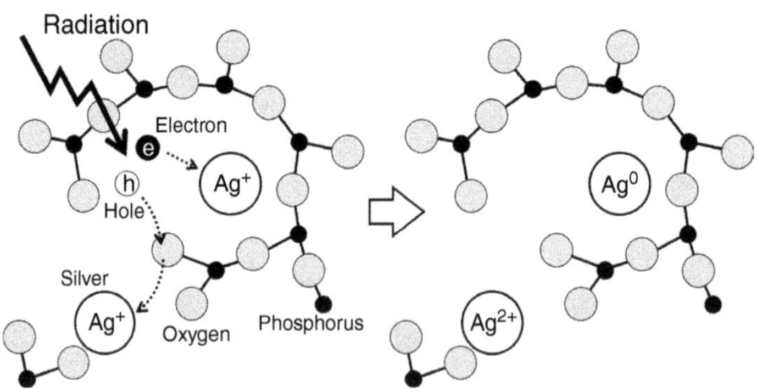

Figure 9.3 Proposed mechanism of the RPL phenomenon in Ag-doped sodium-aluminophosphate glass by Miyamoto et al. Source: Reproduced from Y. Miyamoto et al. [9] by permission of Elsevier.

phosphorus-oxygen hole center (POHC). Then, the hole from POHC is thermally reactivated and then meets Ag^+, forming an Ag^{2+} center.

The thermally-assisted formation of the Ag^{2+} center is well recognized as the so-called build-up effect. The RPL signal (as PL intensity in this context) increases by irradiation, and the signal further increases after irradiation (even at room temperature). This rate of increase can be accelerated when it is heated at elevated temperatures. Figure 9.4 illustrates the thermal response properties of RPL. Figure 9.4a is a transitional RPL response with post-irradiation time at different temperatures. At room temperature (25 °C), it is shown that the RPL signal builds up with time after irradiation. In addition, the build-up rate is accelerated with temperature, and the generation of the Ag^{2+} center is completed almost immediately after the irradiation when the temperature is 100 and 120 °C. Figure 9.4b shows the RPL response as a function of heat-treatment temperature. The treatment time at each temperature was fixed to 10 minutes and each measurement of the signal is performed at room temperature. It is clearly seen that the RPL signal reaches a maximum in the temperature range between about 100 and 200 °C, and the RPL starts to decrease above 200 °C, and becomes completely absent at and above 300 °C. In fact, the RPL response is reproducible even after the removal of the signal by heat. So, this class of RPL materials is reusable for radiation measurements.

McKeever et al. [10] studied the build-up of the RPL signal together with the ESR investigations, and they demonstrated that the main RPL signal is well correlated with the ESR signal of Ag_2^+ (a cluster of two Ag atoms trapping a single hole). Although not exactly the same glass systems, Dmitryuk et al. [52] and Maurel et al. [53] assigned the origin to Ag_3^+. There have been many other studies discussing and trying to identify the origin of Ag_3^+, and the interpretation remains a topic for discussion.

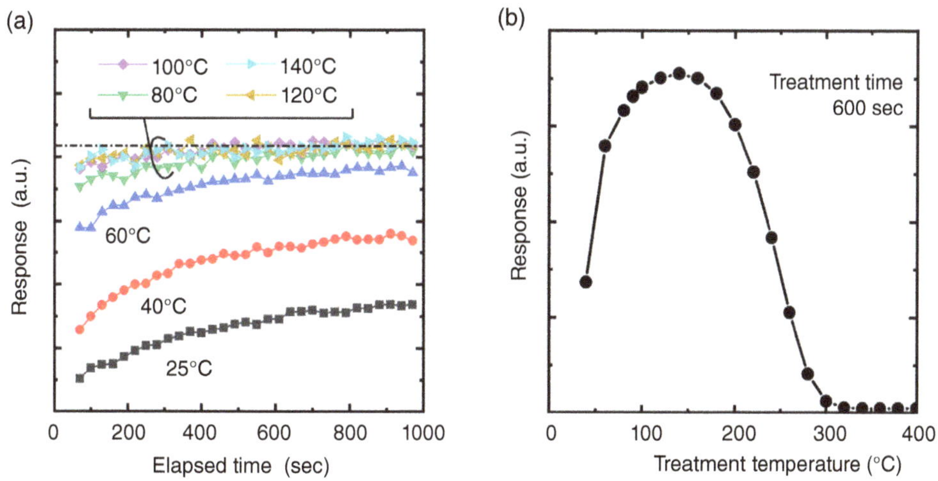

Figure 9.4 *Thermal response properties of RPL. (a) RPL response as a function of elapsed time after irradiation at different temperatures. The response is corrected in terms of the thermal quenching of the luminescent center in order to represent the response value to correspond to normal condition (room temperature). (b) RPL response as a function of treatment temperature. The treatment time is 10 minutes at each temperature.*

9.3.2.3 Application in Personnel and Environmental Dose Monitoring

Ag-doped sodium-aluminophosphate glass is well-known as one of the most successful RPL materials, which have found practical applications. The Chiyoda Technol Corporation is the leading company that commercialize this material system for personnel dose monitoring services. They use Ag-doped sodium-aluminophosphate glass plate as a dose sensing and recording element. Figure 9.5 illustrates the commercial dosimeter badge using the Ag-doped sodium-aluminophosphate glass. Figures 9.5a and b show the appearances of the whole-body-type (general purpose) and ring-type dosimeters, respectively. Figure 9.5c shows the inside of the dosimeter badge. There is much knowledge involved in the design and handling, but one of the most important features to be mentioned is the use of a series of radiation filters aligned in the middle of the badge. As indicated in Table 9.1, the Z_{eff} of Ag-doped sodium-aluminophosphate glass is 12.4, which is considered to be not quite equivalent to that of soft biological tissue (Z_{eff}~7.4). For this, the interactions of radiations with the RPL material (in turn, RPL response signal) are not tissue equivalent, and dose values measured using the RPL glass alone are strongly dependent on the radiation energy. For this reason, the RPL glass detector requires corrections of the energy dependence, and a series of radiation filters are used for this purpose. The radiation filters used in the latest version are Al (0.4 mm), Cu (0.3 mm), Sn (1.4 mm), ABS resin (P1; 0.05 mm), and ABS resin (P2; 0.5 mm). The glass plate is placed along the filters (as shown in Figure 9.5d), and the deposited radiation energy beneath each filter element depends on the filter material and radiation energy. Therefore, with the knowledge of energy dependence of the RPL glass plate, the exposure dose to the human body can be derived.

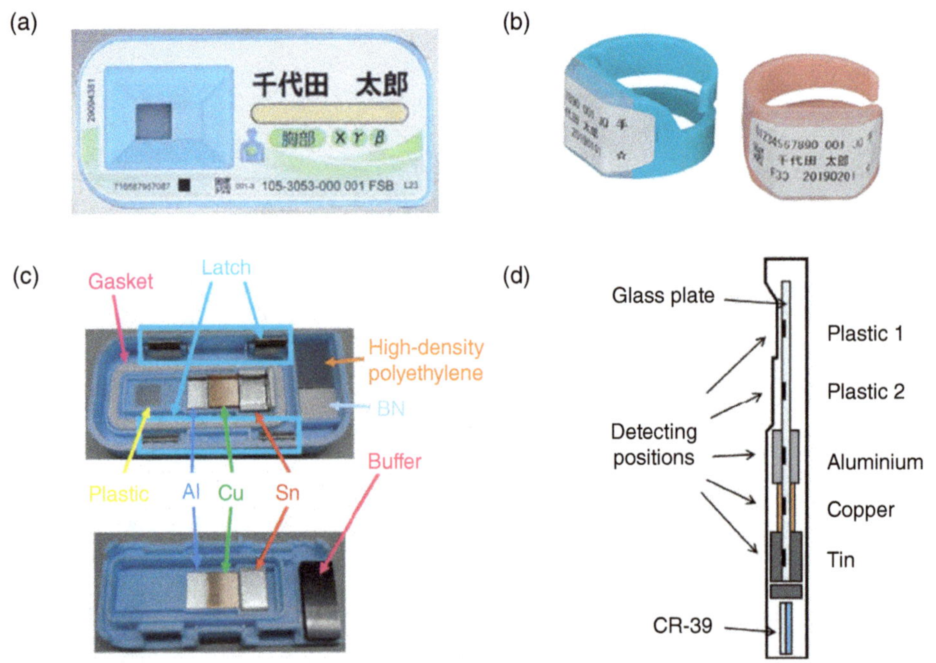

Figure 9.5 *Commercialized personnel dosimeter badges using RPL glass detector (Ag-doped sodium-aluminophosphate glass). Appearances of (a) the whole-body-type dosimeter (general purpose) and (b) ring-type dosimeters. (c) Inside and (d) cross-section image of the dosimeter badge. Source: Courtesy of Chiyoda Technol Corporation.*

The dependence of the RPL signal on radiation energy can be corrected by using a series of radiation filters. Figure 9.6 demonstrates the RPL response signal at different radiation energies with a series of radiation filters. Figure 9.6a shows the RPL glass under a UV lamp after irradiation with 20 Gy of 25 kVp X-rays (top) and 40 Gy of 1.17 and 1.33 MeV ^{60}Co γ-rays (bottom). On the one hand, the low-energy X-rays strongly interact with the filters, especially Cu and Sn, so the RPL signal from the Cu and Sn positions are almost absent by X-ray irradiation. In contrast, the interaction probabilities of high-energy radiation are much lower and not as different by the type of radiation filters. As a result, the RPL signal appears to be almost consistent, regardless of the type of radiation filter; and the response intensity is lower than that for X-ray irradiation of a lower dose. Figure 9.6b compares the RPL response signals with different filters and with different radiation species (X-ray, γ-ray, and β-ray) of irradiation. Since the low-energy X-rays and β-rays have high interaction probabilities with matter and these strongly depend on the type of filter element (or Z_{eff}), the RPL response strongly varies with the

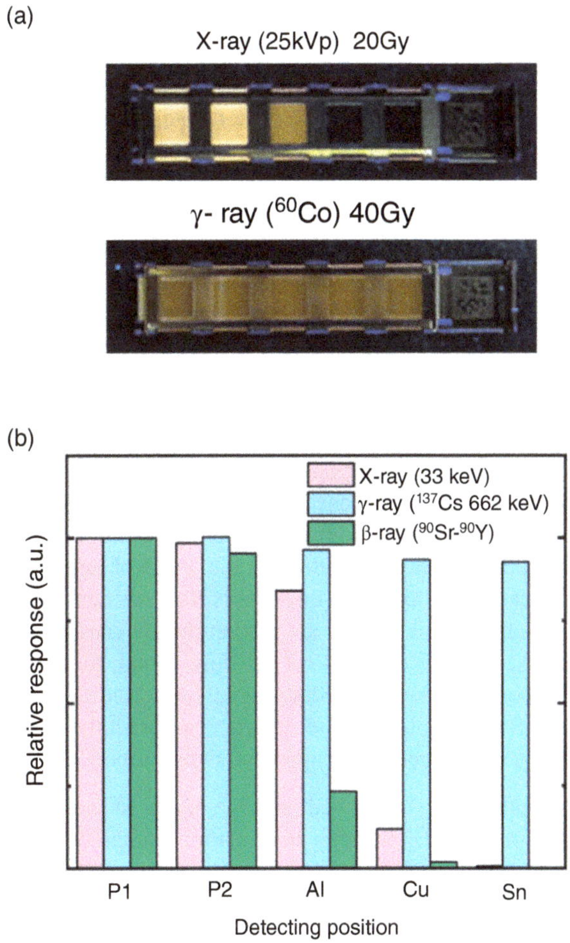

Figure 9.6 Energy correction by using a series of metal filters. (P1 and P2 are defined in the text.) Source: Courtesy of Chiyoda Technol Corporation.

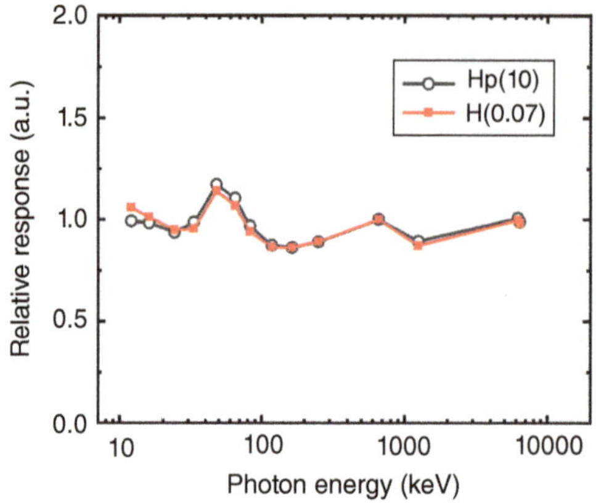

Figure 9.7 *Energy dependence of RPL response after corrections. Source: Courtesy of Chiyoda Technol Corporation.*

filter position. In contrast, the response values are almost consistent when the sample is irradiated with 662 keV ^{137}Cs γ-rays. Even the low energy irradiations by X-rays and β-rays can be distinguished by considering the response signals at the Al position, which are largely different; higher for X-rays and lower for β-rays. With a knowledge of these differences in the response signals at the series of filter positions and the response patterns, we can correct the energy dependence of the RPL glass. Figure 9.7 shows the energy dependence of the dosimeter badge using an RPL glass after the corrections have been applied. The detailed correction protocol for the energy dependence is beyond the scope of this chapter, but can be found elsewhere [54]. With the appropriate correction, the response is reasonably consistent over two orders of magnitude of radiation energy with 17% of error at maximum.

The RPL response corresponds to a PL intensity of Ag^0 and Ag^{2+} centers generated by radiation. In Figure 9.2, it is expressed using the spectral domain for intuitive understanding of the phenomenon, and is a typical way of representing RPL phenomena. For the dosimetric applications in practice, however, the response is not read out in the spectral domain but in the time domain. Figure 9.8 illustrates the concept of the signal reading. The RPL glass detector is excited with a pulse of UV light emitted from a solid-state Nd:YAG laser (355 nm), and then the PL decays with certain lifetimes depending on the origin of luminescence. In the present case, the decay curve can be divided into three parts: fast (<~1 μs), intermediate (<~40 μs), and slow (<~1 ms). The fast and slow parts are predominantly caused by some residual luminescence and some dirt, which are dealt as a pre-dose signal, meaning the residual signal without irradiation. In contrast, the intermediate signal (indicated as *Resp.* in the figure) is the one to be read and corresponding to the dose. In order to extract the intermediate signal from the entire decay curve, a certain geometric approach is performed, as described below:

1. The fast decay signal until t_1 is ignored, and the signal in the intermediate time range between t_1 and t_2 is integrated. The value as a result of integration is F_1.
2. The signal in the slow part between t_3 and t_4 is integrated, and the value is F_2.
3. The integrated value F_2 is multiplied by a certain constant value (fps > 1).
4. The result of 3 (fps × F_2) is subtracted from the integrated value F_1

In this way, the RPL signal is read out and used for dose estimation.

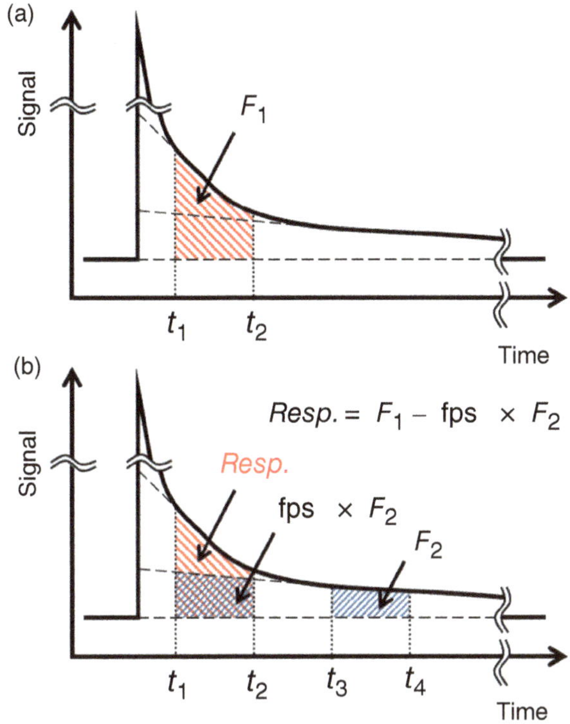

Figure 9.8 *Signal processing techniques of RPL signal for Ag-doped sodium-aluminophosphate glass used in personnel dose monitoring applications. Source: Reproduced from the instruction manual of Dose Ace with permission by AGC Techno Glass and Chiyoda Technol Corporation.*

9.3.2.4 Other Applications in R&D

In the R&D phases, the RPL of Ag-doped sodium-aluminophosphate glasses have been studied for a wide range of applications. For example, Maki et al. [55, 56] introduced a sensitivity to thermal neutrons into the RPL glass by including ^{6}Li impurity in the glass system. The ^{6}Li has a high cross-section to thermal neutrons due to the nuclear reaction ($^6\text{Li} + \text{n} \rightarrow \alpha + {}^3\text{H} + 4.78\,\text{MeV}$), and the produced energy for every reaction is deposited in the glass matrix, which is then ionized followed by the generation of RPL centers. For this approach, the inclusion of Li atoms is of primary importance, and the larger the concentration, the higher the interaction with neutrons. As a general feature of glass materials, the chemical composition (or glass-forming composition) allows a large degree of freedom, so an inclusion of incongruous elements can be easily accepted without the material properties being significantly affected. On the other hand, the chemical composition of crystalline materials is strictly defined by the crystal stoichiometry, so the addition of an impurity element can sometimes be challenging.

The present RPL detector in practice is in the form of a solid glass plate, but the application of RPL could be extended if the detector is flexible. Nanto et al. [57] demonstrated this concept by making the detector into a flexible sheet. In this sheet, micro beads (35 μm in diameter) of Ag-doped sodium aluminophosphate glass (as shown in Figure 9.9a) are

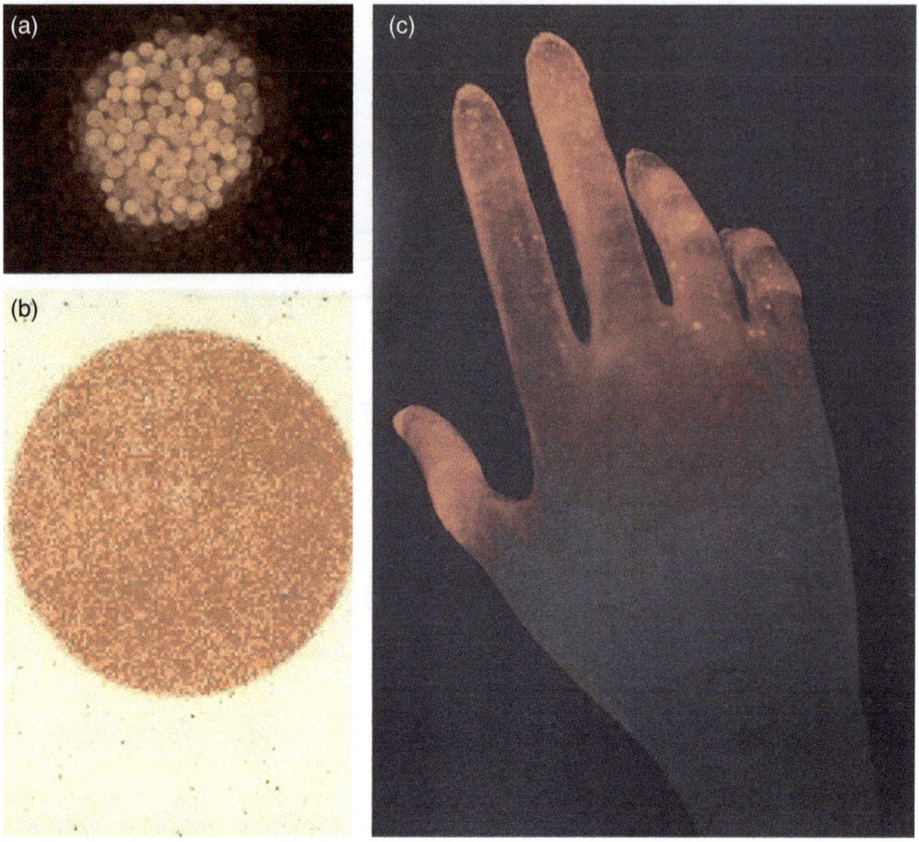

Figure 9.9 A prototype of flexible RPL sheet. (a) Micro beads of RPL glass, (b) γ-ray image acquired by using the RPL sheet, and (c) simulated PL image of an irradiated RPL glove. Source: Reproduced from H. Nanto et al. [57] by permission of Myu KK.

embedded in a polystyrene matrix. In this way, this sheet has both the RPL properties and flexibility, and it can be used to detect 2-D distribution of radiation dose, as demonstrated in Figure 9.9b. This flexible RPL device is potentially used for surgeon gloves in radiosurgery, where the surgeon wears the gloves and supports the patient during the surgery (Figure 9.9c). Also, the flexible device is considered to be used as a material for work suits to be worn during decontamination activities from, for example, a nuclear accident.

9.3.3 Al_2O_3:C,Mg

9.3.3.1 Introduction

RPL of Al_2O_3:C,Mg was first reported in 2003 by Akselrod et al. [58], not for radiation detection at that time but for optical data storage. In this material, an RPL center is generated, not only by ionizing radiation but also via two-photon absorption of blue laser light. Since a very stable RPL center can be generated only at a desired voxel spot and it can be read out by using an optical scanner, this material was proposed as a 3-D optical data storage device. Later, in 2006, the material was reported to be very promising for radiation dosimetry, in particular to be used for nuclear track detections [17], which will be introduced in Section 9.3.3.3. In addition to the contents in this chapter, the reader interested in this topic may also be referred to a review article by Akselrod et al. [59] and the references therein.

9.3.3.2 Principal Properties

Al_2O_3 is probably one of the most well-known inorganic chemical compounds of crystalline material studied in scientific research, as well as attracting attention as a jewelry material from ancient times. It has a trigonal structure with a space group of $R\bar{3}c$, and has many technological applications owing to its desirable mechanical, electronic, and optical properties. Especially, many optical applications such as lasers rely on its notably large band-gap energy, 7–8.8 eV [60–62]. Moreover, the optical properties of Al_2O_3 are very sensitive to the presence of impurity elements. For example, when it includes a small fraction of Cr (usually present as Cr^{3+}), it has a red color. It has a blue color owing to a presence of both titanium and iron. In addition to impurities, an inclusion of defect centers also provide additional optical properties, such as optical absorption, PL, and RPL.

Figure 9.10 shows an as-grown single crystal of Al_2O_3:C,Mg. The inclusion of C and Mg induces the generation of color centers, which cause the visual appearance to be yellow-green in color. More particularly, the generated color centers are denoted as F, F^+, F_2^+, and F_2^{2+} with and without association of included Mg^{2+} ion(s). For example, the F-center corresponds to a single oxygen vacancy with two trapped electrons, which neutralize the charge balance, while the F^+-center is the one with only one trapped electron. The color centers denoted as F_2^+ and F_2^{2+} are two oxygen vacancies with three and two trapped electrons, respectively. In addition to these species of color centers, there are also species associated with Mg^{2+} impurities that are present, and F_2^{2+}(2Mg) and F_2^+(2Mg) are two oxygen vacancies with two Mg^{2+} ions substituting Al^{3+} ions with zero and one trapped electron, respectively.

In addition to strong optical absorptions, these color centers provide luminescent properties to the crystal. Figure 9.11 shows contour graphs of PL excitation and emission spectra of

Figure 9.10 *As-grown Al_2O_3:C,Mg single crystal rod and polished detector plates. Source: Reproduced from G.J. Sykora et al. [63] by permission of Elsevier.*

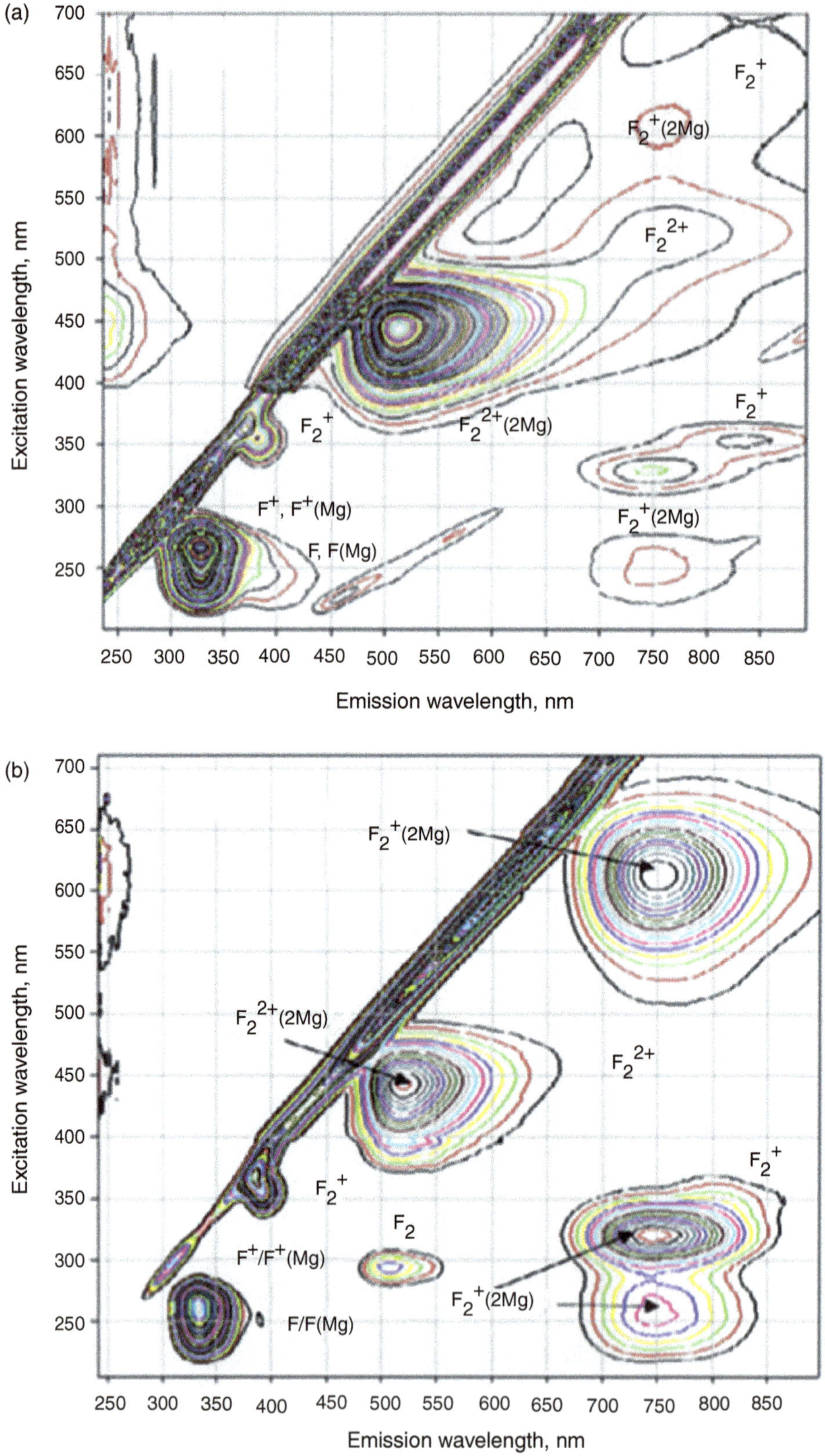

Figure 9.11 *PL excitation-emission contour graphs of Al_2O_3:C,Mg a before and b after irradiation. Source: Reproduced from M.S. Akselrod et al. [64] by permission of Elsevier.*

Al_2O_3:C,Mg. Both the excitation and emission have complex features due to the inclusions of the above-mentioned color centers. In particular, the predominant luminescence of the as-grown crystal is caused by the F_2^{2+}(2Mg) center, which can be excited by blue (~420 nm) light and then emits in the green (~520 nm). In addition, after irradiation with ionizing radiation, a strong emission band appears around 750 nm, which has three excitation bands peaking at 260, 335, and 620 nm. The origin of this emission is attributed to the F_2^{+}(2Mg) center.

The PL emission from the generation of an F_2^{+}(2Mg) luminescent center by ionizing radiation is considered as RPL. The mechanism of RPL in Al_2O_3:C,Mg is well studied and expressed based on charge trapping processes by Sykora [65], which can be described by

$$F_2^{2+}(2Mg)+F_2(2Mg)\xrightarrow{\text{irrad.}}F_2^{2+}(2Mg)+F_2(2Mg)+\left(e^-+h^+\right) \quad (9.6)$$

$$\xrightarrow{\text{migration}}\left(F_2^{2+}(2Mg)+e^-\right)+\left(F_2(2Mg)+h^+\right) \quad (9.7)$$

$$\xrightarrow{\text{trapping}}2F_2^{+}(2Mg) \quad (9.8)$$

Here, electrons and holes are generated by ionizing radiation, and a fraction of electrons is captured by the F_2^{2+}(2Mg) centers and then the latter become F_2^{+}(2Mg) centers. In contrast, the generated holes are captured by F_2(2Mg) centers and these become F_2^{+}(2Mg) and the charge balance is compensated for. Moreover, a similar trapping process occurs with F_2^{2+}(Mg) and F_2^{+}(Mg) centers for electron and hole capture, respectively, although these are not used for dosimetric applications. It should be noted that the charge trapping centers should not be limited to the above-mentioned centers only.

9.3.3.3 Fluorescent Nuclear Track Detector

Al_2O_3:C,Mg has RPL properties, and the RPL can be induced not only by photon irradiation but also by heavy particles. The heavy particles incident on the material deposit their energy as they inelastically migrate in the matrix. The migration distance is typically on a scale of micrometers, and the deposited energy locally ionizes the host along the migration track. As a result of ionization, as described in Section 9.3.3.2, F_2^{+}(2Mg) RPL centers are generated along the nuclear tracks of heavy particles.

The RPL of Al_2O_3:C,Mg was particularly designed and developed for the so-called fluorescent nuclear track detector (FNTD). The FNTD technologies have been developed by using different RPL detectors, but Al_2O_3:C,Mg is the most studied and recognized as the pioneer. The RPL signal cannot only be read out by a conventional steady-state PL technique but also by a confocal fluorescent microscopy technique. The latter microscopy technique is well-established and allows an acquisition of fluorescent image with a diffraction-limited resolution, which is on an order of several hundred nanometers. Figure 9.12a shows a 2-D fluorescent image of recoil proton tracks induced by fast neutrons. The image was obtained by using an Al_2O_3:C,Mg crystal plate as a detector and a confocal fluorescent microscope as a reader. The proton tracks are clearly represented as discrete tracks of fluorescence signal. In addition, the RPL intensity depends on the deposited radiation energy, hence the linear energy transfer (LET) of energetic particles. Therefore, by taking a histogram of the track intensity, we can discriminate the heavy particle species, which is demonstrated in Figure 9.12b.

Recent studies of FNTD [67] have advanced the image resolution by using a microscopy technique called stimulated emission depletion (STED), which breaks the barrier of

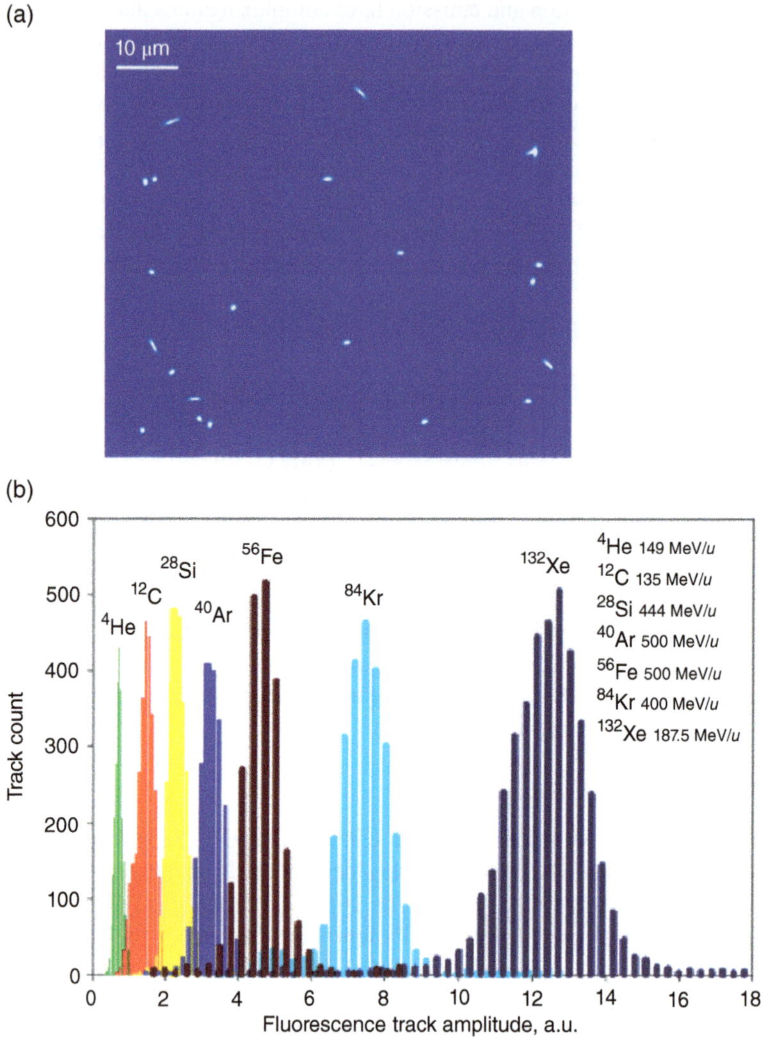

Figure 9.12 *Nuclear track detection demonstrated by using RPL of* Al_2O_3*:C,Mg. (Top) Nuclear track image of fast neutrons. (Bottom) A histogram of nuclear tracks detected. Source: Reproduced from G.J. Sykora et al. [63] and G.J. Sykora et al. [66] by permission of Elsevier.*

diffraction limits and allows the image resolution of several tens of nanometers. With this technology, the use of FNTD can be extended into radiation biology. A notable study was performed by Niklas et al. [68], in which an Al_2O_3:C,Mg RPL detector plate was coated by a layer of biological cells so that a nuclear track and radiation damage in the biological cell (or DNA damage) can be observed together on a per-track basis.

9.3.4 LiF

9.3.4.1 Introduction

LiF is another material well-recognized to show RPL. One of the pioneering works of RPL in LiF may correspond to the study by Levita et al. [20]. In this study, they demonstrated, by using a conventional TSL dosimeter (TLD-100), several fundamental properties such as luminescence spectrum, sensitivity, stability, etc. Compared to other RPL materials, the sensitivity is lower, and this is one of the reasons why it has not found application in personnel dosimetry. However, RPL in LiF is still being actively studied for abnormal dosimetric applications, such as large dose detections for accelerators and micro-dosimetry. Section 9.3.4.2 will review selected works of RPL based on LiF.

9.3.4.2 Principal Properties

Figure 9.13 shows an example PL excitation and emission spectra of LiF measured after irradiation with X-rays. The PL appears in a wide spectral range with two bands peaking around 540 and 630 nm. These emissions can be excited at 440 nm. In this material system, as for Al_2O_3:C,Mg, the emission origins as RPL centers are considered to be color centers, which are denoted as F_3^+ and F_2. The F_3^+ center is considered to be formed with three anion vacancies with two electrons trapped, whereas the F_2 center consists of two anion vacancies with two trapped electrons. Also, LiF contains F_2^+ and F_2^- centers, which are two anion vacancies with one electron and three trapped electrons, respectively. These defect centers are intuitively illustrated in Figure 9.14.

The mechanism of RPL in LiF is considered to be as follows. At low doses (up to 1000 Gy), generation of F-centers as isolated single anion vacancies is dominant. As the irradiation dose increases, the above-mentioned aggregate centers start to populate, while the F-centers start to depopulate. The generated F_2 and F_3^+ centers act as luminescent centers, as shown in Figure 9.13, hence they are recognized as RPL centers [21]. Since the generation of aggregated F-centers require relatively large doses, the sensitivity of RPL in LiF is low compared with other RPL detectors.

The RPL of LiF shows build-up at room temperature and higher. The dominant process involved here is considered to be the following. An electron trapped at the F_2^- center becomes thermally activated and then is transferred to the F_2^+ center at which the electron is captured to form an F_2 center, and the F_2^- center being left by the electron becomes an F_2 center. Therefore, the increase of the number of F_2 centers is thermally assisted.

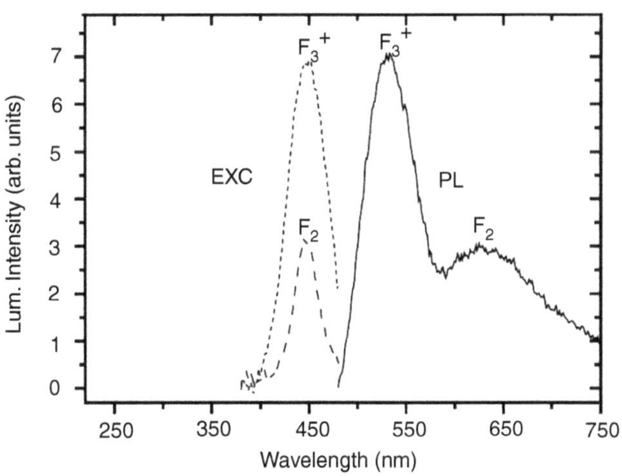

Figure 9.13 *PL excitation and emission spectra of LiF after X-ray irradiation of 126 Gy (absorbed dose). Source: Reproduced from T. Kurobori et al. [69] by permission of Elsevier.*

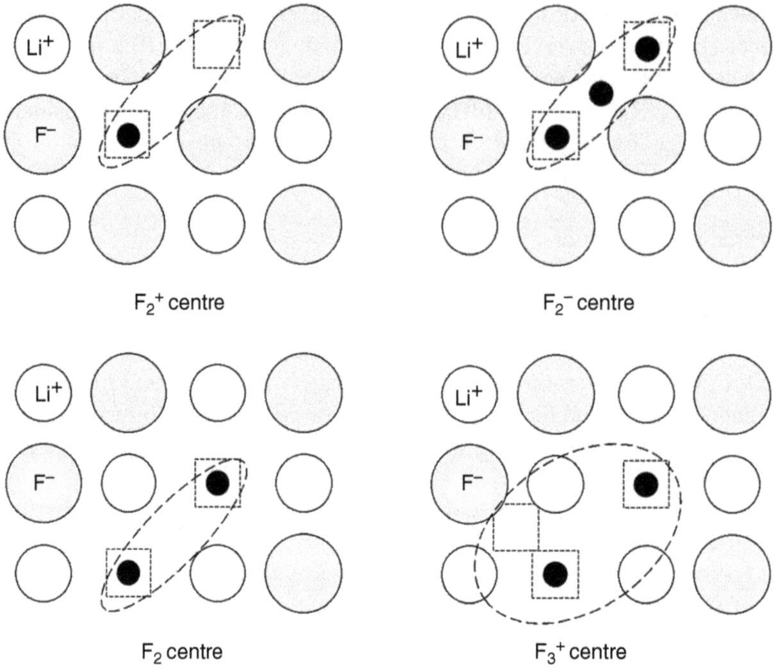

Figure 9.14 *Illustrative representation of a series of defect centers in LiF.*

9.3.4.3 Disk-Type Imaging Plate

Although a large number of studies on RPL of LiF have been performed by using conventional ceramic pellets, which are designed for TSL dosimetry, Kurobori et al. have developed a 2-D dosimetry technique using an RPL disk imaging plate. With the high symmetry of the crystal structure of LiF, it can be easily formed into a high-quality crystal film on a substrate by evaporation techniques. Despite the fact that the signal on a conventional imaging plate in computed radiography is read out by raster scanning, this technique reads the signal as the disk is spinning, just as for the optical disc memory (e.g., CD, DVD, and Blue-ray).

Figure 9.15 shows the schematic diagram of the readout system and the X-ray image obtained. The excitation source is a continuous wave 443 nm laser diode for LiF, whereas the system also allows excitation for an Ag-doped RPL detector (371 nm) as well as a BaFBr:Eu OSL detector (639 nm). The excitation light is guided to the objective lens which focuses it onto the spinning detector disk. The luminescence signal is detected by the PMT, and the signal information is synchronized with the rotation speed and X position in order to reconstruct the 2-D image.

9.3.4.4 Fluorescent Nuclear Track Detector

As for Al_2O_3:C,Mg, there are a number of studies on FNTD by using RPL of LiF. The approach is similar to those done for Al_2O_3:C,Mg, in a sense that the RPL signal induced by incident particles is read out by using a fluorescence microscope. Figure 9.16 shows one of the successful images reported by Bilski et al. [70], demonstrating a detection of nuclear tracks by using RPL of LiF. The incident neutron interacts with ^{6}Li via the $^6Li(n,\alpha)^3H$ nuclear

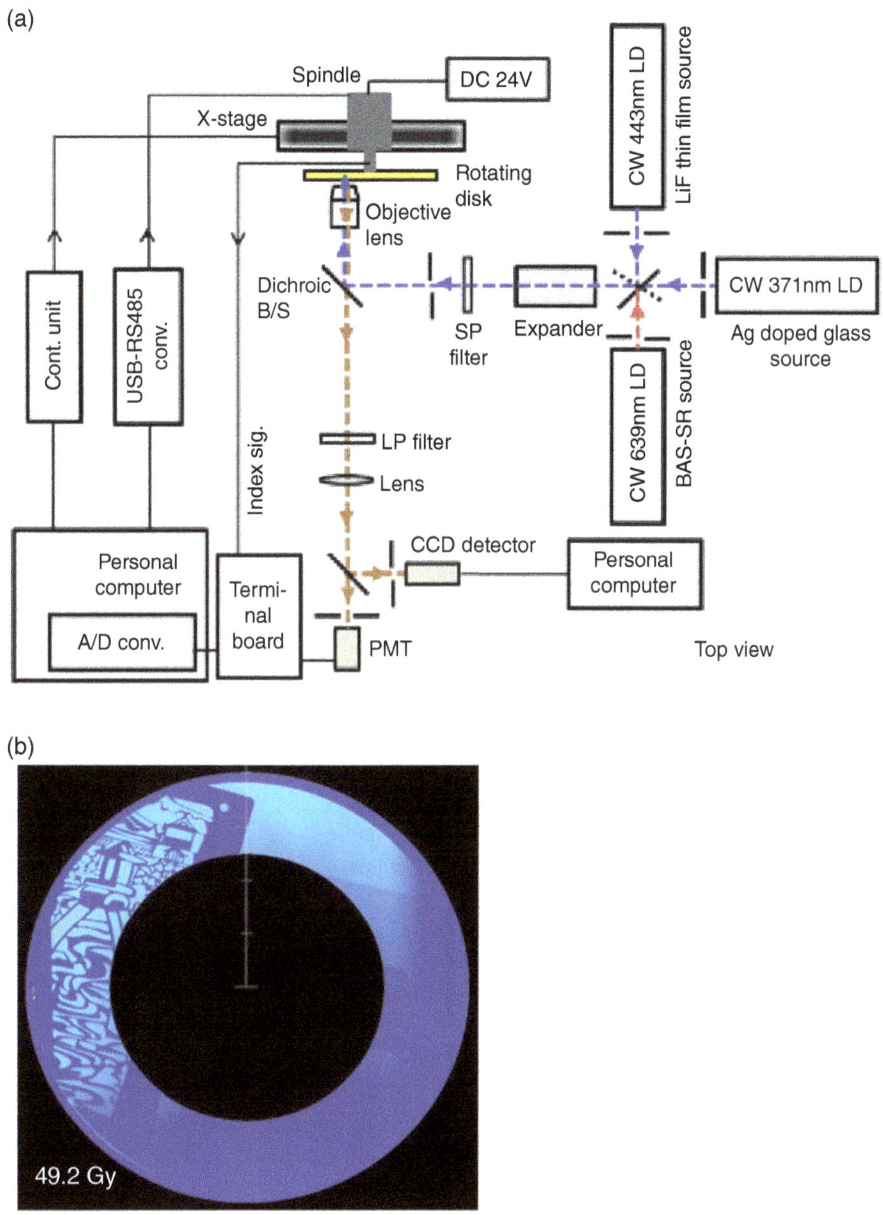

Figure 9.15 Disk-type radiation imaging technique by using a LiF thin-film disk plate. (a) Reader system configuration and (b) obtained X-ray image. The disk diameter is 12 cm. Source: Reproduced from T. Kurobori et al. [69] by permission of Elsevier.

reaction. The result of the latter reaction is a production of α and ^3H nuclei, which deposit their energies along the tracks, which is well represented in the image having the track length of 6 μm for α and 33 μm for ^3H.

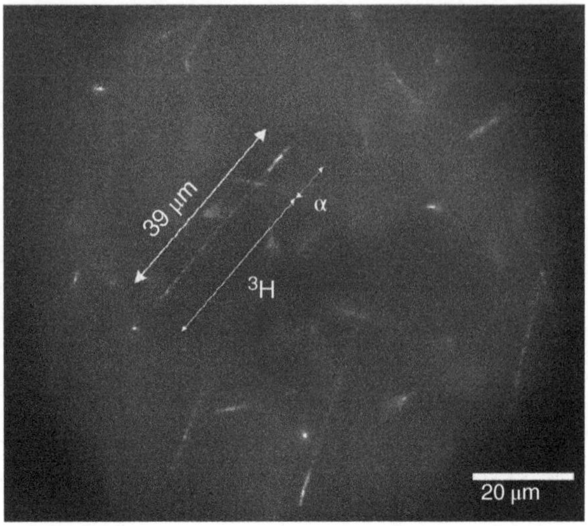

Figure 9.16 *Nuclear track image obtained by using RPL of LiF as FNTD. Source: Reproduced from P. Bilski [70] by permission of Elsevier.*

9.3.5 Sm-Doped Compounds

9.3.5.1 Introduction

The above-mentioned three materials (Ag-doped phosphate glass, Al_2O_3:C,Mg, and LiF) are probably the most well-known RPL materials, and have been studied over several decades. This section, however, introduces another class of RPL materials and their applications which have been intensively studied in recent years and are based on Sm-doped phosphors. The Sm ion has its stable oxidation states in 3+ and 2+. In both the oxidation states, the Sm ion can act as a luminescent center but with different luminescence properties such as luminescence emission wavelength. In some selected host materials, the oxidation state of Sm changes from 3+ to 2+, which is induced by interacting with ionizing radiation, so a luminous Sm^{2+} ion is generated as an RPL center. In the early 2000s, Qiu et al. [71] reported γ-ray induced valence reduction of Sm^{3+} into Sm^{2+} as a treatment process in order to obtain stable Sm^{2+} from Sm^{3+}, as the latter oxidation state is more stable in most materials. The generated Sm^{2+} was to be optically bleached spectral-selectively, or known as spectral hole burning, for high-density optical memory applications, but their interest was not radiation dosimetry. Ten years later, in 2011, Okada et al. [72] applied the phenomenon to microdosimetric applications where the dose distribution of an X-ray microbeam produced by a synchrotron was measured, and this is the main topic in Subsection 9.3.5.3.

9.3.5.2 Principal Properties

Sm is one of the lanthanide elements which have transitional electrons in the 4f orbitals and these play an important role in the optical properties. In general, the Sm ion is stable either in the trivalent or divalent state, and they certainly have different electronic configurations. In the ground state of Sm^{3+}, it has 5 electrons in the 4f orbitals ([Xe]$4f^5$), whereas the Sm^{2+} state has 6 electrons ([Xe]$4f^6$). The Sm ion in both the oxidation states often acts as a luminescent center where the electrons can be brought into the excited state by, for example, UV excitation and then the subsequent relaxation to ground state results in luminescence.

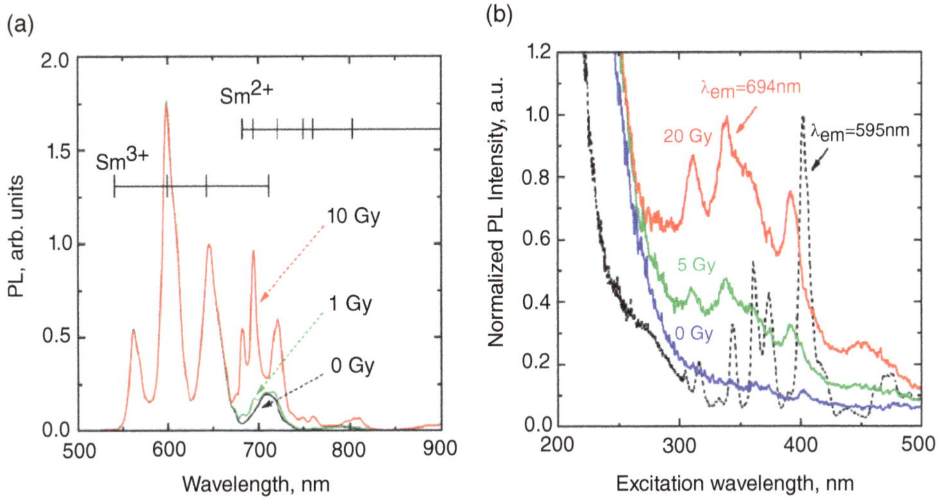

Figure 9.17 *Example PL (a) emission and (b) excitation spectra of an Sm-doped RPL material as a function of radiation dose. The material is Sm-doped BaF_2-Al_2O_3-B_2O_3 glass–ceramic. Source: Reproduced from G. Okada et al. [33] by permission of Elsevier.*

Example PL excitation and emission features of Sm-doped material (BaF_2-Al_2O_3-B_2O_3 glass–ceramic) are represented in Figure 9.17. When excited by UV light (340 nm), it shows luminescence of a set of several sharp peaks at approximately 560, 600, 650, and 710 nm when especially not irradiated by radiation (initial state). These emissions are due to the transitions from the $^4G_{5/2}$ excited state to $^6H_{5/2}$, $^6H_{7/2}$, $^6H_{9/2}$, and $^6H_{11/2}$ ground states ($^4G_{5/2} \rightarrow {}^6H_{J/2}$ [$J = 5, 7, 9, 11$]) of Sm^{3+} ion, respectively. After irradiation, additional emission peaks appear in the range of longer wavelengths with peaks around 682, 694, 721, and 760 nm, and these emissions are attributed to the electronic transitions from the $^5D_0$ excited state to the $^7F_0$, $^7F_1$, $^7F_2$, and $^7F_3$ ground states ($^5D_0 \rightarrow {}^7F_J$ [$J = 0, 1, 2, 3$]) of Sm^{2+}, respectively. Since the number of Sm^{2+} ions generated increases with an increase of the radiation dose, the luminescence intensity of Sm^{2+} also increases with irradiation dose.

The excitation spectrum of the Sm^{3+} emission is complex, and consists of a number of sharp peaks at 304, 316, 331, 343, 361, 373, 389, 402, 418, 438, 462, and 473 nm. These excitation peaks correspond to the transitions from the $^6H_{5/2}$ ground state to the $^3H_{9/2}$, $^4F_{11/2}$, $^4D_{7/2}$, $^3H_{7/2}$, $^4F_{9/2}$, $^4D_{5/2}$, $^6P_{7/2}$, $^4K_{11/2}$, ($^6P_{5/2} + {}^4M_{19/2}$), ($^4G_{9/2} + {}^4I_{15/2}$), ($^4F_{5/2} + {}^4I_{13/2}$), and ($^4I_{11/2} + {}^4M_{15/2}$) excited states of Sm^{3+}, respectively. Without irradiation, no notable signal is detected for the Sm^{2+} emissions. Broad spectral features, however, appear after irradiation, and their intensity increases with irradiation dose as for emission. Considering the broad excitation features, the excitation transitions are attributed to be from the $^7F_0$ ground state to $4f^55d^1$ excitation states.

As above, the Sm^{2+} luminescent center is activated by irradiation, and it is observed as an RPL phenomenon. The process is, in general, considered that electrons and holes are generated by irradiation, and a fraction of the electrons are captured by Sm^{3+} and then become Sm^{2+}. In order to maintain the charge conservation, the generated holes should be neutralized or captured somewhere, and the destination of the holes depends on the host material. For example, in the case of Sm-doped fluorophosphates glass [73], the holes are considered to be captured by phosphorus-oxygen bonds and then POHC is formed. The RPL process of Sm-based phosphors can be expressed as follows:

$$Sm^{3+} + host \xrightarrow{\text{irrad.}} Sm^{3+} + host + \left(e^- + h^+\right) \tag{9.9}$$

$$\xrightarrow{\text{migration}} \left(Sm^{3+} + e^-\right) + \left(host + h^+\right) \tag{9.10}$$

$$\xrightarrow{\text{trapping}} Sm^{2+} + \left(host + h^+\right) \tag{9.11}$$

The RPL response based on Sm is erasable and reproducible after erasure. Figure 9.18 illustrates demonstrations of erasing the RPL response. Figure 9.18a shows the response as a function of optical treatment time with different light sources of wavelength. As soon as exposed to light, the response signal dramatically decreases. Especially, when it is exposed to 285 nm LED light (0.8 mW), the erasure of response is the most effective, and is almost completely erased after 200 minutes. The effectiveness of the erasure seems to depend on the exposure source and also wavelength. For more careful evaluation of the effectiveness, the optical power for each light source and optical absorption by the material should be taken into account. Figure 9.18b illustrates the RPL response signal as a function of heat treatment, or annealing, temperature. The treatment was performed for 30 minutes at each temperature, and the response was measured at room temperature. The response value is consistent and stable after heat treatment of up to 350 °C, whereas it dramatically increases when heat treated around 400–450 °C. In addition, the treatment at higher temperatures results in a decrease of signal, and it completely removes the signal above 510 °C. The increase of the signal when heated to around 400–450 °C can be explained as the thermally stimulated electron trapping process by Sm^{3+} ($Sm^{3+} + e^- \rightarrow Sm^{2+}$), where the origin of the electron is the one generated by ionizing radiation and then trapped by some electron trapping center other than Sm^{3+}.

The RPL signal of some Sm-doped materials is reproducible after erasing the signal, both optically and thermally. Figure 9.19 demonstrates the dose response curves measured with

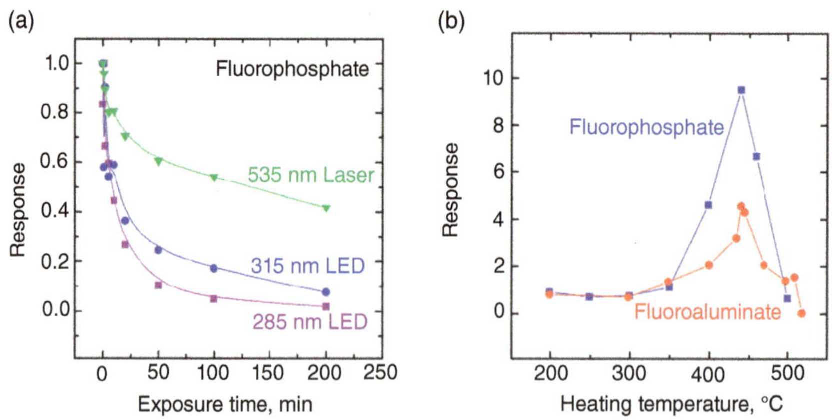

Figure 9.18 *Demonstrations of erasing RPL response by (a) optical means and (b) thermal annealings.*

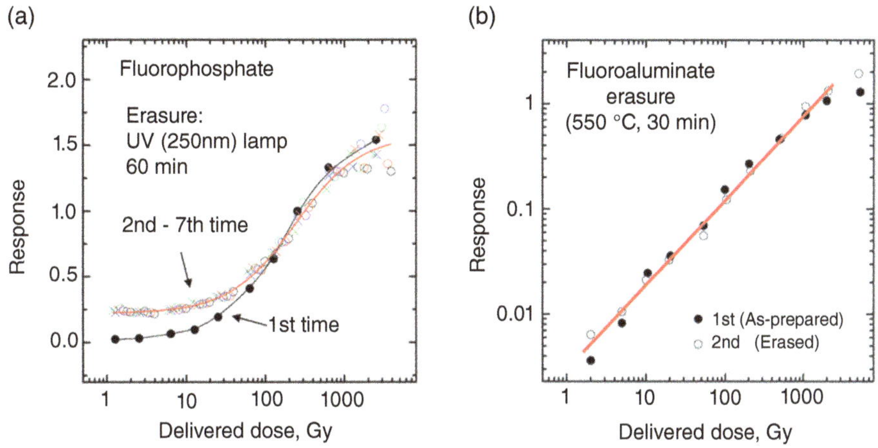

Figure 9.19 *Reusability of Sm-doped RPL detectors after erasure processes (a) by UV exposure and (b) by heat-treatment.*

the erasing processes as examples. In Figure 9.19a, dose response measurements were repeated multiple times, and UV (254 nm) exposure was performed for 60 minutes after each set of response measurements with different doses. The results show that the response curves are reproducible multiple times with UV erasure, which is especially well demonstrated for the second to seventh cycle. The response curves do not agree well with the first time because of insufficient exposure time to completely erase the sample, which requires 200 minutes. Figure 9.19b shows dose response curves measured before and after the thermal erasure process (heating at 550 °C for 30 minutes in air). It well demonstrates that the response curve is almost fully reproducible. The physics involved in the optical and thermal erasure processes are still under discussion, but it can be explained by a charge trapping/detrapping model where the electron trapped at Sm^{2+} is detrapped by UV light and heat, as for optically and thermally stimulated luminescence.

9.3.5.3 Application in Microbeam Radiation Therapy (MRT)

In Microbeam Radiation Therapy (MRT), an array of microplanar X-ray beams is delivered to treat the tumor. Figure 9.20 illustrates the concept of MRT. A highly collimated X-ray beam is generated by a synchrotron (either a bending magnet or wiggler), and the mean X-ray energy is typically 50–150 keV. In the "broad" X-ray beam, a multi-slit collimator is placed to generate an array of microbeams. The collimator used in MRT today is typically the one manufactured by Tecomet (Massachusetts, USA), which is made of tungsten with 8 mm of thickness and more than 100 narrow slits [74]. In the microbeam, each narrow beam has the width of 20–100 µm, and they are accurately aligned by 100–400 µm of separation. With this unique beam geometry, the healthy tissue can tolerate a much higher dose than conventional broad beam therapies (<50 Gy) and, in the R&D of MRT, the X-ray dose delivered at the center of the narrow beam is several hundred Gys or sometimes even higher than 1000 Gy. Also, in between the adjacent beams, a fractional contribution of X-ray dose is present, and it is considered that the ratio of the high dose to the low dose, the so-called peak-to-valley dose ratio, should be large for successful treatment. In order to measure the

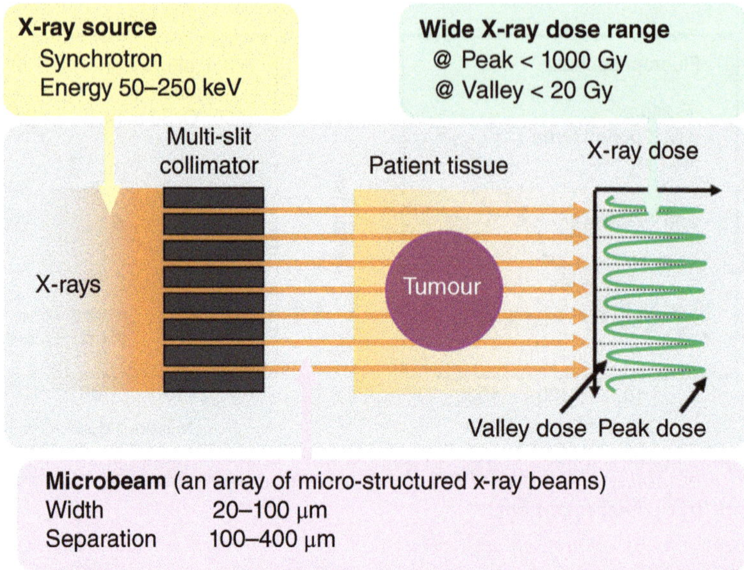

Figure 9.20 *The basic principle of microbeam radiation therapy.*

distribution of X-ray dose in MRT, a detector is required that is capable of detecting doses as low as a few Gys and up to 1000 Gy in conjunction a with a few micrometers of spatial resolution.

In MRT, Sm-doped RPL material can be used to measure the dose distribution. As demonstrated with Al_2O_3:C,Mg (Section 9.3.3.3) and LiF (Section 9.3.4.4), RPL has a great potential for microdosimetry, especially when a fluorescence microscope is used to read out the signal. However, the dynamic range of conventional RPL materials does not satisfy the above requirement (~10 μGy–50 Gy for Ag-doped sodium aluminophosphate glass, ~5 mGy–50 Gy for Al_2O_3:C,Mg, and ~10–over 10 000 Gy for LiF). In addition, other dosimetric techniques do not satisfy both the requirements of dynamic range and spatial resolution together.

Figure 9.21 shows the dose response curves of Sm-doped fluorophosphate and fluoroaluminate glasses with different concentrations of Sm. The response values are normalized to the saturation value for comparison of the dynamic range. All the materials in the graph show a response as low as a few Gys. The response value increases with dose and eventually saturates at large doses. As seen in the figure, the dynamic range can be controlled by the doping concentration of Sm. The higher the concentration, the larger dose required to saturate, and the entire curve shifts toward the higher dose region. The comparison of the two host materials shows that the fluoroaluminate glass has higher sensitivity in the larger dose range than that of fluorophosphate. Particularly, the 1% Sm-doped fluoroaluminate glass has the dynamic range from a few Gys to over 1000 Gy, which covers the dose range for MRT. This composition fluoroaluminate glass plate with a fluorescence confocal microscope was recently used to characterize the MRT beam shape at different energies at the Canadian Light Source (CLS) [75].

The RPL response signal can be read out by using a confocal microscope with a submicrometer of image resolution, as reviewed in [76]. Figure 9.22 represents 2-D fluorescent images obtained by using the 1% Sm-doped fluoroaluminate glass as an imaging plate and a confocal laser scanning fluorescent microscope to read out the signal. The microbeam

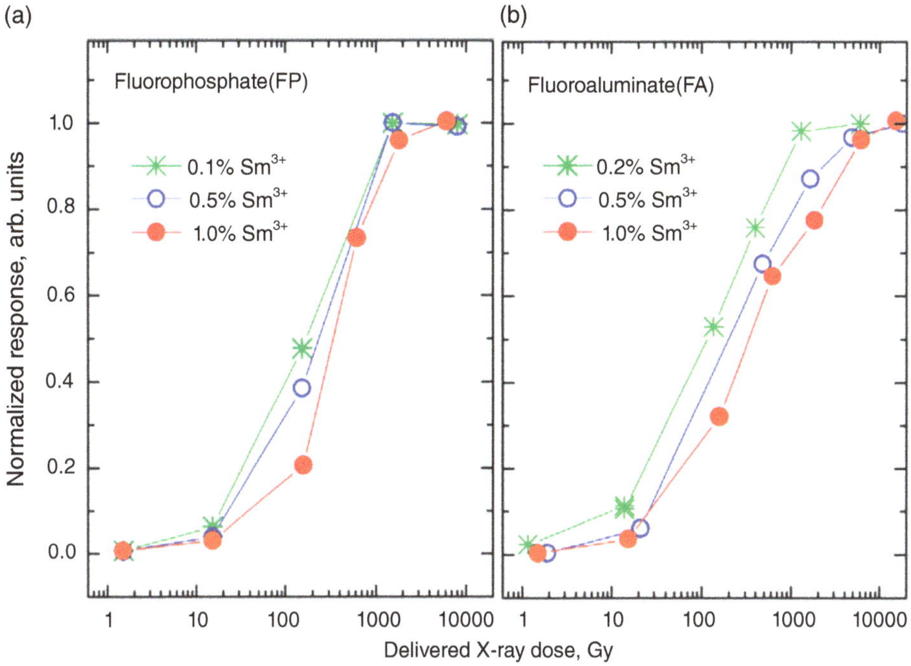

Figure 9.21 Dose response curves of RPL (a) by Sm-doped fluorophosphates glass and (b) by fluoroaluminate glass as a function of concentration of Sm. Source: Reproduced from G. Okada et al. [30] by permission of Elsevier.

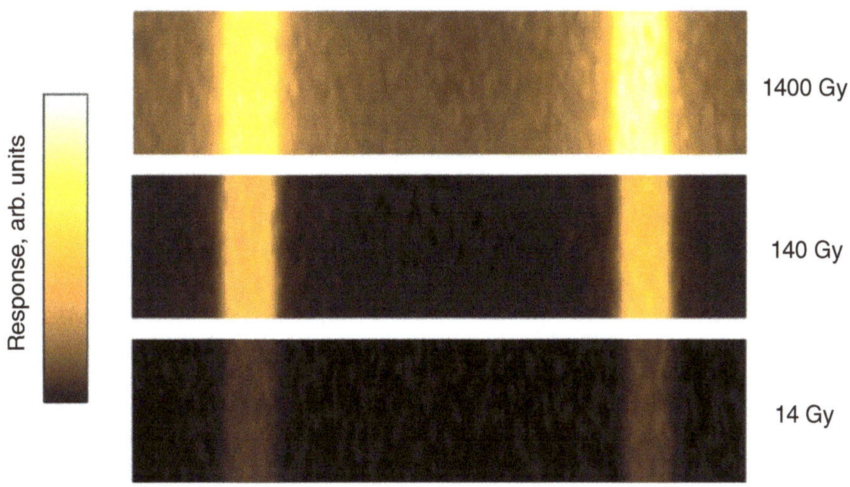

Figure 9.22 Dose distribution of microbeam X-rays measured by using RPL of Sm-doped FA glass detector.

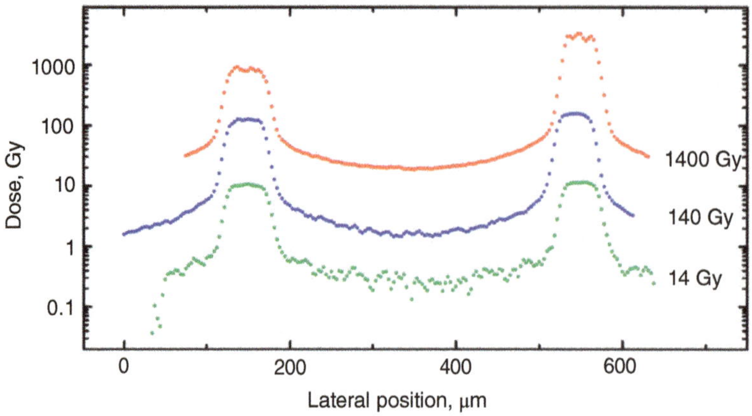

Figure 9.23 *Lateral distribution of X-ray dose in microbeams measured by using RPL of Sm-doped FA glass detector. Source: Reproduced from S. Vahedi et al. [31], with the permission of AIP Publishing.*

irradiation was performed at the BioMedical Imaging and Therapy beamline, CLS. The dose values indicated on the right-hand side are the values at the upstream of the collimator. Two lines of microbeam are clearly illustrated. The width of each beam is 50 μm, and the distance between the centers of the two beams is 400 μm. Figure 9.23 shows the lateral distribution of X-ray dose [31]. It demonstrates that the doses of both the beam centers and between the beams are measured together. In addition, the distribution across the beams is well represented, which carries useful information in radiation physics (e.g., scattering) and therapy (e.g. perfection of alignment). Figure 9.24 compares the microbeam peak shapes measured by using the Sm-doped RPL detector and predicted by Monte Carlo simulation [77]. It is clearly illustrated that both the data well agree with each other, and this fact proves the concept of microbeam measurement by using RPL of Sm-doped detector.

There are several synchrotron facilities worldwide engaged in the R&D of MRT. Each beamline has different characteristics and irradiation geometries (e.g., bending source, fluence, energy, collimation method, etc.). Figure 9.25 compares the simplified irradiation geometries at three different synchrotron beamlines: BioMedical Imaging and Therapy beamline, CLS; Biomedical beamline (ID17); European Synchrotron Radiation Facility (ESRF); and BL28B2 beamline, SPring-8. At the CLS, only a single multi-slit collimator is inserted in the broad beam, whereas a set of two multi-slit collimators as well as horizontal slits are used at the ESRF and the irradiation is performed as the translation stage is moved vertically. The latter geometry allows arbitrary width of narrow beams by adjusting the position. At SPring-8, a single slit collimator is installed horizontally so that one horizontal planar beam is produced to irradiate at a time, and the irradiation is repeated as the sample is translated by the motorized stage. In addition to the irradiation geometries, the X-ray spectrum (mean energy) and dose rate are different; therefore, the dose response of the detector was measured at each facility for calibration. It should be noted that Chicilo et al. recently reported that the response of the Sm-doped fluoroaluminate glass does have energy dependence, but no dose-rate dependence [76]. Figure 9.26 compares the microbeam shapes measured at CLS, ESRF, and SPring-8. The set of measurements were performed by using RPL of 1% Sm-doped fluoroaluminate glass as a recoding element, and the RPL response signal was read out by using a confocal

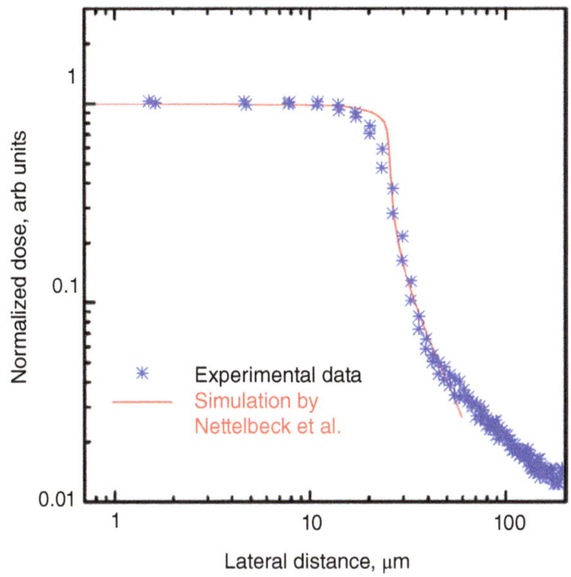

Figure 9.24 *Microbeam peak shapes measured by using Sm-doped RPL detector and simulated by Nettelbeck et al. [77]. Source: Reproduced from G. Okada, PhD Dissertation, University of Saskatchewan (2014) http://hdl.handle.net/10388/ETD-2014-09-1686.*

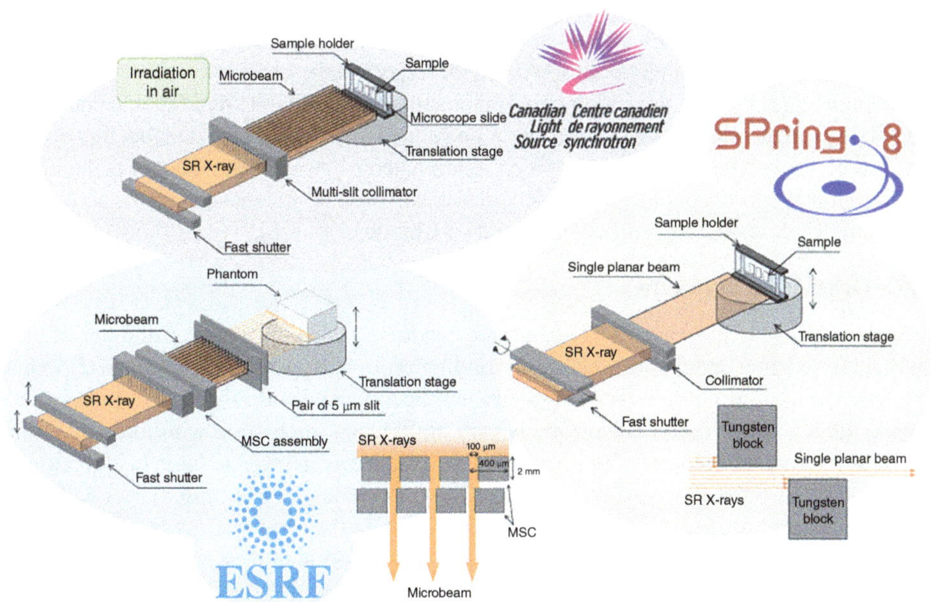

Figure 9.25 *Microbeam irradiation geometries at CLS, ESRF, and SPring-8.*

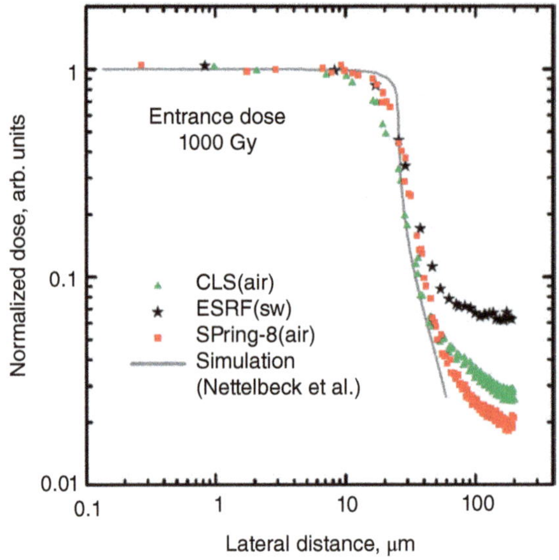

Figure 9.26 *Comparison of microbeam peak shapes measured by using Sm-doped FA glass detector at CLS, ESRF, and SPring-8. Source: Reproduced from G. Okada, PhD Dissertation, University of Saskatchewan (2014) http://hdl.handle.net/10388/ETD-2014-09-1686.*

laser scanning fluorescent microscope. The peak shapes are clearly different between the facilities. Especially at the ESRF, the contrast between the high and low doses (so-called peak-to-valley dose ratio; PVDR) is the lowest, while it is the highest for SPring-8. The reason may be because of significant scattering of X-rays at the multi-slit collimators and horizontal slits. The simpler geometry results in higher contrast, whereas that of CLS is lower than that at SPring-8. In fact, later it was found that the alignment of the collimator at the CLS was imperfect by the rotation angle of only 0.07° with respect to the beam trajectory. This fact, in turn, demonstrates that the present measurement approach is accurate and reliable enough to detect the misalignment as well as the irradiation geometry. It should be noted that this experiment was performed in 2012, and now the irradiation geometry and beam quality at the ESRF have been much improved. Further detailed microbeam dose profile measurements at the CLS with Monte Carlo modeling has been reported by Chicilo et al. [75].

9.3.6 Other RPL Materials

Throughout the search for new RPL materials conducted in recent years, a number of new materials have been found and reported in the literature. Nitrogen doped carbon (C:N) has attracted attention for passive-type dosimetric applications, because the main matrix is carbon and the biological dose equivalency is very high. Upon interaction with ionizing radiation, lattice vacancy is generated in the matrix. The vacancy can migrate upon heat treatment at elevated temperatures, for example, 1000 °C for two hours in a vacuum. Eventually, the migrated vacancy meets a doped nitrogen atom, and then they become a pair, the so-called nitrogen-vacancy (NV) center. The NV center is a class of point defects in carbon and shows fluorescence properties. It can be excited at 450–600 nm and then emits bright red light at 600–800 nm. The use of the NV-center was demonstrated for FNTD by Onoda et al. [40].

Cu-doped aluminoborosilicate glass ($40SiO_2$-$10B_2O_3$-$25Al_2O_3$-$25Na_2O$) was reported to show RPL by Hashikawa et al. [41]. In this material system, the doped Cu ion acts as a luminescent center, and the RPL phenomenon is considered to be due to the valence change of the Cu ion as

$$Cu^{2+} + e^- \rightarrow Cu^+ \tag{9.12}$$

$$Cu^+ + Al-O + h^+ \rightarrow Cu^{2+} + Al-O \text{ or } Cu^+ + AlOHC \tag{9.13}$$

Here, in the initial state, the Cu ion is included in the glass matrix, both in the divalent and monovalent states. When the glass is irradiated with ionizing radiation (X- and γ-rays in the present case), electrons and holes are generated. On the one hand, the electron is captured by the Cu^{2+}, leading to the generation of the Cu^+ center. The Cu^+ center shows a strong luminescence band peaking around 500 nm due to the $3d^{10} \rightarrow 3d^9 4s^1$ transition. On the other hand, the generated hole is captured either by Cu^+ or an Al—O bond. When it is captured by the Cu^+, it becomes Cu^{2+}, whereas AlOHC (aluminum-oxygen hole center) is generated when the hole is captured by the Al—O bond. When the hole trapping by Cu^+ is dominant (meaning that the concentration of Cu^+ is high), the RPL phenomenon hardly occurs. So, one of the key conditions is that the concentration of Cu^+ at the initial state is controlled to be as small as possible. The reported sensitivity to radiation dose is low and not preferred for personnel dosimetric applications but may be acceptable for use in large dose measurements.

Some undoped materials (MgF_2, Mg_2SiO_4, $LiCO_3$, $CaSO_4$, etc.) were reported to show RPL. Unlike the above materials, with the absence of intentional doping, the research especially on finding the luminescence origin is not straightforward and quite challenging. However, without the exact knowledge of the mechanisms, dosimetric properties of some materials were found to be promising. A recent work by Okada et al. [47] revealed that undoped $CaSO_4$ polycrystalline powder demonstrates enhanced RPL properties with higher sensitivity to X-rays than that of a commercial RPL detector (Ag-doped sodium-aluminophosphate glass) by about one order of magnitude. In addition, unlike the conventional detector, it shows no build-up which allows real-time monitoring of radiation dose (Figure 9.27). This material is well-known as a TSL dosimeter material and has been studied over several decades in the field of radiation detection and measurements; however, the

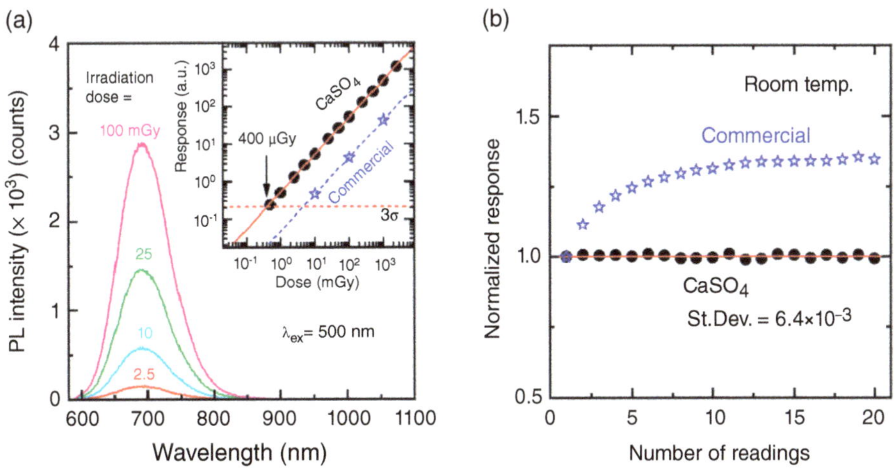

__Figure 9.27__ RPL properties of undoped $CaSO_4$. (a) PL spectra with different irradiation dose and the dose response function compared with the Ag-doped sodium-aluminophosphate. (b) Comparison of RPL stabilities by $CaSO_4$ and Ag-doped sodium-aluminophosphate.
Source: Reproduced from G. Okada et al. [47] by permission of Elsevier.

effect of RPL in this material was only reported recently and it was discovered, in fact, "accidentally." This lesson points to further search for new RPL materials with improved properties, which can lead to future technologies.

9.4 Conclusions

The effect of RPL has been used in radiation dosimetry over the past several decades, and much knowledge, especially on Ag-doped sodium-aluminophosphate glasses, has been accumulated. The major advantage and uniqueness of RPL in contrast to TSL and OSL is the highly stable dosimetric signal, which is not influenced by the history of readout, and this is one of the main reasons that the RPL technology has become one of the most successful dosimetric approaches today. In contrast, despite the history and interest on RPL, there are not so many other materials considered for RPL dosimetry. However, recent studies at length opened the door to new applications, and the recognition of the value of RPL has stimulated a search for new materials. In the near future, together with discoveries of new RPL materials, RPL technology will find new applications, which may not be necessarily only limited to radiation dosimetry.

References

1. Przibram, K. (1921). Über die Phosphoreszenz durch Becquerel Strahlung verfärbter Mineralien. *Wiener Berichte.* 130: 256.
2. Przibram, K. (1923). Verfärbung und Lumineszenz durch Becquerelstrahlen. *Zeitschrift. Für. Phys.* 20: 196–208.
3. Schulman, J.H., Ginther, R.J., Klick, C.C. et al. (1951). Dosimetry of X-rays and gamma-rays by radiophotoluminescence. *J. Appl. Phys.* 22: 1479–1487.
4. Weyl, W.A., Schulman, J.H., Ginther, R.J. et al. (1949). On the fluorescence of atomic silver in glasses and crystals. *J. Electrochem. Soc.* 95: 70–79.
5. Schulman, J.H., Shurcliff, W., Ginther, R.J. et al. (1953) Radiophotoluminescence dosimetry system of the US navy. Nucleonics. 11 OSTI ID: 4426750.
6. Becker, K. (1983). *Solid State Dosimetry*. Cleveland: CRC.
7. Akselrod, G.M.M., Akselrod, M.S.S., Benton, E.R.R. et al. (2006). A novel Al_2O_3 fluorescent nuclear track detector for heavy charged particles and neutrons. *Nucl. Instrum. Methods Phys. Res. Sect. B Beam Interact. Mater. Atoms* 247: 295–306.
8. Miyamoto, Y., Takei, Y., Nanto, H. et al. (2011). Radiophotoluminescence from silver-doped phosphate glass. *Radiat. Meas.* 46: 1480–1483.
9. Miyamoto, Y., Kinoshita, K., Koyama, S. et al. (2010). Emission and excitation mechanism of radiophotoluminescence in Ag^+-activated phosphate glass. *Nucl. Instrum. Methods Phys. Res. Sect. A Accel. Spectrometers Detect. Assoc. Equip.* 619: 71–74.
10. Mckeever, S.W.S., Sholom, S., and Shrestha, N. (2019). Observations regarding the build-up effect in radiophotoluminescence of silver-doped phosphate glasses. *Radiat. Meas.* 123: 13–20.
11. Yokota, R. and Imagawa, H. (1966). Radiophotoluminescence centers in silver-activated phosphate glass. *J. Physiol.* 23: 1038–1048.
12. Huang, D.Y.C. and Hsu, S.-M. (2011). Radio-photoluminescence glass dosimeter (RPLGD). In: *Advanced Cancer Theraphy.*, 553–568. InTech. Hala Gali-Muhtasib https://www.intechopen.com/books/1311
13. Tatsumi, H., Okada, G., Yanagida, T. et al. (2016). Radio-photoluminescence of Ag-doped Li_3PO_4-$Al(PO_3)_3$ glass. *Chem. Lett.* 45: 280–282.
14. Shiratori, D., Isokawa, Y., Samizo, H. et al. (2019). Evaluation of optical and radio-photoluminescence properties in ag-doped $30KPO_3$–$70Al(PO_3)_3$ glasses. *J. Ceram. Soc. Japan* 127: 455–461.
15. Kawamoto, H., Fujimoto, Y., Koshimizu, M. et al. (2019). Temperature dependence of radiophotoluminescence in Ag-doped phosphate glasses containing different alkali metals. *Jpn. J. Appl. Phys.* 58: 062003.

16. Kawamoto, H., Koshimizu, M., Fujimoto, Y. et al. (2019). Formation of radiophotoluminescence centers at room temperature in Ag-doped alkali halides. *Jpn. J. Appl. Phys.* 58: 062004.
17. Akselrod, M.S. and Akselrod, A.E. (2006). New Al_2O_3:C,Mg crystals for radiophotoluminescent dosimetry and optical imaging. *Radiat. Prot. Dosim.* 119: 218–221.
18. Sykora, G.J. and Akselrod, M.S. (2010). Spatial frequency analysis of fluorescent nuclear track detectors irradiated in mixed neutron-photon fields. *Radiat. Meas.* 45: 1197–1200.
19. Sykora, G.J. and Akselrod, M.S. (2010). Photoluminescence study of photochromically and radiochromically transformed Al_2O_3:C,Mg crystals used for fluorescent nuclear track detectors. *Radiat. Meas.* 45: 631–634.
20. Levita, M., Schlesinger, T., and Friedland, S.S. (1976). LiF dosimetry based on radiophotoluminescence (RPL). *IEEE Trans. Nucl. Sci.* 23: 667–674.
21. Nandha Gopal, J., Sanyal, B., and Lakshmanan, A. (2018). Radiophotoluminescence and thermoluminescence characteristics of undoped and Mg doped LiF phosphor in the high dose region. *Radiat. Meas.* 109: 24–34.
22. Mrozik, A., Bilski, P., Marczewska, B. et al. (2014). Radio-photoluminescence of highly irradiated LiF:Mg,Ti and LiF:Mg,Cu,P detectors. *Radiat. Meas.* 71: 31–35.
23. Okada, G., Fujimoto, Y., Tanaka, H. et al. (2017). Dynamics of radio-photoluminescence and thermally-stimulated luminescence in KBr:Sm. *J. Mater. Sci. - Mater. Electron.* 28: 15980–15986.
24. Okada, G., Ueda, J., Tanabe, S. et al. (2014). Samarium-doped oxyfluoride glass-ceramic as a new fast erasable dosimetric detector material for microbeam radiation cancer therapy applications at the Canadian synchrotron. *J. Am. Ceram. Soc.* 97: 2147–2153.
25. Okada, G., Fujimoto, Y., Tanaka, H. et al. (2016). Sm-doped CsBr crystal as a new radio-photoluminescence (RPL) material. *J. Rare Earths* 34: 769–773.
26. Okada, G., Nakamura, F., Kawano, N. et al. (2018). Radiation-induced luminescence centers in Sm:MgF_2 ceramics. *Nucl. Instrum. Methods Phys. Res. Sect. B Beam Interact. Mater. Atoms* 435: 268–272.
27. Nogami, M. and Suzuki, K. (2002). Formation of Sm ions and spectral hole burning in X-ray irradiated glasses. *J. Phys. Chem. B.* 106: 5395–5399.
28. Qiu, J., Miura, K., Suzuki, T. et al. (1999). Permanent photoreduction of Sm^{3+} to Sm^{2+} inside a sodium aluminoborate glass by an infrared femtosecond pulsed laser. *Appl. Phys. Lett.* 74: 10.
29. Morrell, B., Okada, G., Vahedi, S. et al. (2014). Optically erasable samarium-doped fluorophosphate glasses for high-dose measurements in microbeam radiation therapy. *J. Appl. Phys.* 115: 063107.
30. Okada, G., Vahedi, S., Morrell, B. et al. (2013). Examination of the dynamic range of Sm-doped glasses for high-dose and high-resolution dosimetric applications in microbeam radiation therapy at the Canadian synchrotron. *Opt. Mater.* 35: 1976–1980.
31. Vahedi, S., Okada, G., Morrell, B. et al. (2012). X-ray induced Sm^{3+} to Sm^{2+} conversion in fluorophosphate and fluoroaluminate glasses for the monitoring of high-doses in microbeam radiation therapy. *J. Appl. Phys.* 112: 073108.
32. Belev, G., Okada, G., Tonchev, D. et al. (2011). Valency conversion of samarium ions under high dose synchrotron generated X-ray radiation. *Phys. Status. Solidi.* 8: 2822–2825.
33. Okada, G., Shinozaki, K., Komatsu, T. et al. (2017). Radio-photoluminescence in Sm-doped BaF_2-Al_2O_3-B_2O_3 glass-ceramics. *Radiat. Meas.* 106: 73–78.
34. Okada, G., Kawaguchi, N., Kasap, S. et al. (2020). Radio-photoluminescence properties of $LiCaAlF_6$:Sm. *Radiat. Meas.* 132: 106251.
35. Calvert, R.L. and Danby, R.J. (1984). Thermoluminescence and radiophotoluminescence from Eu-and Sm-doped $CaSO_4$. *Phys. Status. Solidi.* 83: 597–604.
36. Okada, G., Hirasawa, K., Kusano, E. et al. (2020). Radio-photoluminescence properties of samarium-doped alkaline earth sulfates. *Nucl. Instrum. Methods Phys. Res. Sect. B Beam Interact. Mater. Atoms* 466: 56–60.
37. Asada, S., Okada, G., Kato, T. et al. (2018). Eu-doped Ca_2SiO_4 as a new radio-photoluminescence phosphor. *Chem. Lett.* 47: 59–61.
38. Okada, G., Shinozaki, K., Shiratori, D. et al. (2019). Radio-photoluminescence observed in Eu-doped BABF glass-ceramics. *Ceram. Int.* 45: 9376–9380.
39. Fujimoto, Y., Okada, G., Sekine, D. et al. (2020). Radiation induced change in the optical properties of NaCl:Yb crystal. *Radiat. Meas.* 133: 106274.
40. Onoda, S., Haruyama, M., Teraji, T. et al. (2015). New application of NV centers in CVD diamonds as a fluorescent nuclear track detector. *Phys. Status Solidi Appl. Mater. Sci.* 212: 2641–2644.
41. Hashikawa, R., Fujii, Y., Kinomura, A. et al. (2018). Radiophotoluminescence phenomenon in copper-doped aluminoborosilicate glass. *J. Am. Ceram. Soc.* 460: 1–10.
42. Nakamura, F., Kato, T., Okada, G. et al. (2017). Scintillation, TSL and RPL properties of MgF_2 transparent ceramic and single crystal. *Ceram. Int.* 43: 7211–7215.

43. Nakamura, F., Kato, T., Okada, G. et al. (2018). Non-doped Li_2CO_3 ceramics as a new radio-photoluminescence material. *Mater. Lett.* 221: 51–53.
44. Nakamura, F., Kato, T. et al. (2017). Non-doped Na_2CO_3 ceramics as a new radio-photoluminescence (RPL) material. *Chem. Lett.* 46: 1383–1385.
45. Nakamura, F., Kato, T., Okada, G. et al. (2018). Radio-photoluminescence in non-doped K_2CO_3 ceramics. *Mater. Lett.* 211: 100–102.
46. Okada, G., Kojima, T., Ushizawa, J. et al. (2017). Radio-photoluminescence observed in non-doped Mg_2SiO_4 single crystal. *Curr. Appl Phys.* 17: 422–426.
47. Okada, G., Koguchi, Y., Yanagida, T. et al. (2020). Undoped $CaSO_4$ showing highly enhanced radio-photoluminescence properties. *Mater. Today Commun.* 24: 101013.
48. Miyamoto, Y., Yamamoto, T., Kinoshita, K. et al. (2008). Emission mechanism of radiophotoluminescence in Ag-doped phosphate glass. *Radiat. Meas.* 431: 017–1023.
49. Miyamoto, Y., Yamamoto, T., Kinoshita, K. et al. (2010). Emission mechanism of radiophotoluminescence in Ag-doped phosphate glass. *Radiat. Meas.* 45: 546–549.
50. Yokota, R. and Imagawa, H. (1966). Radiophotoluminescent centers in silver-activated phosphate glass. *J. Phys. Soc. Jpn.* 23: 1038–1048.
51. Tatsumi, H., Okada, G., Masai, H. et al. (2016). Radiophotoluminescence and thermally-stimulated luminescence of Ag-doped Li_3PO_4–$Al(PO_3)_3$ glass. *J. Lumin.* 179: 545–549.
52. Dmitryuk, A.V., Paramzina, S.E., Perminov, A.S. et al. (1996). The influence of glass composition on the properties of silver-doped radiophotoluminescent phosphate glasses. *J. Non-Cryst. Solids* 202: 173–177.
53. Maurel, C., Cardinal, T., Bellec, M. et al. (2009). Luminescence properties of silver zinc phosphate glasses following different irradiations. *J. Lumin.* 129: 1514–1518.
54. Juto, N. (2002) Glass badge dosimetry system for large scale personal monitoring, in: Proc. Int. Nucl. Conf. 2002 Glob. Trends Perspect., Kuala Lumpur, Malaysia, pp. 142–148.
55. Maki, D., Kobayashi, H., Sato, F. et al. (2011). Development of thermal neutron-sensitive glass dosemeter containing lithium. *Radiat. Prot. Dosim.* 144: 226–230.
56. Maki, D., Sato, F., Murata, I. et al. (2011). Development of neutron-sensitive glass dosimeter containing isotopically enriched boron. *Radiat. Meas.* 46: 1484–1487.
57. Nanto, H., Yanagida, Y., Sugiyama, M. et al. (2017). Visualization of radiation dose distribution utilizing radio-photoluminescence in glass dosimeter. *Sens. Mater.* 29: 1439–1445.
58. Akselrod, M.S.S., Akselrod, A.E.E., Orlov, S.S.S. et al. (2003). Fluorescent aluminum oxide crystals for volumetric optical data storage and imaging applications. *J. Fluoresc.* 13: 503–511.
59. Akselrod, M. and Kouwenberg, J. (2018). Fluorescent nuclear track detectors – review of past, present and future of the technology. *Radiat. Meas.* 117: 35–51.
60. French, R.H. (1990). Electronic band structure of Al_2O_3, with comparison to Alon and AIN. *J. Am. Ceram. Soc.* 73: 477–489.
61. Ealet, B., Elyakhloufi, M.H., Gillet, E. et al. (1994). Electronic and crystallographic structure of γ-alumina thin films. *Thin Solid Films* 250: 92–100.
62. Toyoda, S., Shinohara, T., Kumigashira, H. et al. (2012). Significant increase in conduction band discontinuity due to solid phase epitaxy of Al_2O_3 gate insulator films on GaN semiconductor. *Appl. Phys. Lett.* 101: 1–5.
63. Sykora, G.J., Salaske, M., and Akselrod, M.S. (2008). Properties of novel fluorescent nuclear track detectors for use in passive neutron dosimetry. *Radiat. Meas.* 43: 1017–1023.
64. Akselrod, M.S., Sykora, G.J. et al. (2011). Fluorescent nuclear track detector technology – A new way to do passive solid state dosimetry. *Radiat. Meas.* 46: 1671–1679.
65. Sykora, G.J. (2010). *Photo- and Radiochromic Transformations in Al_2O_3:C,Mg Fluorescent Nuclear Track Detectors and High Resolution Imaging of Radiation Fields*. Oklahoma State University.
66. Sykora, G.L., Akselrod, M.S., Benton, E.R. et al. (2008). Spectroscopic properties of novel fluorescent nuclear track detectors for high and low LET charged particles. *Radiat. Meas.* 43: 422–426.
67. Greilich, S., Osinga, J.-M., Niklas, M. et al. (2013). Fluorescent nuclear track detectors as a tool for ion-beam therapy research. *Radiat. Meas.* 56: 267–272.
68. Niklas, M., Abdollahi, A., Akselrod, M.S. et al. (2013). Subcellular spatial correlation of particle traversal and biological response in clinical ion beams. *Int. J. Radiat. Oncol. Biol. Phys.* 87: 1141–1147.
69. Kurobori, T., Miyamoto, Y., Maruyama, Y. et al. (2014). A comparative study of optical and radiative characteristics of X-ray-induced luminescent defects in Ag-doped glass and LiF thin films and their applications in 2-D imaging. *Nucl. Instrum. Methods Phys. Res. Sect. B.* 326: 76–80.

70. Bilski, P., Marczewska, B., Kłosowski, M. et al. (2018). Detection of neutrons with LiF fluorescent nuclear track detectors. *Radiat. Meas.* 116: 35–39.
71. Qiu, J. and Hirao, K. (2001). γ-Ray induced reduction of Sm^{3+} to Sm^{2+} in sodium aluminoborate glasses. *J. Mater. Sci. Lett.* 20: 691–693.
72. Okada, G., Morrell, B., Koughia, C. et al. (2011). Spatially resolved measurement of high doses in microbeam radiation therapy using samarium doped fluorophosphate glasses. *Appl. Phys. Lett.* 99: 121105.
73. Vahedi, S., Okada, G. et al. (2014). ESR study of samarium doped fluorophosphate glasses for high-dose, high-resolution dosimetry. *Opt. Mater. Exp.* 4: 1244.
74. Bräuer-Krisch, E., Requardt, H., Régnard, P. et al. (2005). New irradiation geometry for microbeam radiation therapy. *Phys. Med. Biol.* 50: 3103–3111.
75. Chicilo, F., Hanson, A.L., Geisler, F.H. et al. (2020). Dose profiles and x-ray energy optimization for microbeam radiation therapy by high-dose, high resolution dosimetry using Sm-doped fluoroaluminate glass plates and Monte Carlo transport simulation. *Phys. Med. Biol.* 65: 075010.
76. Chicilo, F., Okada, G., Belev, G. et al. (2020). Instrumentation for high-dose, high-resolution dosimetry for microbeam radiation therapy using samarium-doped fluoroaluminate and fluorophosphate glass plates. *Meas. Sci. Technol.* 31: 015201.
77. Nettelbeck, H., Takacs, G.J., Lerch, M.L.F. et al. (2009). Microbeam radiation therapy: a Monte Carlo study of the influence of the source, multislit collimator, and beam divergence on microbeams. *Med. Phys.* 36: 447.

10

New Materials for Radiation Detectors: Transparent Ceramics

Takumi Kato, Noriaki Kawaguchi, and Takayuki Yanagida
Nara Institute of Science and Technology, Ikoma, Japan

10.1 Introduction of Transparent Ceramic Materials

10.1.1 Light Scattering Sources in Ceramics

First of all, the words of "Translucent" and "Transparent" are defined. Translucent and Transparent mean light transmittance evaluated by in-line and total light transmittance, respectively. Figure 10.1 is a schematic diagram of measurement methods for in-line and total light transmittance. But, conventionally, ceramics with high transparency are often called "Transparent" ceramics, and those with low transparency are called "Translucent" ceramics.

In general, UV–Vis light cannot be transmitted in ceramics. The reason is that the inside of the material can include light scattering sources. These can include pores and secondary phases due to impurities at the grain boundary, causing scattering sources characteristic of ceramics. In addition, roughness on the surface and macro defects contribute to light loss. These light scattering sources are depicted in Figure 10.2.

Low light propagation in ceramics is approximately represented by Lambert–Beer, as described below. As shown in Figure 10.3, when light intensity I_0 is incident on a material with thickness t and surface reflection R, and taking into account multiple reflections, transmitted light intensity I decays according to Equation 10.1. Here, β and α show decay coefficients of medium and intrinsic absorption based on electronic transition in the crystal of the sintered body, respectively. S_p and S_b are scattering coefficients, where S_p is the scattering coefficient by pore, and S_p is the scattering coefficient by grain boundary, as shown in Figure 10.2.

Phosphors for Radiation Detectors, First Edition. Edited by Takayuki Yanagida and Masanori Koshimizu.

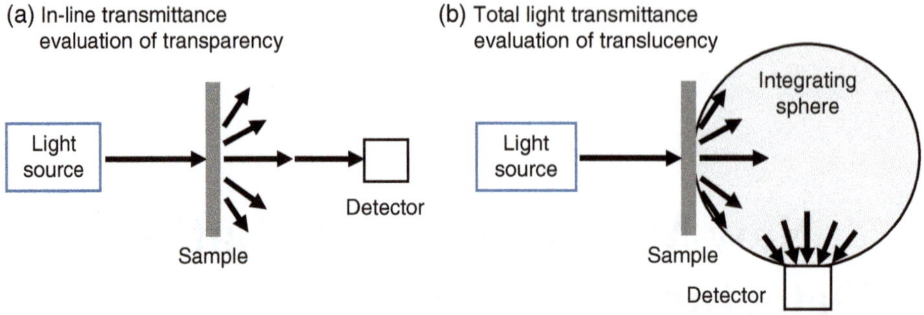

Figure 10.1 *Schematic diagrams of measurement methods for (a) in-line and (b) total light transmittance.*

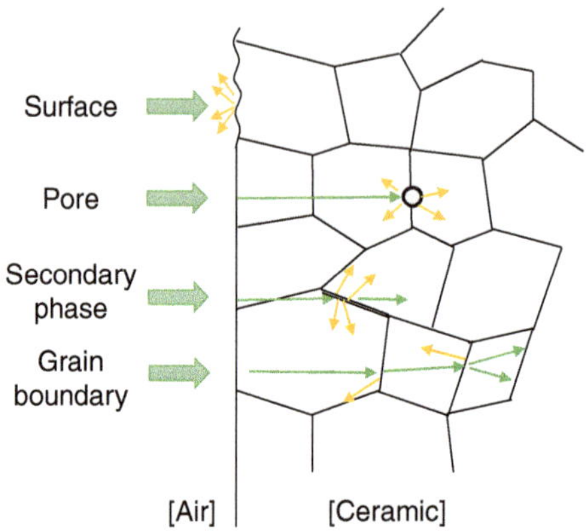

Figure 10.2 *Light scattering sources in ceramics.*

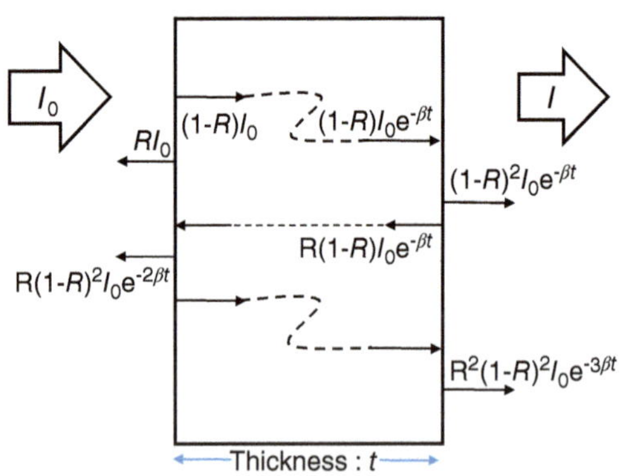

Figure 10.3 *Transmission of light in a medium and multiple reflections at the interface (when there is no light scattering at the interface between the medium and air).*

Moreover, S_b is classified into three scatterings caused by birefringence, secondary phase at grain boundaries, and the grain boundaries themselves:

$$I = I_0 (1-R)^2 e^{-\beta t} \left(1 + R^2 e^{-2\beta t} + R^4 e^{-4\beta t} + \cdots\right)$$

$$\frac{I}{I_0} = \frac{(1-R)^2}{1-R^2 e^{-2\beta t}} e^{-\beta t}$$

$$\beta = \alpha + S_p + S_b \quad (10.1)$$

Ceramics with high transparency can be obtained by decreasing β. R is an intrinsic factor of a material, and α is also an intrinsic factor which depends on the energy band-gap in the UV range and elastic resonance scattering in the IR range. That is why we should select materials without intrinsic absorption in the desired wavelength range to obtain ceramics with high transparency. Moreover, to use high purity powder, mixed powder homogeneously, and to remove any impurities are very important factors.

It is also important that the value of S_p is reduced in the process of making ceramics transparent. To achieve the reduction of S_p, it is required to decrease the number of pores and precipitates during sintering and to reduce the number of grain boundaries by suppressing abnormal grain growth. To decrease the number of pores, raw powders should include fewer pores and be fine-grained with short diffusion lengths. In addition, primary particles (the smallest particles) should not be in an aggregate state, which is essential to obtain ceramics with high transparency. In an aggregate with primary particles, voids are surrounded and go on to form relatively large voids in the sintered body. Sintering is a crucial process to suppress the generation of pores, and there are parameters for control of sintering: sintering temperature, time, atmosphere, pressure, and dopant to promote grain growth. The value of S_b is no problem in the case of a material whose crystal structure is cubic, because birefringence cannot occur in such a material. In the case of a material where the crystal structure is not cubic, the value of S_b, should be taken into consideration, since such a material is optically anisotropic. In the case of optically anisotropic ceramics, the number of grain boundaries can be decreased by moderate growth of crystal grains, achieved by suppressing abnormal grain growth in order to suppress light diffusion phenomenon by birefringence around the grain boundary.

Accordingly, it is theoretically possible that ceramics can have equal transparency to single crystals if the selected material is cubic without being optically anisotropic and by excluding all extrinsic scattering sources.

10.1.2 History and Applications on Transparent Ceramics

The history of transparent ceramics only covers 50 years, with the first transparent ceramics being reported in 1959. R.L. Coble, of General Electric Co. in the United States, was the first to succeed in developing Al_2O_3 transparent ceramics by using MgO as a sintering additive and also being the first to sinter in a hydrogen atmosphere [1]. To obtain denser ceramics, in general, sintering at high temperatures for a long time is needed. Unless the grain growth rate is controlled during the sinter, abnormal grain growth will occur, and a number of pores will remain inside the sintered body. Coble demonstrated that to eliminate pores acting as a light scattering source and suppress abnormal grain growth, by adding the sintering additive of

MgO made it possible to produce dense ceramics with high transparency. Such an Al_2O_3 translucent ceramic has been produced as Lucalox™ and used to produce more than several million high pressure sodium-vapor lamps per year to date. However, this product is not completely transparent, but milky because its crystal structure is hexagonal.

Examples of transparent ceramics developed so far are summarized in Table 10.1, together with their applications. Y_2O_3 basically shows high transmittance from ultraviolet to infrared

Table 10.1 *Example of transparent ceramics and their applications.*

Chemical composition	Application	Reference
Al_2O_3	High temperature window, Laser medium (Cr^{3+} doping), dosimeter	[2–5]
MgO	Window, dosimeter	[6–8]
Y_2O_3	High temperature window, Infrared window, Laser medium	[9–11]
Sc_2O_3	Laser medium (Er^{3+} doping)	[12]
Lu_2O_3	Laser medium (Nd^{3+} doping), Scintillator (Yb^{3+} doping)	[13–15]
$(Y,Gd)_2O_3$	Scintillator (Eu^{3+} doping)	[16–19]
ZrO_2	Infrared window	[20, 21]
ZnO	Scintillator	[22, 23]
$Y_3Al_5O_{12}$ (YAG)	Laser medium (Nd^{3+} doping, etc.), Phosphor for LED (Ce^{3+} doping), Scintillator (Ce^{3+} doping), Persistent phosphor (Ce^{3+}-Cr^{3+} co-doping)	[24–28]
$Lu_3Al_5O_{12}$ (LuAG)	Laser medium (Yb^{3+} doping), Scintillator (Ce^{3+} doping, Pr^{3+} doping)	[29–31]
$Gd_3(Al,Ga)_5O_{12}$ (GAGG)	Scintillator (Ce^{3+} doping)	[32]
$MgAl_2O_4$	Military optical material	[33, 34]
$CaAl_2O_4$	Persistent phosphor (Eu^{2+}-Dy^{3+} co-doping)	[35]
$SrB_xAl_{2-x}O_4$	Persistent phosphor (Eu^{2+}-Dy^{3+} co-doping)	[36]
$ZnAl_2O_4$	High temperature window	[37]
Lu_2SiO_5	Scintillator (Ce^{3+} doping)	[38, 39]
$Sr_2Y_8(SiO_4)_6O_2$	Scintillator (Ce^{3+} doping)	[40]
$Bi_3Ge_4O_{12}$	Scintillator	[41]
$Ca_5(PO_4)_3F$ (FAP)	Laser medium (Yb^{3+} doping)	[42]
$Sr_5(PO_4)_3F$	Laser medium (Yb^{3+} doping)	[43]
ZnS	Laser medium (Cr^{2+} doping)	[44]
Gd_2O_2S (GOS)	Scintillator (Pr^{3+},Ce^{3+},F^{+} co-doping, Tb^{3+} doping)	[45–48]
$Ba(Zr,Mg,Ta)O_3$	Optical lens, Laser medium (Nd^{3+} doping)	[49]
$SrHfO_3$	Scintillator (Ce^{3+} doping)	[50]
$La_2Zr_2O_7$	Scintillator (Bi^{3+} doping)	[51]
$La_2Hf_2O_7$	Scintillator (Pr^{3+} doping)	[52]
$Lu_2Ti_2O_7$	Dielectric	[53]
Lu_3NbO_7	Dielectric	[54]
MgF_2	dosimeter	[55]

Table 10.1 *(Continued)*

CaF_2	Laser medium (Dy^{3+} doping, etc.), Optical lens, dosimeter	[56–58]
BaF_2	Scintillator	[59–61]
CeF_3	Scintillator	[62]
$BaCl_2$	—	[63]
CsCl	Scintillator	[64]
CsBr	Scintillator (Tl^+ doping), Imaging plate (Eu^{2+} doping)	[65–67]
$LaBr_3$	Scintillator (Ce^{3+} doping)	[39]
SrI_2	Scintillator (Eu^{2+} doping)	[68]
SiAlON	Phosphor for LED light (Ce^{3+} doping, Eu^{2+} doping)	[69]
AlN	Heat sink	[70]
$Tb_3Ga_5O_{12}$ (TGG)	Faraday isolator	[71]
$Tb_3Al_5O_{12}$ (TAG)	Faraday isolator	[72]
$Er_3Al_5O_{12}$	Up-conversion phosphor	[73]
(Pb,La)(Ti,Zr)O_3 (PLZT)	Ferroelectric	[74]
(Sr,Ba)Nb_2O_6 (SBN)	Ferroelectric	[75]

and excellent corrosion resistance. Hence, Y_2O_3 is expected to be applied to various purposes, including applications of high-temperature and infrared windows. However, the melting point of Y_2O_3 is as high as 2400 °C or higher, and there is a phase transition (hexagonal ↔ cubic) just below the melting point, so it is difficult to produce large and high-quality single crystals using existing technology. Therefore, Y_2O_3 transparent ceramics have actively been developed since the 1960s by applying sintering technology [9–11].

Similarly, transparent ceramics of Sc_2O_3 and Lu_2O_3, of which melting points are high, have been investigated for a long time [13, 14, 76–78]. Furthermore, transparent ceramics of $MgAl_2O_4$, which have high mechanical strength and excellent infrared transmittance, have been actively researched and developed for military use, mainly in the United States [33, 34, 79].

On the other hand, transparency of various ceramics such as ferroelectric (Pb,La)(Ti,Zr)O_3 (PLZT) and (Sr,Ba)Nb_2O_6 (SBN) and magnetic $Tb_3Al_5O_{12}$ (TAG) has been reported [71, 72, 74, 75, 80]. In recent years, R&D of transparent ceramics with luminescence properties given by a dopant have been actively pursued. Simple oxides and garnet materials represented by $Y_3Al_5O_{12}$ (YAG) have been studied for applications such as scintillators, laser oscillation, wavelength convertors for white LEDs, and persistent phosphors.

10.2 Preparation Methodology

10.2.1 Sintering Mechanism of Ceramics

Sintering of ceramics is defined as a mass transfer phenomenon caused by thermal treatment which reduces surface free energy at a temperature less than the melting point. It is difficult for us to simulate sintering by just one model, because sintering is composed of three steps:

(i) a form of junctions between adjacent partials; (ii) a densification by a disappearance of pores; and (iii) grain growth. Hence, the sintering mechanism of ceramics is described by three further processes divided into initial, intermediate, and final stages [81].

10.2.1.1 Initial Stage

In the initial stage of sintering, the substance transforms from a bond (neck) to a point contact between particles, and a grain boundary is formed between the bonded particles. The initial stage of sintering is a process in which the linear shrinkage becomes 4 and 5%, and the bonding area between particles reaches to about 20% of the average area. When the atoms can be transferred by heating, the overall shape changes in a direction toward a sphere with a smaller free energy, that is, a smaller specific surface area. The driving force at this time is explained as a difference in the activity and chemical potential controlled by the curvature of the surface of each part of the particle. The magnitude of the chemical potential on the particle surface is considered to be the magnitude of the surface stress acting on the surface. As shown in Figure 10.4, if the surface energy on a curved surface is γ, an extra vertical pressure ΔP works in the direction indicated by the arrow in comparison with the plane. The magnitude of ΔP is expressed by the following equation:

$$\Delta P = 2\left\{\gamma dl \sin\left(\frac{\theta}{2}\right) - \gamma dl \sin\left(\frac{\theta'}{2}\right)\right\} / (dl)^2 \cong \gamma\left(\frac{1}{x} - \frac{1}{\rho}\right) \approx -\frac{\gamma}{\rho}(x \gg \rho) \quad (10.2)$$

where γ is the surface energy, x is the radius of the neck portion, and ρ is the radius of curvature.

Here, excessive free energy: ΔF is given by

$$\Delta F = \Delta PV \quad (10.3)$$

where V is volume of 1 mol of particle system. Considering a distribution of the local surface energy in terms of how mass transfer occurs, $\Delta F<0$ is concave because $\Delta P<0$, and $\Delta F>0$ is convex because $\Delta P>0$. Thereby, mass transfer occurs from convex with high energy to concave with low energy. In general, mass transfer paths at the initial stage are regarded as surface diffusion, grain boundary diffusion, and lattice diffusion, as indicated in Figure 10.5 [82].

Surface diffusion, evaporation, and condensation contribute to the growth of the neck, but the shrinkage factor is 0 because no space between particles occurs. In the initial stage, particle diffusion and volume diffusion are mainly involved. Generally, the following equation is

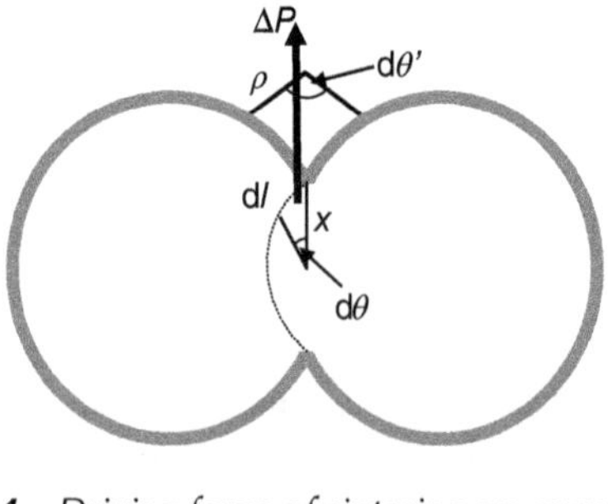

Figure 10.4 *Driving force of sintering on curved surface.*

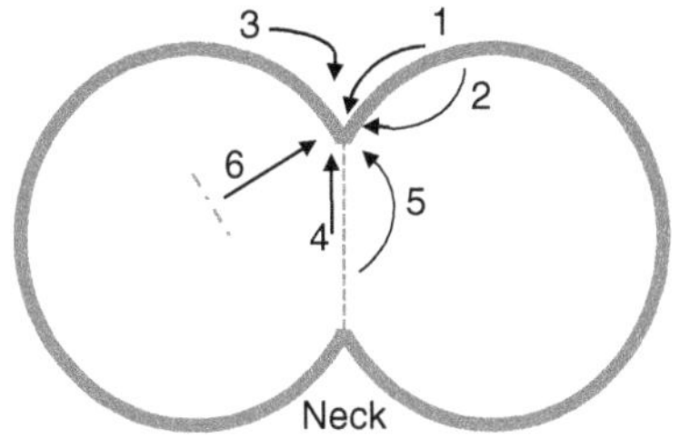

Figure 10.5 *Mass transfer paths during initial stage of sintering.*

widely used as a velocity equation of a mass transfer mechanism by volume diffusion from a grain boundary to a neck portion:

$$\frac{x}{\gamma} = \left(\frac{40\gamma a^3 D}{kT}\right)\gamma^{-3/5}t^{1/5} \tag{10.4}$$

$$\frac{\Delta L}{L_0} = \left(\frac{20\gamma a^3 D}{\sqrt{2}kT}\right)^{2/5}\gamma^{-6/5}t^{2/5} \tag{10.5}$$

where a^3, k, and D are the atomic volume of the diffused vacant, Boltzmann constant, and self-diffusion coefficient, respectively. Here, $\Delta L/L$ represents the shrinkage. Accordingly, it is assumed that the neck growth formed between the particles is proportional to 1/5 power of time, and the shrinkage is proportional to 2/5 power of time.

10.2.1.2 Intermediate Stage

In the intermediate stage, significant shrinkage and densification of the sintered body occurs. Linear shrinkage becomes 15%, and densification process is up to about 95% of a ratio of theoretical density. Grain boundary diffusion is mainly dominant at the intermediate stage, and a point contact of adjacent particles transforms to a surface contact during the procress from initial to intermediate stages of sintering. As a result, the surface gradually changes to a grain boundary, and some voids existing between particles shrink, left behind at grain boundaries as pores, as shown in Figure 10.6. Mass transfer paths during the intermediate stage are addressed, like the model shown in Figure 10.7 [83].

Atoms migrate to the pore surface through mass transfer of I–IV and VI paths, and consequently pores gradually become smaller. Furthermore, grain growth occurs by the migration of the grain boundary through mass transfer of the V path. When migration velocity of grain boundaries is remarkably fast, pores are left behind after grain boundaries migrate because they cannot keep up with such a high migration velocity. On the other hand, in the case where migration velocity of grain boundaries is slow, pores are eliminated to the outside by diffusion along grain boundaries, since pores can migrate at the same time as the grain boundaries. Migration velocity of grain boundaries can be controlled by the heating rate during the sintering process. In addition, since the driving force of mass transfer for sintering is stronger as a particle size is smaller, it is necessary to eliminate pores as much as possible while the grain size is small and before the grain growth process at the end of sintering.

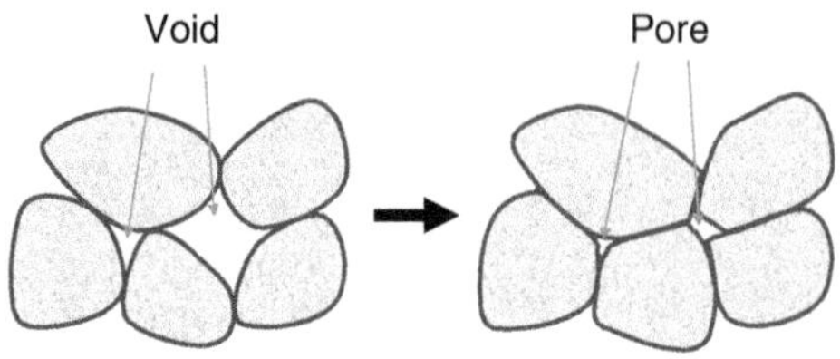

Figure 10.6 Shrinkage of voids between adjacent particles.

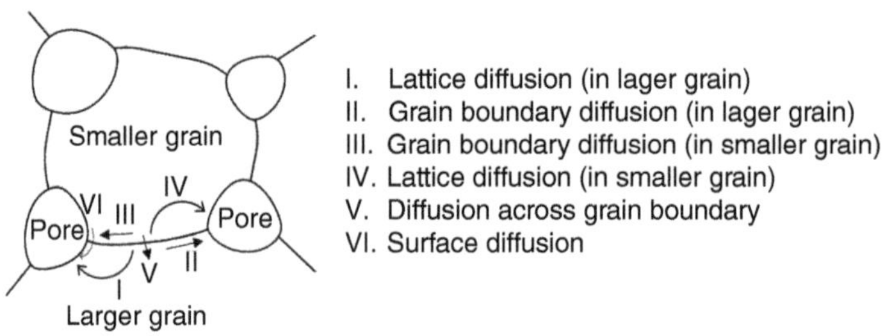

Figure 10.7 Mass transfer paths during intermediate stage in sintering.

10.2.1.3 Final Stage

The final stage is a process that completes densification and grain growth, and main mass transfer is caused by grain boundary diffusion in the same way as at the intermediate stage. In this final stage, the heterogeneous phase, precipitate, and abnormal grain growth are not generated and complete elimination of the pores is essential. The heterogeneous phase and precipitate can be a light scattering source, but can be suppressed by using a high-purity raw powder. Furthermore, abnormal grain growth can also be suppressed by controlling the heating rate; in other words, the migration velocity of the grain boundary. Therefore, the biggest issue is how to completely eliminate the pores. Grain growth progresses at the end of sintering, and the curvature of grain boundaries decreases, so the driving force of mass transfer to remove pores decreases. In a general preparation of transparent ceramics, the average particle size is regarded to be as large as 30 μm or more, in order to reduce pores using diffusion by grain growth. However, when the pores become independent at the final stage of sintering, vacancies diffuse from small pores to large pores, and the small pores eventually disappear, so that the average pore diameter increases. This phenomenon is called Ostwald-ripening [84]. Moreover, there is a possibility that small pores are transformed into large pores and left behind inside the sintered body during the grain growth step (Figure 10.8).

10.2.2 Effect of Residual Pores

The following simulation has been reported on the relationship between pores, which are light scattering sources specific to ceramics, and in-line transmittance. In Ref. [85], it was reported that the effect of residual porosity on the in-line transmittance is very serious, and the porosity needs to be 10 ppm or less in order to obtain the same transmittance as that of a single crystal. It has been found that the in-line transmittance decreases with increase of

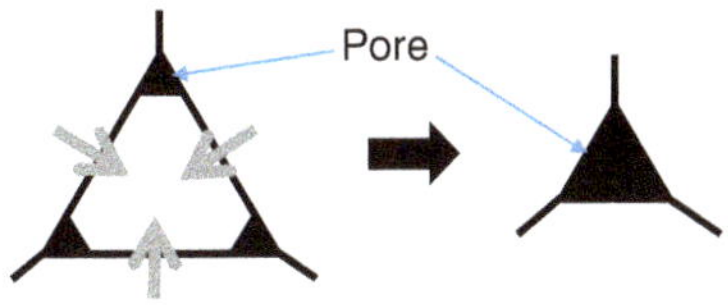

Figure 10.8 Coarsening of pores during grain growth step.

porosity, and ceramics become opaque in the case of 1000 ppm (0.1%) of the porosity. Elimination of the residual pores is the most essential process among other sintering processes and depends on not only sintering processes but the properties of raw powders. Thus, not only molding and sintering processes but a preparation of raw powders is crucial concerning material design for transparent ceramics. Generally, to use a powder with high purity and low aggregation is desirable. The residual pores are mainly intragranular pores left in the grains, and it is important to control the transfer of the grain boundaries to prevent uneven sintering and abnormal grain growth.

The degree of scattering is strongly affected by the size of the scattering source. The in-line transmittance clearly decreases when the scattering source has the same size as visible light. In the case where the scattering source is of larger size, the in-line transmittance simply improves because of a decrease in the number of pores. On the other hand, in a region below 100 nm, the in-line transmittance improves, although the number of pores increases in a certain region. This is due to the effect of Q_{sca}. Although Q_{sca} is about 2 in the region above 100 nm, it decreases in the region of no more than 100 nm and becomes zero in the region below 50 nm. In other words, even if there are some scattering sources, scattering can be suppressed by shrinking the size.

10.2.3 Preparation Methods of Transparent Ceramics

In theory, transparent ceramics can be synthesized; however, it is difficult to theoretically predict the optimal preparation conditions, which can only be obtained experimentally at present. Moreover, such conditions depend on the state of the raw powder (impurity content and particle size) and the sintering method, as well as the chemical composition. Thus, it is difficult to produce transparent ceramics because the optimal preparation conditions are not always determined for one chemical composition.

Similar to that of general ceramics, the process of transparent ceramics also consists of synthesis of precursor powder, compacting, calcination/sintering, and post-treatment (annealing, machining, and polishing) [86]. However, the process of transparent ceramics has its special requirements, especially for sintering methods. Sintering methods to obtain transparent ceramics are mainly classified to the normal sintering, hot-isostatic pressing (HIP), hot pressing (HP), and spark plasma sintering (SPS) methods. Figure 10.9 is a schematic diagram of the normal sintering, HIP, HP, and SPS methods. Normal sintering is a method in which densification is proceeded by heating without actively using an external field. The process is generally performed in air but often under a vacuum or hydrogen atmosphere to obtain transparent ceramics. The HIP method is a manufacturing process used to reduce the porosity of the sintered body and increase the density of various ceramic materials. In the HIP method, the gas pressure is applied isotropically to the sintered body to promote densification in addition to heating during sintering. In the HP method, raw powders are poured into a die made of alumina or carbon, and sintered by uniaxially pressing using punches in an electric furnace. Unlike the HIP method, the

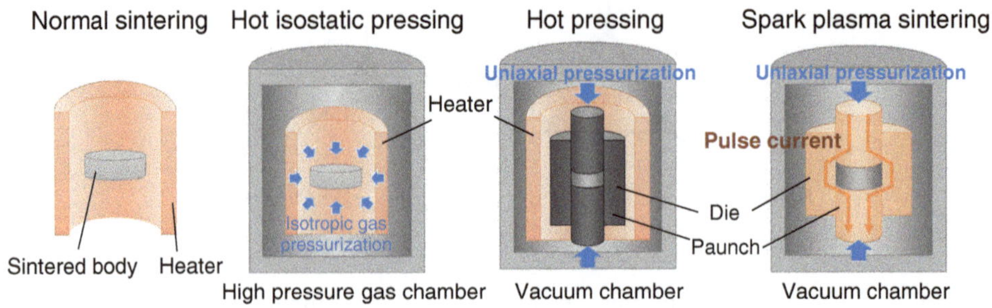

Figure 10.9 *Schematic diagrams of the normal sintering, HIP, HP, and SPS method.*

pressure in the HP method is not isotropic. However, the equipment composition is simple, so it is suitable for preparation of a disc- or plate-like dense body. In the SPS method, sintering is accomplished by a high-density current flux flowing through the sample and the die to cause Joule heating within the sample, while the HP method uses an external heater. Raw powders are uniaxially pressed during sintering, as with the HP method. The SPS technique has a significant advantage over HIP and HP methods, because it can complete the powder densification within a short time duration.

10.3 Transparent Materials

Material forms of inorganic radiation detector are categorized into roughly three types: single crystal, glass, and ceramic. Table 10.2 displays common characteristics of single crystal, glass, and ceramic. Here, the word "ceramic" means not transparent ceramic but opaque ceramic. Glass is the most common optical material because it is advantageous from the viewpoint of upsizing and freedom of shape, and is used for various applications due to its cost-effectiveness; however, thermal properties and mechanical strength of glass are inferior to those of single crystal. On the other hand, single crystal has good thermal and optical properties but is difficult to upsize and is expensive. Although ceramic is inferior to glass and single crystal, from the viewpoint of optical properties, it has good thermal properties, mechanical strength, and freedom of shape and can be synthesized at low cost, even if the melting point is high at over 2000 °C.

Conventionally optical materials such as single crystals and glasses possess both advantages and disadvantages, and neither of them is versatile enough for all applications; therefore, inspired by an invention of a first transparent ceramic, the possibility of using transparent ceramics as an optical material has been attracting attention. If it is possible that ceramics with full high transparency are synthesized by sintering, they can be extremely practical optical materials which have the advantages of both single crystals and glasses. The following are general feature of transparent ceramics.

1. Optical properties of transparent ceramics are almost the same as those of single crystals.
2. Thermal and mechanical properties of transparent ceramics are superior to those of single crystals and glasses.
3. Since the sintered body is homogeneous, there is no unusable portion that occurs in the case of the pulling method for growing a single crystal, and a large amount of material can be manufactured.

Table 10.2 Common characteristics of single crystal, glass, and ceramic.

	Single crystal	Glass	Ceramic
Optical properties	Good	Good	Bad
Upsizing	Bad	Good	Good
Thermal properties	Good	Bad	Good
Mechanical strength	Average	Bad	Good
Cost	Bad	Excellent	Good
Freedom of shape	Bad	Good	Good
High melting point material	Bad	Bad	Good

4. Ceramics can be synthesized at lower temperatures in comparison with single crystals, due to a homogenization and miniaturization of the raw powders.
5. It is possible that ceramics are synthesized in a shorter time than single crystal growth.
6. In the case of ceramics, a material with a desired shape can be synthesized (near net shape).
7. Ceramics can be mass produced at relatively low cost, similar to glasses.
8. Ceramics can be doped with activated ions at a higher concentration than single crystals.
9. In comparison with single crystals, ceramics are difficult to segregate from the dopant.
10. Even activated ions, which are impossible to be doped in single crystals, can be doped in ceramics.
11. It is possible to form composite materials by joining with ceramics, including different emission centers.

10.4 Transparent Ceramic Scintillator

To date, transparent ceramic scintillators with several chemical compositions such as $(Y,Gd)_2O_3$, Gd_2O_2S, and $Y_3Al_5O_{12}$ have been developed. Basically, transparent ceramic scintillators have been developed against specific chemical compositions that have several features; for example, the melting point is relatively high (above 2000 °C), phase transitions occur, and the crystal structure is cubic. In recent years, although there are few reports, transparent ceramics of Lu_2SiO_5, $La_2Zr_2O_7$, $SrHfO_3$, ZnO, CeF_3, BaF_2, CsBr, $LaBr_3$, and SrI_2 have also been synthesized, and their scintillation properties are still being investigated.

10.4.1 Sesquioxide (Such as Y_2O_3, Gd_2O_3, and Lu_2O_3)

Eu-doped $(Y,Gd)_2O_3$ transparent ceramics have been tailor-made for use in both medical and industrial X-ray detectors for CT-scanning applications [16–18]. These ceramics can be fabricated by vacuum sintering and hot pressing methods. Since a phase of Gd_2O_3 transforms from monoclinic to cubic at 1300 °C, it is difficult to turn it into a transparent ceramic. To solve this problem, Gd_2O_3 transparent ceramics were developed, including Y_2O_3, of which the phase transition temperature is 1800 °C, although its effective atomic number is slightly lower than Gd_2O_3. $(Y,Gd)_2O_3$ transparent ceramics maintain a cubic structure of up to 50 mol% or less of the Gd_2O_3 concentration. As the Gd_2O_3 concentration increases, the scintillation intensity decreases, although the absorption coefficient increases, and the optimal concentration is 30–40 mol%.

Eu-doped $(Y,Gd)_2O_3$ showed emission peaks mainly at 610 nm caused by 4f-4f transitions of Eu^{3+}. Pr-codoping in $(Y,Gd)_2O_3$:Eu decreases the afterglow intensity more than two orders of magnitude with increasing Pr. The trivalent Pr acts as a hole trap effectively and competes successfully against the intrinsic hole traps which are reasonable for the afterglow in $(Y,Gd)_2O_3$:Eu.

Yanagida et al. reported the scintillation performance of Yb-doped Re_2O_3 (Re = Sc, Y, Lu) transparent ceramics [13, 15]. Yb:Lu_2O_3 transparent ceramics showed the scintillation due to charge transfer of Yb^{3+}, and these scintillation decay time constants were in the range of 0.5–1.5 ns. The light yield of the fast component was obtained from the ^{137}Cs-excited pulse height spectrum, and the sample doped with Yb 0.3 mol% displayed the maximum light yield of approximately 500 ph/MeV.

10.4.2 Gd_2O_2S (GOS)

Since the 1970s, researchers have tried to develop Gd_2O_2S single crystals. However, good-quality and large-size crystals have not yet been obtained. The reason is because of a loss of S^{2-} ions due to evaporation, and decomposition of the raw materials is inevitable because the melting point of Gd_2O_2S is higher than 2000 °C. As a result, the development of Gd_2O_2S transparent ceramics has been ongoing since the 1980s. Nowadays, a sintering method where Pr, Ce, F-codoped Gd_2O_2S translucent ceramics are prepared by the HIP method has been established [46–48]. The Pr^{3+} ions in Gd_2O_2S emit photons mainly at 510 nm owing to the $^3P_0 \rightarrow {}^3H_J$ transition with a primary decay time of 3 μs. The light output of Pr, Ce, F-codoped Gd_2O_2S translucent ceramics is 1.8 times higher than that of $CdWO_4$ single crystals.

The Ce and F are codoped in this material, mainly to improve afterglow levels. By a very small amount of Ce-doping (<5 ppm), the afterglow intensity reduces by an order of magnitude, and avoids the light output from drastically decreasing (within 10%). It seems that Ce plays a role in suppressing of trapping centers, so that the afterglow intensity is reduced by Ce-doping. Moreover, Ce-doping causes the the light output decrease because an absorption wavelength due to 4f-5d transitions of Ce^{3+} ions overlaps with that due to 4f-4f transitions of Pr^{3+} ions. By the 0.002–0.003 mol% F-doping, the light output increases by 7%, and the afterglow intensity reduces by more than an order of magnitude. A simplified luminescence process of Pr-doped Gd_2O_2S during X-ray excitation is explained as follows: (i) X-rays $\rightarrow$ e^- (electron) + h^+ (hole), carrier formation; and (ii) $Pr^{3+} + h^+ \rightarrow Pr^{4+}$, (3) $Pr^{4+} + e^- \rightarrow (Pr^{3+})^* \rightarrow Pr^{3+} + h\nu$ (luminescence). When the trapping centers capture electrons or holes, a nonradiative combination of electrons and holes is partly created, and this reduces the light output. On the other hand, afterglow appears when electrons or holes captured by the trapping centers during X-ray excitation are released from the traps by thermal activation after switching off X-ray irradiation, and electrons and holes recombine at Pr^{3+} in Gd_2O_2S, which is the emission center. Therefore, these traps reduce the light output and increase afterglow intensity. The F^- ions (ion radius 1.33 Å) are probably substituted for O^{2-} ions (1.32 Å), due to matching of the ion radius. Then, the F^- ions are charged effectively in +1. If the intrinsic defects charge as negative in the neighborhood of the F^- ions, these ions compensate for the charge of the defects. Accordingly, the light output increases and afterglow intensities decrease.

10.4.3 Garnet Materials (Such as YAG, LuAG, and GAGG)

Because the crystal structure of garnet materials is cubic, transparent ceramics can be relatively easily fabricated. Thus, transparent ceramic scintillators of garnet materials have been developed in parallel with single crystals. In the 1990s, a Ce-doped YAG transparent ceramic was first developed by Zych et al. for scintillator use [28]. However, its light yield was inferior to single crystals with the same chemical composition due to trapping phenomena at grain surfaces and interfaces. After this invention, Yanagida et al. succeeded in developing Ce:YAG a transparent ceramic scintillator which showed a higher light yield than that of single crystal [24].

The Ce:YAG transparent ceramic shows a luminescence due to 5d-4f transitions of Ce^{3+} as well as the single crystal. In decay time profiles the single crystal has a slower component than the transparent ceramic. Hence, in Ce:YAG transparent ceramics, the superior scintillation performance is due to the suppression of antisite defects consisting of retrapping of an electron at the anti-site levels in the forbidden gap [87]. In Ce:YAG single crystals, antisite defects caused a slow luminescence and did not accumulate in pulse height measurements. The light yield of Ce:YAG transparent ceramics was about 15% higher than that of the Ce:YAG crystal.

To improve the stopping power to γ-rays, Yanagida et al. also developed $Ce{:}(Gd,Y)_3Al_5O_{12}$ (Ce:GYAG), Ce:GAGG, Ce:LuAG, and Pr:LuAG transparent ceramic scintillators [30–32, 88]. Moreover, some groups follow these developments, and especially, the Ce:GYAG-based scintillator (Ce:GYGAG) shows excellent scintillation performance [89].

The Ce:LuAG transparent ceramic and single crystal have absorption lines at 230, 340, and 460 nm. In wavelengths shorter than 400 nm, absorption depth was stronger in the ceramic due to higher Ce concentration. On the other hand, they showed 80% transparency at wavelengths longer than 500 nm. Thus, it is confirmed that the Ce concentration of ceramics is much denser than that of the single crystal, although the doped Ce amount is the same. In emission spectra, the emission wavelengths of the transparent ceramic and the single crystal were 520 and 510 nm, respectively. Compared with the single crystal, the emission peak slightly shifted to the longer wavelength. Such a phenomenon was also observed in previous work about Ce:YAG [24]. The origin of this phenomenon is still unclear because transmittance around 500 nm is the same in both samples. In ^{137}Cs pulse height spectra, a 662 keV photoabsorption peak and Compton edge were clearly observed. As a result, absolute light yields of the transparent ceramic and single crystal were calculated as $14\,800 \pm 1500$ and $12\,000 \pm 1200$ ph/MeV, respectively. The light yield of the ceramic was 23% larger than that of its single crystalline counterpart.

The Ce:GAGG transparent ceramic showed scintillation due to 5d–4f transition of Ce^{3+} from 500 to 700 nm. The shape of emission spectra resembled those of other Ce-doped garnet materials. In this emission, differences between the transparent ceramic and single crystal were not so high. However, in the transparent ceramic, weak emission peaks clearly appeared at 380 and 415 nm. Based on XRD, PL emission maps, PL decay times, and scintillation decay times, the origin of the weak emissions in the transparent ceramic is attributed to the perovskite phase, because sometimes this phase appears in garnet materials. In pulse height spectra of the Ce:GAGG ceramics under ^{137}Cs exposure, clear photoabsorption peaks were detected. The light yield of Ce:GAGG transparent ceramic was determined as 70 000 ph/MeV. The light yield of the transparent ceramic highly exceeds that of the single crystal, and this Ce:GAGG transparent ceramic is the brightest among transparent ceramic scintillators by far. The enhancement of the light yield will be ascribed to some complicated reasons as follows: (i) difference of the actual Ce concentration; (ii) light scattering near the detection layers of APD; and (iii) an energy migration from the perovskite to garnet phase. Actual Ce concentration of nominally Ce 1% doped GAGG single crystal was around 0.3–0.4%. On the other hand, Ce concentration of the ceramic was close to 1% because it was fabricated by the solid state reaction. Therefore, higher Ce concentration will also affect the light yield enhancement. In addition, the light scattering on ceramic grains and boundaries could cause observed light yield enhancement, where the light is effectively collected only from a layer near the PD, which is much thinner than the sample itself. Generally, a scintillator thickness affects the scintillation light yield; the thinner the sample, the higher the light yield.

10.4.4 Lu_2SiO_5 (LSO)

Ce-doped LSO is considered to be a potential candidate for high performance scintillators that have been applied to detect γ- or X-rays in many applications. Xie et al. reported Ce-doped LSO translucent ceramic scintillator fabricated by the SPS method, and also Wisniewski et al. reported it prepared by the HP method [38, 39]. The non-annealed and annealed Ce:LSO translucent ceramics exhibited a broad main emission peak entered in wavelengths ranging from 400 to 600 nm, which is attributed to 5d-4f transitions of Ce^{3+}. The integrated intensity of the scintillation spectra of Ce:LSO translucent ceramics suggested that the luminescence intensity of annealed samples increased by nearly 3 times. The vacuum-annealed translucent ceramic shows a light yield that is definitely below 50% of the single crystal value. Some improvements in the light yield after additional annealing in water vapor and air is apparent. The well-resolved photoabsorption peak is consistent with an energy resolution similar to that of the single crystal used as a reference, and the light yield of the Ce:LSO transparent ceramic reaches a value of approximately 86% of that of the single crystal.

10.4.5 $SrHfO_3$

$SrHfO_3$ translucent ceramics are one of the candidates for the next-generation scintillator because of its high stopping power against ionizing radiations. Meanwhile, it has a high melting point of over 2500 °C, and its single crystal is hard to grow. Therefore, undoped and Ce-doped $SrHfO_3$ translucent ceramics (codoped with Al and/or Mg) were fabricated using the SPS method [50]. The Al, Ce, Mg-codoped $SrHfO_3$ had the higher light yield of approximately 5000 ph/MeV and better energy resolution than Mg-free ones. Thus, also considering the back scattered electron images, it was reported that the Mg-codoping made grain size larger, and improved the light yield and energy resolution. In other words, this improvement of the scintillation properties resulted mainly from the enhancement of *QY* due to reduction of the pore concentration.

10.4.6 $La_2Zr_2O_7$ and $La_2Hf_2O_7$

$La_2Zr_2O_7$ and $La_2Hf_2O_7$ with their high density and high effective numbers have been attractive as potential scintillators to detect ionizing radiations. $La_2Hf_2O_7$ crystallizes in the cubic pyrochlore phase and exhibits no phase transition. Scintillation performances of Bi:$La_2Zr_2O_7$ and Pr:$La_2Hf_2O_7$ transparent ceramics were investigated [51, 52]. $La_2Zr_2O_7$ and $La_2Hf_2O_7$ transparent ceramics were prepared by conventional sintering under 98% oxygen atmosphere and the low-temperature high-pressure sintering method, respectively. Emission peaks of the undoped and Bi:$La_2Zr_2O_7$ appeared around 460 and 620 nm, respectively. The scintillation intensity was the largest in 0.07 and 0.10% Bi-doped $La_2Zr_2O_7$. The ^{241}Am α-ray-induced pulse height spectra were measured; however, no clear peak was detected. Therefore, it concluded that the absolute light yield of these materials was estimated to be below 100 ph/5.5 MeV-α.

10.4.7 ZnO

The ZnO scintillator can be considered as a suitable candidate for next-generation α-ray detectors and X- and γ-ray detectors in high counting rate environments. The users of processing facilities show a great interest in the ZnO scintillator owing to its similar chemical composition (meaning similar stopping power) with ZnS, medium density (5.6 g/cm^3), potentially high light yield, relatively low cost, and a low sensitivity to background neutrons as well as γ-rays. However, ZnO is available mostly in powder or thin film form, even though large-volume scintillators are required for many purposes.

Rodnyi et al. have prepared undoped, Li-, Zn-, and Ga-doped ZnO transparent ceramics and investigated their scintillation properties [22, 23]. These ceramics emit photons with 2.00–2.75 eV (450–620 nm) under X-ray irradiation, of which decay time constants are typically of several μs. This luminescence is considered to be due to intrinsic defects such as oxygen and zinc vacancies. On the other hand, the Ga:ZnO ceramic emits photons with 3.21 eV (386 nm). Its decay time constant is estimated to be approximately1 ns, which is associated with an emission of the Wannier excitons. Gallium ions replace zinc ions, as well as filling zinc vacancies, in the crystalline structure of the grains, which results in suppression of the intensity of the emission related to defects. As a result of measurements of pulse height spectra of ^{241}Am α-source, the energy resolution of the peak recorded with Li:ZnO ceramics is 62%, and the light yield of Li:ZnO is about 25% of that for Tl:CsI.

10.4.8 BaF_2

BaF_2 crystals have been known as a potential scintillator material since the 1970s. Scintillation from BaF_2 crystals appears around 190–220 and 310 nm due to auger-free luminescence (AFL) and self-trapped excitation (STE), with decay times of ~0.8 and 600–800 ns, respectively. In particular, BaF_2 has a considerably large effective atomic number, so it has the great advantage of being used for γ-ray detections. In 2010s, a research group in Russia first reported scintillation spectra and decays time of BaF_2 transparent ceramics prepared by the HP method [59, 60]. After these reports, a research group in Japan first investigated pulse height spectra of BaF_2 transparent ceramics prepared by the SPS method [61].

Figure 10.10 indicates the scintillation spectra of the BaF_2 transparent ceramics and single crystal in the VUV-UV and UV–VIS spectral regions. When the X-ray was irradiated, all the samples showed an emission at 320 nm due to the STE luminescence. Moreover, in the transparent ceramic sintered in a vacuum, another peak appeared at 500 nm. Supported by the PL analyses, it is thought that the emission is attributed to some kind of defects generated during the sintering processes by the SPS method. Since this emission is not present in the transparent ceramic sintered in N_2 atmosphere, the emission origin would be anion vacancies. Scintillation

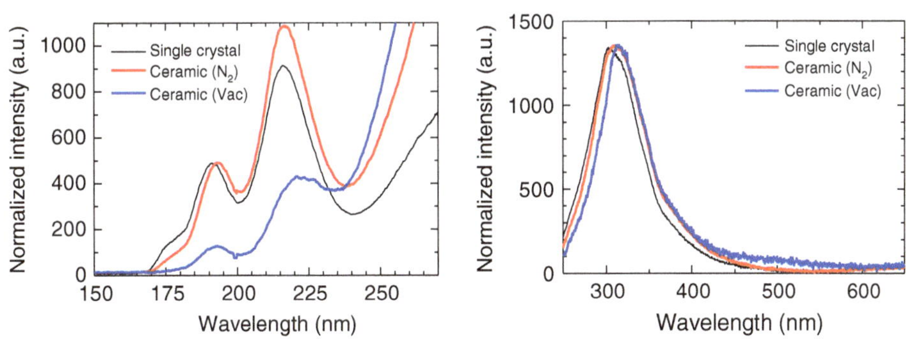

Figure 10.10 X-ray-induced scintillation spectra of all the samples in the VUV-UV (up) and UV–VIS (bottom) spectral regions.

spectra in the VUV range have emission peaks detected at 190 and 220 nm, which are due to the AFL. In the transparent ceramic sintered in a vacuum, the emission intensity of AFL was smaller than that of the ceramic sintered in an N_2 atmosphere, although the sample size was comparable. So, the presence of defects may suppress AFL. Pulse height spectra under ^{137}Cs exposure are shown in Figure 10.11. Clear photoabsorption peaks were detected in the transparent ceramic sintered in N_2 atmosphere and single crystal, while no photoabsorption peak was detected in the transparent ceramic sintered in a vacuum. The absolute light yield of the transparent ceramic sintered in N_2 atmosphere was determined as 6000 ph/MeV. From this result, it was found that green emissions act as negative for scintillation.

10.4.9 CeF_3

CeF_3 single crystal generally shows an emission peak centered at 300 nm with a fast decay time of 10–30 ns. Furthermore, its light yield is insensitive to temperature changes, which makes it suitable for high-rate and high-precision calorimetry. However, CeF_3 single crystal is mainly expensive due to the complicated growth process and low productivity, while the fabrication of transparent ceramics offers a more economical way. X-ray-induced scintillation spectra of CeF_3 translucent ceramic prepared by the HP method were studied by Kou et al. [62].

The CeF_3 translucent ceramic showed a main scintillation emission peak at about 300 nm, emitted by the regular Ce^{3+}. Another weak peak located at 340 nm in the single crystals, corresponding to the perturbed Ce^{3+}, was found to be red-shifted to about 360 nm in the translucent ceramic. The luminescence of CeF_3 is due to the transition from the lowest excited Ce^{3+} 5d level to the ground 4f level. In CeF_3, there are two kinds of Ce^{3+} ions, regular and perturbed Ce^{3+} ions. The regular Ce^{3+} ions were the Ce sites in a perfect crystal structure, whereas the perturbed Ce^{3+} ions were surrounded by some defects. Thus, the origin of the perturbation of these perturbed Ce^{3+} ions was attributed to defects close to them, principally fluorine vacancies or oxygen. A local increase in the crystal field can be produced by these defects which will result in a reduction in the energy gap between 5d and 4f levels of the perturbed Ce^{3+} ions, leading to a longer emission wavelength than regular Ce^{3+} ions. The grain boundary of the translucent ceramic, which is a region where the atom distribution is more disordered and a high concentration of defects is present, may further decrease the energy gap and shift the emission wavelength toward longer wavelengths.

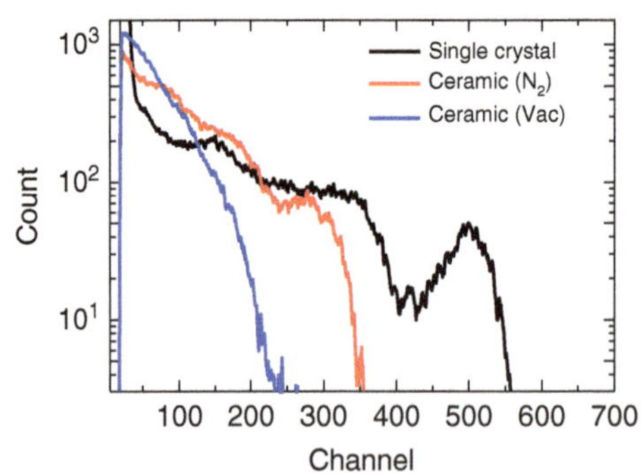

Figure 10.11 Pulse height spectra measured under ^{137}Cs γ-ray irradiation.

10.4.10 CsBr

Eu-doped CsBr is known as a phosphor used for imaging plates. as described in Section 10.5. On the other hand, Tl-doped CsBr transparent ceramics were synthesized for scintillator applications by Kimura et al. [66]. Figure 10.12 shows X-ray-induced scintillation spectra of the single crystal and transparent ceramic. As with the single crystal, the transparent ceramic shows two main emission peaks at around 360 and 500 nm, originating from $^3P_1 \rightarrow {}^1S_0$ transitions of Tl^+ ion and off-center STE, respectively. Pulse height spectra of ^{137}Cs γ-ray were measured using the single crystal and transparent ceramic, as shown in Figure 10.13. The photoabsorption peaks clearly appeared in both samples at almost the same channel. As a result, the calculated light yields of both samples were approximately 23 000 ph/MeV with a typical error of ±10%. The energy resolution at 662 keV of crystal and ceramic samples were approximately 15 and 22%, respectively. The better energy resolution of the single crystal can be understood by better transparency.

10.4.11 $LaBr_3$

Ce:$LaBr_3$ has a hexagonal crystal structure and belongs to the recently discovered group of lanthanide tri-halide scintillators. The recent successes of Ce:$LaBr_3$, that are due to its very high light yield (~60 000 ph/MeV) and outstanding energy resolution (below 3%), are somewhat

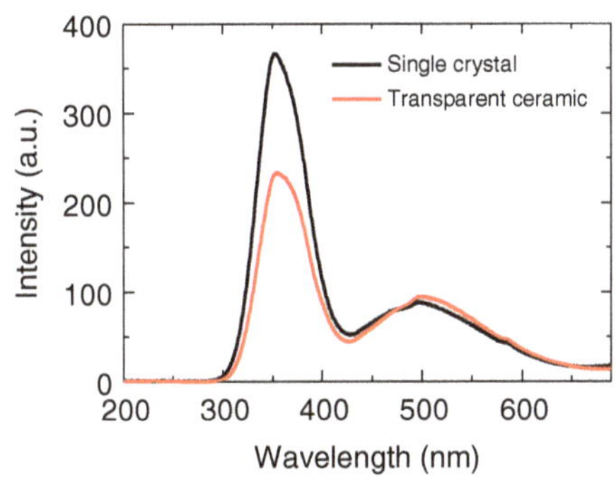

Figure 10.12 X-ray induced scintillation spectra of Tl:CsBr single crystal and transparent ceramic.

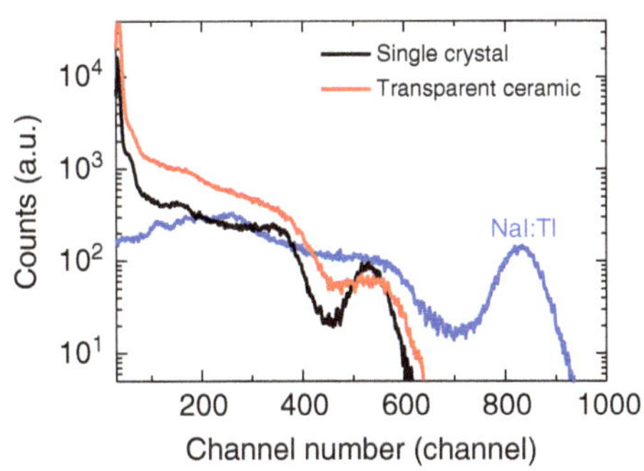

Figure 10.13 Pulse height spectra of ^{137}Cs γ-rays measured using Tl:CsBr single crystal and transparent ceramic. The spectrum obtained using a Tl:NaI scintillator is also plotted as a reference.

mitigated by the difficulties and high cost of producing large, single crystal samples. In contrast, the preparation method for Ce:$LaBr_3$ translucent ceramic and its scintillation characteristics were reported by Wisniewski et al. in 2008 [39]. The light yield of the Ce:$LaBr_3$ translucent ceramic is estimated to approximately 42 000 ph/MeV, that is, 70% of that of Ce:$LaBr_3$ single crystals.

10.4.12 SrI_2

Eu-doped SrI_2 is a scintillator material of interest for γ-ray detection and spectroscopy because of its high density and the effective atomic number as well as the high light yield of the Eu emission. Nowadays, it is well-known that Eu:SrI_2 single crystals can be reproduced with a light yield of approximately 100 000 ph/MeV [23]. Alternatively, Podowitz et al. investigated scintillation properties of Eu:SrI_2 translucent ceramics sintered by the HP method in spite of their orthorhombic crystal structure [68].

The luminescence spectra of Eu:SrI_2 show the characteristic emission band of Eu^{2+} ions in an SrI_2, peaking at 435 nm. A very broad band emission over the entire visible spectrum was also present in Eu-doped SrI_2 with a peak intensity of approximately 4–5% of the Eu emission. This emission is likely defect-mediated and may be due to some form of oxidation or hydration of the iodide. The emission band due to Eu^{2+} ions of the translucent ceramics is red-shifted by 15 nm with respect to the emission reported in single crystals [90]. This shift may be due to either the lower level of transparency in these ceramics, which could cause preferential scattering at shorter wavelengths, or the defect-mediated emission, the intensity of which increases at longer wavelengths. The light yield of Eu:SrI_2 translucent ceramic is calculated to be 21 000 ph/MeV from the photoabsorption peaks under γ-irradiation.

10.5 Transparent Ceramics for Dosimeter

In comparison with scintillators, there are few studies on the dosimetric properties of transparent ceramics. In this section, TSL, OSL, and RPL properties of transparent ceramics with some chemical compositions are introduced. These materials are expected to apply to personal dosimeters and imaging plates. In addition, persistent luminescence properties of transparent ceramics are also introduced in this section, since its luminescent process is similar to TSL. The advantage as to why transparent ceramics are applied to dosimetric materials includes "volume effects." If the dosimetric material is transparent, we can detect TSL and OSL, not only on the surface but also inside the materials. That is to say, it is expected that transparent ceramics show higher TSL, OSL, or RPL intensity than opaque ceramics.

10.5.1 Al_2O_3

Al_2O_3 was one of the classical materials studied for its possible application as a radiation dosimeter, owing to its superior thermal stability, chemical stability, and low effective atomic number close to the human body. Today, Al_2O_3 doped with C crystals as TSL and OSL dosimeters are now well-established in personal dosimetry, having been already commercially available for almost two decades. Kato et al. investigated TSL properties of undoped and C-doped Al_2O_3 transparent ceramics sintered by the SPS method [4, 5]. Figure 10.14 shows TSL glow curves of the undoped Al_2O_3 transparent ceramic and single crystal measured

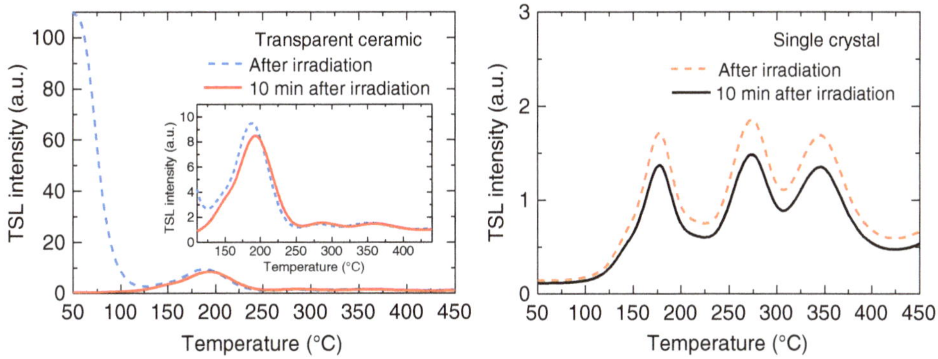

Figure 10.14 *TSL glow curves of the Al_2O_3 transparent ceramic and single crystal.*

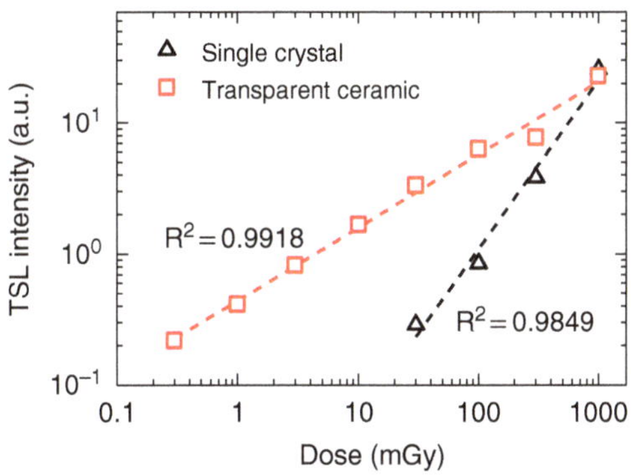

Figure 10.15 *TSL dose response curves of the Al_2O_3 transparent ceramic and single crystal.*

immediately and 10 minutes after X-ray irradiation (100 mGy) at room temperature [4]. Both materials showed four separate glow peaks at 50, 190, 290, and 360 °C. Despite the similarities, the measured intensities were very different between the transparent ceramic and single crystal. The intensity at 50 °C in the transparent ceramic was much larger than that in single crystal by a factor of approximately 100. Moreover, the peak intensity at 190 °C in the transparent ceramic was about 4.5 times higher than that in the single crystal. The main reason is that the transparent ceramic was synthesized in a highly reductive atmosphere by SPS. TSL dose response curves of the Al_2O_3 transparent ceramic and single crystal are compared in Figure 10.15. The transparent ceramic showed a good linearity from 0.3 to 1000 mGy, while the single crystal showed a good linearity from 30 to 1000 mGy. As a result, the sensitivity of the Al_2O_3 transparent ceramic was higher than that of single crystal

Furthermore, Liu et al. reported TSL glow curves and OSL decay curves of Cr-doped and Ti,Mg co-doped Al_2O_3 transparent ceramics fabricated by the vacuum sintering method [91, 92]. The Cr:Al_2O_3 transparent ceramics show two different TSL glow peaks, a small peak at 124 °C and a dominating one at 230 °C, which basically agrees well with the behavior of Cr:Al_2O_3 single crystals. TSL intensities of glow peaks increased with increasing Cr_2O_3 content. It had a main peak at 230 °C of very high intensity and linear concentration dependence up to high concentration, which is related to Cr^{3+} ions.

10.5.2 CaF_2

CaF_2 doped with an impurity is one of the well-known phosphor materials for radiation dosimetric methods, owing to high sensitivity and low effective atomic number. For example, Dy-, Tm-, and Mn-doped CaF_2 are commercially known as TLD-200, -300, and -400, respectively.

A comparative study on TSL properties of undoped CaF_2 transparent ceramic and single crystal was reported by Nakamura et al. [58]. The CaF_2 transparent ceramic was prepared by the SPS technique. Figure 10.16 presents TSL glow curves of CaF_2 ceramic and single crystal samples after 100 mGy X-ray irradiation. Both samples exhibited glow peaks from 100 to 150 °C. In addition, only the single crystal sample showed a peak at 400 °C, which may be attributed to defects that were introduced while synthesizing the single-crystal sample at a higher temperature than melting point. The TSL intensity of the ceramic is much greater than that of crystal, which means that the ceramic sample shows much higher TSL sensitivity.

Singh et al. investigated the TSL properties of Mn-doped CaF_2 transparent ceramics [93]. These ceramics were prepared by the hot pressing method using nano-powders. The single glow peak at 260 °C was detected to be the same as single crystal with the same chemical composition. This ceramic was able to show low-level doses down to about 3 mGy and exhibited a linear output up to 100 mGy. To compare the TSL response of the transparent ceramic with an opaque Mn-doped CaF_2 pallet, the as-prepared nano-powder (with same Mn doping concentration) was cold pressed into a pallet and annealed at 300 °C. After exposure to a dose of 10 Gy, a glow peak could not be recorded for the opaque pellet. This suggests that the transparent ceramic is more sensitive compared to an opaque pellet and so can be used to measure low doses.

10.5.3 MgO

MgO transparent ceramics were fabricated by the SPS technique to investigate these TSL properties in comparison with MgO single crystal [8]. In addition to undoped MgO transparent ceramics, MgO doped with rare earth or transition metal ions were prepared as transparent ceramics, and these dosimetric properties were then studied [94, 95]. Figure 10.17 shows TSL glow curves of the MgO transparent ceramic and single crystal after 1 Gy X-ray exposure. Both materials showed a strong single glow peak at around 140 °C. Furthermore, there are some additional glow peaks observed at around 240 and 350 °C for the transparent ceramic, and at 320 and 400 °C for the single crystal. This indicates that there exist three

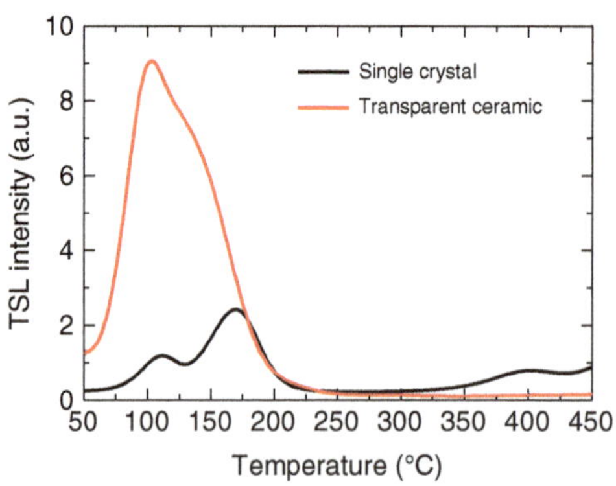

Figure 10.16 TSL glow curves of CaF_2 transparent ceramic and single crystal.

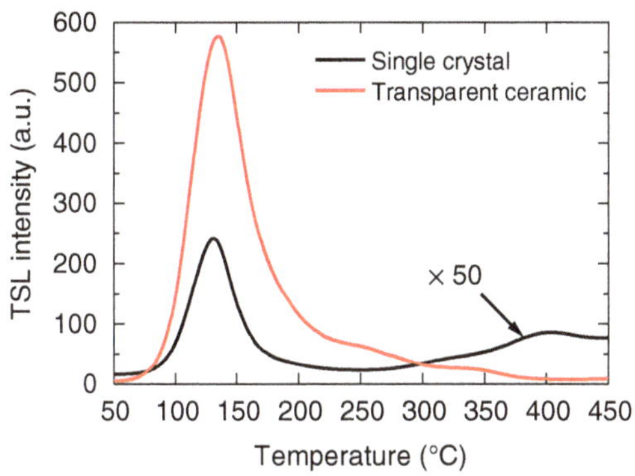

Figure 10.17 *TSL glow curves of the MgO transparent ceramic and single crystal.*

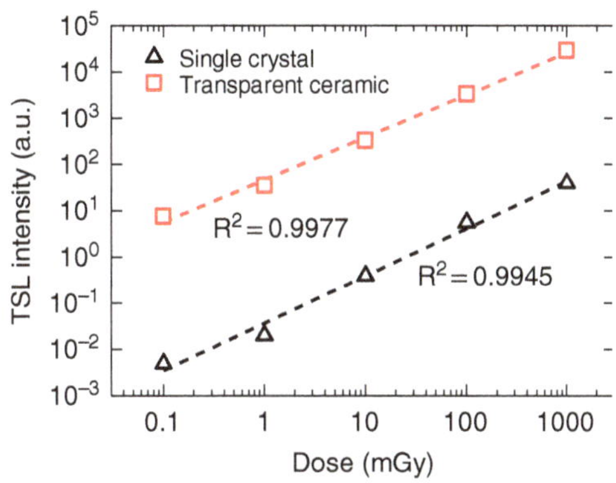

Figure 10.18 *Dose response curves. The TSL signal used here is the peak intensity observed at 140 °C.*

different trapping centers in both transparent ceramic and single crystal. Overall, the TSL intensity is much stronger for the transparent ceramic than single crystal. Figure 10.18 indicates TSL dose response curves of both materials. Both materials have a linear response to the incident X-ray dose over the dose range from 0.1 to 1000 mGy.

10.5.4 MgF_2

TSL glow curves and RPL properties of MgF_2 transparent ceramics prepared by the SPS technique have been researched [55]. Figure 10.19 illustrates PL spectra of the MgF_2 transparent ceramic and single crystal measured before and after irradiations with varying X-ray doses. Only the transparent ceramic showed new emission bands around 415 and 700 nm after irradiation, while the single crystal did not. It should be emphasized that this appearance of new PL emission bands was clear evidence of RPL. The transparent ceramic showed an emission peak of around 415 nm under 370 nm excitation, and the excitation and emission wavelengths were consistent with those of the M center (aggregation of two anion vacancies trapping two electrons). The emission peak at around 700 nm could not be measured because the signal was not stable under PL excitation and degraded during measurement. It was

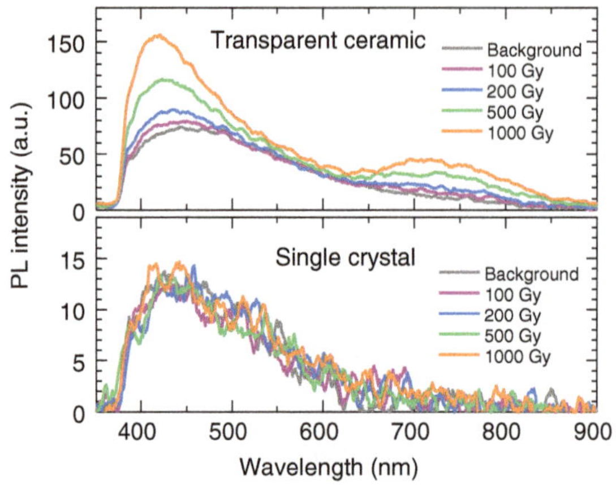

Figure 10.19 *PL emission spectra of MgF_2 transparent ceramic (top) and single crystal (bottom) irradiated by X-ray of varying dose from 100 to 1000 Gy under 340 ± 40 nm excitation.*

anticipated that there were more defect centers, particularly at the M^+ and M^- centers, included in the transparent ceramic than the single crystal because the ceramic was synthesized in a highly reductive atmosphere by the SPS technique. These M^+ or M^- centers captured electrons or holes and then became the M center.

10.5.5 CsBr

Since the late 1990s, Eu-doped CsBr has attracted much attention as imaging plates because of its outstanding properties such as efficient X-ray absorption, relatively short lifetime (~10 μs), high luminescence output, and linear dose dependence. In Eu-doped CsBr, F centers are known to act as electron traps, while holes are captured at V_K centers located in the vicinity of Eu^{2+} activators. The Eu^{2+} ions replaced in the Cs ions show an emission peak of around 440 nm. In practice, the emission can be observed as OSL under He—Ne laser (633 nm) stimulation. In addition to Eu-doped CsBr, it has been shown that undoped CsBr also exhibits notable OSL properties.

Kimura et al. researched TSL and OSL properties of undoped and Eu-doped CsBr transparent ceramics sintered by the SPS technique [65, 67]. Figure 10.20 shows TSL glow curves of undoped CsBr transparent ceramic and single crystal measured after X-ray irradiation. Both materials exhibited major glow peaks at around 125 and 220 °C. The glow peak can be related to anion defects because the intensity of the ceramic sample is much higher than that of the single crystal sample. The inset shows the TSL dose responses of the CsBr transparent ceramic and single crystal samples. Both samples showed good linearity from 0.01 to 100 mGy. Figure 10.21 shows OSL spectra of the CsBr transparent ceramic and single crystal samples. Both materials exhibited OSL with broad emission peaking at around 390 and 450 nm. These origins can be attributed to oxygen impurities. The dose responses of both samples showed good linearity from 1 to 1000 mGy. Because the transparent ceramic sample was synthesized by SPS performed in a highly reductive atmosphere, the ceramic sample generated more trapping centers, and the dosimetric properties could be enhanced.

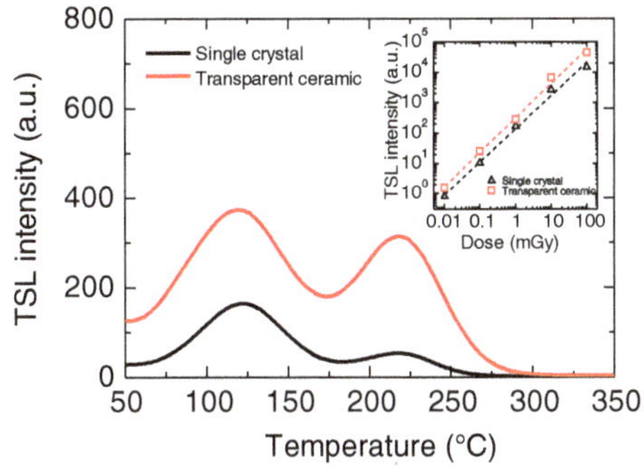

Figure 10.20 *TSL glow curves of CsBr transparent ceramic and single crystal measured after 100 mGy X-ray irradiation. The inset shows TSL dose response curves from 0.01 to 100 mGy.*

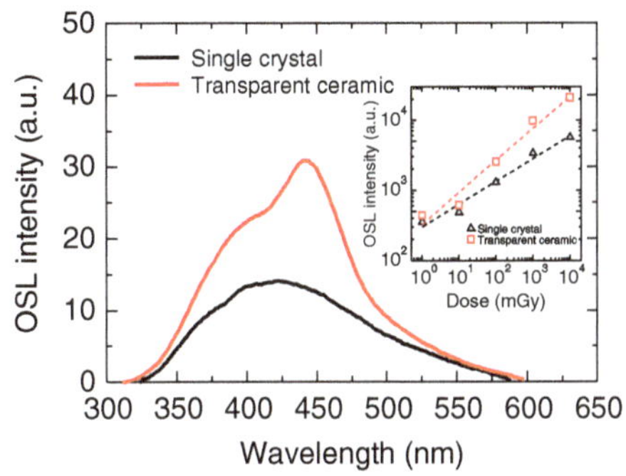

Figure 10.21 *OSL emission spectra of CsBr transparent ceramic and single crystal measured under 630 nm stimulation after the samples were irradiated by X-rays (10 Gy). The inset shows OSL dose response curves.*

Figure 10.22 shows OSL spectra of the Eu-doped CsBr transparent ceramic and single crystal. The inset indicates the OSL dose response function. The dose response curves showed a good linearity from 0.1 to 100 mGy. The intensity of the transparent ceramic was significantly higher than that of the single crystal by a factor of approximately 8.

10.5.6 Y_3Al_5-xGaxO12 (YAGG)

Persistent luminescence represents the light emission lasting from several minutes to a few hours after switching off excitation sources. General phosphors that possess persistent luminescence properties show TSL glow peaks at around room temperature (15 °C). Persistent luminescence is caused by thermally stimulating electrons or holes captured at a relatively shallow trapping level that is equivalent to room temperature.

Persistent luminescence of Ce:YAGG and Ce, Cr:YAGG transparent ceramics with different thicknesses (1.4, 2.1, 2.8 mm) has been reported [26]. When the thickness of the Ce, Cr:YAGG transparent ceramic is further increased from 1.4 to 2.8 mm, the persistent behavior increases notably as well, especially for the 2.8 mm thickness sample. Compared with the Ce, Cr:YAGG opaque ceramic, the corresponding transparent ceramic seems to show brighter persistent luminescence. In the transparent ceramics, not only the surface but also the inside of the

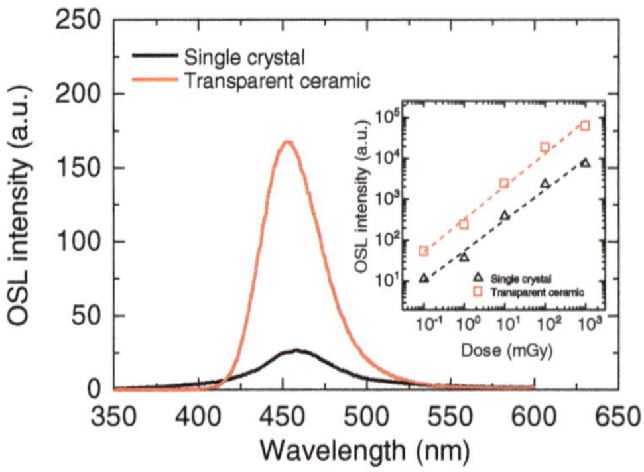

Figure 10.22 *OSL emission spectra of Eu-doped CsBr transparent ceramic and single crystal measured under 630 nm stimulation. Prior to the measurement, the samples were irradiated by X-rays (~10 Gy). The inset shows OSL dose response curves.*

ceramic can be excited under UV light. owing to their lower optical scattering coefficient than that of conventional opaque ceramics. In opaque ceramics, only the surface area can efficiently be excited due to the strong scattering of the excitation light. For a transparent ceramic persistent phosphor, the persistent luminance can be increased with increasing sample thickness, which is regarded as the "volume effect."

References

1. Coble, R. (1957) Transparent Al_2O_3 and Method of Preparation, 3, 026210. https://patents.google.com/patent/US3026210A/en.
2. Alvarez-Clemares, I., Mata-Osoro, G., Fernández, A. et al. (2010). Transparent alumina/ceria nanocomposites by spark plasma sintering. *Adv. Eng. Mater.* 12: 1154–1160. doi: 10.1002/adem.201000176.
3. Wang, D., Shen, L., Ran, S. et al. (2014). Transparent alumina fabricated by SPS sintering with AlF_3 doping. *Scr. Mater.* 92: 31–34. doi: 10.1016/j.scriptamat.2014.08.007.
4. Kato, T., Kawano, N., Okada, G. et al. (2017). Comparative study of dosimeter properties between Al_2O_3 transparent ceramic and single crystal. *Radiat. Meas.* 107: 43–47. doi: 10.1016/j.radmeas.2017.09.006.
5. Kato, T., Kawano, N., Okada, G. et al. (2017). Dosimetric properties of Al_2O_3 transparent ceramics doped with C. *Nucl. Instrum. Methods Phys. Res. Sect. B Beam Interact. Mater. Atoms*: 0–1. doi: 10.1016/j.nimb.2017.12.013.
6. Chen, D., Jordan, E.H., and Gell, M. (2008). Pressureless sintering of translucent MgO ceramics. *Scr. Mater.* 59: 757–759. doi: 10.1016/j.scriptamat.2008.06.007.
7. Fang, Y., Agrawal, D., Skandan, G. et al. (2004). Fabrication of translucent MgO ceramics using nanopowders. *Mater. Lett.* 58: 551–554. doi: 10.1016/S0167-577X(03)00560-3.
8. Kato, T., Okada, G., and Yanagida, T. (2016). Optical, scintillation and dosimeter properties of MgO transparent ceramic and single crystal. *Ceram. Int.* 42: 5617–5622. doi: 10.1016/j.ceramint.2015.12.070.
9. Dutta, S.K. and Gazza, G.E. (1969). Transparent Y_2O_3 by hot-pressing. *Mater. Res. Bull.* 4: 791–796. doi: 10.1016/0025-5408(69)90001-4.
10. Ikegami, T., Li, J.-G., Mori, T. et al. (2004). Fabrication of transparent Yttria ceramics by the low-temperature synthesis of yttrium hydroxide. *J. Am. Ceram. Soc.* 85: 1725–1729. doi: 10.1111/j.1151-2916.2002.tb00342.x.
11. Saito, N., Matsuda, S., and Ikegami, T. (1998). Fabrication of transparent Yttria ceramics at low temperature using carbonate-derived powder. *J. Am. Ceram. Soc.* 81: 2023–2028. doi: 10.1111/j.1151-2916.1998.tb02583.x.
12. Lupei, A., Lupei, V., Gheorghe, C. et al. (2008). Excited states dynamics of Er^{3+} in Sc_2O_3 ceramic. *J. Lumin.* 128: 918–920. doi: 10.1016/j.jlumin.2007.11.032.

13. Yanagida, T., Fujimoto, Y., Yagi, H. et al. (2014). Optical and scintillation properties of transparent ceramic Yb:Lu_2O_3 with different Yb concentrations. *Opt. Mater.* 36: 1044–1048. doi: 10.1016/j.optmat.2014.01.022.
14. Lu, J., Takaichi, K., Uematsu, T. et al. (2002). Promising ceramic laser material: highly transparent Nd^{3+}:Lu_2O_3 ceramic. *Appl. Phys. Lett.* 81: 4324–4326. doi: 10.1063/1.1527234.
15. Yanagida, T., Fujimoto, Y., Kurosawa, S. et al. (2011). Ultrafast transparent ceramic scintillators using the Yb3+ charge transfer luminescence in RE_2O_3 host. *Appl. Phys. Express.* 4: 126402. doi: 10.1143/APEX.4.126402.
16. Kim, Y.K., Kim, H.K., Kim, D.K. et al. (2004). Synthesis of Eu-doped $(Gd,Y)_2O_3$ transparent optical ceramic scintillator. *J. Mater. Res.* 19: 413–416. doi: 10.1557/jmr.2004.19.2.413.
17. Grabmaier, B.C. and Rossner, W. (1993). New scintillators for X-ray computed tomography. *Nucl. Tracks Radiat. Meas.* 21: 43–45. doi: 10.1016/1359-0189(93)90043-9.
18. Winnacker, A., Grabmaiert, B.C., and Ktjstler, W. (1995). Effect of Pr-codoping on the X-ray induced afterglow of $(Y, Gd)_2O_3$:Eu. *J. Phys. Chem. Solid* 56: 907–913. doi: 10.1016/0022-3697(95)00023-2.
19. Greskovich, C. and Duclos, S. (1997). Ceramic scintillators. *Annu. Rev. Mater. Sci.* 27: 69–88. doi: 10.1146/annurev.matsci.27.1.69.
20. Anselmi-Tamburini, U., Woolman, J.N., and Munir, Z.A. (2007). Transparent nanometric cubic and tetragonal zirconia obtained by high-pressure pulsed electric current sintering. *Adv. Funct. Mater.* 17: 3267–3273. doi: 10.1002/adfm.200600959.
21. Zhang, H., Kim, B.N., Morita, K. et al. (2011). Effect of sintering temperature on optical properties and microstructure of translucent zirconia prepared by high-pressure spark plasma sintering. *Sci. Technol. Adv. Mater.* 12 doi: 10.1088/1468-6996/12/5/055003.
22. Rodnyi, P.A., Chernenko, K.A., Gorokhova, E.I. et al. (2012). Novel scintillation material ZnO transparent ceramics. *IEEE Trans. Nucl. Sci.* 59: 2152–2155. doi: 10.1109/TNS.2012.2189896.
23. Rodnyi, P., Chernenko, K., Klimova, O. et al. (2016). Influence of annealing on the scintillation properties of zinc oxide powders and ceramics. *Radiat. Meas.* 90: 136–139. doi: 10.1016/j.radmeas.2016.01.021.
24. Yanagida, T., Takahashi, H., Ito, T. et al. (2005). Evaluation of properties of YAG (Ce) ceramic scintillators. *IEEE Trans. Nucl. Sci.* 52: 1836–1841. doi: 10.1109/TNS.2005.856757.
25. Ikesue, A., Furusato, I., and Kumata, K. (1995). Fabrication and optical properties of high-perfrormance polycrystalline Nd:YAG ceramics for solid-state lasers. *J. Am. Ceram. Soc.* 78: 1033–1040. doi: 10.1111/j.1151-2916.1995.tb08433.x.
26. Xu, J., Ueda, J., Kuroishi, K. et al. (2015). Fabrication of Ce^{3+}-Cr^{3+} co-doped yttrium aluminium gallium garnet transparent ceramic phosphors with super long persistent luminescence. *Scr. Mater.* 102: 47–50. doi: 10.1016/j.scriptamat.2015.01.029.
27. Nishiura, S., Tanabe, S., Fujioka, K. et al. (2011). Properties of transparent Ce:YAG ceramic phosphors for white LED. *Opt. Mater.* 33: 688–691. doi: 10.1016/j.optmat.2010.06.005.
28. Zycha, E., Brecher, C., Wojtowicz, A.J. et al. (1997). Luminescence properties of Ce-activated YAG optical ceramic scintillator materials. *J. Lumin.* 75: 193–203. doi: 10.1016/S0022-2313(97)00103-8.
29. Luo, D., Zhang, J., Xu, C. et al. (2012). Yb:LuAG laser ceramics: a promising high power laser gain medium. *Opt. Mater. Express* 2: 1355–1358. doi: 10.1364/OME.2.001425.
30. Yanagida, T., Fujimoto, Y., Yokota, Y. et al. (2011). Comparative study of transparent ceramic and single crystal Ce doped LuAG scintillators. *Radiat. Meas.* 46: 1503–1505. doi: 10.1016/j.radmeas.2011.03.039.
31. Yanagida, T., Yoshikawa, A., Ikesue, A. et al. (2009). Basic properties of ceramic Pr:LuAG scintillators. *IEEE Trans. Nucl. Sci.* 56: 2955–2959. doi: 10.1109/TNS.2009.2026475.
32. Yanagida, T., Kamada, K., Fujimoto, Y. et al. (2013). Comparative study of ceramic and single crystal Ce:GAGG scintillator. *Opt. Mater.* 35: 2480–2485. doi: 10.1016/j.optmat.2013.07.002.
33. Morita, K., Kim, B.N., Hiraga, K. et al. (2008). Fabrication of transparent $MgAl_2O_4$ spinel polycrystal by spark plasma sintering processing. *Scr. Mater.* 58: 1114–1117. doi: 10.1016/j.scriptamat.2008.02.008.
34. Morita, K., Kim, B.N., Yoshida, H. et al. (2009)). Spark-plasma-sintering condition optimization for producing transparent $MgAl_2O_4$ spinel polycrystal. *J. Am. Ceram. Soc.* 92: 1208–1216. doi: 10.1111/j.1551-2916.2009.03074.x.
35. Souza, N.R.S., Silva, D.C., Sampaio, D.V. et al. (2017). Laser sintering of persistent luminescent $CaAl_2O$:$Eu^{2+}Dy^{3+}$ ceramics. *Opt. Mater.* 68: 2–6. doi: 10.1016/j.optmat.2016.10.050.
36. Alves, Y.G.S., Sampaio, D.V., Souza, N.R.S. et al. (2017). Persistent luminescence properties of $SrB_XAl_{2-X}O_4$:Eu,Dy laser-sintered ceramics. *Opt. Mater.* 70: 63–68. doi: 10.1016/j.optmat.2017.05.017.
37. Kim, B.-N., Hiraga, K., Jeong, A. et al. (2014). Transparent $ZnAl_2O_4$ ceramics fabricated by spark plasma sintering. *J. Ceram. Soc. Japan* 122: 784–787. doi: 10.2109/jcersj2.122.784.

38. Xie, J., Shi, Y., Fan, L. et al. (2013). Microstructure and luminescent properties of Ce:Lu_2SiO_5 ceramic scintillator by spark plasma sintering. *Opt. Mater.* 35: 744–747. doi: 10.1016/j.optmat.2012.08.013.
39. Wisniewski, D.J., Boatner, L.A., Neal, J.S. et al. (2008). Development of novel polycrystalline ceramic scintillators. *IEEE Trans. Nucl. Sci.* 55: 1501–1508. doi: 10.1109/TNS.2008.919259.
40. Shen, Y., Chen, R., Gurzadyan, G.G. et al. (2012). Fabrication and spectroscopic characterization of Ce^{3+} doped $Sr_2Y_8(SiO_4)_6O_2$ translucent ceramics. *Opt. Mater.* 34: 1155–1160. doi: 10.1016/j.optmat.2012.01.020.
41. Macedo, Z.S., Silva, R.S., Valerio, M.E.G. et al. (2004). Laser-sintered bismuth germanate ceramics as scintillator devices. *J. Am. Ceram. Soc.* 87: 1076–1081. doi: 10.1111/j.1551-2916.2004.01076.x.
42. Sato, Y., Arzakantsyan, M., Akiyama, J. et al. (2014). Anisotropic Yb:FAP laser ceramics by micro-domain control. *Opt. Mater. Express* 4: 2006. doi: 10.1364/OME.4.002006.
43. Chen, S., Wu, Y., and Yang, Y. (2013). Spark plasma sintering of hexagonal structure Yb^{3+}-doped $Sr_5(PO_4)_3F$ transparent ceramics. *J. Am. Ceram. Soc.* 96: 1694–1697. doi: 10.1111/jace.12362.
44. Chlique, C., Merdrignac-Conanec, O., Hakmeh, N. et al. (2013). Transparent ZnS ceramics by sintering of high purity monodisperse nanopowders. *J. Am. Ceram. Soc.* 96: 3070–3074. doi: 10.1111/jace.12570.
45. Gorokhova, E.I., Demidenko, V.A., Eron'ko, S.B. et al. (2005). Spectrokinetic characteristics of the emission of Gd_2O_2S-Tb(Ce) ceramics. *J. Opt. Technol.* 72: 53. doi: 10.1364/JOT.72.000053.
46. Nakamura, R. and Yamada, N. (1997). Effect of Ce radiation damage of Gd_2O_2S:Pr ceramics scintillator. *J. Ceram. Soc. Japan* 850: 847–850. doi: 10.2109/jcersj.105.847.
47. Ito, Y. Yamada, H. Yoshida, M. Fujii, H. Toda, G. Takeuchi H. and Tsukuda Y. (1988). Hot isostatic pressed Gd_2O_2S:Pr, Ce, F translucent scintillator ceramics for X-ray computed tomography detectors. *Jpn. J. Appl. Phys.* 27: 1371–1373. doi:10.1143/JJAP.27.L1371.
48. Nakamura, R., Yamada, N., and Ishii, M. (1999). Effects of halogen ions on the X-ray characteristics of Gd_2O_2S:Pr ceramic scintillators. *Jpn. J. Appl. Phys.* 38: 6923–6925. doi: 10.1143/JJAP.38.6923.
49. Kuretake, S., Tanaka, N., Kintaka, Y. et al. (2014). Nd-doped $Ba(Zr,Mg,Ta)O_3$ ceramics as laser materials. *Opt. Mater.* 36: 645–649. doi: 10.1016/j.optmat.2013.11.001.
50. Chiba, H., Kurosawa, S., Harata, K. et al. (2016). Luminescence properties of the Mg co-doped Ce:$SrHfO_3$ ceramics prepared by the spark plasma sintering method. *Radiat. Meas.* 90: 287–291. doi: 10.1016/j.radmeas.2016.02.025.
51. Yanagida, T. and Okada, G. (2016). Characterizations of optical properties and radiation induced luminescence of Bi-doped $La_2Zr_2O_7$ transparent ceramics. *J. Ceram. Soc. Japan* 124: 564–568. doi: 10.2109/jcersj2.15237.
52. Trojan-Piegza, J., Gierlotka, S., Zych, E. et al. (2014). Spectroscopic studies of nanopowder and nanoceramics $La_2Hf_2O_7$:Pr scintillator. *J. Am. Ceram. Soc.* 97: 1595–1601. doi: 10.1111/jace.12788.
53. An, L., Ito, A., and Goto, T. (2011). Highly transparent lutetium titanium oxide produced by spark plasma sintering. *J. Eur. Ceram. Soc.* 31: 237–240. doi: 10.1016/j.jeurceramsoc.2010.09.010.
54. An, L., Ito, A., and Goto, T. (2011). Fabrication of transparent Lu_3NbO_7 by spark plasma sintering. *Mater. Lett.* 65: 3167–3169. doi: 10.1016/j.matlet.2011.07.010.
55. Nakamura, F., Kato, T., Okada, G. et al. (2017). Scintillation, TSL and RPL properties of MgF_2 transparent ceramic and single crystal. *Ceram. Int.* 43: 7211–7215. doi: 10.1016/j.ceramint.2017.03.009.
56. Liu, Z., Mei, B., Song, J. et al. (2015). Microstructure and optical properties of hot-pressed Er:CaF_2 transparent ceramics. *J. Alloys Compd.* 646: 760–765. doi: 10.1016/j.jallcom.2015.05.272.
57. Aubry, P., Bensalah, A., Gredin, P. et al. (2009). Synthesis and optical characterizations of Yb-doped CaF_2 ceramics. *Opt. Mater.* 31: 750–753. doi: 10.1016/j.optmat.2008.03.022.
58. Nakamura, F., Kato, T., Okada, G. et al. (2017). Scintillation and dosimeter properties of CaF_2 translucent ceramic produced by SPS. *J. Eur. Ceram. Soc.* 37: 1707–1711. doi: 10.1016/j.jeurceramsoc.2016.11.016.
59. Demidenko, A.A., Garibin, E.A., Gain, S.D. et al. (2010). Scintillation parameters of BaF_2 and BaF_2:Ce^{3+} ceramics. *Opt. Mater.* 32: 1291–1293. doi: 10.1016/j.optmat.2010.05.003.
60. Garibin, E.A., Gain, S.D., Gusev, P.E. et al. (2011). New scintillators based on barium fluoride crystals and ceramics. *Bull. Russ. Acad. Sci. Phys.* 75: 1066–1069. doi: 10.3103/S106287381107015X.
61. Kato, T., Okada, G., Fukuda, K. et al. (2017). Development of BaF_2 transparent ceramics and evaluation of the scintillation properties. *Radiat. Meas.* 106: 140–145. doi: 10.1016/j.radmeas.2017.03.032.
62. Li, W., Chen, M., Kou, H. et al. (2015). Translucent cerium fluoride ceramics fabricated by hot-pressing for scintillation application. *Int. J. Appl. Ceram. Technol.* 12: 638–642. doi: 10.1111/ijac.12215.
63. Shoulders, W.T. and Gaume, R.M. (2017). Phase-change sintering of $BaCl_2$ transparent ceramics. *J. Alloys Compd.* 705: 517–523. doi: 10.18494/SAM.2019.2186.
64. Kimura, H., Kato, T., Nakauchi, D. et al. (2019). Vacuum-UV-excited photoluminescence and scintillation properties of CsCl transparent ceramics and single crystal. *Sens. Mater.* 31: 1265–1271.

65. Kimura, H., Nakamura, F., Kato, T. et al. (2018). Comparative study of radiation-induced luminescence between non-doped CsBr transparent ceramic and crystal. *J. Ceram. Soc. Japan* 126: 184–189. doi: 10.2109/jcersj2.17212.
66. Kimura, H., Kato, T., Nakauchi, D. et al. (2020). Comparative study of CsBr:Tl transparent ceramic and single crystal for radiation detector applications. *Nucl. Inst. Methods Phys. Res. A.* 954: 161226. doi: 10.1016/j.nima.2018.09.061.
67. Kimura, H., Nakamura, F., Kato, T. et al. (2018). Comparative study of dosimeter properties of Eu-doped CsBr transparent ceramic and single crystal. *Optik* 157: 421–428. doi: 10.1016/j.ijleo.2017.11.104.
68. Podowitz, S.R., Gaumé, R.M., Hong, W.T. et al. (2010). Fabrication and properties of translucent SrI_2 and Eu:SrI_2 scintillator ceramics. *IEEE Trans. Nucl. Sci.* 57: 3827–3835. doi: 10.1109/TNS.2010.2084590.
69. Joshi, B., Kshetri, Y.K., Gyawali, G. et al. (2015). Transparent Mg-α/β-Sialon:Eu^{2+} ceramics as a yellow phosphor for pc-WLED. *J. Alloys Compd.* 631: 38–45. doi: 10.1016/j.jallcom.2015.01.081.
70. Xiong, Y., Fu, Z., Wang, Y. et al. (2006). Fabrication of transparent AIN ceramics. *J. Mater. Sci.* 41: 2537–2539. doi: 10.1007/s10853-006-5314-8.
71. Yue Feng, S.Z., Lin, H., Chen, C. et al. (2015). Fabrication of transparent $Tb_3Ga_5O_{12}$ ceramic. *Chin. Opt. Lett.* 13: 0.31602. doi: 10.3788/COL201513.031602.Faraday.
72. Lin, H., Zhou, S., and Teng, H. (2011). Synthesis of $Tb_3Al_5O_{12}$(TAG) transparent ceramics for potential magneto-optical applications. *Opt. Mater.* 33: 1833–1836. doi: 10.1016/j.optmat.2011.06.017.
73. Hu, S., Qin, X., Liu, X. et al. (2017). Fabrication and luminescent properties of highly transparent $Er_3Al_5O_{12}$ ceramics. *Opt. Mater.* 71: 86–89. doi: 10.1016/j.optmat.2016.05.018.
74. Sun, P., Xu, C.-N., Akiyama, M. et al. (1999). Controlled oxygen partial pressure sintering of (Pb,La)(Zr,Ti)O_3 ceramics. *J. Am. Ceram. Soc.* 82: 1447–1450. doi: 10.1016/S1872-1508(06)60057-0.
75. Kubota, T., Tanaka, N., Kageyama, K. et al. (2009). Preparation of highly oriented transparent (Sr, Ba)Nb_2O_6 ceramics and their ferroelectric properties. *Jpn. J. Appl. Phys.* 48: 031405. doi: 10.1143/JJAP.48.031405.
76. Prakasam, M., Viraphong, O., Michau, D. et al. (2014). Critical parameters to obtain Yb^{3+} doped Lu_2O_3 and ZnO transparent ceramics. *Ceram. Int.* 40: 1859–1864. doi: 10.1016/j.ceramint.2013.07.088.
77. Gazza, G.E. and Roderick, D. (1971). Transparent Sc_2O_3 by hot-pressing. *J. Mater. Sci.* 6: 1137–1139. doi: 10.1007/BF00980612.
78. Li, J.G., Ikegami, T., and Mori, T. (2005). Fabrication of transparent, sintered Sc_2O_3 ceramics. *J. Am. Ceram. Soc.* 88: 817–821. doi: 10.1111/j.1551-2916.2005.00162.x.
79. Sokol, M., Kalabukhov, S., Dariel, M.P. et al. (2014). High-pressure spark plasma sintering (SPS) of transparent polycrystalline magnesium aluminate spinel (PMAS). *J. Eur. Ceram. Soc.* 34: 4305–4310. doi: 10.1016/j.jeurceramsoc.2014.07.022.
80. Carl, K. and Hardtl, K.H. (1977). Electrical after-effects in Pb(Ti, Zr)O_3 ceramics. *Ferroelectrics* 17: 473–486. doi: 10.1080/00150197808236770.
81. Kingery, W.D., Bowen, H.K., Uhlmann, D.R. et al. (1976). *Introduction to Ceramics*. Wiley.
82. Fang, Z.Z. (2010). *Sintering of Advanced Materials*. Elsevier.
83. Ikegami, T. and Moriyoshi, Y. (1984). Intermediate-stage sintering of a homogeneously packed compact connect. *J. Am. Ceram. Soc.* 67: 19–20. doi: 10.1111/j.1151-2916.1984.tb19736.x.
84. Voorhees, P.W. (1985). The theory of Ostwald ripening. *J. Stat. Phys.* 38: 231–252. doi: 10.1007/BF01017860.
85. Tsukuma, K., Yamashita, I., and Kusunose, T. (2008). Transparent 8 mol% Y_2O_3-ZrO_2 (8Y) ceramics. *J. Am. Ceram. Soc.* 91: 813–818. doi: 10.1111/j.1551-2916.2007.02202.x.
86. Wang, S.F., Zhang, J., Luo, D.W. et al. (2013). Progress in solid state chemistry transparent ceramics: processing, materials and applications. *Prog. Solid State Chem. J.* 41: 20–54. doi: 10.1016/j.progsolidstchem.2012.12.002.
87. Blaz, K. and Ambrosio, C.D. (2007). Luminescence and scintillation properties of YAG:Ce single crystal and optical ceramics. *J. Lumin.* 126: 77–80. doi: 10.1016/j.jlumin.2006.05.004.
88. Yanagida, T., Itoh, T., Takahashi, H. et al. (2007). Improvement of ceramic YAG(Ce) scintillators to $(YGd)_3Al_5O_{12}$(Ce) for gamma-ray detectors. *Nucl. Instrum. Methods Phys. Res. A.* 579: 23–26. doi: 10.1016/j.nima.2007.04.173.
89. Cherepy, N.J., Payne, S.A., Asztalos, S.J. et al. (2010). Scintillators with potential to supersede lanthanum bromide. *IEEE Trans. Nucl. Sci.* 56: 873–880. doi: 10.1109/TNS.2009.2020165.
90. Wilson, C.M., Van Loef, E.V., Glodo, J. et al. (2008). Strontium iodide scintillators for high energy resolution gamma ray spectroscopy. *Proc. SPIE* 7079: 707917. doi: 10.1117/12.806291.
91. Liu, Q., Yang, Q., Zhao, G. et al. (2014). Titanium effect on the thermoluminescence and optically stimulated luminescence of Ti,Mg:α-Al_2O_3 transparent ceramics. *J. Alloys Compd.* 582: 754–758. doi: 10.1016/j.jallcom.2013.07.189.

92. Liu, Q., Yang, Q.H., Zhao, G.G. et al. (2013). The thermoluminescence and optically stimulated luminescence properties of Cr-doped alpha alumina transparent ceramics. *J. Alloys Compd.* 579: 259–262. doi: 10.1016/j.jallcom.2013.06.070.
93. Singh, S.G., Sen, S., Patra, G.D. et al. (2012). Fabrication, properties and thermo-luminescent dosimetric application of CaF_2:Mn transparent ceramic. *Nucl. Instrum. Methods Phys. Res. Sect. B Beam Interact. Mater. Atoms* 287: 51–55. doi: 10.1016/j.nimb.2012.06.002.
94. Kato, T., Okada, G., and Yanagida, T. (2016). Optical, scintillation and dosimeter properties of MgO translucent ceramic doped with Cr^{3+}. *Opt. Mater.* 54: 134–138. doi: 10.1016/j.optmat.2016.02.030.
95. Kato, T., Okada, G., Kawaguchi, N. et al. (2017). Dosimeter properties of Ce-doped MgO transparent ceramics. *J. Lumin.* 192: 316–320. doi: 10.1016/j.jlumin.2017.06.067.

11

Luminescence in Glass-Based Materials by Ionizing Radiation

Hirokazu Masai and Kenji Shinozaki
National Institute of Advanced Industrial Science and Technology, Ikeda, Japan

11.1 Introduction

Glasses are abundant in nature and have been made by humankind for over 3000 years. Since glass possesses various excellent characteristics, such as the transparency, chemical durability, thermal stability, hardness, and a wide range of chemical compositions, it is one of the fundamental materials in our daily lives. Glass has evolved from a basic structural material, such as window glass, to an enabling material for use in advanced electronic, biological, and photonic products, for example optical fibers and glass covers for smartphones. The notable advantage of glass materials is their ability to be mass produced, with relatively cheaper fabrication costs compared with other single crystalline substances. On the other hand, because glass lacks a long-range structural order without grain boundary, various shapes of optical materials can be prepared. Good formability is also the reason why glass materials are so widely used in industry. It could be said that our life cannot be established without glass materials.

One of the applications of glass materials is that of optical functional materials, owing to its good transparency and good formability. Conventional oxide glass, such SiO_2 glass, exhibits a good transparency window from the visible to UV region. On the other hand, chalcogenide glass, in which oxygens are substituted by other elements in the group 16, such as S and Se, the transparency window is open in the IR region. These transparent windows have been used in optical applications, such as for fibers, lenses, and filters. Although glass has a very long history, various kinds of glass are still studied worldwide. This is because glass is in a thermodynamically metastable state whose structure and physical properties can

Phosphors for Radiation Detectors, First Edition. Edited by Takayuki Yanagida and Masanori Koshimizu.

be widely tailored, and various kinds of compositions can be introduced into a monolithic glass. From these points of view, glass materials will surely continue to be used in the future.

In the field of glass research, the interaction with a quantum beam is roughly classified into three parts: (i) radiation shielding typified by lead-containing glass; (ii) defect generation by quantum beam irradiation; and (iii) prompt and storage emission by quantum beam irradiation. Since some glass exhibits dosimetric properties, as discussed in Chapters 7 to 9, please refer to these chapters for the detailed mechanism. On the other hand, fiber-type elements are important for radiation therapy dosimetry applications, and oxide glass with high durability has been used in such applications. Therefore, the preparation process of glass-based materials is also briefly discussed in this chapter.

11.2 Structural and Physical Properties of Glass

If a material having an ordered structure is defined as a solid, a glass having no ordered structure will be defined as a liquid. One of the characteristics of glass as a liquid is a network connectivity with no periodicity and the absence of grain boundaries. The "randomly" formed network allows glass to assume complex shapes, such as fibers and craft materials. In addition, the glass network allows doping of impurities with higher concentration compared to crystals. Since the prefix "*a-*" in *amorphous* means not or without, the "amorphous" material is positioned as "a material that has no crystalline morphology." Glasses are amorphous solids that exhibit the glass transition phenomenon [1–4]. The glass transition definition by the National Research Council Ad Hoc Committee is "phenomenon in which a solid amorphous phase exhibits with changing temperature a more or less sudden change in the derivative thermodynamic properties from crystal" [1]. The *glass transition* phenomena that defines "glass" is one of the physical phenomena that have not yet been elucidated upon. The glass transition and the supercooled liquid state of homogeneous melt has been actively discussed [2–4]. The cooling process from a supercooled liquid is a behavior unique to glass, making it different from crystals. Here, it should be noted that not all materials can be vitrified. Zachariasen proposed several requirements for oxide to form a glass [5]. These simple rules are: (i) an oxygen atom is linked to not more than two cations; (ii) the number of oxygen atoms surrounding cations must be small; and (iii) the oxygen polyhedra share corners with each other, not edges or faces. Based on this rule, there is a classification of constituents of glass. The first group is able to form glass, known as the network forming oxide, or glass former. SiO_2, B_2O_3, P_2O_5, GeO_2, and As_2O_3 belong to the network forming oxide, which can vitrify by itself. The second group mainly consists of alkaline metal oxide and alkaline earth metal oxide, whose network formabilities are too poor to be vitrified. This group is called the network modifying oxide, or network modifier. Sun approved this categorization and establish another category "intermediate," which depends on the single bond strength between metal and oxygen [6]. The single bond strength of each metal oxide is obtained by the dissociation energy of the metal oxide divided by the coordination number. The intermediate group can work either as a network former or as a network modifier, which depends on the chemical composition of the glass. Although such a category is roughly adaptable to metal oxides, the glass forming ability of each oxide glass is different.

The glass forming ability is also affected by the viscosity of the melt. Since the thermodynamic change from the melt to solid occurs in a finite time, a more ordered structure is formed depending on the cooling conditions. In other words, a structural rearrangement to a thermodynamically stable crystal occurs during the vitrification (solidification) process. If the entropy difference (difference in free energy) between the crystal and the glass is small, liquid (glass melt) is generally vitrified. There is an empirical relationship between viscosity and temperature, called the Vogel–Tammann–Fulcher (VTF) equation [7–9].

$$\eta = A\exp.\left[B/\left(T-T_{o}\right)\right] \tag{11.1}$$

where *A* and *B* are temperature-independent constants. If the viscosity of a glass behaves in a nearly Arrhenius fashion, the glass is called "strong glass" [2, 10]. SiO_2 glass is a prototype of this strong glass (Figure 11.1). However, if the temperature dependency of viscosity is apart from Arrhenius behavior, the glass is called "fragile glass" [2, 10]. Although it is hard to fabricate fragile glass, whose vitrification ability is poor, by a conventional melt-quenching method, a fragile glass can be obtained by using an ultra-fast quenching process [11, 12].

On the other hand, another name for amorphous materials is non-crystalline materials, exhibiting that there is no characteristic atomistic ordering. Crystals are defined by their periodicity, that is, manifested by distinct diffraction peaks in X-ray and neutron diffraction data. In contrast to crystalline materials, there is no distinct signature of ordering, especially in the longer distance range [13] in non-crystalline materials. In the case of a crystal, such as SiO_2, there several crystalline phases, depending on the preparation temperature and pressure [14]. In contrast to ordered crystals, whose structure can be clarified using X-ray and

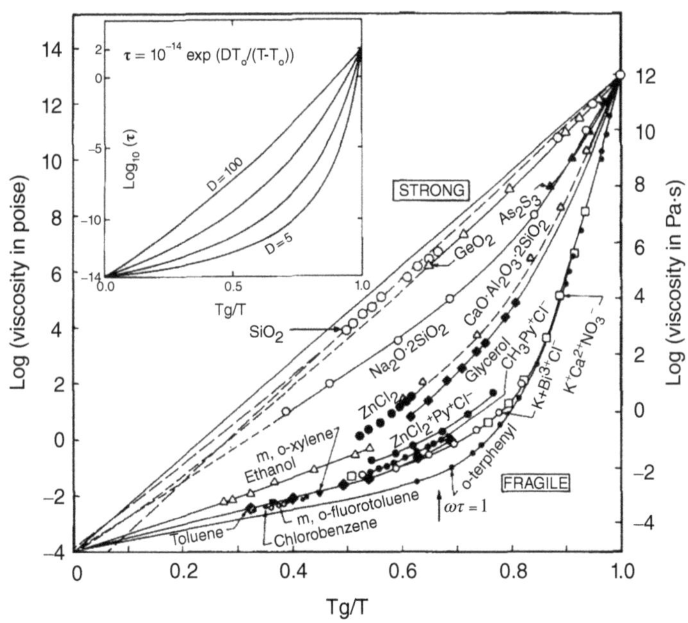

Figure 11.1 *Strong–fragile Arrhenius representation of liquid viscosities scaled by T_g. Insert shows how the pattern may be generated by variation of the D parameter in the equation* $\tau = 10^{-14}\exp\left(\frac{DT_0}{T-T_0}\right)$. *(Source: Reproduced from C. A. Angell [10] by permission of Elsevier [10].)*

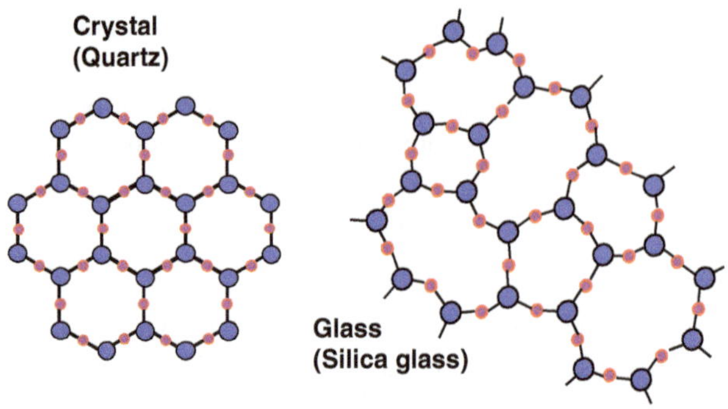

Figure 11.2 *Schematic illustration of crystal (quartz) and glass (silica glass). Although there is no long-range periodic structure in glass, there is no remarkable difference between the local* SiO_4 *structure in glass and that in crystal.*

neutron diffraction analyses, the structure of glass should be revealed with several measurement techniques, because of the lack of long-range periodicity [13, 15, 16]. Indeed, the local structures in glasses, such as bonding distances, bonding angles, coordination numbers, or defects, have extremely wide structural deviations. Figure 11.2 shows a schematic illustration of the structure of a crystal (quartz) and glass (silica glass). Glass is considered to be a material that has no periodic structure on the order of several nanometers. However, on the atomic scale, i.e., short-range structure (bond lengths and bond angles), glass with a stoichiometric composition of crystal has the similar structure or density to those of the specific crystal. Such characteristics are easily understood from diffraction data.

For amorphous materials, there is no periodical structure that satisfies the diffraction condition of Bragg's law. Therefore, diffraction phenomena will be discussed by scattering and interference of wave vector. Here, we assume that the coherent scattering lengths and position vectors of the i-th, j-th atoms are b_i, b_j and $\mathbf{r}_i$, $\mathbf{r}_j$, respectively. If the scattering angle is represented as 2θ, the scattering vector $Q = |\mathbf{Q}| = 4\pi\sin\theta/\lambda$. The coherent scattering intensity $I(\mathbf{Q})$, which is scattered by atoms in the system, is expressed as the square of the conjugate of the sum of the combined scattering amplitudes by N atoms as follows:

$$I(Q)=\left\langle \sum_{i=1}^{N}\sum_{j=1}^{N} b_i b_j \exp\left|i\mathbf{Q}(r_i - r_j)\right|\right\rangle \tag{11.2}$$

If we calculate the average angle between $\mathbf{Q}$ and $\mathbf{r}_i - \mathbf{r}_j = \mathbf{r}_{ij}$, by addition of equal weight,

$$I(Q)=\sum_{i=1}^{N}\sum_{j=1}^{N} b_i b_j \frac{sinQr_{ij}}{Qr_{ij}} \tag{11.3}$$

From Equations (11.2) and (11.3), the following is obtained:

$$S(Q)=\frac{\frac{I(Q)}{N}-\left(\langle b^2\rangle-\langle b\rangle^2\right)}{\langle b\rangle^2} \tag{11.4}$$

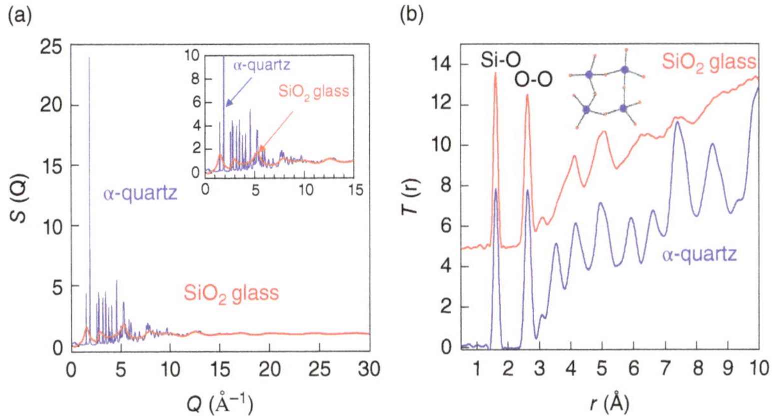

Figure 11.3 *(a) Neutron total structural factor S(Q) of α-quartz and SiO_2 glass. (b) Neutron total correlation function Tr of α-quartz and SiO_2 glass. (Source: reproduced by permission of Kohara and Onodera).*

Figure 11.3a shows the neutron total structural factor $S(Q)$ of α-quartz and SiO_2 glass. Although clear diffraction peaks are observed in α-quartz, SiO_2 glass exhibits the halo pattern. It is notable that the halo pattern of SiO_2 glass is located at the similar Q values of α-quartz, suggesting that the structural distribution of glass spreads from that of a crystal.

The pair distribution function $g(r)$, which is obtained by Fourier transformation of $S(Q)$, is represented with average atomic density ρ_0 as:

$$g(r)=1+\frac{1}{2\pi^2\rho_0 r}\int_0^\infty Q\left[S(Q)-1\right] \tag{11.5}$$

There is a relationship between $g(r)$ and total correlation function $T(r)$: $T(r) = 4\pi r\rho_0 g(r)$. In $T(r)$, correlations of atoms can be fitted with the Gaussian function, because $T(r)$ is based on an assumption that atomic correlations fluctuate at an atomic distance r_i. Figure 11.3b shows neutron total correlation function $T(r)$ of α-quartz and SiO_2 glass. The atomic distances of Si—O (1.61 Å) and O—O (2.63 Å) in SiO_2 glass are almost the same as that in α-quartz. It can be said that a crystal is a singular point of chemical composition and structure. On the other hand, the structure at the intermediate range suggested by X-ray or neutron diffraction is one of the unresolved issues in glass science [17, 18].

The general public thinks that glass is transparent, like window glass or drinking glasses. For transparent glass, the optical absorption property is one of the important parameters. The detailed experimental data of amorphous chalcogenide and α-Si are described in [19]. Conventional oxide glasses have a wide transmission window in the visible region, because most of network forming metal oxides, such as SiO_2, B_2O_3, and P_2O_5, possess a band-gap larger than 5 eV. Figure 11.4 is schematic representation of electronic band structures of perfectly ordered crystal (a) and amorphous material (b). Compared with crystalline material, glass has diversity of structure. The tail states of optical absorption, therefore, reflect

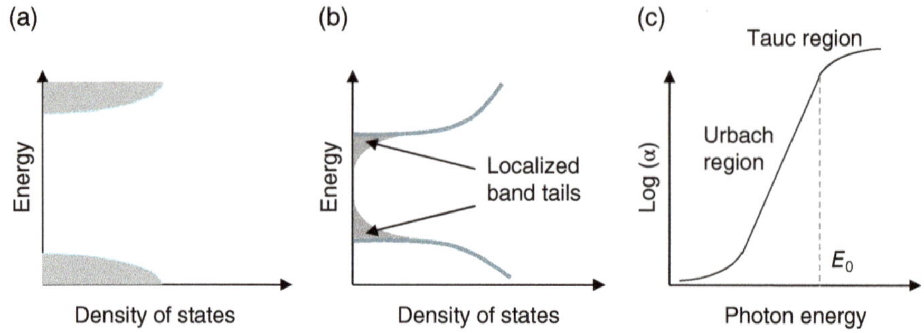

Figure 11.4 *Schematic representation of electronic band structure of perfectly ordered crystal (a) and amorphous material (b). (c) Optical absorption spectra of amorphous materials [20].*

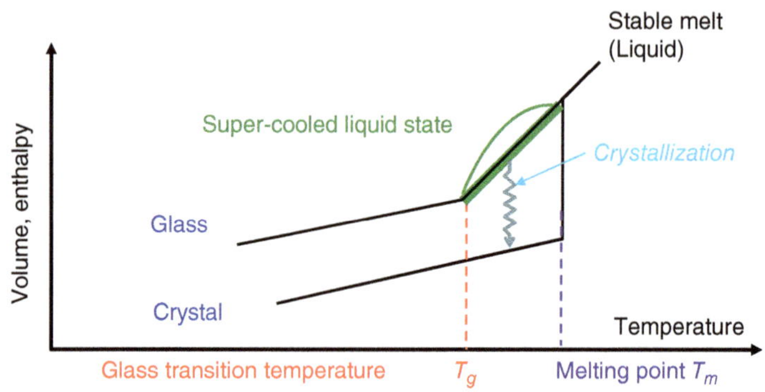

Figure 11.5 *A schematic relationship between volume (or enthalpy) and temperature of glass and crystal.*

such diversity of the electronic state in glass. The tail of the optical absorption is empirically expressed as

$$\alpha(E,T) = \alpha_0 \exp\left[(E - E_0) \cdot \frac{\sigma}{k_B T}\right] \tag{11.6}$$

where, α_0 and E_0 are almost independent of the incident photon energy and the temperature T [21, 22]. Equation (11.6) suggests that the logarithms of the absorption coefficient are focused to α_0 at E_0 (see Figure 11.4c). The defects in glass in these photon energy regions, which also affect the optical and emission properties of the glass, are shown in Section 11.4.

On the other hand, crystals are precipitated from a glass by heat-treatment at a temperature equal to or higher than the glass transition temperature, because glass is in a thermodynamically metastable state. This precipitation is the so-called "crystallization of glass (Figure 11.5). Generally, crystallization behavior, which is the thermodynamically stabilization of a super-cooled liquid, is not preferable for conventional window glass or fiber application. However, there are glass-based materials in which the precipitated crystalline phase has a positive effect on the physical properties of the matrix. The glass-based materials with precipitated crystals after heat-treatment are known as "glass-ceramics." The concept of glass-ceramics was firsty proposed by Stookey [23]. The most famous glass-ceramic is an Li_2O-Al_2O_3-SiO_2 (LAS)-based glass-ceramic, in which β-quartz and β-eucryptite, $LiAlSiO_4$, are precipitated, with an almost zero thermal expansion coefficient. Since LAS glass-ceramics possess optical transparencies in addition to near zero thermal expansion, many industrial products, such as telescope mirrors and cookware fire doors, have been made [24]. On the other hand, Na_2O–CaO–SiO_2–P_2O_5 glass-ceramics with precipitation of apatite has been used as bioactive materials, such as in bone surgical applications [25].

It is noted that different glass-ceramics can be prepared from the same glass by controlling the thermal history for crystallization. On the other hand, in addition to conventional precipitation of dielectric crystals by heat-treatment, a change in the refractive index or a different phase structure such as a metal can be introduced into a material by applying a strong electric field, i.e., laser irradiation [26, 27]. The radiation-induced defect formation is shown in Section 11.4. If a limited space can be heated by irradiation with a laser, space selective crystallization can be induced [28, 29]. If laser scanning conditions, such as wavelength, power, and scanning speed, are well controlled, a continuous crystalline line can be drawn by CW laser scanning (Figure 11.6, [28]). This is because the metastable state of the supercooled liquid has been changed by the heat history. For example, when the preferential growth direction of the precipitated crystal matches the spontaneous polarization of the crystal, a characteristic precipitation morphology of crystal in glass is observed [29, 30]. Such a structure unique to glass is very interesting from a material sciences perspective. It is reported that nano-crystalline dielectrics nanocrystallites were precipitated after heat-treatment of metal-precipitated glass at around glass transition temperatures (Figure 11.7) [31]. This behavior can be said to be the formation of a heterogeneous nano-structure in a monolithic glass-based material.

The valence state and the local coordination field of the activator is an important feature that dominates the performance of the material [32–34]. Although activators are often added as a dopant to amorphous materials, it is extremely difficult to visualize the local

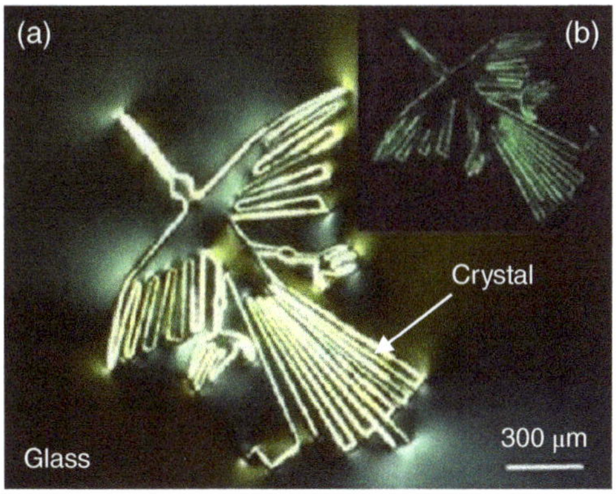

Figure 11.6 Polarization optical micrograph (a) and SHG microscope observation (b) for the curved line (designed as a ground picture [bird] in Nazca) at the surface of glass written by YAG laser irradiation [28]. (Source: Reproduced from R. Ihara et al. [28] by permission of Elsevier.)

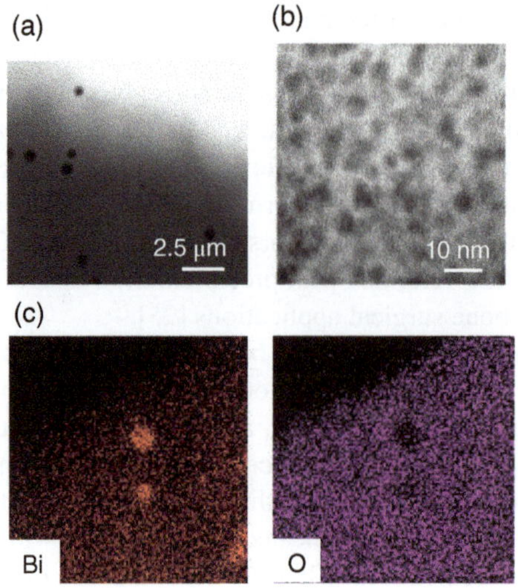

Figure 11.7 (a) TEM image of an AlN-doped aluminoborate glass-ceramic heat-treated at 590°C for 3 hours. (b) HRTEM image of the glass-ceramic. (c) EDX maps of bismuth and oxygen in the vicinity of Bi nanoparticles [31].

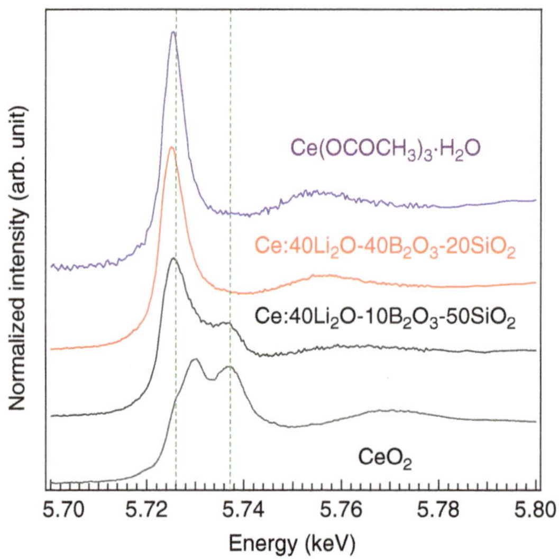

Figure 11.8 Ce L_3-edge XANES spectra of Ce-doped lithium borosilicate glasses along with those references: $Ce(OCOCH_3)_3{\cdot}H_2O$ and CeO_2 [36].

coordination states in glasses. In these cases, X-ray absorption fine structure (XAFS) measurement is often one of the techniques for evaluation of the local coordination [35–38]. Examination of heavier cations, such as rare earth cations, is generally performed using L-edge XAFS measurements, [35, 36] as shown in Figure 11.8. By comparison with the references, the average valence states of dopants are evaluated. On the other hand, recent measurement techniques and equipment using K-edge XAFS analysis have been performed for even heavier cations. By using the K-edge XAFS, the EXAFS region can be obtained in a wide *k* range, which is different from the L-edge analysis, in which the EXAFS region is restricted by each L-edge. However, we should note the validity of the K-edge spectra by consideration of the resolution of the measurements. Figure 11.9a shows the mass attenuation coefficient of tin as a function of energy [39]. There are large absorptions of K-edge and L-edge, which depends on each transition in the tin atom. Figures 11.9b and c show Sn K-edge XANES and L_2-edge XANES of SnO-doped ZnO-P_2O_5 glass, respectively [40]. Since a higher absorption edge generally indicates a higher oxidation state of the cation, we discuss the valence state of activator (Sn^{2+}) using the absorption edge energy E_0. In the case of K-edge XANES (Figure 11.9b), we take the E_0 of the SnO-doped ZnO-P_2O_5 glass prepared in air as lower than that of the glass melted in Ar. However, this evaluation is wrong, which is the opposite result when considering the results of ^{119}Sn Mössbauer spectra. However, as shown in Figure 11.9c, the spectrum of the glass prepared in Ar is similar to that of SnO, while the spectrum of the glass prepared in air is similar to that of SnO_2, indicating that each preparation atmosphere has a clear effect on the L_2-edge XANES spectra. Since Mössbauer spectroscopy is not adaptable to all elements, such evaluation of the validity of XANES spectra is important. A similar result is also observed in Ag-XANES spectra [41]. Although XANES analysis seems to be the strongest tool for evaluation of dopant cations in

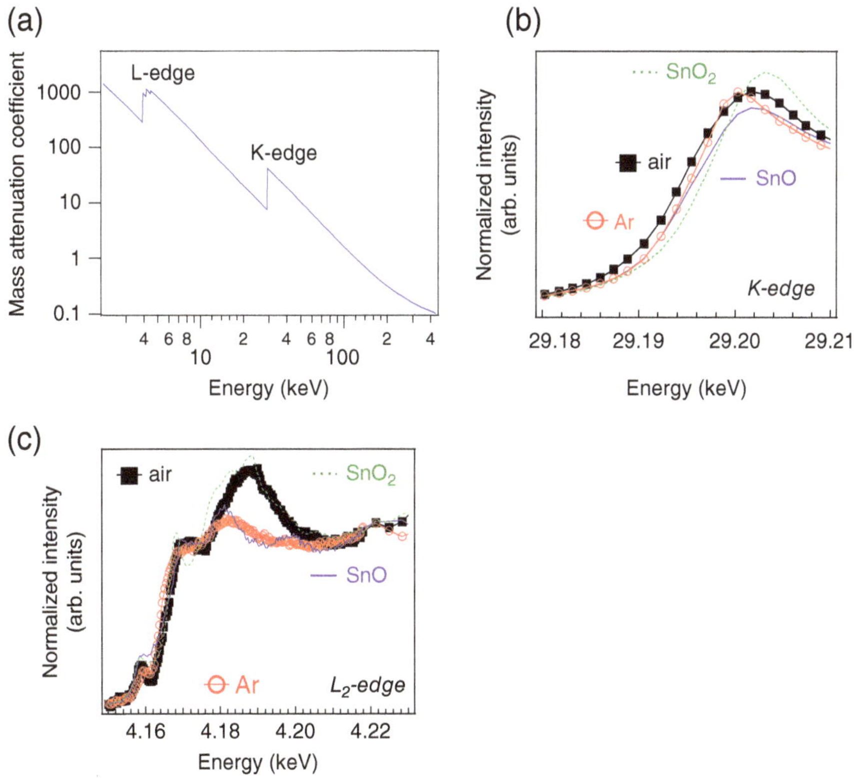

Figure 11.9 *(a) Mass attenuation coefficient of tin as a function of energy [39]. Comparison between Sn K-edge XANES (b) and L_2-edge XANES (c) of SnO-doped ZnO-P_2O_5 glass [40]. Source: Data from X-Ray Form Factor, Attenuation, and Scattering Tables, NIST.*

glass, the researchers should take account of the fact that the observed XANES spectra is the average of the target element. If a trace amount of the species affects the luminescence properties, the XANES data do not show a linear relationship between luminescent properties and local coordination states. Despite XANES analysis of glass not always giving a final answer, the importance of using XANES will become more important when combining it with other measurement techniques [42].

11.3 Attenuation of Quantum Beam as Shielding Materials

Bulk glass possesses both good transparency in the visible region and homogeneity of density and chemical composition, in addition to excellent formability of large-size plates. Therefore, it has been used as window glass for a long time ago. If such glasses can also exhibit absorbance of X-rays, and remain transparent, they will be favorable as radiation shielding materials. This shielding application is important not only for basic scientific experiment but also for medical applications such as in the radiotherapy room, and in iodine injection, etc.

In the case of X-ray shielding applications, effective atomic number (Z_{eff}) is one of the important parameters. PbO-containing glasses have been widely used as shielding glass, because PbOs are useful materials that exhibit both good transparency in the visible region and relatively higher glass forming ability (large PbO fraction in glass), in addition to higher Z_{eff} due to the high atomic number of lead [43–48]. Although lead-containing glasses are still widely used as protective materials, alternative lead-free shielding glasses have been recently reported [47–49]. For precise calculations, several parameters, such as mass attenuation coefficients (μm) [50], total atomic cross-sections (σ_t) [51], molar extinction coefficients (ε), electronic cross-sections (σ_e), and electron densities (N_{el}) should be considered in addition to Z_{eff} [52].

11.4 Defect Formation in Oxide Glass by Quantum Beam Irradiation

As mentioned above, short-range structures (at the atomic scale) of glass are similar to those of crystal. Therefore, defects, which are defined as an atomic structure with local disorder in the crystal, can also exist in glass. However, since glass has no ordered structure in the long-range, there are many defects in glass, even in simple SiO_2 glass [53–55], since SiO_2 glass is the most commonly and widely used glass. In addition, it is a very important optical material in industry, such as in telecommunications and medical applications. Therefore, the defects in SiO_2 glass have been well studied among oxide glasses (see Figures 11.10 and 11.11). The defects of SiO_2 originate from both the starting materials and the fabrication processes. Famous defects originating from the starting materials include hydrogen, hydroxyl group, chlorine, and so on. In the case of GeO_2-doped SiO_2 fiber, which has been used as an optical telecommunication fiber, defects of Ge species are also examined [56, 57]. For these glasses, irradiation light sources for generation of defects are mainly the UV or VUV laser, because quantification is easier compared with X-ray or γ-ray irradiation.

Such defects in glass have been examined by optical absorption measurement and photoluminescence, which can detect trace components. The correlation between optical absorption and emission band in photoluminescence in SiO_2 glass is summarized in Figure 11.10 [54]. The assignments of defects in glass have been carried out by a

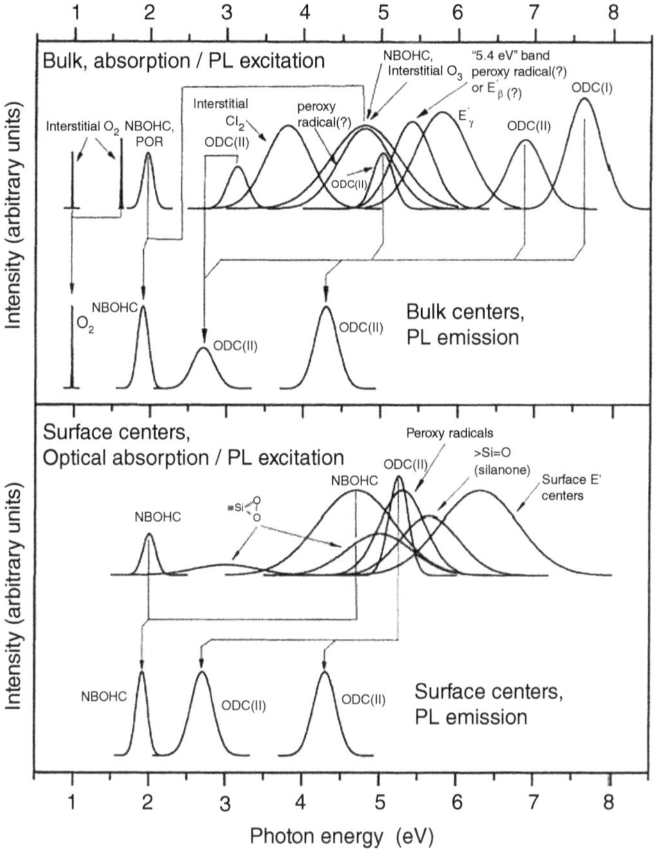

Figure 11.10 *PL excitation and emission band of SiO_2 glass. Top and bottom figures show defect centers in synthetic silica and of centers on the surfaces of SiO_2, respectively. (Source: Ref. [54] reproduced from L. Skuja et al. (1998) by permission of Elsevier.)*

combination of electron spin resonance (ESR) spectroscopy that can detect trace amounts of unpaired electrons, such as the paramagnetic center, quantitatively [58].

SiO_2 glass fiber is used not only for optical telecommunications but also for the diagnostics systems of nuclear sites. In the latter case, pure SiO_2 optical fiber, which has a resistance to high dose irradiation, is desired. There are several reports on γ-irradiated silica and silica glass fiber [59–63]. The defects in SiO_2 glasses are summarized in Table 11.1 [64].

The defect formations by γ-ray are also reported in the borate [65, 66] and phosphate [67, 68] systems. In the case of zinc phosphate glass [68], there is a linear relationship between the generated optical absorption and thermoluminescence. It is clarified that certain generated traps by X-ray irradiation can function as storage luminescent centers below the T_g. The generated defects serve as emission centers of photoluminescence, which has been confirmed by comparison between the optical absorption and PL excitation spectra.

By irradiation of laser or X-ray, a light-induced red-shift in the optical absorption edge is sometimes observed. This behavior is known as photodarkening, which is observed in both chalcogenide [69] and oxide glasses [70].

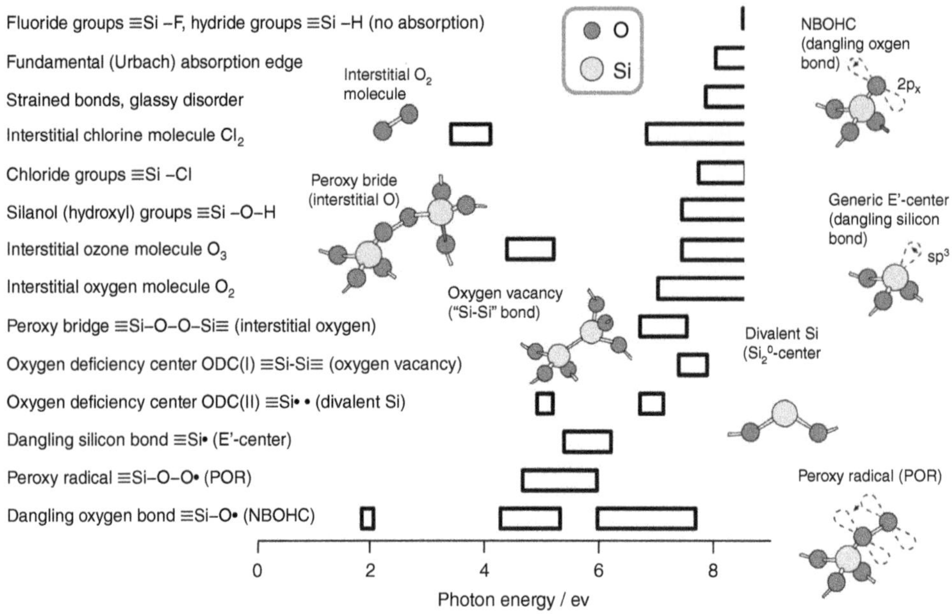

Figure 11.11 *Optical absorption bands of intrinsic defects and impurity in synthetic SiO_2 glass. Source: Absorption bands and the illustrations of defects are modified from Ref. [55].*

Table 11.1 *Representative defects in SiO_2 glasses [64].*

Type	Defects
Oxygen deficiency center	≡Si-Si≡(Oxygen deficiency center ODC(I)) =Si: (divalent Si, ODC(II)) ≡Si• (Dangling silicon bond, E′-center)
Peroxy linkage	≡Si–O–O• (Peroxy radical, POR) ≡Si–O• (Dangling oxygen bond, non-bridging oxygen hole center, NBOHC) ≡Si–O–O–Si≡(Peroxy bridge)
Charged defect	≡Si–O–Si≡(self-trapped hole, STH)
Extrinsic defect, Extrinsic charged defect	≡Si–OH, ≡Si–F, ≡Si–H, ≡Si–Cl ≡Si–O–Al≡(Aluminum oxygen hole center) ≡Ge• (Germanium E′-center), ≡Ge•–(Germanium electron center, GEC) ≡P•–(Phosphorus electron center, PEC) =PO_2• (Phosphorus oxygen hole center, POHC)
Interstitial chemicals	H_2, O_3, O_2, Cl_2, N_2, H_2O, HCl, ClClO, H•, Cl•, ClO_2•, ClO_3•, HO_2•

On the other hand, radiation-induced valence change of the dopant cation is sometimes observed [27, 71, 72]. The radiation-induced change is generally a reduction of valence state of cation. In the case of Sm^{3+}, the generated Sm^{2+} species by irradiation work as an emission center [71, 72]. On the other hand, it has been reported that the Ag^+ center in phosphate glass is changed into Ag^0 and Ag^{2+} by X-ray irradiation [73]. Recently, the valence states of Ag cations after irradiation were discussed with Ag L_3-edge XANES analysis, and it is suggested that generated Ag species by X-ray irradiation possess a valence state higher than 2 [41]. The phenomenon, in which electronic transitions are observed within luminescent centers, which are generated in phosphors by ionizing radiation, belongs to radiophotoluminescence. The details of this mechanism is described in Chapter 9.

11.5 Scintillation in Oxide Glass

Phosphors are a kind of energy converter that generate light in a broad range of wavelengths, from ultraviolet (UV) to infrared (IR). Conventional phosphors are classified into two types: (i) phosphors excited by UV or visible light; and (ii) and phosphors excited by ionizing radiation. In addition, radiation-induced phosphors are categorized into two systems: (i) scintillation (prompt luminescence); and (ii) dosimetry (storage luminescence). Studies of glass scintillator have a long history, particularly those of non-oxide systems. The details of non-oxide glasses are shown in Section 11.7. On the other hand, scintillation behaviors of oxide glasses have recently attracted attention, because oxide glasses are superior to non-oxide glasses in terms of chemical stability. In Section 1.5.1, the scintillation from oxide-based glasses is described.

11.5.1 Glass Scintillators for X-Ray and γ-Ray

Although there is no commercially available glass scintillator for X-ray detection, the study of the glass scintillator has been performed since the 1960s. On the other hand, scintillators possessing higher quantum efficiency are still being investigated worldwide, especially from a security point of view. Since such scintillators are commonly used in bulk shapes, whose size can be of several meters, design of both chemical composition and fabrication processes is important for preparation of large practical scintillators. Also, fabrication cost will be important if the materials become widespread. From this viewpoint, glass materials have an advantage in industry.

Since the physics of scintillation are also described in other chapters, the mechanism is only briefly introduced here. Basically, as the atomic number of the metal cation increases (i.e., the interaction coefficient with X-rays increases), the total absorption energy of X-rays increases. Here, we use the total absorption energy using the following expression:

$$\zeta = \int EN_0(E)\frac{\mu_{\mathrm{EA}}(E)}{\mu_{\mathrm{T}}(E)}\left[1-\exp\{\mu_T(E)\cdot t\}\right]dE, \tag{11.7}$$

where ζ is the absorbed energy in the sample along the irradiation axis per unit area, E is the incident radiation energy, N_0 is the number of incident photons per unit area, $\mu_{\mathrm{T}}(E)$ is the total attenuation coefficient of the sample, μ_{EA} is the energy absorption coefficient of sample, and t is the thickness of the sample.

In scintillation, radiation-induced luminescence is brought about by energy transfer from the host matrix to activators [74–77]. Since the photon energy is far beyond the band-gap of materials, X-ray-induced scintillation is a complex process involving absorption of X-rays in the matrix and scintillation in the activators. The absorption of X-rays by a material, i.e., the total attenuation of ionizing radiation, which can be expressed in terms of an absorption cross-section, is proportional to the density of the material, ρ, and the fourth-power of effective atomic number of the material, Z_{eff} [74, 75]. On the other hand, I_{scinti} is a product of the total absorption energy ζ and the scintillation efficiency $\eta = \beta_{e-h} \cdot S_{trans} \cdot Q_{eff}$, and is given by

$$I_{scinti} = \zeta \cdot \beta_{e-h} \cdot S_{trans} \cdot Q_{eff}. \tag{11.8}$$

where β_{e-h}, S_{trans}, and Q_{eff} are the efficiencies of the processes for generating electron–hole pairs (generation of the secondary particles), transferring the energies of the secondary particles to luminescent centers, and exciting and emitting light at luminescent centers, respectively. The value of Q_{eff} is conventionally referred to as the internal quantum efficiency of photoluminescence (PL). Although the expression looks different, the physical meaning is basically the same with (1.11) and (1.13). Generally, the development of scintillators mainly focuses on the values of Z_{eff} and Q_{eff}, because it is difficult to discuss quantitatively the efficiencies for the electron–hole generation or energy transfer processes.

The fundamental unit of an oxide glass consists of cation and oxygen. Since the Z_{eff} value of glasses depends on the atomic number, heavier cations are preferable. Although the most well-known heavier element is Pb, it is expected that Pb-rich oxide glasses are not suitable for this scintillation application. It is because Pb^{2+} belongs to the ns^2-type emission centers [32]. The ns^2-type emission center is categorized by the electron transition of ns^2-ns^1np^1. Table 11.2 shows several ns^2-type emission centers in glass materials. It is noted that these species are the metastable cations of each element, which suggests that the valence state of these cations can be changed during the melting of the glass. Since most of attenuation glasses contain a large PbO fraction, it might be said that heavy concentration quenching occurs in PbO-rich glass to avoid increase in the Q_{eff} value. It is also noted that prepared $2Bi_2O_3$-$3GeO_2$ glass, which has the stoichiometric chemical composition of $Bi_4Ge_3O_{12}$, exhibits low scintillation efficiency. It is speculated that the random network of $2Bi_2O_3$-$3GeO_2$ glass allows the existence of many defects or color centers, which prevent effective luminescence efficiency. Therefore, lanthanide cations are sometimes used to increase the Z_{eff} values of oxide glasses (see Table 11.2).

As mentioned in Section 11.2, tailoring of the valence state of activators is important to attain high Q_{eff}. When investigating the valence states without XANES analysis, the concept of the "optical basicity," as defined by Duffy, [94, 95] is sometimes used for estimations. Optical basicity, such as the average basicity of oxides in the glass, is a concept based on the

Table 11.2 *Typical ns^2-type emission centers reported in oxide glass system.*

ns^2-type center	Host glass
In^+	Phosphate [37]
Sn^{2+}	Borate [78–82] Phosphate [82–86] Germanate [82]
Sb^{3+}	Borate [77, 82] Phosphate [82, 87]
Te^{4+}	Borate [77, 88]
Tl^+	Alkali (alumino)silicate [89] Aluminoborate [90]
Pb^{2+}	Borate [78, 82] Phosphate [82, 91] Borosilicate [91]
Bi^{3+}	Aluminosilicate [92] Borate [82, 93] Phosphate [82]

polarization of electrons. The idea of basicity of glasses is sometimes useful for evaluation of the physical properties of bulk glasses. For example, the optical basicity of Li_2O, B_2O_3, and SiO_2 are reported to be 1, 0.42, and 0.48, respectively [94]. Therefore, when the optical basicity of glass increases by substitution of SiO_2 for B_2O_3, it is expected that an oxidation reaction of activators occurs even in an inert atmosphere. However, this concept is usable for conventional speculation of activators and valence state of transition metal in the original glass. There is no doubt that the XANES analysis or X-ray photoelectron spectroscopy (XPS) are the best for quantification of the valence states.

Table 11.3 shows a list of glass scintillators from the literature. Although there are many chemical compositions, a unifying concept about the luminescence efficiency and the chemical composition has not yet been established. Masai et al. reported a correlation between the optical basicity and scintillation properties of Ce-doped Li_2O-B_2O_3-SiO_2 glasses, based on the results of photoluminescence and Ce L_3-edge XANES analysis. Figure 11.12a shows the X-ray induced scintillation properties of 0.5 mol% Ce-doped 40Li_2O-xB_2O_3-(60–x)SiO_2 glass [35]. Although the *Ce*$^{3+}$ *ratio* (=$Ce^{3+}/(Ce^{3+} + Ce^{4+})$) and internal quantum efficiency (Q_{eff}) of SiO_2-rich 40Li_2O-10B_2O_3-50SiO_2 glass are the smallest (see Figure 11.8), the scintillation intensity is the highest among these glasses (Figure 11.12b). On the other hand, the proportion of absorbed X-rays and secondary electrons generated and the energy transfer processes are still largely unknown. Since we have no quantitative information about the values of β_{e-h} and S_{trans}, their product, ($\beta_{e-h} \cdot S_{trans}$), is treated as a coefficient that can be evaluated using I_{scinti}, $\zeta_{relative}$, and Q_{eff}, and represents the efficiency for generating electron–hole pairs followed by energy transfer to luminescent centers in each glass. Figure 11.12c shows relative total attenuation and ($\beta_{e-h} \times S_{trans}$) values of 0.5 mol% Ce-doped 40Li_2O-xB_2O_3-(60–x)SiO_2 glass as a function of B_2O_3 fraction. There is no linear correlation between the total attenuation and ($\beta_{e-h} \times S_{trans}$) values, indicating that the energy transfer process of the generated charged secondary particles to activators depends on the chemical composition, and it is important for attaining high scintillation efficiency. Therefore, tailoring the energy transfer process is expected to enable fabrication of high-performance scintillators. The results illustrate the difficulty in understanding the luminescence induced by ionizing radiation, including the energy absorption and subsequent energy transfer. However, understanding of such energy transfer paths will be important for further materials design of X-ray or neutron detectors.

However, in order to attain effective energy transfer processes, glass-based scintillators prepared by the glass-ceramics route are reported. Since crystallization of glass is a kind of thermodynamic stabilization, phase separation behavior is sometimes observed. Using this analogy, halide crystallites can be precipitated from oxide-rich host glass. From this approach, several glass scintillators were reported [103–106]. The details are shown in the latter part of Section 11.6.

11.5.2 Glass Scintillators for Neutrons

A neutron with no electric charge mainly interacts with the nucleus of an atom. For addressing the helium-3 shortage, alternative neutron detection materials have been examined worldwide [141]. If a glass material shows neutron-induced scintillation comparable to the crystalline phosphor, it will be considered as a scintillator capable of good formability and productivity, which is important in industrial manufacturing processes. Because the neutron cross-sections of each element are not proportional to the atomic number, a specific species

Table 11.3 *A selection of glasses for X-ray glass scintillators.*

Glass system	Other components	Activator[a]	Light yield ph/MeV[b]	References
Al_2O_3-B_2O_3	MgO	Ce^{3+}		[96]
Al_2O_3-B_2O_3	CaF_2	Ce^{3+}	7200[c]	[97]
Al_2O_3-B_2O_3	SrF_2	Ce^{3+}	240[d]	[98]
Al_2O_3-B_2O_3	BaF_2	Ce^{3+}	1800[e]	[99]
Al_2O_3-B_2O_3-SiO_2	Gd_2O_3, GeO_2, AlF_3	Ce^{3+}	600	[100]
Al_2O_3-B_2O_3-SiO_2	Gd_2O_3	Ce^{3+}		[101]
Al_2O_3-P_2O_5	MgO, Li_2O	Ce^{3+}, Tb^{3+}		[102]
Al_2O_3-P_2O_5		Ce^{3+}		[36]
Al_2O_3-P_2O_5	Li_2O	Ce^{3+}		[103]
Al_2O_3-P_2O_5	Cs_2O	Ce^{3+}	530[e]	[104]
Al_2O_3-P_2O_5	Cs_2O, CsCl	Ce^{3+}	2100[e]	[105]
Al_2O_3-P_2O_5	CaF_2, AlF_3	Ce^{3+}	93[f]	[106]
Al_2O_3-SiO_2	K_2O, KF, LuF_3	Tb^{3+}:KLu_2F_7[a]		[107]
Al_2O_3-SiO_2	Li_2O, LaF_3	Tb^{3+}:TbF_3[a]		[108]
Al_2O_3-SiO_2	K_2O, CaO, BaF_2, LuF_3	Tb^{3+}:$BaLuF_5$[a]		[109]
Al_2O_3-SiO_2	CaO, CaF_2	Tb^{3+}:CaF_2[a]		[110]
Al_2O_3-SiO_2	Li_2O, BaF_2	Ce^{3+}, Tb^{3+}		[111]
Al_2O_3-SiO_2	Li_2O-MgO	Ce^{3+}	4000	[112, 113]
B_2O_3	Li_2O, Y_2O_3	Dy^{3+}		[114]
B_2O_3	SrO	Sn^{2+}		[81]
B_2O_3	SrO	Ce^{3+}		[39, 115]
B_2O_3	BaO	Ce^{3+}		[115, 116]
B_2O_3-GeO_2	La_2O_3, Gd_2O_3	Eu^{3+}		[117]
B_2O_3-GeO_2	Gd_2O_3	Tb^{3+}		[118]
B_2O_3-P_2O_5	Li_2O	Ce^{3+}		[119]
B_2O_3-SiO_2	Li_2O	Ce^{3+}		[38, 120]
B_2O_3-SiO_2	Ga_2O_3, Ln_2O_3 (Ln = Y, Gd, Lu), P_2O_5	Tb^{3+}	2000	[121]
B_2O_3-SiO_2	CaO, Gd_2O_3	Sm^{3+}		[122]
B_2O_3-SiO_2	CaO, Gd_2O_3	Dy^{3+}		[123]
B_2O_3-SiO_2	AlF_3, Ln_2O_3 (Ln = Y, Gd, Lu)	Eu^{3+}		[124]
B_2O_3-SiO_2	AlF_3, Ln_2O_3 (Ln = Y, Gd, Lu)	Ce^{3+}		[125]
B_2O_3-SiO_2	BaO, Gd_2O_3	Ce^{3+}		[126]
P_2O_5	ZnO	Sn^{2+}		[83] [84]
P_2O_5	ZnO	Sn^{2+}, Ce^{3+}		[127]
P_2O_5	Na_2O, Gd_2O_3	Ce^{3+}		[128, 129]
P_2O_5	Na_2O, Gd_2O_3	Tb^{3+}		[129, 130]
SiO_2	BaO, ZnO, CaO	Ce^{3+}		[131]
SiO_2[d]		Ce^{3+}		[132]
SiO_2	BaO, MgO, Li_2O, K_2O Al_2O_3	Ce^{3+}	400	[133]
SiO_2		Sn^{2+}	1648[f]	[134]
SiO_2		Tl^{+}	1100[f]	[135]

Table 11.3 *(Continued)*

Glass system	Other components	Activator[a]	Light yield ph/MeV[b]	References
SiO_2		Eu^{2+}, Eu^{3+}	407[f]	[136]
ZrF_4-$BaCl_2$-NaF	LaF_3, AlF_3	Eu^{2+}		[137]
ZrF_4-$BaCl_2$-NaF	LaF_3, AlF_3 + α	Sm^{3+}		[138]
ZrF_4-BaF_2-NaF	LaF_3, AlF_3, HfO_2 + α	Ce^{3+}	150	[139]

[a] Glass-ceramics, please refer to Section 11.6
[b] Light yield by ^{241}Am α-ray irradiation.
[c] Light yield by ^{137}Cs gamma ray irradiation.
[d] Light yield by ^{252}Cf neutron irradiation.
[e] Light yield by ^{57}Co γ-ray irradiation.
[f] Estimated from light yield of BGO crystal (8000 ph/MeV) or Li-Glass (4000 ph/MeV) [140]

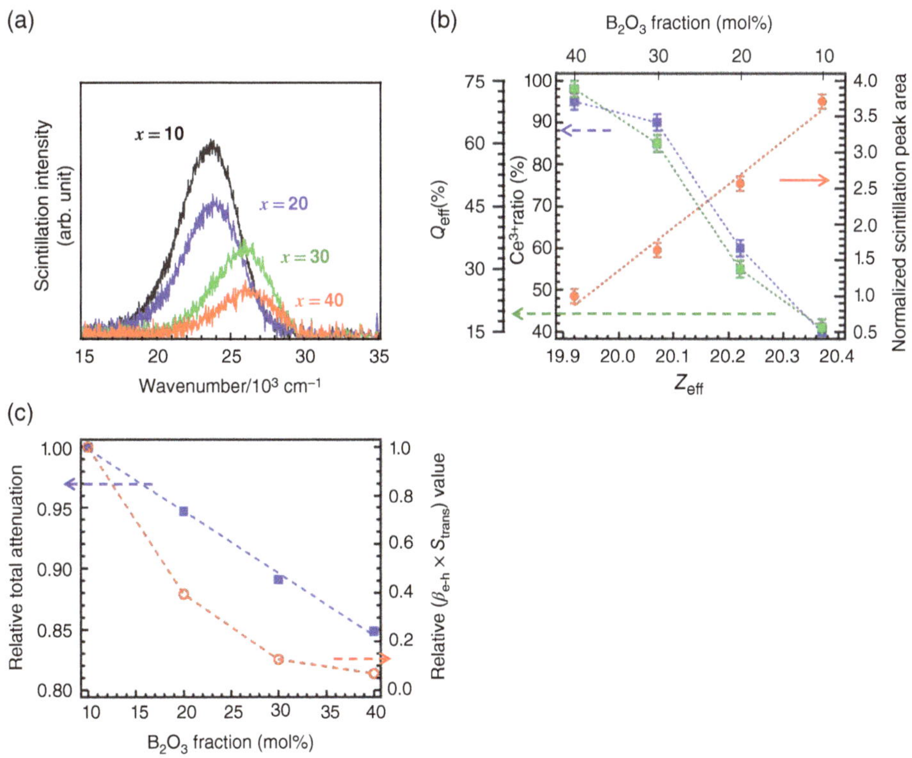

Figure 11.12 *(a) X-ray induced scintillation property of 0.5 mol% Ce-doped 40Li_2O-xB_2O_3-(60-x)SiO_2 glasses. (b) Values of Q_{eff}, the Ce^{3+} ratio (= $Ce^{3+}/(Ce^{3+}+Ce^{4+})$) and the normalized scintillation peak area as a function of Z_{eff} and the B_2O_3 fraction. (c) Total attenuation and ($\beta_{e-h} \times S_{trans}$) relative to the values for the Ce-doped glass [35].*

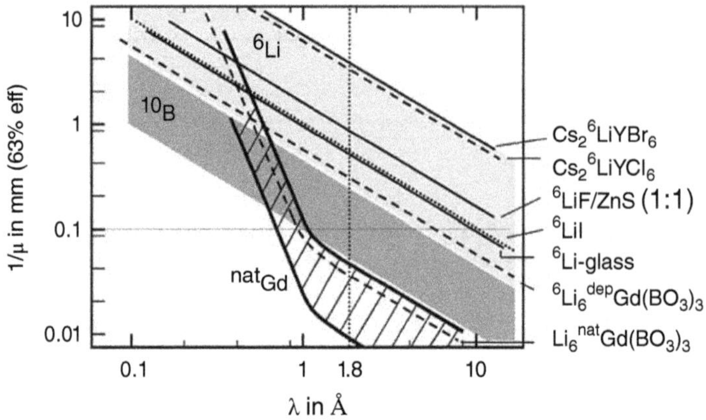

Figure 11.13 Absorption lengths of typical inorganic scintillators based on ^{6}Li, ^{10}B, and ^{nat}Gd as a function of neutron wavelength [143]. (Source: Reproduced from C.W.E. van Eijk, A. Bessiére, P. Dorenbos et al. [143]. by permission of Elsevier.)

Table 11.4 A selection of glasses for scintillator applications [136–143]. * means glass-ceramics, please refer to Section 11.6

Glass system	Other components	Activator*	ref
Al_2O_3-SiO_2	Na_2O, B_2O_3, Li_2O, MgO	Ce^{3+}	[143]
Al_2O_3-SiO_2	MgO, Li_2O	Ce^{3+}	[144, 146, 148–151]
Al_2O_3-SiO_2	Li_2O, MgO, BaO	Tb^{3+}	[153]
Al_2O_3-SiO_2	Li_2O, MgO	Pr^{3+}	[154]
Al_2O_3-SiO_2	LiF, CaF_2	Eu^{2+}:CaF_2*	[155]
Al_2O_3-P_2O_5	LiF	Pr^{3+}	[156]
B_2O_3	Li_2O	Ce^{3+}	[157]

should be selected for neutron detection. Considering the neutron cross-section, plausible nuclei for glass are ^{6}Li, ^{10}B, and ^{155}Gd, whose natural abundances are approximately 7.6, 19.9, 14.8%, respectively [142]. Based on these values of cross-section, several glass scintillators have been developed since the 1960s [143–157]. Figure 11.13 shows absorption lengths of typical inorganic neutron scintillators containing ^{6}Li, ^{10}B, and ^{nat}Gd species [142]. Among oxide glasses, the ^{6}Li-containing glass system is the most investigated [142–151]. Li-glasses (Saint-Gobain K.K.) mainly consist of the Ce-doped Li_2O-Al_2O_3-SiO_2 system, and the physical properties are tuned by selecting the chemical composition of glass [151]. Similar to X-ray induced glass scintillators, a relationship between the chemical composition and neutron scintillation efficiency has not yet been clarified. It is expected that detailed structural analysis of host glasses will hint to designing high efficiency. Representative glasses for neutron scintillator application are listed in Table 11.4.

11.5.3 Storage Luminescence in Glass

Since there are so many defects in glass, storage type phosphor, whose origins are defects or impurities of materials, might be one of the targets of glass-based material. So far, it is expected that the most famous storage phosphor is Eu^{2+}, Dy^{3+}:$SrAl_2O_4$ [158]. The long

Table 11.5 *Chemical composition of Ag-doped phosphate glasses for RPL applications [41].*

ID	Chemical composition (mol%)						T_g (°C)
	Ag_2O	Al_2O_3	Li_2O	P_2O_5	MgO	Na_2O	
FD-3	1.64	10.24	27.87	60.25	0	0	426
FD-5	0.28	13.35	0	63.36	0	23.01	480
FD-6	0.75	11.73	0	61.72	9.25	16.55	491
FD-7	0.09	13.14	0	59.04	0	27.73	458

phosphorescent phosphor at room temperature has been widely used in industry. Based on this report, there are roughly two approaches for possessing the long phosphorescence to glass-based material. One is a method of mixing the phosphor with the starting materials of glass [159], and another is the glass-ceramics route with precipitation of the $SrAl_2O_4$ phase [160], or other phases [161]. This storage emission is a kind of thermoluminescence, in which electrons or holes are captured at trap sites followed by the relaxation from the trap sites with thermal assistance.

On the other hand, there is OSL, in which UV irradiation is used as a stimulated source from the trap sites. The detailed mechanism is described in Sections 11.7–11.9. Here, radio-photoluminescence (RPL) in glasses is briefly described from the compositional viewpoint. RPL is a luminescence owing to electronic transitions within luminescent centers which are generated in phosphors by ionizing radiation [73, 162–167]. In other words, additional emission centers are generated by ionizing radiation in RPL. There are several reports on RPL behavior of glass containing Ag^+ or Sm^{3+}. Table 11.4 shows chemical compositions of Ag-doped phosphate glasses for RPL applications, which is reported by Yokota et al. [164], The base composition is aluminophosphate, and alkaline metal oxide is added. Since the glass transition temperature T_g is low, Al_2O_3 is added in order to improve the chemical durability of glasses. As shown in Table 11.5, the concentration of Ag_2O depends on the chemical composition of glass, indicating that the spatial distribution or local coordination state of Ag might be different depending on the chemical composition. Detailed studies will be necessary for further functionalization.

11.6 Scintillation and Dosimetry in Non-oxide Glass

On the constituents of glass, most of practical glasses are oxides, but non-oxide and mixed anion glasses show unique features because replacing the cation replaces the second neighboring ion, but replacing the anion replaces the nearest ion, therefore substitution of an ion greatly affects the properties. Especially glasses containing halogen, chalcogen, nitrogen, or carbon, have received much attention because of their unique functions, such as a wide transmission wavelength range, emission characteristics, ionic conduction, and hardness [168–172]. Halide glasses (fluoride, chloride, etc.) [168] and chalcogenide (sulfide, selenide, etc.) [169] are especially focused upon as photonic materials. The phonon energy (ν) is given by the following equation:

$$2\pi(\nu) = \sqrt{\frac{k}{m}} \tag{11.9}$$

where k is the spring constant and m is the mass. From this equation, two strategies for reducing phonon energy is given, one of which is suppressing the spring constant. Since the Coulomb force decreases as the valence decreases, the halide composed of the monovalent anions F^-, Cl^-, Br^-, and I^- exhibits lower phonon energy than the oxide (O^{2-}). The other strategy is an increase in the mass number of the anion. The chalcogenide composed of S^{2-}, Se^{2-}, and Te^{2-} has a larger m than that of O^{2-}, therefore chalcogenide glasses exhibit lower phonon energy than oxides. Thermal oscillations in halide and chalcogenide glasses are less likely to be induced by infrared light, and transmit infrared light more efficiently up to longer wavelengths than oxides; in particular, chalcogenide glasses are a candidate for infrared transmission window materials. Table 11.6 shows the typical phonon energy values of glasses composed of oxides and non-oxides [168,173–179]. As you can be seen from the table, halide and chalcogenide glasses have much smaller phonon energy than oxide. Since chalcogenides usually have absorption in the visible regions and low resistance to radiation, there is some interest as a dosimeter but few studies as scintillators. Halide glasses have low phonon energy, and the phonon energy decreases as the period of the halogen increases. Halides other than fluoride have very low phonon energies, but most of these glasses show a low glass transition temperature of less than 100 °C [180], low thermal stability, and extremely poor hygroscopicity. There is little research on radiation detection due to practical drawbacks in use and experimental constraints, such as the fact that all the procedures in glass synthesis and evaluation must be performed in an inert atmosphere due to their terrible hygroscopicity and poor thermal stability.

Fluoride glass is an important material as a special glass because of its various optical properties which are not found in oxide glass. Since fluoride ions have a smaller binding energy than oxides due to their smaller charge, fluoride does not form a strong network structure, unlike in conventional glasses. It is known that BeF_2 vitrifies as a single component and has been studied mainly because of its structural interest. As oxide glass forms a network by sharing the vertices of an SiO_2 tetrahedral unit, BeF_2 also forms a tetrahedral

Table 11.6 *Maximum phonon energy of typical glasses and crystals.*

Host	Maximum phonon energy (cm^{-1})	References
Borate glass	1400	[168]
Phosphate glass	1200	[168]
Silicate glass	1100	[168]
Germanate glass	900	[168]
Tellurite glass	800	[168]
YAG crystal	700	[173]
Fluoride glass (HfF_4-base)	590	[174]
Fluoride glass (ZrF_4-base)	580	[175]
Fluoride glass (AlF_3-base)	530	[175]
Fluoride glass (ThF_4-base)	400	[175]
Sulfide glass (As_2S_3-base)	350	[175]
LaF_3 crystal	350	[177]
Chloride glass ($ZnCl_2$-base)	300	[178]
Bromide glass ($ZnBr_2$-base)	174	[179]
Iodide glass (ZnI_2-base)	124	[179]

structure in a BeF_2-based glass and forms a network by corner sharing. Vitrification of the BeF_2-system has been known for some time, but has problems such as toxicity, deliquescence, and low glass-forming ability. Practical attention was drawn from the report of ZrF_4-based glass by Poulain et al. in 1975 [181]. Fluoride heavy metal (HMG) glass does not contain highly toxic elements, has high glass formation properties, and has a practical level of water resistance. Some compositions of the fluoride glass are thermally stable to the extent that the fiber can be drawn. Ultra-low loss optical fibers beyond silica glass were expected and were actively studied around the 1980s [182, 183]. Although the development of less loss glass than silica glass was not successful, these investigations resulted in new applications such as fiber lasers [171]. Due to their low phonon energy, application to optical amplifiers and fiber lasers using 1.3 μm band or mid-infrared light emission mainly by Pr^{3+} were studied [175, 184]. ZrF_4-BaF_2-LaF_3-AlF_3-NaF (ZBLAN) glass is particularly famous in fluoride glasses, but other feasible fluoride glass systems for photonics of HfF_4, ThF_4, AlF_3, and InF_3 have also been developed. Nowadays, so many glass formers are known, including divalent ones such as BeF_2, ZnF_2, MnF_2, and CdF_2, trivalent ones such as InF_3 and AlF_3, and tetravalent ones such as ZrF_4, HfF_4, and ThF_4. Fluoride glasses can contain a cation with a large atomic number as a network forming component [182]. Table 11.7 shows the characteristics of typical fluoride glasses. Fluoride reduces optical basicity resulting a low refractive index [185]. It has been pointed out that the properties of fluoride glass are remarkably lost due to contamination of OH and O. Generally, $NH_4F \cdot HF$ is added to the raw material and heated to remove oxides that become contaminants, thereby oxidizing oxide impurities [186]. Contamination oxides and transition metal ions impair optical characteristics and radiation resistance, so that purity is extremely important.

As a fluoride glass scintillator, d-f transition of Ce^{3+} as a luminescence center is well researched. For optical applications, ZrF_4 systems such as ZBLAN are well studied, but radiation resistance is not high; therefore, other fluorides, especially HfF_4, ThF_4, and AlF_3 systems, are often studied for scintillator applications. These glass systems also have excellent glass-forming ability depending on the composition, and glass of a centimeter square has been easily synthesized. For radiation applications, hafnium fluoride glasses (HFG) based on HfF_4, which has high density and high resistance to radiation, have been well studied. These glasses have high scintillation characteristics and radiation hardness. For example, a typical 5% Ce-doped fluorohafnate glass reported by Auffray (1996) has a density of 5.95 g/cm³ [139]. Figure 11.14 shows the transmission spectrum and the scintillation spectrum of the glass. There is an absorption edge derived from the Ce ion at 290 nm, and scintillation derived from Ce^{3+} appears at 325 nm. The lifetime of Ce^{3+} is 25 ns, and the light yield

Table 11.7 *Composition and properties of typical fluoride glasses [182].*

Composition	Density (g/cm³)	Refractive index	Glass transition, T_g (°C)	Thermal stability, T_g–T_x (°C)
ZBLA(ZrF_4-BaF_2-LaF_3-AlF_3)	4.54	1.519	320	72
ZBLAN(ZrF_4-BaF_2-LaF_3-AlF_3-NaF)	4.35	1.498	262	90
HBLA(HfF_4-BaF_2-LaF_3-AlF_3)	5.88	1.504	312	78
BATY(BaF_2-AlF_3-ThF_4-YF_3)	5.10	1.487	446	98
YABC(YF_3-AlF_3-BaF_2-CaF_2)	4.0	1.440	430	105

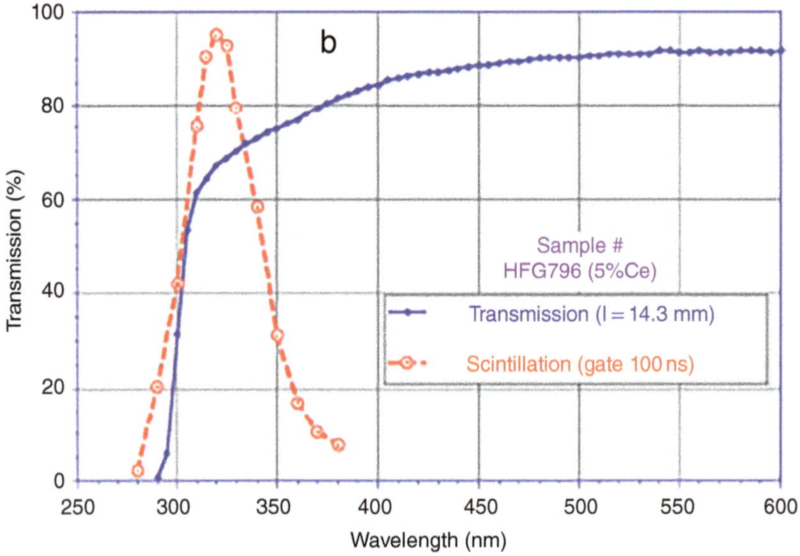

Figure 11.14 *Transmittance and scintillation spectra of a Ce-doped hafnium fluoride glass [187]. (Source: reproduced from E. Auffray et al. [187].)*

is, for example, about 100 photons/MeV (by ^{137}Cs gamma ray) in 5% Ce-doped HFG [188]. Attempts have been made to codope halide to enhance the scintillation properties of fluoride glass. Dmitruk (1997) reported the scintillation properties when CeF_3 and $CeCl_3$ were added to the HfF_4-BaF_2-CeF_3-AlF_3-NaF system [189]. Light yield of the 5% CeF_3-doped sample was 0.62% compared to NaI (Tl), but the chloride codoped sample increased the light yield, e.g., 4.3% CeF_3–8% $CeCl_3$ increased to 0.95% (by ^{137}Cs gamma ray). The scintillation properties may be improved by preparing a transparent nanocrystallized glass by heat-treating fluoride glasses or fluorochloride glasses to precipitate a fluoride nanocrystal or a halide nanocrystal. For example, Cheng reported $BaCl_2$: Eu^{2+} nanocrystals with 10–20 nm from ZBLAN-based glass by heat-treatment, and the spatial resolution of X-ray CT was higher than a CdWO4 single crystal [190].

Fluoride glasses containing polyanion such as metaphosphoric acid (e.g., $NaPO_3$), NbO_2F, and TiO_2F are also worth introducing. By adding a small amount of phosphoric acid to fluoride, the thermal stability and glass forming ability dramatically improved by formation of a network structure composed of PO_4 and $Al(O,F)_6$ [191–193]. They are especially important materials in optics, such as for lenses with low wavelength dispersion and anomalous partial dispersion. Attention has also been paid to them as a host for luminescent material. Addition of an oxide remarkably improves glass forming ability, so that a large amount of rare earth ions can be doped. Poulain (1998) studied $NaPO_3$-BaF_2-CeF_3 glass and succeeded in doping 25% CeF_3 to the glass [194]. The scintillation wavelength was 370 nm and the emission lifetime was 30 ns, and this glass was reported to exhibit a 60% light yield of CeF_3 [194]. In addition, a rare earth ion-doped $20Al(PO_3)_3$–80LiF glass containing a large amount of ^{6}Li for detecting neutron beams has been reported [195]. The thermal neutron beam shows 110 photon/neutron for the Pr^{3+}-doped sample and 570 photon/neutron for the Ce doped sample. Halophosphate glasses containing halides and fluorophosphate have also been reported [105].

$CeCl_3$-CsCl-$CsPO_3$-$Al(PO_3)_3$ glass shows a gamma-ray response of 2100 photons/MeV. It has also been reported that fluoroborate glasses exhibit high scintillation properties by adding rare earth ions [196, 197]. For example, BaF_2-Al_2O_3-B_2O_3 glass has been reported to have 1800 photons/MeV (gamma ray of ^{137}Cs) by doping with 5% Ce [99]. CaF_2-Al_2O_3-B_2O_3 glasses show good dosimetric properties with linear responses in the range of 0.01 to 1000 mGy, covering the same range as commercial glass scintillators [97].

Fluorophosphate glasses are also known to exhibit good RPL properties. When irradiated, in a glass containing earth ions or transition metals having multiple oxidation states such as Sm^{3+}/Sm^{2+} and Eu^{3+}/Eu^{2+}, the valence state changes. For example, when Sm^{3+}-containing fluorophosphate glass was irradiated by γ-rays, part of the Sm^{3+} reduced to Sm^{2+}, and a new PL peak appeared (Figure 11.15). Since the emission intensity ratio of Sm^{3+} and Sm^{2+} corresponds to the dose of irradiated radiation, high sensitivity imaging can be performed using

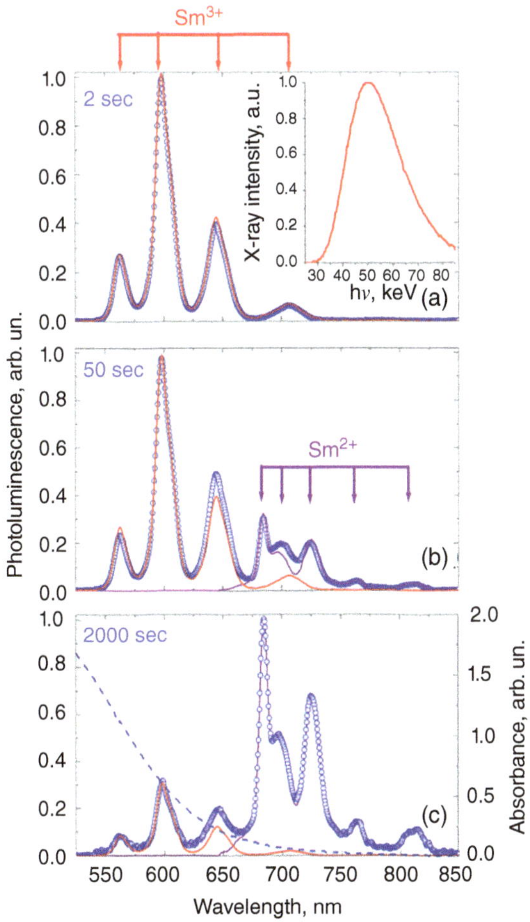

Figure 11.15 *The evolution of PL spectra under synchrotron x-ray irradiation. The irradiations times are (a) 2 seconds, (b) 50 seconds, and (c) 2000 seconds, respectively [191]. (Source: reproduced from G. Okada et al. [191].)*

this emission intensity ratio (Figure 11.16) [191, 192, 198]. Reading is not a detrap mechanism like TSL or OSL, but a change in PL response due to a change in valence in the glass, so reading can be performed many times and not be affected by temperature and light; further details are described in Chapter 9. In glass materials, this phenomenon is not only reported for fluorophosphate glass but also for fluoride glass and crystallized glass [192, 198]. Fluorophosphate glass is one of the feasible RPL materials because of its good radiation response characteristics, large size productivity, and chemical stability.

Oxyfluoride glass containing halide nanocrystals is an excellent material for photonics. Halide nanocrystals having excellent luminescence properties are precipitated in an oxide glass having high thermal stability, and chemical and mechanical durability [199–202]. After the report of the precipitation of Er^{3+}, Yb^{3+} co-doped $Pb_xCd_{1-x}F_2$ nanocrystals from oxyfluoride glass by Wang and Ohwaki [199] in 1993, research on oxyfluoride nanocrystallized glasses was actively conducted, and precipitation of various fluoride nanocrystals such as PbF_2, CdF_2, LaF_3, BaF_2, CaF_2, $NaYF_4$, $KNiF_4$ has since been reported [200, 201]. Since fluoride crystals are the emitting host, the phonon energy can be further reduced as compared with fluoride glass. Figure 11.17 shows a TEM image of BaF_2 nanocrystallized glass as an

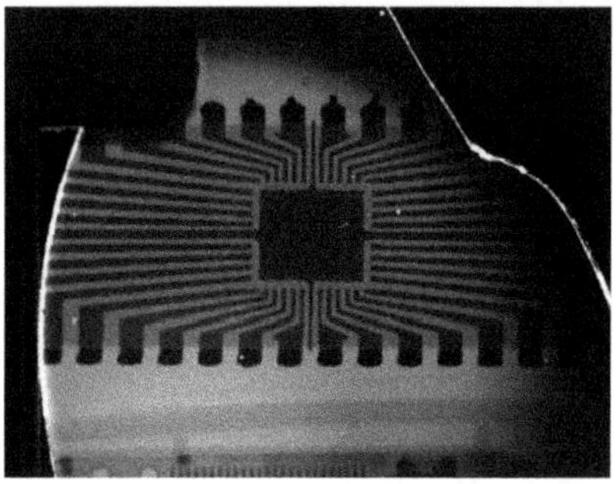

Figure 11.16 *X-ray image of an integrated circuit taken by RPL of Sm-doped fluorophosphate glass [192]. (Source: reproduced from A. Edgar et al. [192].)*

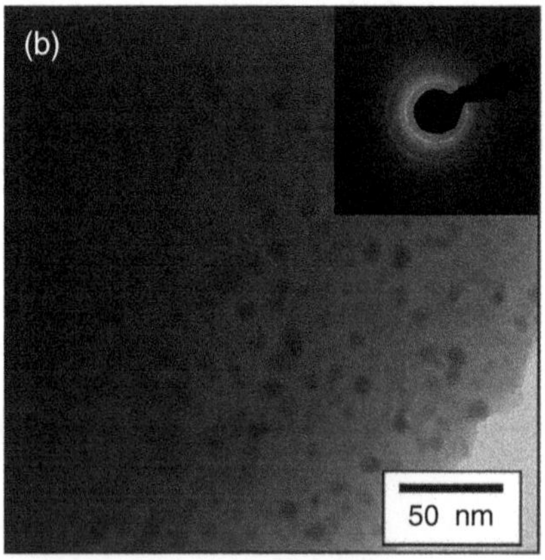

Figure 11.17 *TEM image of oxyfluoride nanocrystallized glass. 5 nm of BaF_2 nanocrystals were precipitated by heat-treatment [203]. (Source: reproduced from K. Shinozaki et al. [203].)*

example. By controlling the crystallization conditions, the precipitation grain size can be reduced to 30 nm or less, and light scattering can also be reduced, so that high transparency can be maintained. In addition, efforts have been made to adapt the laser-induced local heating method to induce crystallization on only one of the glass surfaces to crystallize on the waveguide [202]. Solving issues such as low water resistance, mechanical strength, and glass forming ability of fluoride glass, and exhibiting higher luminescence characteristics, the oxyfluoride crystallized glass is excellent material for fiber lasers, optical amplifiers, solar lighting, lighting, sensing, etc.

These transparent nano-crystallized glasses are synthesized by heat-treatment crystallization of base glass. The base glass may not only be a glass produced by melting and quenching but also an oxyfluoride glass produced by the sol–gel method. Two-step heat-treatment is often attempted to control nucleation and crystal growth rate, e.g., for forming nuclei by heat-treatment at a temperature around the glass transition temperature and then growing the crystals at a much higher temperature. In some cases, fluoride glass is used as a base glass, but an oxide is often used as the matrix for improving chemical durability and mechanical strength. Since the emission center is a nanocrystal, it is considered that the property of the glass matrix does not significantly affect the emission property. In oxyfluoride glass, the silicate or aluminosilicate system, which is a highly viscous component, is selected as the base glass composition [201]. For example, in the case of CaF_2 nanocrystals precipitation, CaF_2-Al_2O_3-SiO_2-based glass is often selected because it forms a viscous barrier around fluoride nanocrystals, and suppress the crystal growth rate [202–204]. Recently, transparent nanocrystallization in the fluoroborate system has been reported [203]. The nanocrystallized glass exhibits good scintillation and dosimetry characteristics derived from the precipitated crystals. It is reported that $17SiO_2$–$11B_2O_3$–$10Al_2O_3$–$6Cs_2O$–$32BaF_2$–$10GdF_3$–$7GdBr_3$–$7TbF_3$ precipitates $BaGdF_5$: Tb nanocrystals after heat-treatment and exhibit 80% of light yield to GOS:Tb [205]. $50SiO_2$-$16Al_2O_3$-$19NaF$-$13GdBr_3$-$2CeBr_3$ glass with $GdBr_3$:Ce nanocrystals exhibits 1700 photons/MeV (^{137}Cs gamma ray) of Ce^{3+} emission [206]. It is expected that halides having poor water resistance and durability can be handled stably by precipitating them in glass.

11.7 Preparation of Glass

In this section, the preparation of glassy materials is briefly discussed. Although there are various glass manufacturing methods, they are roughly classified into a melt process and other processes. In a conventional melt process, solid raw materials are heated to a homogenous glass melt at the melting temperature, and then cooled into glass, which is the most general method for manufacturing glass. On the other hand, the other processes are roughly divided into a vapor phase method and a liquid phase method.

11.7.1 Melt Process

The solid raw material is heated to obtain a uniform liquid phase and then vitrified. At that time, there are various parameters such as temperature for melting, heating duration, heating method, time for holding the melt, and atmosphere, and so on. The network former (NWF) oxides (see Section 11.2), which easily form glass, are used as the main chemical component, and generally other different oxides are also added. Alkali metals, alkaline earth

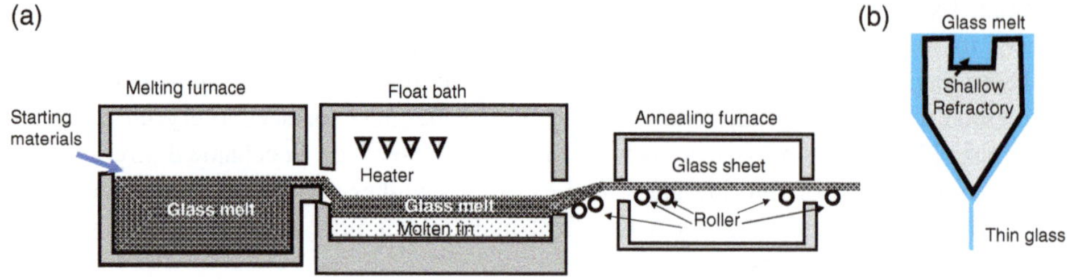

Figure 11.18 *Tentative preparation process of sheet glasses. (a) A schematic illustration of float process. (b) A schematic illustration of fusion process.*

metals, halides, etc. are sometimes added to the melt as components that bring about a low melting point. As the raw material, not only common oxides but also carbonate, nitrates, sulfates, hydroxides, etc. are used with a wide variety of chemical compositions.

One of the characteristics of glass is transparency. Since existence of bubbles, stria, cords, etc. decrease the transparency, these impurities need to be removed. Generally, most modern glass sheet is produced by the *float process*, in which glass melt flows onto the surface of a bath of molten tin. The sheet on the tin surface is slightly contaminated by molten tin, while the air side is clean with an excellent surface (Figure 11.18a). In the case of production of thin sheet glass, the fusion method is also used, in which overfilled glass melt from the tank allows a shallow flow over each side to form a single thin sheet (Figure 11.18b). The advantage of this process is that both sides of the glass sheet have flawless surfaces.

Important in the glass manufacturing process is control of trace amounts of impurities from the manufacturing process as well as the starting material. As shown in Section 11.4 impurities work as defects or emission centers in glasses. A conventional preparation method for bulk glass is the melt-quenching method using crucibles. Using this preparation method, powdered starting materials are melted in an electric furnace. When a glass is prepared using crucibles at a laboratory, impurities are contaminated from crucibles during the melting. In the case of melting with oxide-based crucibles, such as porcelain, alumina, and silica crucibles, considerable amounts of contaminants enter the glass melt from the crucible, depending on the temperature. On the other hand, even when melting with a platinum crucible, a gold crucible, or a carbon crucible, it is considered that trace amounts of contaminants from the crucible is unavoidable [207]. We should consider that the effect of impurities on the physical properties and structure of the glass can be negligible or not for each application.

As a method of avoiding mixing from the crucible, a containerless method of producing glass by laser irradiation is also used [208, 209]. The schematic illustration of the aerodynamic levitation technique is shown in Figure 11.19. A solidified material is melted during laser irradiation. After the laser is switched off, the temperature rapidly decreases, which is effective to prevent crystallization. Since a gas flows during laser irradiation to keep the glass melt sphere levitated in the air, the heterogeneous nucleation from the surface of the crucible is prevented. In order to suppress a change in the cation-oxygen fraction due to evaporation of oxygen, detailed irradiation (temperature) conditions are being studied. Furthermore, the production atmosphere for controlling the redox reaction of the melt and the melting temperature are also factors that greatly affect the physical properties.

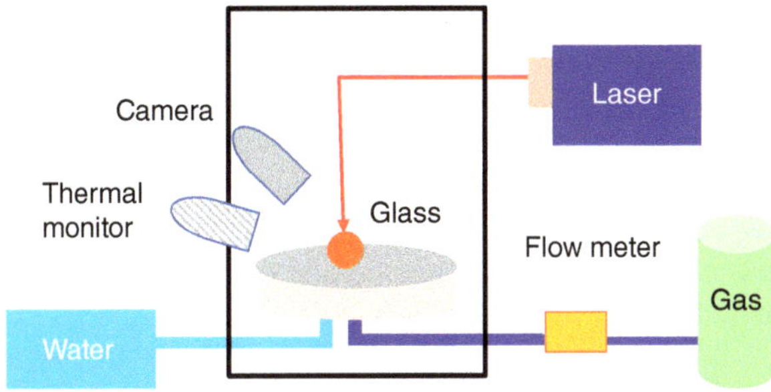

Figure 11.19 *Schematic illustration of aerodynamic levitation technique. Transparent glass spheres with diameters of several mm are commonly obtained.*

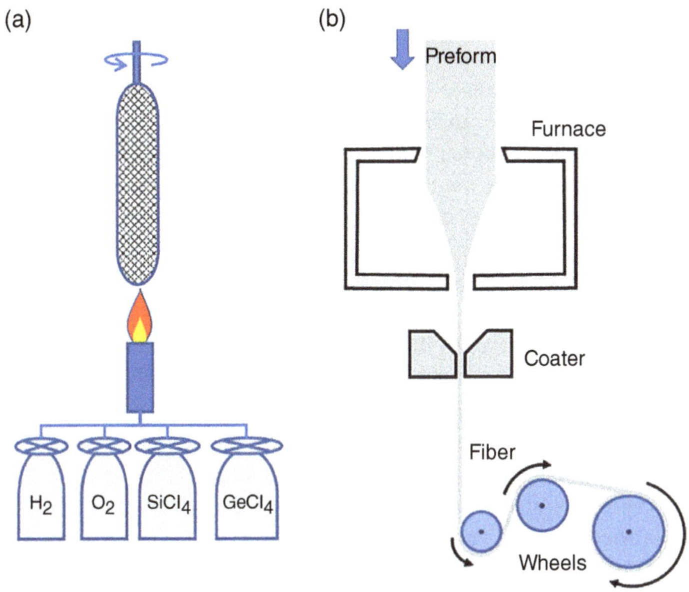

Figure 11.20 *Glass fiber processes. (a) Schematic illustration of VAD process. (b) Schematic illustration of fiber redraw method.*

11.7.2 Vapor Process and Fiber Drawing

There have been many efforts to tailor the fiber drawing process. In the case of SiO_2 glass fiber, the vapor phase axial deposition (VAD) method, which is very effective to produce high purity SiO_2 glass, was invented in 1980 [210]. The schematic illustration is shown in Figure 11.20a. Using reactive gases (O_2 and H_2), high purified chemicals (conventionally metal halides) are solidified by burning. There are other deposition processes for fiber preform. One is outer vaper deposition (OVD), in which oxygen and metal halide vapors react on the outside of a rotating silica rod to form the *soot* particles. After the sintering process using porous soot particles, a solid rod is obtained. Another method is the modified chemical vapor deposition (MCVD). The deposition process is carried out in a silica tube in which vapors react inside the tube without contamination from the flame. After deposited soot is obtained, the entire tube is heated to complete the preform. After the fiber preform is obtained, the fiber drawing process is applied. Figure 11.20b shows a schematic illustration of the fiber redrawing method. Both moving speed of fiber preform and drawing speed are controlled in order to obtain excellent fiber with small propagation loss. In the case of the drawing process of multi-clad fiber, the rod-in-tube method is commonly used. In this method, tubes (clads) and the rod (core) are heat-treated to obtain a multi-clad fiber preform. This drawing process can be adopted for preparation of fibers with the complicated shape, such as panda fibers [211].

11.7.3 Liquid Process

As a liquid process for fabrication of glass materials, the sol–gel process is the most famous approach. In the sol–gel process, the basic part of the glass network is formed by a liquid phase reaction at relatively low temperatures. The name derives from the material being prepared via a *sol* in which fine particles are suspended in solution, and a *gel* is obtained by solidification of the sol. Although there have been many scientific papers published since the 1980s, practical application is limited due to the high cost and long time required to obtain the monolithic glass. Therefore, in industry, fiber-type or film-type materials are mainly fabricated using the sol–gel process [212]. The characteristics of this process are: (i) various shapes, such as bulk, fiber, film, particles, can be prepared; (ii) low-temperature synthesis compared with the conventional melt process; and (iii) a high purification process is available.

The process of sol–gel synthesis consists of three parts: (i) preparation of sol and the solidified gel after drying; (ii) heat-treatment to obtain the porous precursor; and (iii) sintering process at higher temperature. At the first stage, selection of starting chemicals and various reaction conditions, such as temperature, pH, catalyst, atmosphere, purity, are important. The conventional starting chemicals for sol–gel are metal alkoxides, whose hydrolysis and polycondensation rates are different. Therefore, obtaining a homogenous gel consisting of a multicomponent system is not an easy task at the first stage. The important point at the second stage is prevention of cracking in the precursor. Because conventional gels contain solvent in the solidified matrix, cracking is often observed when the residual solvent is evaporated. When the sintering process is applied as the last stage, heat-treatment conditions should be carefully monitored for densification of matrix. If the conditions are not optimial, residual OH groups or small bubbles will remain in the matrix. Because considerable shrinkage is observed through these three stages, handling of materials is also important. Recently, fabrication of bulk SiO_2 glass by sol–gel synthesis is reported. It is notable that the optical properties of SiO_2 glass prepared at 1200 °C is comparable to conventional pure SiO_2 glass [213]. Although synthesis with large scaling is not easy, there is no doubt that such basic research will contribute to great improvement of this research field.

11.8 Future Prospectives for Glass-Based Materials

For optical (phosphor) applications, a doping of an activator into glass is widely used, because the intrinsic emission intensity of amorphous materials is generally low. However, in an amorphous material (glass), it is very hard to assume that all the activators exhibit the same local coordination states and luminescence characteristics. Rather, it should be considered that there are local sites exhibiting high conversion efficiency and with low efficiency in a monolithic glass. For example, even when evaluated using X-ray absorption fine structure analysis (XANES), the result obtained is the average of all atoms, and it is not clear whether each local structure is important for efficient light emission. It has been reported that the rare-earth (Er) cations in Er-doped fiber amplifiers can be directly observed by a scanning transmission electron microscope [214]. However, it should be noted that this technique is effective when there is a contrast in atomic number of elements. For examination of local structures of doped activators, not only experimental results but also theoretical calculation are necessary. At the same time, in order to examine the surrounding glass network structure, an investigation of the average 3-D structure of glass [39], which is established by a combination of structural analysis methods, will be important.

Acknowledgement

The author (H. M.) is appreciative of Dr Shinji Kohara and Prof Yohei Onodera for many useful discussions.

References

1. Wong, J. and Angell, C.A. (1976). *Glass Structure by Spectroscopy*. New York: Marcel Dekker Inc.
2. Angell, C.A. (1995). Formation of glasses from liquids and biopolymers. *Science* 267: 1924.
3. Greaves, G.N. and Sen, S. (2007). Inorganic glasses, glass-forming liquids and amorphizing solids. *Adv. Phys.* 56: 1.
4. Debenedetti, P.G. and Stillinger, F.H. (2001). Supercooled liquids and the glass transition. *Nature* 410: 259.
5. Zachariasen, W.H. (1932). The atomic arrangement in glass. *J. Am. Chem. Soc.* 54: 3841–3851.
6. Sun, K.H. (1947). Fundamental condition of glass formation. *J. Am. Ceram. Soc.* 30: 277.
7. Vogel, H. (1921). Das Temperaturabhangigkeitsgesetz der Viskositat von Flussigkeiten. *Phys. Z.* 22: 645.
8. Fulcher, G.S. (1925). Analysis of recent measurements of the viscosity of glasses. *J. Am. Ceram. Soc.* 8: 339.
9. Tammann., G. and Hesse, W. (1926). Die abhängigkeit der viscosität von der temperatur bie unterkühlten flüssigkeiten. *Z. Anorg. Allg. Chem.* 156: 245.
10. Angell, C.A. (1988). Structural instability and relaxation in liquid and glassy phases near the fragile liquid limit. *J. Non Cryst. Solids* 102: 205.
11. Velikov, V., Borick, S., and Angell, C.A. (2001). The glass transition of water, based on hyperquenching experiments. *Science* 294: 2335.
12. Kim, S.W., Shimoyama, T., and Hosono, H. (2011). Solvated electrons in high-temperature melts and glasses of the room-temperature stable electride $[Ca_{24}Al_{28}O_{64}]^{4+}\cdot 4e^-$. *Science* 333: 71.
13. Warren, B.E. and Biscoe, J. (1938). The structure of silica glass by X-ray diffraction studies. *J. Am. Ceram. Soc.* 21: 49–54.
14. Demuth, T., Jeanvoine, Y., Hafner, J. et al. (1999). Polymorphism in silica studied in the local density and generalized-gradient approximations. *J. Phys. Condens. Matter* 11: 3833.
15. Brow, R.K. (2000). Review: the structure of simple phosphate glasses. *J. Non Cryst. Solids* 263–264: 1.
16. McGreevy, R.L. (2001). Reverse Monte Carlo modelling. *J. Phys. Condens. Matter* 13: R877–R913.
17. Elliott, S.R. (1991). Medium-range structural order in covalent amorphous solids. *Nature* 354: 445–452.
18. Salmon, P.S. (2018). *Magmas Under Pressure* (eds. Y. Kono and C. Sanloup). Amsterdam, Netherlands: Elsevier.
19. Shimkawa, K., Singh, J., and O'Leary, S.K. ((2006). *Optical Properties of Disordered Condensed Matter. Edited by J. Singh* (ed. J. Singh), 47. West Sussex, UK: Wiley.
20. Saito, K. and Ikushima, A.J. (2000). Absorption edge in silica glass. *Phys. Rev. B.* 62: 8584.
21. Urbach, F. (1953). The long-wavelength edge of photographic sensitivity and of the electronic absorption of solids. *Phys. Rev.* 92: 1324.

22. Sumi, H. and Toyozawa, Y. (1971). Urbach-martienseen rule and exciton trapped momentarily by lattice vibrations. *J. Physical Soc. Japan* 31: 342.
23. Stookey, S.D. (1959). Catalyzed crystallization of glass in theory and practice. *Ind. Eng. Chem.* 51: 805.
24. Beall, G.H. and Pinckney, L.R. (1999). Nanophase glass-ceramics. *J. Am. Ceram. Soc.* 82: 5.
25. Kokubo, T. and Takadama, H. (2006). How useful is SBF in predicting *in vivo* bone bioactivity? How useful is SBF in predicting *in vivo* bone bioactivity? *Biomater* 27: 2907.
26. Davis, K.M., Miura, K., Sugimoto, N. et al. (1996). Writing waveguides in glass with a femtosecond laser. *Opt. Lett.* 21: 1729–1731.
27. Qiu, J.R., Shirai, M., Nakaya, T. et al. (2002). Space-selective precipitation of metal nanoparticles inside glasses. *Appl. Phys. Lett.* 81: 3040–3042.
28. Ihara, R., Honma, T., Benino, Y. et al. (2005). Writing of two-dimensional crystal curved lines at the surface of Sm_2O_3–Bi_2O_3–B_2O_3 glass by samarium atom heat processing. *Solid State Commun.* 136: 273–277.
29. Komatsu, T. and Honma, T. (2019). Laser patterning and growth mechanism of orientation designed crystals in oxide glasses: A review. *J. Solid State Chem.* 275: 210–222.
30. Takahashi, Y., Yamazaki, Y., Ihara, R. et al. (2013). Parasitic amorphous on single-domain crystal: Structural observations of silicate glass-ceramics. *Sci. Rep.* 3: 1147.
31. Masai, H., Takahashi, Y., Fujiwara, T. et al. (2010). Precipitation of heterogeneous nanostructures: Metal nanoparticles and dielectric nanocrystallites. *J. Appl. Phys.* 108: 023503.
32. Yen, W.M., Shionoya, S., and Yamamoto, H. (2007). *Phosphor Handbook*, 2e. Boca Raton: CRC Press.
33. Feldmann, C., Justel, T., Ronda, C.R. et al. (2003). Inorganic luminescent materials: 100 years of research and application. *Adv. Funct. Mater.* 13: 511–516.
34. Blasse, G. (1988). Luminescence of inorganic solids: From isolated centres to concentrated systems. *Prog. Solid State Chem.* 18: 79–171.
35. Masai, H., Okada, G., Torimoto, A. et al. (2018). X-ray-induced scintillation governed by energy transfer process in glasses. *Sci. Rep.* 8: 623.
36. Masai, H., Shinozaki, K., Okada, G. et al. (2018). Luminescence of Ce^{3+} in aluminophosphate glasses prepared in air. *J. Lumin.* 195: 413.
37. Masai, H., Yamada, Y., Okumura, S. et al. (2015). Photoluminescence of monovalent indium centres in phosphate glass. *Sci. Rep* 5: 13646.
38. Peters, P.M. and Houde-Walter, S.N. (1997). X-ray absorption fine structure determination of the local environment of Er^{3+} in glass. *Appl. Phys. Let.* 70: 541.
39. https://physics.nist.gov/PhysRefData/FFast/html/form.html
40. Masai, H., Ina, T., Okumura, S. et al. (2017). Validity of valence estimation of dopants in glasses using XANES analysis. *Sci. Rep.* 8: 415.
41. Masai, H., Koshimizu, M., Kawamoto, H. et al. (2019). X-ray absorption near-edge structure of Ag cations in phosphate glasses for radiophotoluminescence applications. *J. Cerma. Soc. Jpn.* 127: 924.
42. Onodera, Y., Kohara, S., Masai, H. et al. (2017). Formation of metallic cation–oxygen network for anomalous thermal expansion coefficients in binary phosphate glass. *Nat. Commun.* 8: 15449.
43. Mostafa, A.M.A., Issa, S.A.M., and Sayyed, M.I. (2017). Gamma ray shielding properties of PbO-B_2O_3-P_2O_5 doped with WO_3. *J. Alloys Comp* 708: 294–300.
44. Matori, K.A., Sayyed, M.I., Sidek, H.A.A. et al. (2017). Comprehensive study on physical, elastic and shielding properties of lead zinc phosphate glasses. *J. Non Cryst. Solids* 457: 97–103.
45. Limkitjaroenporn, P., Kaewkhao, J., Limsuwan, P. et al. (2011). Physical, optical, structural and gamma-ray shielding properties of lead sodium borate glasses. *J. Phys. Chem. Solid* 72: 245–251.
46. Archer, B.R., Fewell, T.R., Conway, B.J. et al. (1994). Attenuation properties of diagnostic X-ray shielding materials. *Med. Phys.* 21: 1499–1507.
47. Singh, N., Singh, K.J., Singh, K. et al. (2004). Comparative study of lead borate and bismuth lead borate glass systems as gamma-radiation shielding materials. *Nucl. Instrum. Methods B* 225: 305–309.
48. Singh, V.P., Badiger, N.M., and Kaewkhao, J. (2014). Radiation shielding competence of silicate and borate heavy metal oxide glasses: Comparative study. *J. Non Cryst. Solids* 404: 167–173.
49. Issa, S.A.M. and Mostafa, A.M.A. (2017). Effect of Bi_2O_3 in borate-tellurite-silicate glass system for development of gamma-rays shielding materials. *J. Alloys Compd.* 695: 302–310.
50. https://dx.doi.org/10.18434/T4D01F
51. https://physics.nist.gov/PhysRefData/Xcom/html/xcom1.html
52. https://dbshino.nifs.ac.jp/nifsdb

53. Pacchioni, G., Skuja, L., and Griscom, D.L. (eds.) (2000). *Defects in SiO_2 and Related Dielectrics: Science and Technology*. Dordrecht, Netherlands: Kluwer Academic Publishers.
54. Skuja, L. (1998). Optically active oxygen-deficiency-related centers in amorphous silicon dioxide. *J. Non Cryst. Solid* 239: 16–48.
55. Skuja, L., Hirano, M., Hosono, H. et al. (2005). Defects in oxide glasses. *Phys. Stat. Sol. C* 2: 15–24.
56. Hosono, H., Abe, Y., Kinser, D.L. et al. (1992). Nature and origin of the 5-eV band in SiO_2:GeO_2 glasses. *Phys. Rev. B.* 46: 11445.
57. Ikushima, A.J., Fujiwara, T., and Saito, K. (2000). Silica glass: A material for photonics. *J. Appl. Phys.* 88: 1201.
58. Griscom, D.L., Friebele, E.J., and Sigel, G.H. Jr. (1974). Observation and analysis of the primary ^{29}Si hyperfine structure of the *E'* center in non-crystalline SiO_2. *Solid State Commun.* 15: 479.
59. Griscom, D.L. (2011). Trapped-electron centers in pure and doped glassy silica: A review and synthesis. *J. Non-Cryst. Solids* 357: 1945–1962.
60. Nishikawa, H., Watanabe, E., Ito, D. et al. (1996). Visible photoluminescence from Si clusters in γ-irradiated amorphous SiO_2. *Appl. Phys.* 80: 3513–3517.
61. Griscom, D.L., Friebele, E.J., Long, K.J. et al. (1983). Fundamental defect centers in glass: Electron spin resonance and optical absorption studies of irradiated phosphorus-doped silica glass and optical fibers. *J. Appl. Phys.* 54: 3743–3762.
62. Griscom, D.L. (1997). Radiation hardening of pure-silica-core optical fibers: Reduction of induced absorption bands associated with self-trapped holes. *Appl. Phys. Lett.* 71: 175.
63. Kashaykin, P.F., Tomashuk, A.L., Salgansky, M.Y. et al. (2017). Anomalies and peculiarities of radiationinduced light absorption in pure silica optical fibers at different temperatures. *J. Appl. Phys.* 121: 213104.
64. Kajihara, K. (2018). Point defects in silica glass I. *New Glass* 124: 27. (in Japanese).
65. Griscom, D.L., Taylor, P.C., Ware, D.A. et al. (1968). ESR studies of lithium borate glasses and compounds γ irradiated at 77°K: evidence for a new interpretation of the trapped-hole centers associated with boron. *J. Chem. Phys.* 48: 5158.
66. Lee, S. and Bray, P.J. (1963). Electron spin resonance studies of irradiated glasses containing boron. *J. Chem. Phys.* 39: 2863.
67. Weeks, R.A. and Bray, P.J. (1968). Electron spin resonance spectra of gamma-ray-irradiated phosphate glasses and compounds: oxygen vacancies. *J. Chem. Phys.* 48: 5.
68. Masai, H., Okada, G., Kawaguchi, N. et al. (2019). Relationship between defect formation by X-ray irradiation and thermally stimulated luminescence of binary zinc phosphate glasses. *Opt. Mater. Expres* 9: 2037.
69. Pfeiffer, G., Paesler, M.A., and Agarwal, S.C. (1991). Reversible photodarkening of amorphous arsenic chalcogens. *J. Non Cryst. Solids J. Non Cryst. Solids* 130: 111.
70. Chiesa, M., Mattsson, K., Taccheo, S. et al. (2014). Defects induced in Yb^{3+}/Ce^{3+} co-doped aluminosilicate fiber glass preforms under UV and γ-ray irradiation. *J. Non Cryst. Solids* 403: 97.
71. Qiu, J. and Hirao, K. (2001). γ-ray induced reduction of Sm^{3+} to Sm^{2+} in sodium aluminoborate glasses. *J. Mater. Sci. Lett.* 20: 691.
72. Malchukova, E., Boizot, B., Petite, G. et al. (2007). Optical properties and valence state of Sm ions in aluminoborosilicate glass under β-irradiation. *J. Non Cryst. Solids* 353: 2397.
73. Yokota, R. and Imagawa, H. (1965). ESR studies of radiophotoluminescent centers in silver-activated phosphate glass. *J. Physical Soc. Japan* 20: 1537.
74. Jackson, D. and Hawkes, D.J. (1981). X-ray attenuation coefficients of elements and mixtures. *Phys. Rep.* 70: 169–233.
75. Knoll, G.F. (2012). *Radiation Detection and Measurement*, 4e. Wiley.
76. Robbins, D.J. (1980). On predicting the maximum efficiency of phosphor systems excited by ionizing radiation. *J. Electrochem. Soc.* 127: 2694–2702.
77. Lempicki, A., Wojtowicz, A.J., and Berman, E. (1993). Fundamental limits of scintillator performance. *Nucl. Instrum. Methods Phys. Res. A.* 333: 304–311.
78. Masai, H. and Yanagida, T. (2016). Photoluminescence of ns^2-type center-containing zinc borate glasses. *J. Non Cryst. Solids* 431: 83–87.
79. Leskelä, M., Koskentalo, T., and Blasse, G. (1985). Luminescence properties of Eu^{2+}, Sn^{2+}, and Pb^{2+} in SrB6O10 and Sr1−*x*Mn*x*B6O10. *J. Solid State Chem.* 59: 272–279.
80. Masai, H., Suzuki, Y., Yanagida, T. et al. (2015)). Luminescence of Sn^{2+} center in ZnO-B_2O_3 glasses melted in air and Ar condition. *Bull. Chem. Soc. Jpn.* 88: 1047–1053.

81. Masai, H., Koreeda, A., Fujii, Y. et al. (2016). Photoluminescence of Sn^{2+}centre as probe of transient state of supercooled liquid. *Opt. Mater. Express* 6: 1827–1836.
82. Reisfeld, R., Boehm, L., and Barnett, B. (1975). Luminescence and nonradiative relaxation of Pb^{2+}, Sn^{2+}, Sb^{3+}, and Bi^{3+} in oxide glasses. *J. Solid State Chem.* 15: 140–150.
83. Masai, H., Yanagida, T., Fujimoto, Y. et al. (2012). Scintillation property of rare earth-free SnO-doped oxide glass. *Appl. Phys. Lett* 101: 191906.
84. Masai, H., Takahashi, Y., Fujiwara, T. et al. (2010). High photoluminescent property of low-melting Sn-doped phosphate glass. *Appl. Phys. Express* 3: 082102.
85. Masai, H., Tanimoto, T., Fujiwara, T. et al. (2012). Correlation between emission property and concentration of Sn^{2+} center in the SnO-ZnO-P_2O_5 glass. *Opt. Express* 20: 27319.
86. Masai, H., Tanimoto, T., Okumura, S. et al. (2014). Correlation between preparation conditions and the photoluminescence properties of Sn^{2+} centers in ZnO-P_2O_5 glasses. *J. Mater. Chem. C* 2: 2137.
87. Masai, H., Matsumoto, S., Fujiwara, T. et al. (2012). Photoluminescent properties of Sb-doped phosphate glass. *J. Am. Ceram. Soc.* 95: 862.
88. Masai, H., Yamada, Y., Okumura, S. et al. (2013). Photoluminescence of Te^{4+} center in zinc borate glass. *Opt. Lett.* 38: 3780.
89. Alexander, M.N., Onorato, P.I.K., Struck, C.W. et al. (1987). Structure of alkali (alumino) silicate glasses II. Luminescence of thallium-doped sodium aluminosilicates, and implications for optical basicity theories. *J. Non Cryst. Solids* 91: 63–82.
90. Ghosh, A.K. (1968). The nature of trapping and luminescence centers due to thallium in glass. *J. Phys. Chem. Solid* 29: 1387–1393.
91. Ehrt, D. (2004). Photoluminescence in the UV–VIS region of polyvalent ions in glasses. *J. Non Cryst. Solids* 348: 22–29.
92. Fujimoto, Y. and Nakatsuka, M. (2001). Infrared luminescence from bismuth-doped silica glass. *Jpn. J. Appl. Phys.* 40: L279.
93. Sontakke, A., Tarafder, K., Biswas, A.D. et al. (2003). Sensitized red luminescence from Bi^{3+} co-doped Eu^{3+}: ZnO–B_2O_3 glasses. *Phys. B-Condensed Matter* 404: 3525.
94. Duffy, J.A. and Ingram, M.D. (1976). An interpretation of glass chemistry in terms of the optical basicity concept. *J. Non Cryst. Solids* 21: 373–410.
95. Duffy, J.A. (1993). A review of optical basicity and its applications to oxidic systems. *Geochim. Cosmochim. Acta* 57: 3961–3970.
96. Nakauchi, D., Okada, G., Fujimoto, Y. et al. (2017). Optical and radiation-induced luminescence properties of Ce-doped magnesium aluminoborate glasses. *Opt. Mater.* 72: 90–194. 96.
97. Takumi Kato, S., Hirano, H., Samizo, G. et al. (2019). Dosimetric, luminescence and scintillation properties of Ce-doped CaF_2-Al_2O_3-B_2O_3 glasses. *J. Non Cryst. Solids* 509: 60–64.
98. Kimura, H., Shinozaki, K., Okada, G. et al. (2019). Scintillation properties of Ce-doped SrF_2-Al_2O_3-B_2O_3 glasses. *J. Non-Cryst. Solids* 508: 46–50.
99. Samizo, H., Shinozaki, K., Kato, T. et al. (2019). X-ray induced luminescence properties of Ce-doped BaF_2-Al_2O_3-B_2O_3 glasses. *Opt. Mater.* 90: 64–69.
100. Fu, J., Parker, J.M., Brown, R.M. et al. (2003). Compositional dependence of scintillation yield of glasses with high Gd2O3 concentrations. *J. Non Cryst. Solids* 326–327: 335–338.
101. Chewpraditkul, W., He, D., Chen, X. et al. (2011). Luminescence and scintillation of Ce^{3+}-doped oxide glass with high Gd_2O_3 concentration. *Phys. Status Solidi A* 208: 2830–2832.98.
102. He, D., Yu, C., Cheng, J. et al. (2011). Effect of Tb^{3+}concentration and sensitization of Ce^{3+}on luminescence properties of terbium doped phosphate scintillating glass. *J. Alloys Compd.* 509: 1906–1909.
103. Tatsumi, H., Okada, G., Yanagida, T. et al. (2016). Scintillation and dosimeter properties of Ce-doped Li_3PO_4–$Al(PO_3)_3$ glasses. *J. Cerma. Soc. Jpn.* 124: 550–553.
104. Kagami, K., Fujimoto, Y., Koshimizu, M. et al. (2019). Photoluminescence and radiation response properties of $CsPO_3$-$Al(PO_3)_3$-$CeCl_3$-based glass scintillators. *Opt. Mater.* 87: 127–131.
105. Kagami, K., Fujimoto, Y., Koshimizu, M. et al. (2020). Photoluminescence and scintillation properties of $Al(PO_3)_3$-$CeCl_3$-CsCl-$CsPO_3$ glass scintillators. *J. Mater. Sci. Mater. Electron.* 31: 4488–4493.
106. Akatsuka, M., Shinozaki, K., Nakauchi, D. et al. (2019). Scintillator and dosimeter properties of Ce^{3+}doped CaF_2-AlF_3-$AlPO_4$ glasses. *Opt. Mater.* 94: 86–91.
107. Fu, J., Kobayashi, M., Sugimoto, S. et al. (2009). Scintillation from Eu^{2+}in nanocrystallized glass. *Ceram. Soc.* 92: 2119–2121.
108. Cao, J., Wang, X., Li, X. et al. (2001). A role of Gd3+ in scintillating processes in Tb-doped Na–Gd phosphate glasses. *J. Lumin.* 94–95: 321–324.

109. Pan, Z., James, K., Cui, Y. et al. (2008). Terbium-activated lithium–lanthanum–aluminosilicate oxyfluoride scintillating glass and glass-ceramic. *Nucl. Instrum. Methods A* 594: 215–219.
110. Cao, J., Chen, W., Chen, L. et al. (2016). Synthesis and characterization of $BaLuF_5$: Tb^{3+} oxyfluoride glass ceramics as nanocomposite scintillator for X-ray imaging. *Ceram. Int.* 42: 17834–17838.
111. Sun, X. and Huang, S. (2010). Tb^{3+}-activated SiO_2–Al_2O_3–CaO–CaF_2 oxyfluoride scintillating glass ceramics. *Nucl. Instrum. Methods A*. 621: 322–325.
112. Yanagida, T., Ueda, J., Masai, H. et al. (2016). Optical and scintillation properties of Ce-doped $34Li_2O$–5MgO–$10Al_2O_3$–$51SiO_2$ glass. *J. Non Cryst. Solids* 31: 140–144.
113. Zarrella, G., Zarrnoni, R., Dall'Igna, R. et al. (1994). A new cerium scintillating glass for X-ray detection. *Nucl. Instrum. Methods A* 345: 198–201.
114. Kaewnuam, E., Wantana, N., Kim, H.J. et al. (2017). Development of lithium yttrium borate glass doped with Dy^{3+} for laser medium, W-LEDs and scintillation materials applications. *J. Non Cryst. Solids* 464: 96–103.
115. Torimoto, A., Masai, H., Okada, G. et al. (2017). Emission properties of Ce-doped alkaline earth borate glasses for scintillator applications. *Opt. Mater.* 73: 517–522.
116. Torimoto, A., Masai, H., Okada, G. et al. (2017)). Emission properties of Ce^{3+} centers in barium borate glasses prepared from different precursor materials. *Opt. Mater.* 72: 52–57.
117. Sun, X., Jiang, D., Chen, S. et al. (2013). Eu^{3+}-Activated borogermanate scintillating glass with a high Gd_2O_3 content. *J. Am. Ceram. Soc.* 96: 1483–1489.
118. Sun, X., Yu, X., Wang, W. et al. (2013). Luminescent properties of Tb^{3+}-activated B_2O_3–GeO_2–Gd_2O_3 scintillating glasses. *J. Non Cryst. Solids* 379: 127–130.
119. Isokawa, Y., Hirano, S., Kawano, N. et al. (2018). Dosimetric and scintillation properties of Ce-doped Li_3PO_4-B_2O_3 glasses. *J. Non Cryst. Solids* 487: 1–6.
120. Masai, H. and Yanagida, T. (2015). Emission property of Ce^{3+}-doped Li_2O-B_2O_3-SiO_2 glasses. *Opt. Mater. Express* 5: 1851–1858. 114.
121. Fu, J., Kobayashi, M., and Parker, J.M. (2008). Terbium-activated heavy scintillating glasses. *J. Lumin.* 128: 99–104. 115.
122. Wantana, N., Kaewjaeng, S., Kothan, S. et al. (2017). Energy transfer from Gd^{3+} to Sm^{3+} and luminescence characteristics of CaO–Gd_2O_3–SiO_2–B_2O_3 scintillating glasses. *J. Lumin.* 181: 382–386. 116.
123. Kaewkhao, J., Wantana, N., Kaewjaeng, S. et al. (2016). Luminescence characteristics of Dy^{3+} doped Gd_2O_3-CaO-SiO_2-B_2O_3 scintillating glasses. *J. Rare Earths* 34: 583–589.117.
124. Fu, J., Kobayashi, M., Sugimoto, S. et al. (2008). Eu^{3+}-activated heavy scintillating glasses. *Mater. Res. Bull.* 43: 1502–1508. 118.
125. Wang, Q., Yang, B., Zhang, Y. et al. (2013). High light yield Ce^{3+}-doped dense scintillating glasses. *J. Alloys Compd.* 581: 801–804. 19.
126. Jiang, C., Deng, P., Zhang, J. et al. (2004). Radioluminescence of Ce^{3+}-doped B_2O_3–SiO_2–Gd_2O_3–BaO glass. *Phys. Lett. A.* 323: 323–328. 120.
127. Masai, H., Hino, Y., Yanagida, T. et al. (2013). Excitation light source dependence of emission in Sn^{2+}-Ce^{3+} codoped ZnO-P_2O_5 glasses. *J. Appl. Phys.* 114 (083502): 121.
128. Nikl, M., Nitsch, K., Mihokova, E. et al. (2000). Efficient radioluminescence of the Ce^{3+}-doped Na–Gd phosphate glasses. *Appl. Phys. Lett.* 77: 2159, 122–2161.
129. Baccaro, S., Dall'Igna, R., Fabeni, P. et al. (2000). Ce^{3+} or Tb^{3+}-doped phosphate and silicate scintillating glasses. *J. Lumin.* 87–89: 673–675.
130. Maresa, J.A., Nikla, M., Nitscha, K. et al. (2001). A role of Gd^{3+} in scintillating processes in Tb-doped Na–Gd phosphate glasses. *J. Lumin* 94–95: 321–324.
131. Qian, G., Baccaro, S., Guerra, A. et al. (2007). Gamma irradiation effects on ZnO-based scintillating glasses containing CeO_2 and/or TiO_2. *Nucl. Instrum. Methods A* 262: 276–280.
132. Chiodini, N., Fasoli, M., Martini, M. et al. (2002). High-efficiency SiO_2:Ce^{3+} glass scintillators. *Appl. Phys. Lett.* 81: 4374–4376.
133. Kobayashi, M., Yoshimura, Y., Chiba, M. et al. (1981). Performance of a scintillator glass shower detector for 20–120 MeV electrons. *Nucl. Instrum. Methods* 184 (399): 127.
134. Shiratori, D., Kimura, H., Nakauchi, D. et al. (2004)). Dosimetric properties of Zn-doped SiO_2 glasses synthesized by the spark plasma sintering method. *Radiat. Meas.* 134 (106297): 131.
135. Hashimoto, K., Shiratori, D., Nakauchi, D. et al. (2020). Evaluation of radiation-induced luminescence properties in Tl-doped SiO_2 glasses prepared by the spark plasma sintering method. *J. Ceram. Soc. Jpn* **128**: 267–272.
136. Isokawa, Y., Kimura, H., Kato, T. et al. (2019). Radiation induced luminescence properties of Eu-doped SiO_2 glass synthesized by spark plasma sintering. *Opt. Mater.* 90: 187–193.

137. Johnson, J.A., Schweizer, S., Henke, B. et al. (2006). Eu-activated fluorochlorozirconate glass-ceramic scintillators. *J. Appl. Phys.* 100 (034701): 28.
138. Okada, G., Egar, A., Kasap, S. et al. (2016). Radioluminescence properties of Sm-doped fluorochlorozirconate glasses and glass-ceramics. *Jpn. J. Appl. Phys.* 55: 02BC07.
139. Auffray, E., Bouttet, D., Dafinei, I. et al. (1996). Cerium doped heavy metal fluoride glasses, a possible alternative for electromagnetic calorimetry. *Nucl. Instrum. Methods A* 380: 524–536.
140. Kobayashi, M. (2014). Sciti-glass. In: *Radiation Measurement with Scintillators, (V2-solution)*, 310. *(in Japanese)*.
141. Kouzes, R.T., Lintereur, A.T., and Siciliano, E.R. (2010). Cerium doped heavy metal fluoride glasses, a possible alternative for electromagnetic calorimetry. *Nucl. Instrum. Methods A* 623: 1035.
142. Lide, D.R. (ed.) (2002). *CRC Handbook of Chemistry and Physics*, 83e. Boca Raton FL: CRC Press.
143. van Eijk, C.W.E., Bessiére, A., and Dorenbos, P. (2004). Inorganic thermal-neutron scintillators. *Nucl. Instrum. Methods A* 529: 260–267.
144. Bollinger, L.M., Thomas, G.E., and Ginther, R.J. (1962). Neutron detection with glass scintillators. *Nucl. Instum. Methods* 17: 97.
145. Fairley, E.J. and Spowart, A.R. (1978). Neutron scintillating glasses. Part III: Pulse decay time measurements at room temperature. *Nucl. Instum. Methods* 150: 159.
146. Wraight, L.A., Harris, D.H.C., and Egelstaff, P.A. (1965). Improvements in thermal neutron scintillation detectors for time-of-flight studies. *Nucl. Instrum. Methods* 33: 181.
147. Zetterstrom, H., Schwarz, S., and Strömberg, L.G. (1966). Multiple scattering of fast neutrons in ^{6}Li-glass scintillators. *Nucl. Instrum. Methods* 42: 277.
148. Cameron, J., Harrison, L.M., and Parker, J.B. (1967). Monte Carlo and *Sn* calculations of the efficiency of ^{6}Li loaded glass detectors. *Nucl. Instrum. Methods* 56: 45.
149. Spowart, A.R. (1969). Measurement of the absolute scintillation efficiency of granular and glass neutron scintillators. *Nucl. Instrum. Methods* 75: 35.
150. Spowart, A.R. (1970). Measurement of the gamma sensitivity of granular and glass neutron scintillators and films. *Nucl. Instrum. Methods* 82: 1.
151. Spowart, A.R. (1977). Neutron scintillating glasses. Part II: The effects of temperature on pulse height and conductivity. *Nucl. Instrum. Methods* 140: 19–28.
152. http://www.apace-science.com/ast/g_scint.htm.
153. Zanella, G., Zannoni, R., Dall'Igna, R. et al. (1995). Development of a terbium-lithium glass for slow neutron detection. *Nucl. Instrum. Methods A.* 359: 547–555.
154. Masai, H., Okada, G., Kawaguchi, N. et al. (2019). Optical and luminescent properties of Pr-doped Li_2O-MgO-Al_2O_3-SiO_2 glasses. *Opt. Mater* 88: 1–6.
155. Struebing, C., Chong, J., Lee, G. et al. (2016). A neutron scintillator based on transparent nanocrystalline CaF_2:Eu glass ceramic. *Appl. Phys. Lett.* 108: 153106.
156. Arikawa, Y., Yamanoi, K., Nakazato, T. et al. (2009). Pr^{3+}-doped fluoro-oxide lithium glass as scintillator for nuclear fusion diagnostics. *Rev. Sci. Instrum.* 80: 113504.
157. Ishii, M., Kuwano, Y., Asaba, S. et al. (2004). Luminescence of doped lithium tetraborate single crystals and glass. *Radiat. Meas* 38: 571–574.
158. Matsuzawa, T., Aoki, T., and Takeuchi, N. (1996). A new long phosphorescent phosphor with high brightness, $SrAl_2O_4$:Eu^{2+}, Dy^{3}. *Electrochem. Soc.* 143: 2670–2673.
159. Massera, J., Gaussiran, M., Gluchowski, P. et al. (2015). Processing and characterization of phosphate glasses containing $CaAl_2O_4$:Eu^{2+}, Nd^{3+} and $SrAl_2O_4$:Eu^{2+},Dy^{3+} microparticles. *J. Eur. Ceram. Soc.* 35: 3863.
160. Xiao, L., Xiao, Q., Liu, Y. et al. (2010). A transparent surface-crystallized Eu^{2+}, Dy^{3+} co-doped strontium aluminate long-lasting phosphorescent glass-ceramic. *Alloys Compds.* 495: 72–75.
161. Chen, D. (2014). Near-infrared long-lasting phosphorescence in transparent glass ceramics embedding Cr^{3+}-doped LiGa5O8 nanocrystals. *J. Eur. Ceram. Soc.* 34: 4069.
162. Schulman, J.H., Ginter, R.J., and Klick, C.C. (1951). Dosimetry of X-rays and Gamma-rays by radiophotoluminescence. *J. Appl. Phys* 22: 1479.
163. Schulman, J.H. and Etzel, H.W. (1953). Small-volume dosimeter for X-rays and Gamma-rays. *Science* 118: 184.
164. Yokota, R. and Imagawa, H. (1966). Radiophotoluminescent centers in silver-activated phosphate glass. *J. Physical Soc. Japan* 23: 1038.
165. Miyamoto, Y., Takei, Y., Nanto, H. et al. (2011). Radiophotoluminescence from silver-doped phosphate glass. *Radiat. Meas.* 46: 1480.

166. Vahedi, S., Okada, G., Morrell, B. et al. (2012). X-ray induced Sm^{3+} to Sm^{2+} conversion in fluorophosphate and fluoroaluminate glasses for the monitoring of high-doses in microbeam radiation therapy. *J. Appl. Phys.* 112: 073108.
167. Okada, G., Morrell, B., Koughia, C. et al. (2011). Spatially resolved measurement of high doses in microbeam radiation therapy using samarium doped fluorophosphate glasses. *Appl. Phys. Lett.* 99: 121105.
168. Layne, C.B., Lowdermilk, W.H., and Weber, M.J. (1977). Multiphonon relaxation of rare-earth ions in oxide glasses. *Phys. Rev. B.* 16: 10.
169. Snopatin, G.E., Shiryaev, V.S., Plotnichenko, V.G. et al. (2009). High-purity chalcogenide glasses for fiber optics. *Inorg. Mater.* 45: 1439.
170. Pantano, C.G., Singh, A.K., and Zhang, H. (1999). Silicon oxycarbide glasses. *J. Sol-Gel Sci. Technol.* 14: 7.
171. Lucas, J., Smektala, F., and Adam, J.-L. (2002). Fluorine in optics. *J. Fluor. Chem.* 114: 113.
172. Adam, J.-L. (2002). Lanthanides in non-oxide glasses. *Chem. Rev.* 102: 2461.
173. Zverev, G., Kolodnyĭ, G., and Onishchenko, A. (1971). Nonradiative transitions between levels of trivalent rare-earth ions in yttrium-aluminum garnet crystals. *Sov. J. Exp. Theor. Phys.* 33: 497.
174. Bendow, B., Drexhage, M.G., Banerjee, P.K. et al. (1981). Vibrational spectra of fluorohafnate glass. *Solid State Commun.* 37: 485.
175. Yeh, D.C., Sibley, W.A., Suscavage, M. et al. (1987). Multiphonon relaxation and infrared-to-visible conversion of Er^{3+} and Yb^{3+} ions in barium-horium fluoride glass. *J. Appl. Phys.* 62: 266.
176. Hewak, D.W., Medeiros Neto, J.A., Samson, B. et al. (1994). Quantum-efficiency of praseodymium doped Ga:La:S glass for 1.3 mm optical fibre amplifiers. *IEEE Photonics Technol. Lett.* 6: 609.
177. Lage, M.M., Righi, A., Matinaga, F.M. et al. (2004). Raman-spectroscopic study of lanthanide trifluorides with the ⊠-YF_3 structure, J. Phys. Condens. Matter. 16: 3207.
178. Shojiya, M., Takahashi, M., Kanno, R. et al. (1997). Optical transitions of Er^{3+} ions in $ZnCl_2$-based glass. *J. Appl. Phys.* 82: 6259.
179. Kadono, K., Kageyama, H., Kamijo, N. et al. (1992). Structure of zinc halide-based glasses. *J. Non Cryst. Solids* 140: 98.
180. Hu, H., Ma, F., and Mackenzie, J.D. (1983)). New halide glasses – bromide glasses based on $ZnBr_2$ I: Vitreous zinc bromide. *J. Non Cryst. Solids* 55: 169.
181. Poulain, M., Poulain, M., Lucas, J. et al. (1975). Fluorine-containing glass with ZrF_4-optical properties of a glass doped with Nd^{3+}. *Mater. Res. Bull.* 10: 243.
182. Aggarwal, I.D. and Lu, G. (eds.) (1991). *Fluoride Glass Fiber Optics*. Academic Press.
183. Shibata, S., Horiguchi, M., Jinguji, K. et al. (1981). Prediction of loss minima in infra-red optical fibres. *Electron. Lett.* 17: 775.
184. Ohishi, Y., Snitzer, E., Sigel, G.H. et al. (1991)). Pr^{3+}-doped fluoride fiber amplifier operating at 1.31 μm. *Opt. Lett.* 16: 1747.
185. Duffy, J.A. (1989). A common optical basicity scale for oxide and fluoride glasses. *J. Non Cryst. Solids J. Non Cryst. Solids* 109: 35.
186. Kotsar, M.L., Bateev, V.B., Baskov, P.B. et al. (2001). Preparation of high-purity ZrF_4 and HfF_4 for optical fibres and radiation-resistant glasses. *Inorg. Mater.* 37: 1085.
187. Auffray, E., Bouttet, D., Dafinei, I. et al. (1996). Nuclear instrument methods physics research section A. Accelerated spectrometers. *Detect. Assoc. Equip.* 380: 524.
188. Shaukat, S.F.F., McKinlay, K.J.J., Flower, P.S.S. et al. (1999). Optical and physical characteristics of HBLAN fluoride glasses containing cerium. *J. Non Cryst. Solids* 244: 197.
189. Dmitruk, L., Vinogradova, N., Kozlov, V. et al. (1997). Scintillating HfF4-based glasses doped cerium chloride and cerium oxide compounds. *J. Non Cryst. Solids* 213–214: 311.
190. Chen, G., Johnson, J., Schweizer, S. et al. (2015) in Med. Imaging 2006 Phys. Med. Imaging, (Flynn, M.J. and Hsieh, J. eds), (2006), p. 61422X.
191. Okada, G., Edgar, A., Kasap, S. et al. (2016). Radioluminescence properties of Sm-doped fluorochlorozirconate glasses and glass-ceramics. *Jpn. J. Appl. Phys.* 55: 02BC07.
192. Edgar, A., Varoy, C.R., Koughia, C. et al. (2009). Optical properties of divalent samarium-doped fluorochlorozirconate glasses and glass ceramics. *Opt. Mater. (Amst)* 31: 1459.
193. Ehrt, D. (2015). Phosphate and fluoride phosphate optical glasses – properties, structure and applications. *Phys. Chem. Glas. Eur. J. Glas. Sci. Technol. Part B.* 56: 217.
194. Jiang, J., Zhang, G., and Poulain, M. (1998). Cerium-containing glasses for fast scintillators. *J. Alloys Compd.* 275–277: 733.

195. Arikawa, Y., Yamanoi, K., Nagai, T. et al. (2010). Light output enhanced fast response and low afterglow ^{6}Li glass scintillator as potential down-scattered neutron diagnostics for inertial confinement fusion. *Rev. Sci. Instrum.* 81: 1.
196. Shinozaki, K., Honma, T., and Komatsu, T. (2014). High quantum yield and low concentration quenching of Eu^{3+} emission in oxyfluoride glass with high BaF_2 and Al_2O_3 contents. *Opt. Mater. (Amst)* 36: 1384.
197. Shinozaki, K., Sukenaga, S., Shibata, H. et al. (2018). Effect of Mg^{2+} and fluorine on the network and highly efficient photoluminescence of Eu^{3+} ion in MgF_2–BaO–B_2O_3 glasses. *J. Am. Ceram. Soc.* 102: 16165.
198. Okada, G., Vahedi, S., Morrell, B. et al. (2013). Examination of the dynamic range of Sm-doped glasses for high-dose and high-resolution dosimetric applications in microbeam radiation therapy at the Canadian synchrotron. *Opt. Mater. (Amst)* 35: 1976.
199. Wang, Y. and Ohwaki, J. (1993). New transparent vitroceramics codoped with Er^{3+} and Yb^{3+} for efficient frequency upconversion. *Appl. Phys. Lett.* 63: 3268.
200. Dejneka, M.J. (1998). The luminescence and structure of novel transparent oxyfluoride glass-ceramics. *J. Non Cryst. Solids* 239: 149.
201. Fedorov, P.P., Luginina, A.A., and Popov, A.I. (2015). Transparent oxyfluoride glass ceramics. *J. Fluor. Chem.* 172: 22.
202. Shinozaki, K., Noji, A., Honma, T. et al. (2013)). Morphology and photoluminescence properties of Er^{3+}-doped CaF_2 nanocrystals patterned by laser irradiation in oxyfluoride glasses. *J. Fluor. Chem.* 145: 81.
203. Shinozaki, K., Konaka, R., and Akai, T. (2019). Synthesis of new transparent borate-based BaF_2 nanocrystallized glass by formation of nucleation sites induced by rare earth ions. *J. Eur. Ceram. Soc.* 39: 1735.
204. Raghuwanshi, V.S., Hoell, A., Bocker, C. et al. (2012). Experimental evidence of a diffusion barrier around BaF2 nanocrystals in a silicate glass system by ASAXS. *Crst.. Eng. Comm.* 14: 5215.
205. Beckert, M., Gallego, S., Ding, Y. et al. (2016). Medical imaging scintillators from glass-ceramics using mixed rare-earth halides. *Opt. Mater.* 60: 513.
206. Kang, Z.T.T., Rosson, R., Barta, B. et al. (2013). $GdBr_3$:Ce in glass matrix as nuclear spectroscopy detector. *Radiat. Meas.* 48: 7.
207. Masai, H., Hamaguchi, K., Suzuki, Y. et al. (2012). Precipitation of Pt nano-crystallites from BaO-TiO_2-GeO_2 re-melted glass. *J. Appl. Phys.* 111: 123513.
208. Ansell, S., Krishnan, S., and Weber, J.K.R. (1997). Structure of liquid aluminum oxide. *Phys. Rev. Lett.* 78: 464–466.
209. Weber, J.K.R., Hampton, D.S., Merkley, D.R. et al. (1994). Aero-acoustic levitation: A method for containerless liquid-phase processing at high temperatures. *Rev. Sci. Instrum.* 65: 456–465.
210. Hanawa, F., Sudo, S., Kawachi, M. et al. (1980). Fabrication of completely OH-free vad fibre. *Electron. Lett.* 16: 699.
211. Ballato, J. and Snitzer, E. (1995). Fabrication of fibers with high rare-earth concentrations for Faraday isolator applications. *Appl. Optics* 34: 6848.
212. Brinker, C.J., Hurd, A.J., Frye, G.C. et al. (1991). Sol-Gel thin film formation. *Cerma. Soc. Jpn.* 99: 862.
213. Kajihara, K., Hirano, M., and Hosono, H. (2009). Sol-gel synthesis of monolithic silica gels and glasses from phase-separating tetraethoxysilane–water binary system. *Chem. Commun.*: 2580–2582.
214. Mizoguchi, T., Findlay, S.D., Masuno, A. et al. (2013). Atomic-scale identification of individual lanthanide dopants in optical glass fiber. *ACS Nano* 7: 5058.

12

Detectors Using Radiation Induced Luminescence

Kenichi Watanabe
Department of Applied Quantum Physics and Nuclear Engineering, Kyushu University, Fukuoka, Japan

12.1 Introduction

Radiation detectors using radiation induced luminescence materials are widely used in various fields, such as medical, security, and nuclear science and technology. Radiation induced luminescence materials are categorized into prompt and delayed luminescence materials. The former are scintillators. The latter are thermally stimulated luminescence (TL), optically stimulated luminescence (OSL), and radio-photoluminescence (RPL) materials. In this chapter, various detectors are described based on this categorization. Figure 12.1 shows the flow of measurement using prompt and delayed luminescence materials. When using prompt luminescence materials, signals are simultaneously read out during irradiation. In cases of delayed luminescence materials, signals are separately read out by thermal or optical stimulation after irradiation.

In this section, a general description will be given for readers to easily understand to how manufacture radiation detectors using radiation induced luminescence materials. As shown in Figure 12.2, the radiation detector using radiation induced luminescence materials basically consists of a luminescence material, photon transport and/or collection assembly, and photodetector and signal processing unit. For reading signals from the delayed luminescence material, the detector needs a stimulation mechanism to produce the delayed luminescence phenomenon. For the TL materials, a resistive heater is used. On the other hand, a stimulation light source, such as a laser or light emitting diode, is generally used.

The radiation detector is the equipment used to converting radiation quanta to an electrical signal. In gaseous detectors or semiconductor detectors, electrical charge carriers are directly created from high-energy radiation quanta. When using radiation induced luminescence

Phosphors for Radiation Detectors, First Edition. Edited by Takayuki Yanagida and Masanori Koshimizu.

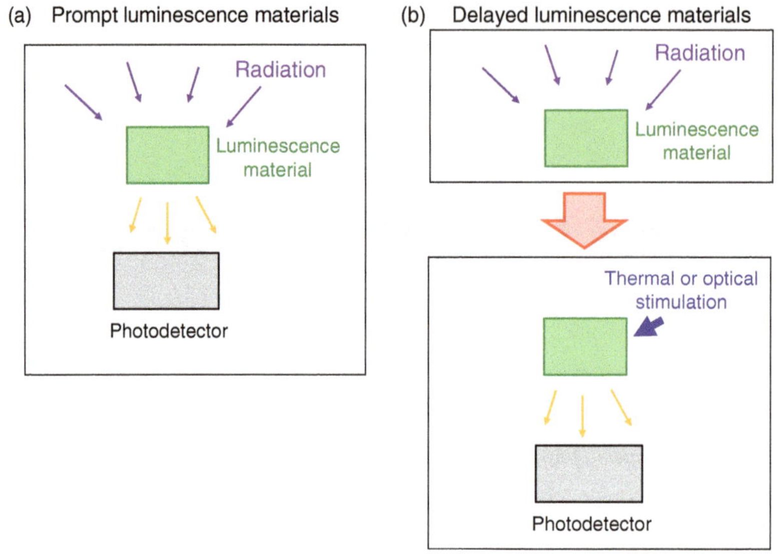

Figure 12.1 *Flow of measurements using (a) prompt and (b) delayed luminescence materials.*

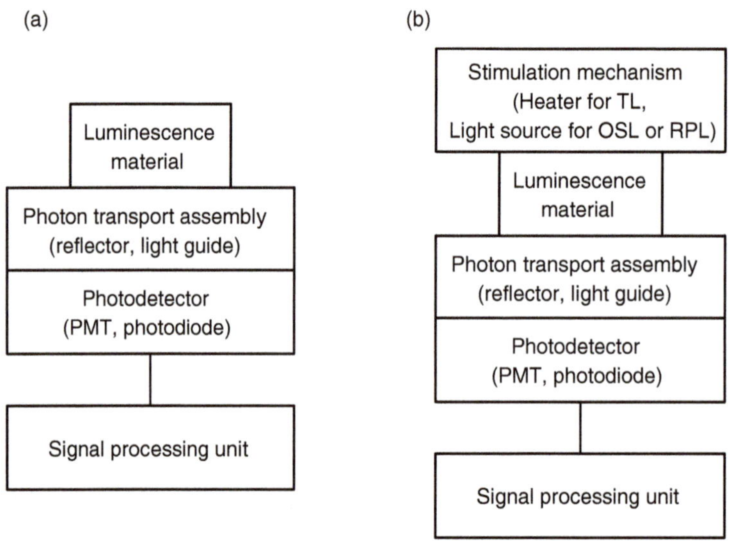

Figure 12.2 *Basic configuration of detectors using radiation induced luminescence materials. (a) Scintillator-based detectors, (b) Delayed luminescence material-based detectors.*

material for radiation detectors, a high-energy radiation quantum at once creates relatively low-energy photons (usually visible photons, sometimes ultraviolet (UV) or infrared (IR) photons). Then, these photons are converted into an electrical signal. The number of information carriers directly created from each radiation quantum is not so large. Therefore, in many cases, radiation detectors have to fight against noise problems. However, fortunately, this kind of problem with the luminescence-based radiation detectors might be reduced. This is because the photomultiplier tube (PMT) can help to overcome such noise problems. The PMT is a highly sensitive photodetector. This photodetector consists of a photoelectric converter from visible photons to electrons and an electron multiplier. The PMT can generate relatively large electrical signals compared with those emitted from gaseous and semiconductor detectors. The noise problems for PMT-based detectors might be reduced owing to relatively large signals. This is one of the advantages of luminesence-based detectors compared with other types of radiation detectors, such as gaseous and semiconductor detectors.

In Section 12.2, we will describe the general issues in manufacturing the radiation detector using luminescence materials. This will be followed by introducing individual detector types.

12.2 General Issues to Manufacturing the Detector

One of the important points when manufacturing fabrication detectors using radiation induced luminescence materials is the design of the signal readout system. As described above, the radiation induced luminescence materials are categorized into prompt and delayed luminescence materials. The prompt luminescence material is generally known as a scintillator. To manufacture a scintillator-based radiation detector, we have two choices: pulse counting and current mode detectors. Figure 12.3 shows a conceptual drawing of these two operation modes.

In the pulse counting mode detectors, signals generated by each ionizing radiation quantum are independently recorded, or counted. In this measurement mode, since only events exceeding the discrimination level are recorded, noise lower than the discrimination level can be eliminated. Therefore, the pulse counting mode measurement can generally have a higher signal to noise ratio compared with the current mode. In addition to this noise elimination ability, the pulse counting mode has another advantage. Individual signals not

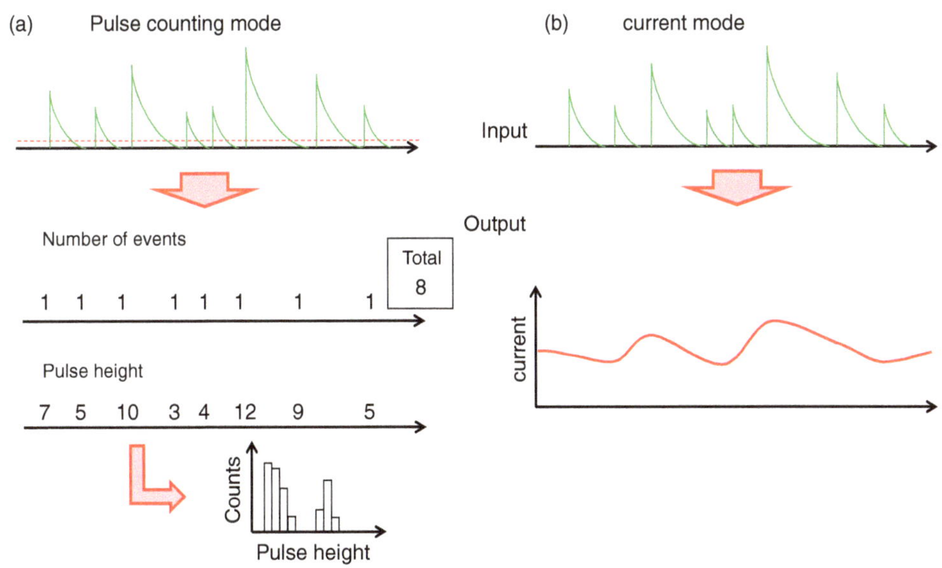

Figure 12.3 *(a) Pulse counting mode detector and (b) current mode detector.*

only contain information on the number of events but also additional information on each event, such as deposition energy, occurrence, timing, and type of radiation, etc. A typical example is the energy spectrum measurement, in which a histogram on deposition energy in the scintillator can be created.

In the current mode, information of each event is no longer recorded. In this mode, the deposition energy or emission light yield averaged over the time constant in the signal processing unit is recorded. The current mode operation is generally used in relatively intense radiation fields, such as an X-ray irradiation field. This operating mode is similar to the integrated mode operation, in which the time constant is almost infinite and the accumulated information can be reset at arbitrary timing. An electrical circuit for signal processing in this mode is simpler than that in the pulse counting mode. Therefore, this operation mode is useful in the detector with many pixels, such as the 1-D line sensor and 2-D flat panel detector (FPD).

In delayed luminescence materials, the relatively stable trapped electrons and holes are latently accumulated by irradiating with ionizing radiation. Since the trapped electrons and holes are stable, the signals from detectors using delayed luminescence materials can be read out after irradiation. This is different from scintillator detectors, in which signals are generated only during irradiation. In the TL detector, the trapped electrons and holes are thermally re-excited by heating the luminescence material. Therefore, the TL detector has a heater and photon detection system. The heater temperature is controlled to increase in line with the arbitrary temperature rise. The photon detection system is operated to synchronize with the heater. The TL glow curve, which is a time trend of TL luminescence yield, can be recorded by some TL detectors. From the TL glow curve, the TL temperature dependence, from which we can extract information on the depth of the trapping site, can be obtained. In the OSL, the trapped electrons and holes are optically re-excited by irradiating with stimulation light. The re-excitation efficiency depends on the wavelength of the light and the depth of the trapping site. As a stimulation light source, a laser diode (LD) or a light emitting diode (LED) are applied. The wavelength of the light source should be selected depending on characteristics of the luminescence material. Because the stimulation light might disturb the measurement of the luminescence intensity, the photon detection system has to eliminate stimulation light entering the photodetector, for example, by using an optical filter. The wavelength of the OSL stimulation light is generally longer than that of the luminescence one. In the TL and OSL detectors, the accumulated information will reduce and finally disappear after readout processes. The RPL detector has basically similar configuration to the OSL detector. In the RPL material, emission centers for photoluminescence are created by irradiating with ionizing radiation. In the case of the Ag-doped phosphate glass, doped Ag^{+} ions change to Ag^{0} or Ag^{2+} ions by irradiating with radiation. The Ag^{0} or Ag^{2+} ions are stable after readout or irradiating with stimulation light. Therefore, this type of detector can repeatedly be read out. This feature is completely different from OSL-based detectors.

Another important point when manufacturing luminescence-based radiation detectors is consideration on matching wavelength characteristics in a luminescence material, photodetectors, and other optical components. In addition, a design of a photon collection assembly, such as a reflector surrounding the luminescence material, light guide, lens, etc., is important because the number of photons emitted from the luminescence materials is very few. When considering transport of photons from luminescence materials, we have to consider the famous Snell's law. When a photon enters from region A to B, Snell's law can be written as

$$n_A \sin\theta_A = n_B \sin\theta_B, \tag{12.1}$$

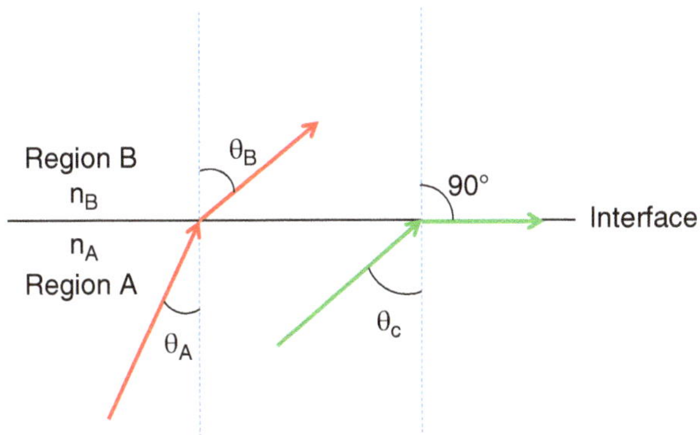

Figure 12.4 *Snell's law.*

where n_A and n_B are refractive indices of the region *A* and *B*, respectively. θ_A and θ_B are the incident and refractive angles, which are photon direction angles from the normal vector of the interface between region A and B, as shown in Figure 12.4. When we are considering photon transport, one of the important phenomena is the total reflection. Since the maximum angle of refraction is 90°, the total reflection occurs when the incident angle is larger than the critical angle θ_c given by

$$\sin\theta_c = \frac{n_B}{n_A}. \tag{12.2}$$

In addition to Snell's law, the Fresnel reflection is also important. The Fresnel reflection occurs when photons pass through an interface between regions having different refractive indices. The Fresnel reflection increases when the difference in the refractive index is large. If the refractive index of the luminescence material is large and the surrounding region is air (refractive index $n_{air} \sim 1.0$), a relatively large fraction of luminescence photons cannot escape from the luminescence materials due to the total reflection and Fresnel reflection. In this case, the air gap between the luminescence material and photodetector should be eliminated by filling with an optical buffer, which ideally has an intermediate refractive index between the luminescence material and photodetector window. Optical grease is often used as the optical buffer. As in other cases, when using a light guide, such as an optical fiber, you have to consider Snell's law to transport luminescence photons to a photodetector.

When using a transparent luminescence material with a well-featured shape, such as a cylinder or a rectangle, you should consider the optimal reflector surrounding the luminescence material. Although a metallic mirror reflector might be considered as the best option, in many cases, it is not the optimal solution. In this case, photons can undergo only a few types of incident angles to the interface. If photons cannot escape efficiently with these incident angles, a mirror reflector cannot work as a valid photon collection component. For this kind of situation, a diffusive reflector will be more effective. The diffusive reflector can change angles randomly and increase the possibility for photons to escape from the luminescence material. White powder, which is basically composed of a transparent material and can cause multiple refractions and reflections at the interface between powder materials and air gaps, is frequently used as the diffusive reflector. White fluororesin sheets are also used as the diffusive reflector. A numerical simulation code for ray tracing of the luminescence photons might be useful for design of light collection assembly.

12.3 Scintillation Detectors for Gamma-Rays and X-Rays

The scintillation detectors can be applied for gamma- and X-ray detection. These types of detectors are widely used in various fields, such as nuclear engineering, security, and medical applications. In this section, the individual detectors will be introduced in detail.

12.3.1 Gamma-Ray Spectrometer

The gamma-ray spectrometer is used to measure gamma-ray energy distribution. Since gamma rays emitted from nuclides have characteristic energies, which can show their origins or parent nuclides, we can gain some information of these parent nuclides by measuring the gamma-ray spectrum. The ideal output from the gamma-ray spectrometer is, of course, the gamma-ray energy distribution, in which the horizontal and vertical axes represent the gamma-ray energy and the amount of gamma-rays having the corresponding energy. However, unfortunately, all the gamma-ray spectrometers cannot directly measure real gamma-ray energy distribution. The radiation detectors can only generate signals corresponding to the amount of electron–ion or electron–hole pairs created by interactions between a radiation quantum and detector material. The amount of electron–ion or electron–hole pairs does not directly correspond to the gamma-ray energy itself, but rather the deposition energy in the detector. This means that gamma rays might not deposit their entire energy, especially in Compton scattering, in which a part of the gamma-ray's energy will be emitted from the detector. Although the ideal gamma-ray spectrometer should show an output completely corresponding to an incident gamma-ray energy, that is a delta function response, the real signal output from the gamma-ray spectrometer is distorted from the original gamma-ray energy distribution.

The signal output from scintillation detectors is the number of scintillation photons. Figure 12.5 shows the general configuration of the gamma-ray spectrometer. The scintillation photons are converted into electrical signals by a photodetector, such as PMT. The electrical signals are processed by signal processing circuits. The number of scintillation photons emitted in each gamma-ray event basically corresponds to the number of electron–hole pairs created or deposition energy in the scintillator. Note that the electron–hole pairs

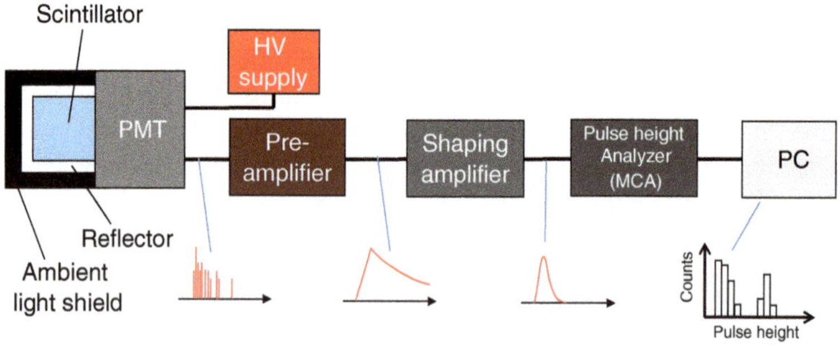

Figure 12.5 *General configuration of the scintillator-based gamma-ray spectrometer.*

are not directly created by the gamma-ray itself but by fast electrons created by the gamma-ray interactions, such as photoelectric absorption, Compton scattering, and pair production. If the decay constant of the scintillator is sufficiently short, a signal pulse having the amplitude corresponding to the number of scintillation photons emitted in each gamma-ray event will be created. The signal amplitude or signal pulse height is extracted and recorded by a signal pulse height analyzer, such as a multichannel analyzer (MCA). As a result, the scintillator-based gamma-ray spectrometer creates the signal pulse height spectrum, shown in Figure 12.6, instead of the real gamma-ray energy spectrum. Therefore, the scintillator-based gamma-ray spectrometer also has a distorted energy response from an ideal delta function response. Figure 12.6 shows a typical pulse height spectrum or energy response from a scintillator-based gamma-ray spectrometer.

As shown in Figure 12.6, the gamma-ray spectrometer shows a complex response [1]. When a gamma ray is absorbed by the photoelectric effect, almost all the gamma-ray energy is transferred into an orbital electron bound to an atom and then the high energy electron escapes from the atom and creates additional electron–hole pairs in the scintillator. These electron–hole pairs generate scintillation photons. After the photoelectric absorption, the residual atom is excited and emits an Auger electron or characteristic X-ray. Generally, since the Auger electrons or characteristic X-rays are re-absorbed in the scintillator material, the entire gamma-ray energy is fully absorbed. As a result, the gamma-ray events induced by the photoelectric absorption create the full-energy peak. If the atomic number of an atom absorbing a gamma-ray is relatively high, the characteristic X-ray might have high energy. Since the high-energy X-rays have relatively high penetrating powers, they can escape from the scintillator. As a result, these types of events show the characteristic X-ray escape peak, which is at a position lower than the energy of a characteristic X-ray. The characteristic

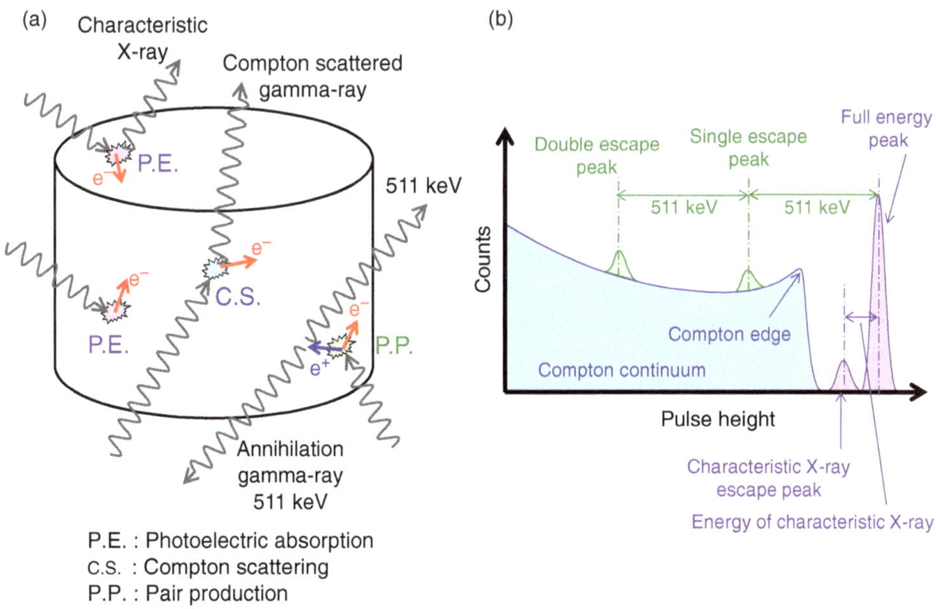

Figure 12.6 *Typical response from the scintillator-based gamma-ray spectrometer. (a) Conceptual drawing of each interaction. (b) Typical response in the pulse height spectrum.*

X-ray escape peak cannot be seen in a scintillator consisting of only light elements, because, in many cases, the energy of the characteristic X-ray is lower than the energy resolution of the scintillation detector.

When a gamma ray undergoes Compton scattering, the scattered gamma rays might escape from the scintillator. In this case, the energy of an incident gamma ray is partially absorbed into the scintillator and a part of the energy is taken out by the scattered gamma ray. As a result, these kinds of events are recorded at the lower energy side compared with the full energy peak. The energy of the scattered gamma rays depends on the scattering angle θ as written by [1]

$$h\nu' = \frac{h\nu}{1 + \frac{h\nu}{mc^2}(1 - \cos\theta)}, \tag{12.3}$$

where $h\nu'$, $h\nu$, and mc^2 are the scattered and incident gamma-ray energy and the rest mass energy of an electron. Since the recoil electron, which creates electron–hole pairs in the scintillator, shows the difference in energy between the incident and scattered gamma rays, the deposition energy in these kinds of events changes depending on the scattering angle. As a result, the Compton scattering events create the Compton continuum. From Equation (12.3), it can be seen that the minimum energy of the Compton scattered gamma-ray $h\nu'_{min}$ is obtained at $\theta = \pi$ (backscattering) and written by

$$h\nu'_{min} = \frac{h\nu}{1 + 2\frac{h\nu}{mc^2}}. \tag{12.4}$$

Conversely, the recoil electron in the Compton scattering has the maximum value ($=h\nu - h\nu'_{min}$). At this maximum energy, the Compton continuum has a higher energy side edge. This is known as the Compton edge.

When a gamma ray has energy of more than 1.022 MeV, it might cause the pair production phenomenon. After this interaction, an electron–positon pair is created. The energy of the incident gamma ray is shared by an electron and a positron. Since they have sufficiently high energy, they create electron–hole pairs in the scintillator and generate scintillation photons. When the positron loses its energy, it undergoes the annihilation phenomenon with a surrounding electron and generates two annihilation gamma rays with energy of 511 keV, which is the rest mass energy of an electron. If one of the annihilation gamma rays escapes from the scintillator, the absorbed energy in the scintillator is lowered by 511 keV, which is the energy of the annihilated gamma ray, compared with the incident gamma-ray energy. If the both of them escape, the energy deposition is lower by 1.022 MeV than the full energy peak.

As with the gamma-ray spectrometer, it is ideal to show only the full energy peak. From this viewpoint, the scintillator for the gamma-ray spectrometer is desired to include heavy elements, which have a high fraction of the photoelectric absorption compared with other interactions, such as Compton scattering and pair production. In addition, when undergoing multiple interactions, the gamma rays might deposit their entire energy into the scintillation. In these cases, these events are also recorded at the full energy peak. Therefore, the large scintillator also has an advantage from the viewpoint of the ideal response, especially for high energy gamma-ray measurements.

One of the most important performances of the gamma-ray spectrometer is the energy resolution. The energy resolution is defined as the energy width of the peak. Generally, the energy resolution is shown as the full width at half maximum (FWHM) W of the peak. The relative energy resolution $W(E)/E$ of the gamma-ray energy of E is frequently used. The relationship between the FWHM W and the standard deviation σ of the Gaussian distribution is given as

$$W = 2\sqrt{2\ln 2}\,\sigma = 2.355\sigma. \tag{12.5}$$

In the scintillation detector, the total standard deviation σ_{total} at the gamma-ray peak energy E can be written by

$$\sigma^2_{total} = \sigma^2_{statistic} + \sigma^2_{light\ collection} + \sigma^2_{non-proportionality} + \sigma^2_{electric\ noise}, \tag{12.6}$$

where $\sigma^2_{statistic}$, $\sigma^2_{light\ collection}$, $\sigma^2_{non\text{-}proportionality}$, and $\sigma^2_{electric\ noise}$ are the standard deviations on statistics of information carrier, light collection, non-proportionality, and electrical noise, respectively. The statistical and non-proportionality components are considered to contribute to the intrinsic energy resolution, which is defined as the resolution originated only from scintillator material properties. In the scintillation detector, the statistical standard deviation $\sigma^2_{statistic}$ plays the most important role, because contribution of the electrical noise is relatively small owing to its large signal generated by the PMT. The statistical standard deviation $\sigma^2_{statistic}$ is based on a statistical fluctuation of the number of information carriers created in the scintillation detector. If we ignore the light collection term assuming a thin sample coupled with an enough large area photodetector, Equation (12.6) will be physically identical to Equation (1.19). When the average number of created information carriers is N, since the fluctuation of the number of created information carriers is derived from the Poisson statistics, the relative standard deviation is written by

$$\frac{\sigma_{statistic}}{N} = \frac{\sqrt{N}}{N} = \frac{1}{\sqrt{N}}. \tag{12.7}$$

For general radiation detectors, the statistical component should be considered based on Equation (1.18). Since the Fano factor is 1 for scintillation detectors, Equation (12.7) is equivalent to Equation (1.18). From this equation, a large number of the information carriers cause good energy resolution. The fundamental information carriers in the scintillation detector are, of course, scintillation photons. Thus, scintillators showing high light yield have a tendency to show high energy resolution. Therefore, it is important to use a bright scintillator in a high energy resolution spectrometer. In the case of scintillation, the other candidates as information carriers are photoelectrons created at the photoelectric conversion section in the PMT and multiplied electrons in the PMT, etc. From Equation (12.7), the statistical fluctuation is dominant when the number of the information carriers is at minimum value. Therefore, the statistical fluctuation is dominantly determined by the number of photoelectrons created in the PMT. So the photoelectric conversion efficiency at the PMT is also important to achieve high energy resolution.

Uniform light collection from the whole part of the scintillator is also important. The signal pulse height is determined by the number of scintillation photons reaching the photodetector. When using a poor light collection system, in which light collection efficiency is not uniform, we cannot achieve high energy resolution, even if the scintillator emits a constant number of scintillation photons. In order to achieve uniform light collection, a highly efficient diffusive reflection coating, such as Al_2O_3, MgO, and $BaSO_4$ powder, is applied.

The non-proportionality of the scintillation light yield is also an important issue. The scintillation light yield is considered to be proportional to the deposition energy in the scintillator. However, practically, the light yield per unit energy often changes depending on the energy of fast electrons created in gamma-ray interactions, especially in a relatively low energy region of less than 100 keV. This means that the average energy for creating one scintillation photon changes, depending on the fast electron energy. When full energy deposition events are caused through multiple interactions, including Compton scattering, the energy division ratio of fast electrons created at each interaction points is randomly distributed. Therefore, the total scintillation light yield also fluctuates, even if the total deposition energy remains the same. In the conventional Tl:NaI scintillator, the light yield at 10 keV is 20% higher than that at 1 MeV. The information of non-proportionality in some scintillators can be found in the literature [1, 2–4].

As described above, the energy resolution depends on the number of scintillation photons or gamma-ray energy. Therefore, the resolution at the energy of 662 keV, which is the gamma-ray energy of ^{137}Cs, is frequently evaluated as one of the specifications.

Here, we discuss the decay time of scintillators. In the pulse counting mode operation, the signal created by scintillation photons emitted in each event should be individually integrated into the signal processing circuit. Therefore, the time constant of the circuit should be matched with or longer than the decay time of the scintillator. As a result, the time width of scintillation pulse signals is comparable to scintillation decay. The detector using a slow scintillator suffers from problems of counting loss and/or signal pile-up, when the counting rate or gamma-ray intensity is high. In order to measure an intense gamma-ray field, a fast scintillator should be applied.

The scintillation light yield and/or sensitivity of the photodetector often depend on temperature. If we need high energy resolution, we need to control room temperature. Additionally, when conducting a long measurement, we should take care of ambient temperature.

Finally, we describe some typical gamma-ray spectrometers using a scintillator. The most conventional and representative scintillator for the gamma-ray spectrometer is the Tl:NaI. The light yield, decay time, and peak emission wavelength are 40 000 photons/MeV, 230 ns and 415 nm [5, 6]. Since this material shows hydroscopicity, the scintillator should be completely sealed to prevent it from exposure to ambient air. The Tl:NaI shows the energy resolution of 7% for 662 keV gamma rays emitted from a ^{137}Cs source [7]. Recently, Ce:$LaBr_3$ and Eu:SrI_2 have been developed as high energy resolution gamma-ray spectrometers. The Ce:$LaBr_3$ shows a light yield of 61 000 photons/MeV, a decay time of 30 ns, and a peak emission wavelength of 387 nm. The energy resolution of the Ce:$LaBr_3$ was reported to be 3% [8]. The Eu:SrI_2 shows the light yield of 120 000 photons/MeV, a decay time of 1.2 μs, and a peak emission wavelength of 435 nm. The energy resolution of the Eu:SrI_2 was reported to be 3% [9].

12.3.2 Survey Meter and Area Monitor

The scintillation survey meter and area monitor are used to measure gamma- or X-ray dosage. The survey meter is a portable device and the area monitor is a stationary or mounted one.

Although there are various definitions of the radiation dose, the dose measured by the scintillation survey meter and area monitor is the ambient dose equivalent. The ambient dose equivalent is one of the operational quantities for radiological protection. Radiological protection requires the quantification of the extent of exposure of the human body to ionizing radiation. For this purpose, the concept of two types of the quantities; "protection quantities" and "operational quantities," has been introduced by the International Commission of Radiological Protection (ICRP) and International Commission on Radiation Units and Measurements (ICRU) [10, 11]. The protection quantities, such as mean absorbed dose, equivalent dose in an organ or tissue, and effective dose, are used for implementation of dose limits and for guiding quantitatively the optimization principles. However, the protection quantities in an organ or tissue cannot be measured in practice. Therefore, in practical cases, the operational quantities are used in radiation measurements for external exposure to estimate the protection quantities for radiation management. Again, the survey meter and area monitor are equipment to measure one of the operational quantities; the ambient dose equivalent.

The ambient dose equivalent is defined as the dose equivalent at a point in a radiation field that would be produced at a 10 mm depth in the ICRU sphere, which is a sphere of 300 mm in diameter made of tissue equivalent material, at a 10 mm depth from the surface. The ambient dose equivalent practically provides a conservative estimate of the limiting quantities of the effective dose, which is one of the protection quantities. In order to determine the ambient dose equivalent, the conversion coefficients from the fluence (photons/cm^2) of high energy photons, such as gamma- or X-rays, are used determine to the ambient dose equivalent. The conversion coefficient can be calculated by the Monte Carlo transport simulation code.

The detector configuration of the survey meters and area monitors is basically the same as the gamma-ray spectrometer. The final output of these monitors is not the signal pulse height spectrum but the ambient dose equivalent. If gamma rays have a mono-energetic spectrum, of course, the ambient dose equivalent is proportional to the number of gamma-ray photons. However, the conversion coefficient is energy-dependent. In order to measure precise dose, this energy dependence should be considered in the survey meters and area monitors. One of the energy-response improvement methods is the G-function method [12–14]. The ambient dose equivalent H*(10) is determined as

$$H^*(10) = \int h(E) \frac{d\Phi(E)}{dE} dE, \tag{12.8}$$

where $h(E)$ is the conversion coefficient from the photon fluence to the ambient dose equivalent at an energy of E and $d\Phi(E)/dE$ is the photon fluence with energy from E to $E+dE$. The gamma-ray detector has response function $R(E, L)$, which is the expected mean signal count with a pulse height of L when irradiating with gamma rays with an energy of E and unit photon fluence. From the relationship between the response function and conversion coefficient, the G-function $G(L)$ is defined by

$$h(E) = \int R(E, L) G(L) dL. \tag{12.9}$$

Meanwhile, the signal pulse height distribution $P(L)$ can be written by

$$P(L) = \int R(E, L) \frac{d\Phi(E)}{dE} dE. \tag{12.10}$$

Consequently, by using the signal pulse distribution and G-function, the ambient dose equivalent $H^*(10)$ can be determined as

$$H^*(10) = \int P(L)G(L)dL. \tag{12.11}$$

The G-function can also be calculated by Monte Carlo transport simulation.

One of the important issues of the survey meter is portability. Therefore, the survey meter is, generally, designed to show only the ambient dose equivalent rate (Sv/h) to remove unnecessary functions. Considerations on consumption of electrical power are also important for design of the survey meter. On the other hand, since the area monitors are free for these kinds of problems, some of them, especially in the monitoring station, can provide the pulse height spectrum to identify radioactive isotopes. Almost all survey meters and area monitors use Tl:NaI as the scintillator, owing to its low cost and relatively high performance, such as for detection efficiency and energy resolution.

12.3.3 Scintillation Detectors for Medical Applications

Scintillation detectors are also used for various medical applications. Especially, in medical diagnostic techniques, the scintillator is one of the essential elements. In this section, some applications will be introduced in detail.

12.3.3.1 X-Ray Flat Panel Detectors (FPDs)

The X-ray radiography, which is transmission imaging using X-rays, is frequently used in the medical diagnostic field. Traditionally, X-ray films were used as the imager in X-ray radiography. Instead of the film, the X-ray FPDs can be used. The X-ray FPDs are categorized into two types; direct and indirect conversion FPDs. The direct conversion type FPD applies to semiconductor detectors as a sensitive volume and directly converts X-ray photons into an electrical charge. On the other hand, the indirect conversion type uses a scintillator sheet as a sensitive volume. In this type of detector, X-ray photons are converted into visible photons in the scintillator and then converted into an electrical charge by a photodetector. Here, the indirect conversion type or scintillator-based FPDs will be described. Figure 12.7 shows the fundamental structure of the scintillator-based X-ray FPD. The X-ray FPD consists

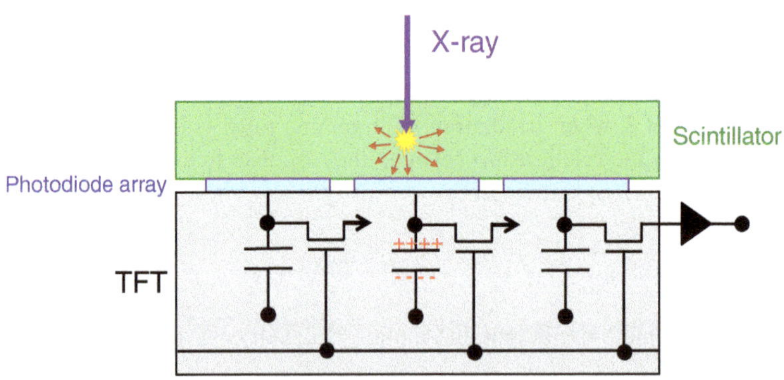

Figure 12.7 Schematic view of the scintillator-based X-ray flat panel detector (FPD).

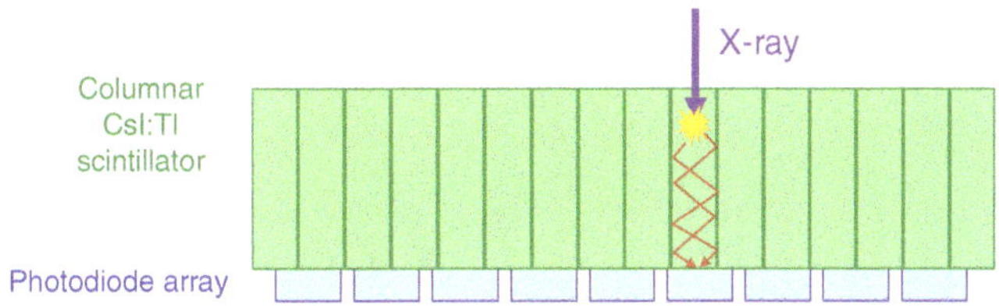

Figure 12.8 *Schematic drawing of the columnar CsI:Tl based FPD.*

of a scintillator, photodetector, and signal readout unit. The 2-D photodiode array is generally used as a photodetector. To read out the signal from the photodiode array, a thin film transistor (TFT) array is applied in many cases. As a scintillator, Gd_2O_2S:Tb (GOS) or CsI:Tl are frequently used. These scintillators are bright and their emission wavelength is matched with the wavelength response of the photodiode. The sensitivity of the FPD directly depends on the thickness of the scintillator. On the other hand, a thick scintillator makes spatial resolution worse because of diffusion of scintillation photons. The scintillator thickness is decided based on a trade-off between sensitivity and spatial resolution.

The CsI:Tl scintillator film produced by a physical vapor deposition process is known to display a micro-columnar structure [15]. This type of scintillator film has an excellent feature. Figure 12.8 shows a schematic drawing of the columnar CsI:Tl-based FPD. As described above, although the sensitivity increases by using a thick scintillator film, the spatial resolution is sacrificed. However, in this kind of columnar structure, scintillation photons are roughly confined by reflection on the column wall. This effect can reduce degradation in the spatial resolution when increasing the scintillator thickness.

12.3.3.2 X-Ray Tomography (CT)

X-ray computed tomography (CT) is widely used in medical diagnostics. Figure 12.9 shows a basic configuration of the X-ray CT system. The CT system consists of rotation mechanism mounting with a fan-beam X-ray generator and line sensors. In the X-ray line sensor, scintillators are applied as radiation sensitive medium. Transmission line images of a subject are acquired from all directions by using rotation mechanism. The X-ray transmission probability is inversely related with X-ray attenuation along the beam axis. The X-ray transmission T is given by

$$T = \frac{I}{I_0} = \exp(-\mu d). \tag{12.12}$$

where I and I_0 are X-ray intensity before and after passing through the subject, respectively. μ and d are an attenuation coefficient and thickness of the subject. Therefore, X-ray attenuation A along an X-ray pathway S is represented as

$$A = -\ln T = \int \mu\, d\, dS. \tag{12.13}$$

The X-ray transmission shows information integrating the attenuation strength along each X-ray pathway. From the transmission line data, the cross-sectional image of the X-ray

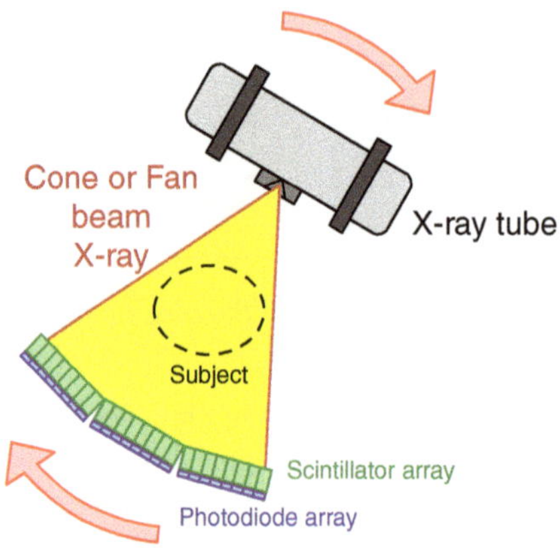

Figure 12.9 *Basic configuration of the X-ray CT system.*

attenuation coefficients can be reconstructed by mathematical procedures, such as Radon transform.

In advanced CT systems, the X-ray generator emits a cone-shaped beam, instead of a fan beam. Accordingly, the line sensor is replaced with a multi-detector row. This multi-detector row CT system can reconstruct multi-layered cross-sections simultaneously. This system is also known as a multi-slice CT system.

12.3.3.3 Single Photon Emission Computed Tomography (SPECT)

The medical diagnostic techniques based on ionizing radiation imaging are categorized into two groups; morphological and functional imaging technique. The X-ray CT, described above, is a morphological imaging technique. The functional imaging is realized by nuclear medicine imaging. In nuclear medicine imaging, pharmaceutical drugs containing radioactive isotopes are taken into a patient's body. Then, external detectors detect gamma-rays emitted from the radioactive drugs, from which the spatial distribution of radioactive isotopes or drugs is reconstructed. Since spatial distribution of the drugs reflects biological function, the nuclear medicine imaging can acquire biological functional images of patients. The single photon emission computed tomography (SPECT) is one of the nuclear medicine imaging techniques.

The SPECT system consists of a 2-D gamma camera mounted on a rotation mechanism. A typical planar gamma camera is shown in Figure 12.10. The planar gamma camera is generally used in nuclear medicine 2-D imaging. This type of camera has a collimator in front of the scintillator to collimate gamma-ray photons. Since a position-sensitive PMT is applied as a photodetector, a gamma-ray interaction positon in the scintillator can be determined by using a position identification technique, such as the center of gravity calculation shown in Figure 12.11.

Owing to the collimator, an integral amount of radioactive isotopes can be determined. Since total scintillation signal intensity in each event represents gamma-ray energy or deposition energy in the event, this kind of detector can acquire information about incident gamma-ray energy. Therefore, the gamma camera has abilities not only to obtain spatial distribution but also to distinguish radionuclides emitting gamma rays. Again, this kind of gamma camera can obtain the integral amount of radioisotopes along a line of sight of the

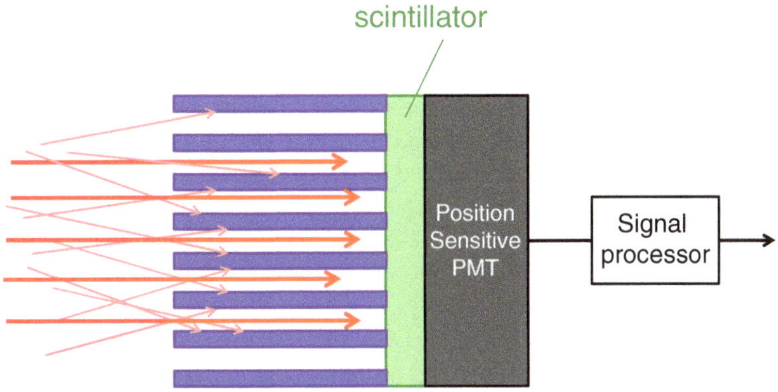

Figure 12.10 *Gamma camera with a collimator.*

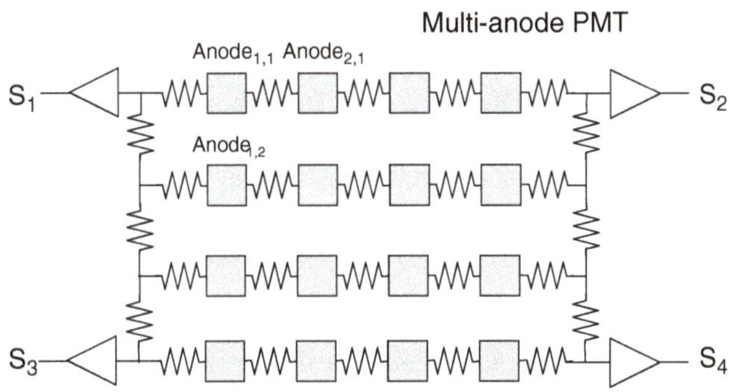

$$X = \frac{S_2 + S_4}{S_1 + S_2 + S_3 + S_4}, \quad Y = \frac{S_3 + S_4}{S_1 + S_2 + S_3 + S_4}$$

Figure 12.11 *Position identification based on a center of gravity calculation.*

collimator. In the SPECT system, 2-D images of the gamma camera are acquired from all directions by using rotation mechanism. Therefore, 3-D spatial distribution of radioisotopes contained in the drug can be reconstructed in the same way as the X-ray CT.

12.3.3.4 Emission Tomography (PET)

Positon emission tomography (PET) is a similar modality to SPECT. In PET diagnostic, drugs, including radioisotopes emitting a positron, which is an antiparticle of the electron, are administrated into a patient body. The most frequent application of PET is for cancer screening or exploration of cancer material using fluorodeoxyglucose (FDG), an analogue of glucose, labeled with ^{18}F ($T_{1/2}$: 110 minutes). Other typical positron emitters used in the PET diagnostic are ^{11}C ($T_{1/2}$: 20.4 minutes), ^{13}N ($T_{1/2}$: 9.97 minutes), and ^{15}O ($T_{1/2}$: 2.04 minutes). These short-lived radioisotopes are generally created by nuclear bombardment reactions using a cyclotron proton accelerator. A position emitted from these radioisotopes travels in tissue for a short distance until it interacts with an electron. The encounter annihilates both electron and positron, producing a pair of 511 eV annihilation gamma rays moving in approximately opposite directions. If a positron has no momentum when annihilating, these directions are exactly the opposite. However, in many cases, since a positron has a slight momentum, these directions will be deviated from exactly opposite directions.

Figure 12.12 shows a basic configuration of the PET system [16]. A pair of annihilation gamma rays with 511 keV will be detected when they reach scintillators surrounding a patient. The technique is based on simultaneous or coincident detection of the pair of gamma-rays moving in approximately opposite directions. The annihilation point of a positron can possibly be localized along a straight line, which connects two scintillators coincidently detecting 511 keV annihilation gamma rays. This line is known as the line of response (LOR). By accumulating information of the LOR, the 3-D distribution of radioisotopes emitting positrons can be reconstructed by using a similar algorithm to SPECT and X-ray CT.

In practice, the LOR has a non-zero width because the directions of annihilation gamma-rays are not exactly opposite. In addition, as shown in Figure 12.13a, especially in the peripheral field of view, the parallax error when defining the LOR will be large. This is because scintillators used in the PET system have a long and narrow shape to satisfy high sensitivity and high spatial resolution. In order to reduce the parallax error of the LOR, information of depth of interactions (DOI) is useful. The depth of interactions is a longitudinal position along the long and narrow scintillator. Figure 12.13b shows the effect of the parallax error correction by information of the DOI in peripheral field of view. In order to obtain information of the DOI, long scintillators are generally segmented in a longitudinal direction and the DOI is located by some sophisticated position-coding methods using a position sensitive photodetector [17].

The advanced technique in the PET is the time-of-flight (TOF) method [18]. The TOF information, which is a time difference in detecting timing of a coincident detection event, can localize the event position along the LOR. Figure 12.14 shows the effect of TOF information. The signal-to-noise ratio of the PET image will improve by using the TOF information.

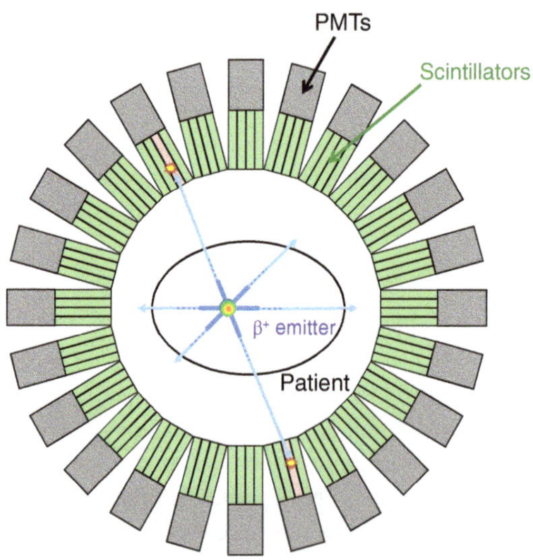

Figure 12.12 *Basic configuration of the PET system.*

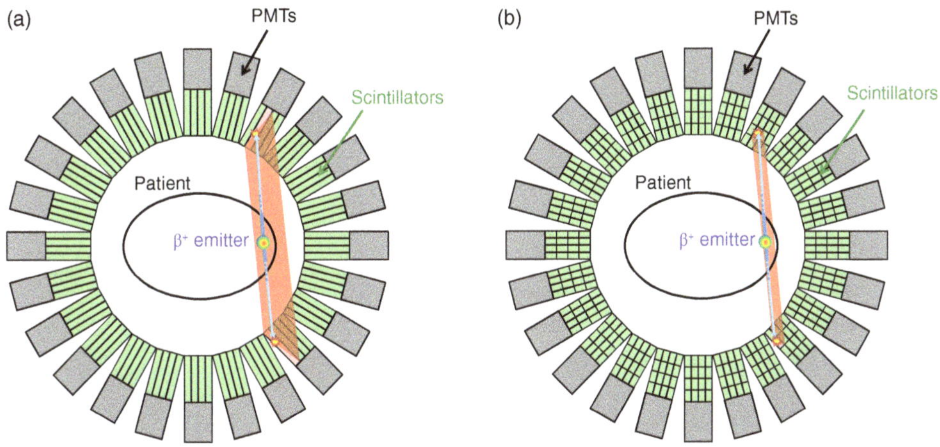

Figure 12.13 *Effect of parallax error correction by information of the depth of interactions (DOI) in peripheral field of view. The PET system (a) without and (b) with the DOI information.*

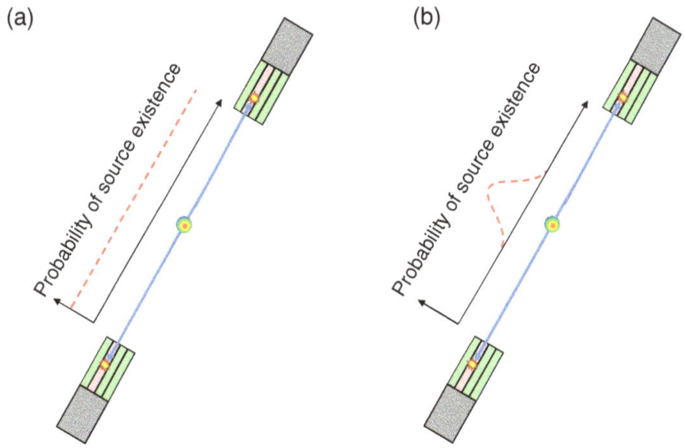

Figure 12.14 *Effect of time-of-flight information. The PET system (a) without and (b) with the PET information.*

A high timing resolution of approximately 650 ps, at least, is required to apply the TOF technique. In the TOF-PET system, the timing resolution is one of the most important specifications. Even if flight lengths of annihilation gamma-ray photons emitted in opposite directions are the same, the arrival time difference between two signals follows the Gaussian distribution, as shown in Figure 12.15a. Uncertainty of the signal arrival time is considered to be attributed to not only finite scintillator size or variety in the flight length, but also the timing jitter in the signal pick-up process. Although various signal-pick-up methods are applied in the TOF measurements, the simplest one is the leading-edge method. As shown in Figure 12.15b, real signal shapes could show some fluctuation. In the leading-edge method, the system picks up the timing when the signal pulse crosses over a fixed threshold level. The

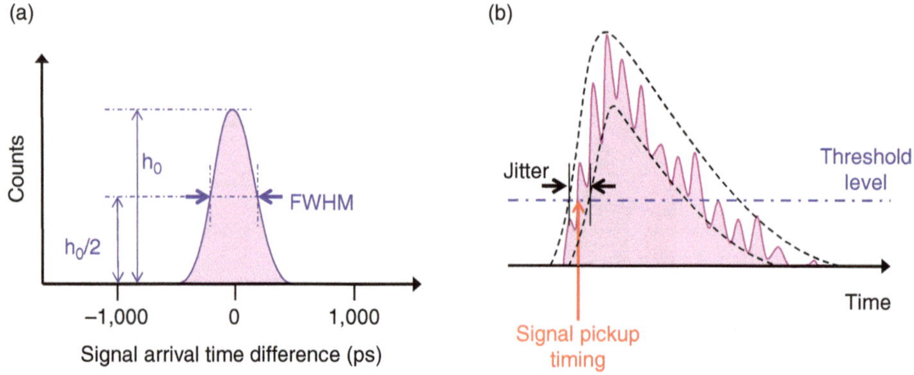

Figure 12.15 (a) Distribution of the arrival time difference between two signals. (b) Schematic drawing of the signal pickup process of the leading-edge method.

pick-up timing could be changed by signal fluctuation. This is one of the reasons why the distribution of the signal arrival time difference, as shown in Figure 12.15a, has a finite width. The time resolution is generally defined as the full width at half maximum (FWHM) of the distribution. The time resolution depends on scintillator properties, such as the light yield, rise, and decay time. In order to reduce the timing jitter width, it is desired that the signal leading edge has a steep slope. To achieve a steep leading edge, it is important to use a scintillator showing not only a short decay time but also a high maximum instantaneous scintillation intensity, which is achieved by both the high light yield and short rise and decay time. The LYSO scintillator is sometimes used in the TOF-PET system because of its short decay time and high light yield.

12.3.4 Scintillation Detectors for Other Applications

Scintillators are also used in other various applications. Many scintillation detectors are applied in high energy physics experiments. However, these applications are beyond the scope of this text. In this section, some other applications will be introduced.

12.3.4.1 Nuclear Well Logging

Nuclear well-logging is based on nuclear measurements in boreholes for mining exploration. This can provide some information on rock characteristics. Nuclear-logging tools apply two types of radiation: gamma rays and neutrons [19]. Figure 12.16 shows various types of nuclear well-logging techniques. The well-logging tools using gamma rays are categorized into two types. One of them is based on measurement of gamma rays emitted from natural radioisotopes, such as ^{40}K, ^{232}Th, ^{238}U, and their decay products, etc. By continuously measuring these compositions, we can gain some information on the geological formation. This is a passive gamma-ray logging method. Another type of gamma-ray logging is the active gamma-ray logging, which is used to measure the density of the geological formation. In this method, a gamma-ray source and a detector unit are inserted into a borehole. As a gamma-ray source, ^{137}Cs or ^{60}Co sources are frequently used. By measuring back-scattered gamma rays, bulk density of the geological formation can be ascertained.

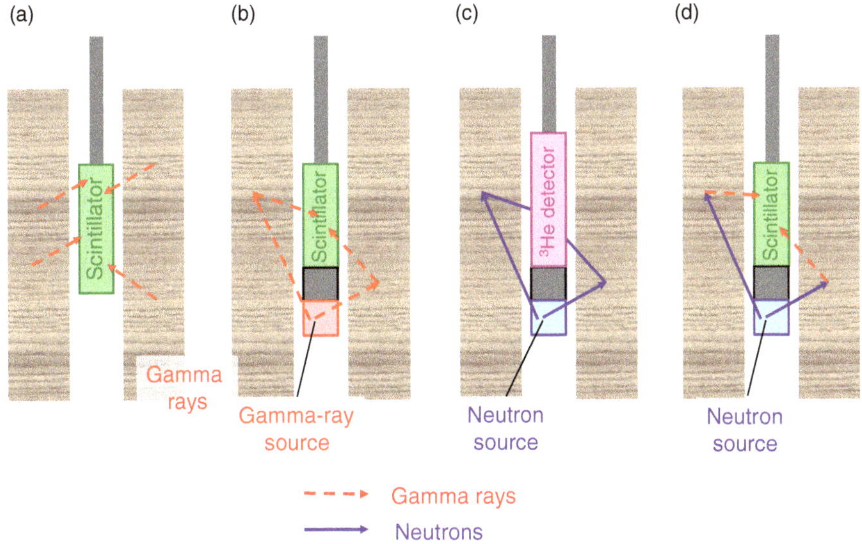

Figure 12.16 *Various types of nuclear well-logging techniques. (a) Passive gamma-ray logging. (b) Active gamma-ray logging. (c) Back-scattered neutron logging. (d) Neutron induced prompt gamma-ray logging.*

The well-logging techniques using neutrons are also categorized into two types. One of them is based on back-scattered neutron detection. This means that the neutron-logging tools consist of a neutron source and neutron detector, such as the ^{3}He gaseous detector. Another type of neutron logging exploits neutron induced prompt gamma rays, which are emitted from nuclides excited by neutron absorption. These kind of gamma rays can be used to identify a parent nuclide composing geological formation. Gamma rays used in nuclear well-logging are detected by scintillation-detector-based gamma-ray spectrometers. Since neutron induced prompt gamma rays sometimes have high energy of more than 3 MeV, which are rare in the delayed gamma rays emitted from radioactive isotopes, large volume scintillators are useful in this application.

One of the unique properties in the well-logging is the high temperature environment. Sometimes, the temperature in a borehole can be about 200 °C or higher. Therefore, scintillators for this application are desired to have small temperature dependence. In addition, the high-temperature PMT should be used in the well-logging system.

12.3.4.2 Security

Scintillators are also used for some security purposes. The most famous application in security is X-ray non-destructive inspection: for example, baggage inspection at airports. This is fundamentally based on X-ray radiography or X-ray transmission imaging, which was described in the previous sections on medical applications. However, a difference from medical applications is X-ray energy. In medical applications, X-ray energy is relatively low, because the subject is a biological tissue composed of light elements, such as H, C, N, and O. On the other hand, in security applications, since the subject would have a higher density than that in medical applications, higher energy X-rays are used. Therefore, scintillator thickness for X-ray inspection is thicker than that of scintillator films used in medical applications. Since the subjects in security inspection, such as baggage, do not move, the inspection device combining a 1-D line scanner and belt conveyer system is generally used instead of 2-D detectors, such as FPDs. For large containers, high energy X-rays (of the order of ~MeV) are applied. For this kind of application, heavy scintillators are used, such as $Bi_4Ge_3O_{12}$ (BGO) and $CdWO_4$ (CWO).

Another type of security application using scintillators is in nuclear security. A fundamental purpose of nuclear security is to prevent the spread of nuclear weapons and to ensure that nuclear technology is only made available for peaceful purposes [20]. The nuclear security system is based on detection of nuclear and other radioactive materials out of regulatory control. Nuclear materials are plutonium, uranium, thorium, and any material containing one or more of the above. Radioactive materials are defined as any matter which emits ionizing radiation. Measurement system of gamma rays and neutrons emitted from these materials are essential tools for nuclear security. Especially, the gamma-ray spectrometer is a powerful tool used to identify radioactive and nuclear materials.

For nuclear security purposes, radiation portal monitors (RPMs) are also applied. The RPMs are passive radiation detection devices used for the screening of individuals, vehicles, or cargo for detection of threat sources at borders or secure facilities. The RPM is designed to detect traces of radiation emitted from an object passing through an RPM. Gamma-ray detection is generally applied in RPMs, and in some cases neutron detection complements it when nuclear materials should be detected. As a gamma-ray detection technique in the RPMs, scintillation detectors are widely used. Some RPMs exploit plastic scintillators for gamma-ray detection because they can easily be manufactured into large volume detectors. RPMs using plastic scintillators sometimes suffer from an inability to discriminate gamma rays from nuclear materials and naturally occurring radioactive materials (NORMs), because of their limited energy resolution. Although RPMs using Tl:NaI scintillators were also developed as the advanced RPMs, they have not been able to show totally better performance than the RPMs using plastic scintillators because of cost.

12.4 Scintillation Detectors for Charged Particles

High energetic charged particles, such as alpha- and beta-rays, are a directly ionizing radiation group, unlike X-ray, gamma-ray, or neutrons, which are an indirectly ionizing group. Once a directly ionizing radiation particle enters matter, ionization and excitation immediately occur. From these features, the intrinsic detection efficiency, which is defined as the ratio of the number of detected particles to the number of incident particles entering into the sensitizing area of a detector, is almost 100% and penetrating power of these particles is limited to relatively short lengths. Note that directly ionizing radiation particles might stop at an entrance window of a detector because of their low penetrating power or short range. Therefore, in the design of the radiation detector for charged particles, consideration of an entrance window or dead layer is essential.

The charged particle detectors are commonly used for radioactive contamination monitoring. The decay modes of almost all radioisotopes are the alpha and beta decay. This means that almost all radioactive materials emit alpha and beta (or beta plus) particles. As described above, since energetically charged particles entering a detector are almost perfectly detected, detection of radioactive contamination using alpha and/or beta particle detectors is highly efficient compared with that using gamma-ray detectors, especially for surface contamination. For gamma-ray detection, large volume or thickness of a detector is an important factor to increase the detection efficiency. However, for alpha- or beta-particle detection, a large solid angle or large sensitive area is more important than the volume or thickness. As a result, a large and thin scintillator sheet is frequently applied for radioactive contamination monitoring.

For contamination monitoring of alpha-ray emitting radioisotopes, a ZnS:Ag scintillator sheet is generally used. The ZnS:Ag scintillator is well known as one of the brightest scintillator materials. Since the ZnS:Ag scintillator is used in the form of a white compressed powder sheet, scintillation photons can penetrate only a very thin thickness into the scintillator sheet. For contamination monitoring of beta-ray emitting radioisotopes, a plastic scintillator plate is also used owing to its low cost. Since a plastic scintillator is transparent, the scintillator thickness can be controlled according to its purpose. As a radioactive contamination monitor for both alpha- and beta-particle emitting radioisotopes, a thin sheet type of ZnS:Ag/plastic dual scintillator is used [21]. Figure 12.17 shows the basic configuration of the radioactive contamination monitor using a ZnS:Ag/plastic dual scintillator.

Alpha particles emitted from radioactive isotopes have relatively higher energy (~4 or 5 MeV) compared with beta particles (maximum 3 MeV). Since alpha particles have high stopping power dE/dx, they cannot penetrate the ZnS:Ag scintillator sheet and deposit all of their energy in the ZnS:Ag scintillator. On the other hand, relatively high energy beta particles have small stopping power. This means that these beta particles deposit only small fractions of their energy into the ZnS:Ag scintillator sheet and pass through it. As a result, a large fraction of energy of a high energy beta particle will be deposited in the plastic scintillator. The light yield per unit deposition energy of the ZnS:Ag scintillator is approximately five times higher than that of the plastic scintillator. As described above, higher energy alpha particles deposit their energy in the bright ZnS:Ag scintillator and smaller energy beta particles deposit almost all of their energy in the plastic scintillator, which has a smaller light yield than ZnS:Ag. Consequently, signal pulse heights obtained in alpha particle events are higher than that obtained in beta particle events. Therefore, alpha and beta particle events can be separated by simple pulse height discrimination. This type of detector is widely used as the radioactive contamination monitor with alpha/beta discrimination ability. Since the plastic scintillator has a shorter decay time than the ZnS:Ag scintillator, alpha and beta particle

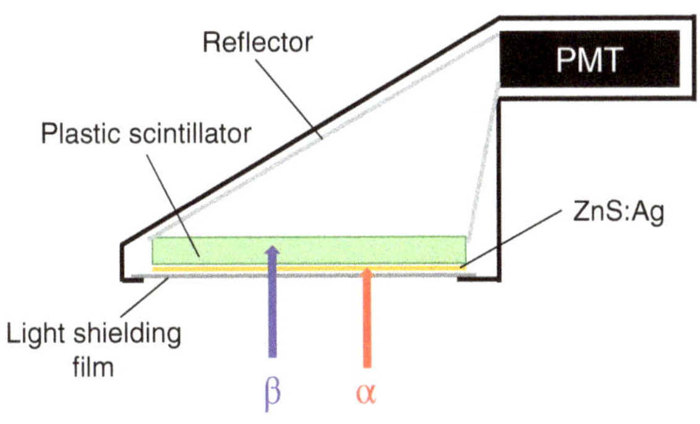

Figure 12.17 Basic configuration of the radioactive contamination monitor using a ZnS:Ag/plastic dual scintillator.

events can also be discriminated by the pulse shape discrimination technique. If combining the pulse height and pulse shape discrimination techniques, the radioactive contamination monitor using a ZnS:Ag/plastic dual scintillator might be improved in terms of alpha/beta discrimination ability.

12.5 Scintillation Detectors for Neutrons

Neutron detection techniques are widely used in various fields, such as security, medicine, and material science. Since neutrons are indirectly ionizing radiation particles, they have to be converted into charged particles, which can directly ionize and excite a scintillation material, for neutron detection. In the case of high-energy photons, such as gamma- and X-rays, they are converted into energetic electrons through electromagnetic interactions. On the other hand, neutrons are converted into charged particles through nuclear reactions. This is the most important difference between high-energy photon and neutron detection. Based on interaction processes, interaction of high-energy photons depends on the atomic number of a target material. On the other hand, neutron interaction depends on properties of each isotope or nuclide of a target material. Anyway, in order to detect neutrons, the detector must have the neutron converter, in which neutrons are converted into energetic charged particles.

Neutrons are generally categorized into two groups: thermal and fast neutrons, from the viewpoint of their energy. In this section, general descriptions of the detectors categorized by neutron energy will be shown.

12.5.1 Thermal Neutron Detectors

The energy of thermal neutrons is low and comparable to that of the thermal motion of ambient atoms. The average velocity and energy of thermal neutrons, with the temperature of 293 K, are 2200 m/s and 25 meV, respectively. Generally, thermal neutrons show a Maxwellian energy distribution, because they reach equilibrium thermally with surrounding atoms through elastic and inelastic scattering. These thermal neutrons may be absorbed by surrounding atoms when nearing the working region of nuclear force of the atoms. Since the probability of this kind of interaction depends on a time remaining in the working region of the nuclear force, the neutron absorption cross-section is inversely proportional to the neutron velocity. This is known as the 1/v law.

After absorbing a neutron, almost all atoms emit one or more gamma rays. This means that the neutron field generally includes gamma rays. Therefore, one of the essential properties in neutron detectors is gamma-ray suppression or discrimination ability. Almost all nuclides or isotopes can absorb neutrons, but they have a large variation in the reaction cross-section. Although the neutron detection efficiency might depend on the cross-section, almost all isotopes might be used as neutron sensitive isotopes. However, almost all isotopes can emit only gamma rays and/or internal conversion electrons in the immediate processes following neutron absorption. These secondary emitted particles are also generated in gamma-ray interactions. If we want to reduce the gamma-ray sensitivity, we cannot generally apply these reactions for neutron detection. However, several limited isotopes, such as ^{3}He, ^{6}Li, ^{10}B, ^{14}N, and ^{35}Cl, can promptly emit energetic heavy charge particles, such as proton, triton, and alpha particles, after neutron absorption. These kinds of isotopes can be useful for neutron detection media. The golden standard of thermal neutron detectors has been the ^{3}He gaseous detector, which has high detection efficiency and low gamma-ray sensitivity. However, an alternative to the ^{3}He detector needs to be developed because of its precarious supply. As the alternative neutron detector, neutron sensitive scintillators have been developed.

In scintillators sensitive to thermal neutrons, ^{6}Li and ^{10}B are considered to be candidates as neutron converters because they have sufficiently high cross-sections (^{10}B:3800 b, ^{6}Li:940 b) for thermal neutrons and can form a solid compound [22]. They interact with neutrons through the following exothermic nuclear reactions:

$$ {}^{6}_{3}Li + {}^{1}_{0}n \rightarrow {}^{4}_{2}\alpha + {}^{3}_{1}t + 4.78\,MeV\ \left(\alpha : 2.05\,MeV, t : 2.73\,MeV\right), $$

$$ {}^{10}_{5}B + {}^{1}_{0}n \rightarrow {}^{7}_{3}Li + {}^{4}_{2}\alpha + 2.79\,MeV\ \ (6\%)\left({}^{7}Li : 1.01\,MeV, \alpha : 1.78\,MeV\right) $$
$$ \rightarrow {}^{7}_{3}Li^{*} + {}^{4}_{2}\alpha + 2.31\,MeV\ \ (94\%)\left({}^{7}Li : 084\,MeV, \alpha : 1.47\,MeV\right). $$

Although ^{10}B has a higher cross-section than ^{6}Li, ^{6}Li is widely used in the neutron sensitive scintillator because its reaction products, such as triton and alpha particles, have smaller charge and higher kinetic energy than these of ^{10}B. Since low energy and highly-charged ions have high linear energy transfer (LET), which induces severe quenching in a scintillator, ^{6}Li is considered to be a preferential isotope as the neutron converter in scintillators.

Table 12.1 shows the properties of ^{6}Li-based scintillators. As described above, a ^{6}Li nuclide disintegrates into triton (^{3}H) and alpha particle (^{4}He) and releases energy of 4.78 MeV after absorbing a neutron. Since thermal neutron energy is small compared with the energy released, total kinetic energy shared into a triton and alpha particle is 4.78 MeV. In addition, since the range or path length of a triton and alpha particle in these scintillators is in the order of micrometers, the entire energy released in this reaction is deposited in a scintillator in almost all cases. When using a transparent scintillator, we can uniformly collect scintillation photons induced in this reaction from the entire region of the scintillator. As a result, the signal pulse height should be uniform for neutron scintillation events. This means that the peak structure will be created in the pulse height spectrum for neutron events. Therefore, transparency is an advantageous point for neutron scintillators. The relative peak width depends on the scintillation light yield because it is based on the Poisson statistics described in the above section on the gamma-ray spectrometers (see Equation 12.6). Therefore, the light yield is also an important property. The emission wavelength and decay time are also important from the viewpoint of the wavelength matching with the photodetector and the counting loss problem, respectively. Of course, the Li content has a large impact on neutron detection efficiency.

One of the unique properties for neutron scintillators is the α/β ratio, which is defined as the ratio of the light yields for an alpha particle and electron per unit energy. Generally, the released energy in ^{6}Li reactions is sufficiently high compared with typical energy of gamma rays detected in neutron scintillators, which has a relatively low effective atomic number to reduce gamma-ray sensitivity. so shows low sensitivity, especially for high-energy gamma rays. Therefore, neutron events can easily be distinguished from gamma-ray events by pulse height discrimination if the scintillation efficiency is the same for electrons and high LET particles, such as tritons and alpha particles. However, as described above, the scintillation efficiency generally decreases for high LET particles. because a part of densely excited

Table 12.1 *Properties of ^{6}Li-based scintillators [23]*

Scintillator	Light yield (photons/n)	Decay time (ns)	Density (g/cm^3)	Li content (atoms/cm^3)	Emission wavelength (nm)	α/β ratio	Hygroscopicity	Transparency
Ce:Li glass	6000	75	2.5	~1.6×10^{22}	395	0.23	No	Yes
Eu:LiI	49 500	1400	4.06	1.8×10^{22}	470	0.9	Yes	Yes
LiF-Ag:ZnS[a]	~150 000	200	3.06	~2.0×10^{22}	450	–	No	Opaque
Ce:$LiCaAlF_6$	5000	40	2.99	9.6×10^{21}	285	0.4	No	Yes
Eu:$LiCaAlF_6$	40 000	1600	2.99	9.6×10^{21}	375	0.2	No	Yes
LiF/Eu:CaF_2 eutectics	~10 000	~700	2.85	3.8×10^{22}	428	–	No	Yes
Ce:Cs_2LiYCl_6	64 000	1/50/1000	3.31	3.5×10^{21}	370	0.67	Yes	Yes

[a] LiF:ZnS = 2 : 1 (weight ratio), Grain packing ratio 75%

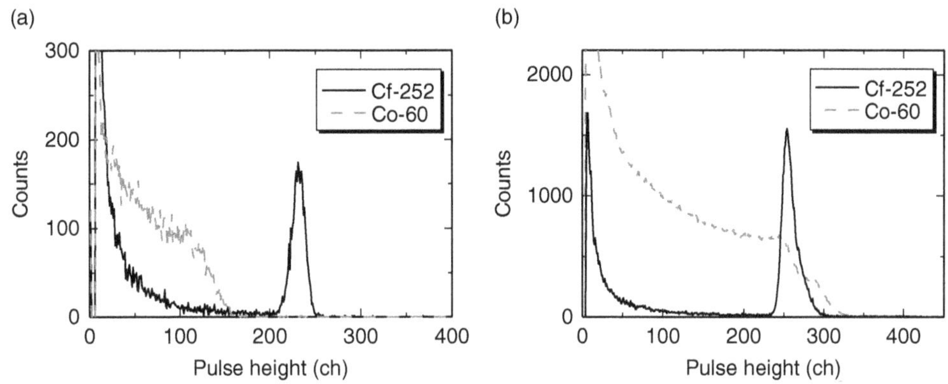

Figure 12.18 *Pulse height spectra obtained from (a) Ce:LiCaAlF$_6$ and (b) Eu:LiCaAlF$_6$ scintillators when irradiated with neutrons and gamma rays. The neutron and gamma-ray sources were ^{252}Cf surrounded by polyethylene moderator bricks and ^{60}Co sources, respectively.*

electrons cannot easily find emission sites. The α/β ratio is an effective index showing the scintillation efficiency for high LET particles produced by neutron reactions. The ^{6}Li-based scintillators showing high α/β ratio have an advantage from the viewpoint of gamma-ray suppression ability. Figure 12.18 shows the pulse height spectra obtained from Ce:LiCaAlF$_6$ and Eu:LiCaAlF$_6$ scintillators when irradiated with neutrons and gamma rays [24, 25]. The neutron and gamma-ray sources were ^{252}Cf surrounded by polyethylene moderator bricks and ^{60}Co sources, respectively. The gamma-ray energies of ^{60}Co are 1.17 and 1.33 MeV. Since these scintillators are composed of only light elements, Compton scattering is a dominant interaction and the full energy deposition peak cannot be observed. Consequently, only the Compton continuum is seen. The energy of the Compton edge created by 1.33 MeV gamma rays is 1.12 MeV. If the scintillation efficiency is the same for both the high LET heavy charged particles and electrons, the neutron peak position (4.78 MeV) should be roughly 4 times higher than the position of the ^{60}Co Compton edge (1.12 MeV). However, the Compton edge position in the spectrum obtained from the Eu:LiCaAlF$_6$ scintillator is a little higher than the neutron peak position. This is because the α/β ratio is 0.2 for the Eu:LiCaAlF$_6$ scintillator. On the other hand, in the case of the Ce:LiCaAlF$_6$ scintillator, the ^{60}Co gamma-ray Compton edge position is lower than the neutron peak position, owing to its relatively high α/β ratio. In this scintillator, ^{60}Co gamma-ray events can be discriminated from neutron events by simple pulse height discrimination.

Some possible applications using scintillators sensitive to thermal neutrons will be introduced in the rest of this section.

12.5.1.1 Medical Applications: Boron Neutron Capture Therapy

In medical applications, a thermal neutron scintillator can be used. One of the radiation therapies using neutrons is the boron neutron capture therapy (BNCT). The BNCT is an efficient tumor selective form of radiotherapy. In the BNCT, first, a boron drug is administrated into a patient and carried preferentially into tumor cells. Then, the patient is externally irradiated with thermal or epi-thermal neutrons and then ^{10}B(n,α) reactions are induced into the tumor cells. The tumor cells are damaged by attack with energetic charged particles: an alpha particle and

^{7}Li nuclide. The ranges or path lengths of these charged particles are comparable to the size of typical cells. Therefore, the damage is limited within the cells accumulating the boron drug. The BNCT can selectively attack tumor cells by using a tumor selective boron drug. Since the BNCT is a radiotherapy using neutrons, the neutron monitoring has an important role. For BNCT neutron monitoring, the detector must be able to measure an intense neutron field up to the neutron flux of 10^9 n/cm^2/s, which is the typical neutron flux used in the BNCT, to suppress the gamma-ray sensitivity and to have little interference with the irradiation field.

One of the candidates for the BNCT neutron monitor is an optical fiber-based neutron scintillation detector [26, 27]. In this detector, a small piece of Eu:$LiCaAlF_6$ scintillator crystal is mounted at the tip of an optical fiber. Scintillation photons are transmitted in the optical fiber and detected by a photodetector mounted at the other end of the optical fiber. Figure 12.19 shows the geometrical relation on pathways of scintillation photons in the optical fiber.

As described in Section 12.2, we have to take care of efficient light collection and/or light transmission. An optical fiber is an efficient light guide utilizing the total reflection. As shown in Figure 12.19a, photons in the transmission mode, in which photons with a larger reflection angle than the critical angle are reflected by total reflection, can be transmitted with little loss. However, the acceptable angle into the transmission mode of the optical fiber is usually small for scintillation photons. Therefore, the optical fiber light guide has low attenuation but small light collection efficiency. From this viewpoint, the important parameter of the optical fiber is the numerical aperture (NA). The NA is defined as

$$(NA) = n \sin\theta_{\max} = n_{core} \sin\left(\frac{\pi}{2} - \theta_c\right) = \sqrt{n_{core}^2 - n_{clad}^2}, \tag{12.14}$$

where n is the refractive index of the medium around the fiber (in this case, the scintillator), $\theta_{\max}$ is the maximum acceptable angle into the fiber, n_{core} and n_{clad} are the refractive indices of the optical fiber core and cladding, and θ_c is the critical angle. This parameter is an index of the acceptable angle into the optical fiber. Therefore, it is important to select a high NA optical fiber for efficient light collection. On the other hand, when the optical fiber is used in an

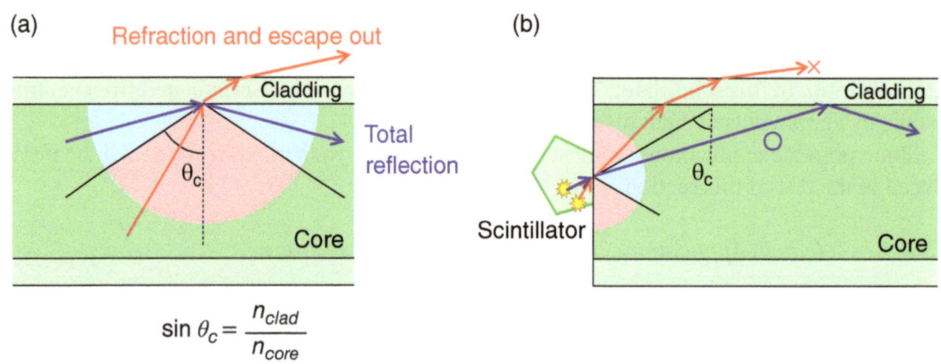

Figure 12.19 Geometrical relation on pathways of scintillation photons in the optical fiber. (a) Light transmission mode in the optical fiber. (b) Acceptable angle of scintillation photons into the optical fiber.

intense radiation field, the radiation damage of the fiber should be taken care of, because the high NA fibers might have a long path length in the optical fiber transmission mode.

In order to efficiently collect scintillation photons, it is necessary to place the scintillator within the core area on the end surface of an optical fiber. This is because it is difficult to place scintillation photons into the transmission mode from outside the core region if the scintillator is surrounded by a reflector. This means that the desired scintillator size is smaller than that of the optical fiber core. When using a large size scintillator crystal, a lens optics assembly should be considered to be applied to put scintillation photons into the transmission mode from a large volume scintillator. Of course, since the lens assembly has a relatively large volume, it is not appropriate to use these kind of optics when not required to interfere with the measured field by the detector itself.

The optical fiber-based BNCT neutron detector uses a small neutron scintillator. The small size of scintillator has an additional advantage. By controlling the scintillator size, gamma-ray events can be suppressed selectively by pulse height discrimination. Figure 12.20a shows the concept of gamma-ray suppression by controlling the scintillator size.

^{6}Li(n,t) reaction products, alpha and triton, have a short range in the order of 10 μm. On the other hand, a fast electron created by gamma-ray interaction has a relatively long range in the order of mm. By controlling the scintillator size into a few hundreds of micrometers, ^{6}L(n,t) reaction products can deposit their entire energy. However, fast electrons can easily escape from the scintillator before significantly depositing its energy. As a result, the signal pulse height induced in the gamma-ray event will be smaller compared with that of the neutron event. We, therefore, can easily distinguish neutron induced events from gamma-ray events in the pulse height spectrum. Figure 12.20 shows an example of pulse height spectra obtained from the neutron detector using a small Eu:LiCaAlF$_6$ scintillator. Although the scintillator material Eu:LiCaAlF^6 used in this detector was the same as that shown in Figure 12.18b, the pulse height in gamma-ray events is suppressed by controlling the scintillator size. This feature is useful from the viewpoint of gamma-ray suppression ability. The optical fiber-based neutron scintillation detector is one of the promising candidates as the neutron monitor in the BNCT irradiation field.

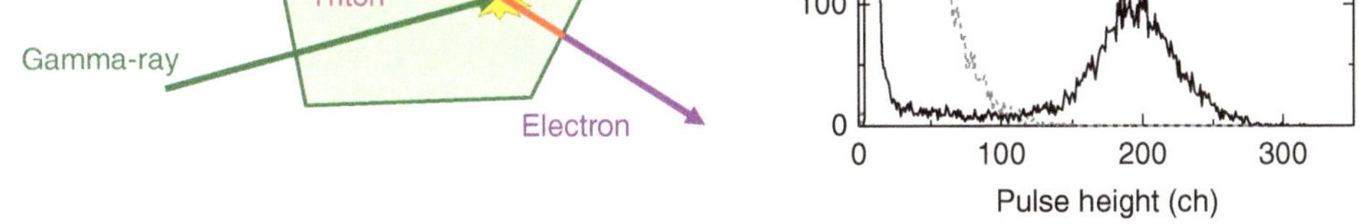

Figure 12.20 (a) Conceptual drawing of gamma-ray suppression by controlling the scintillator size. (b) Pulse height spectra obtained from the optical fiber-based neutron detector using a small piece of the Eu:LiCaAlF$_6$ scintillator. The neutron and gamma-ray sources were ^{252}Cf surrounded by polyethylene moderator bricks and ^{60}Co sources, respectively.

12.5.1.2 Security Applications: Radiation Portal Monitor

Neutron scintillators are also used in nuclear security applications. The purpose of the nuclear security is to prevent the spread of nuclear weapons and to ensure that nuclear technology is made available for peaceful purposes. The nuclear security is based on control of nuclear materials. Nuclear materials are plutonium, uranium, thorium, and their compounds. Many isotopes in these elements could be neutron sources through spontaneous fission or (α, n) reactions in the surrounding light elements, such as oxygen. Therefore, for nuclear security purposes, we can apply not only gamma-ray spectrometers but also neutron detectors. Since neutron background is generally lower than gamma-ray one, the neutron detection for nuclear security has an advantage compared with the gamma-ray detection from the viewpoint of the signal-to-noise ratio.

The RPMs (radiation portal monitors) are the instruments for nuclear security, as described above. In RPMs, the neutron detectors are frequently used. Since the RPM is designed to detect trace amounts of neutrons emitted from nuclear materials, the detector should provide a large solid angle. This means that a large size detector is required for this purpose. The ^{3}He gaseous detectors surrounded by a polyethylene moderator are generally used in RPMs. The ^{3}He detector is ideal for this purpose, owing to it low gamma-ray sensitivity because the advantage of the neutron detection is low background. However, since the ^{3}He detectors sometimes suffer from an unstable supply, alternatives to the ^{3}He detector needs to be developed.

The Li-based neutron scintillators are developed as an alternative to the ^{3}He detectors. For RPMs, the detectors should have a large area or volume and low gamma-ray sensitivity. Scintillators are generally solid phase and have much higher atomic density compared with gaseous detector. Since gamma rays basically interact with electrons bonded to the atoms, scintillator-based neutron detectors have some sensitivity to gamma rays and generally suffer from gamma-ray interference. As one of the candidates for alternatives to ^{3}He-based RPMs, a Transparent RUbber SheeT (TRUST)-type neutron scintillator has been developed [23]. As described in the previous section, a small piece of an Li-based scintillator has a large advantage for gamma-ray suppression. However, of course, a small scintillator has low sensitivity. To increase the neutron sensitivity by using small scintillators, a large number of small scintillators have to be used. It is not easy to uniformly collect scintillation photons from a lot of randomly arranged small scintillators, because photons are refracted or reflected by the interface of the scintillators. In order to reduce deposition energy from fast electrons induced by gamma rays, scintillators have to be reduced in size and be physically separated from each other to prevent fast electrons from re-entering into a neighboring scintillator. On the other hand, to efficiently collect scintillation photons, it is important to eliminate optical interfaces between a scintillation region and a photodetector. A way to satisfy both requirements is to fill in between the scintillators with transparent resin or rubber, which has the same or similar refractive index to the scintillators. According to the Fresnel equations, refraction and reflection occur on the interface having differences in the refractive index. In other words, on the interface with no difference in the refractive index, photons cannot be refracted or reflected and consequently go straight through the interface. In this case, scintillation photons can be collected efficiently and the deposition energy from fast electrons can be reduced. Figure 12.21 shows photographs of single crystal and TRUST Eu:$LiCaAlF_6$ scintillators and the pulse height spectra obtained from the TRUST Eu:$LiCaAlF_6$ scintillator.

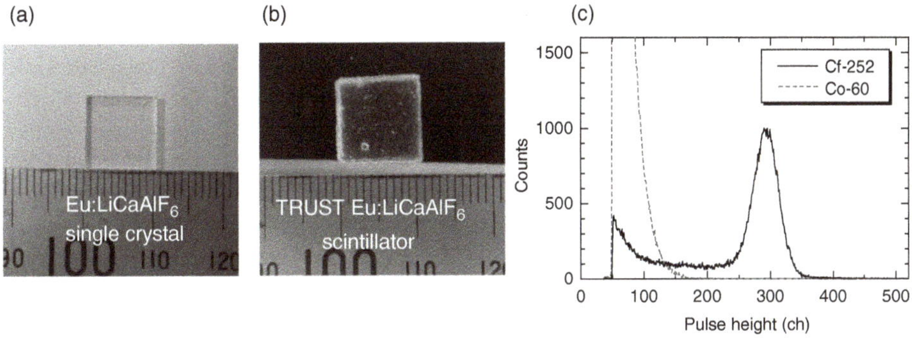

Figure 12.21 *Photographs of (a) a Eu:LiCaAlF_6 single crystal and (b) TRUST Eu:LiCaAlF_6 scintillator. (c) Pulse height spectra obtained from TRUST Eu:LiCaAlF_6 scintillator.*

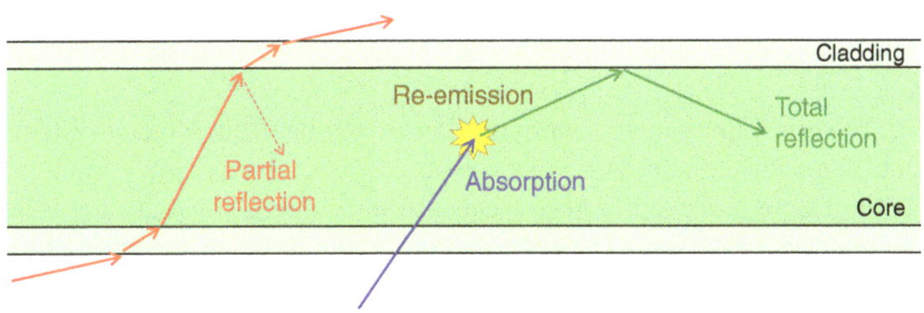

Figure 12.22 *Schematic drawing of the wavelength-shifting fiber (WLSF).*

Since the TRUST scintillator is formed by the low temperature molding process, it can easily be formed into a large-sized sheet. This is an advantage to increase the neutron sensitivity. However, we have to consider how to efficiently and uniformly collect scintillation photons from a large area and/or volume scintillator. A useful way to collect photons from a large area scintillator is to use a wavelength-shifting fiber (WLSF) [28]. Figure 12.22 shows a schematic drawing of the WLSF. Generally, an incident photon from the side surface of the optical fiber cannot enter into the transmission mode, in which photons will repeatedly undergo the total reflection and can transmit in the optical fiber with little loss. Although partial reflections will occur on the interface between the fiber core and cladding, transmitted photons using this reflection will be strongly attenuated. Therefore, generally, the optical fiber cannot collect photons into the transmission mode from the side surface. In the case of the WLSF, an incident photon from the side surface will be absorbed in the core and then a wavelength-shifted photon will be re-emitted and can be transmitted in the transmission mode with little loss. The WLSFs are generally designed to pass through the wavelength-shifted photons in the core. The core material of the WLSF has strong absorption for incident scintillation photons and small attenuation for wavelength-shifted photons. Consequently, the WLSF can collect and transmit photons incident from the side surface. The WLSFs are generally used as a light collection and transmission optics from a large area scintillator. The large area neutron detector using the TRUST Eu:LiCaAlF_6 scintillator and the WLSFs is one of the candidates as a neutron detector in the RPMs.

12.5.1.3 Material Science Applications: Neutron Imaging

The neutron imaging technique is widely used for non-destructive analysis in various fields, for example, material science. Since neutrons have complementary properties to X-rays, neutron imaging can extract some interesting information. Although there are various types of neutron imagers, the most general way is to combine the neutron scintillator sheet and optical camera system. In this type of neutron camera, the LiF and Ag:ZnS mixed powder scintillator formed into a thin sheet is widely used [29]. Lithium atoms work as a neutron converter. Energetically charged particles produced in ^{6}Li(n,t) reactions will enter the Ag:ZnS scintillator powder and induce the scintillation process. The LiF-Ag:ZnS scintillator is one of the brightest neutron scintillators. The Ag:ZnS is not colored but has a high refractive index. Therefore, the mixed powder is white and opaque because photons are refracted and reflected many times before escaping. Consequently, the effective or available thickness of this kind of scintillator is not very thick: a few 100 μm at maximum. The sensitivity depends on the scintillator sheet thickness but has a maximum limited by transparency of the scintillator sheet. Additionally, when using the thick scintillator sheet, the spatial resolution also deteriorates as the neutron camera. The scintillator sheet thickness should be determined by balancing between the sensitivity and spatial resolution.

Figure 12.23 shows the general configuration of the neutron camera system using the neutron scintillation sheet. Scintillation images are acquired by an optical camera, such as a CCD or CMOS camera, through a mirror because the camera generally has relatively low resistance to neutron irradiation. The optical camera should be protected from neutron and gamma-ray irradiation to reduce noise signals. The spatial resolution depends not only on scintillator thickness but also on the number of pixels in the camera, magnification of a lens, and emittance of neutron beam.

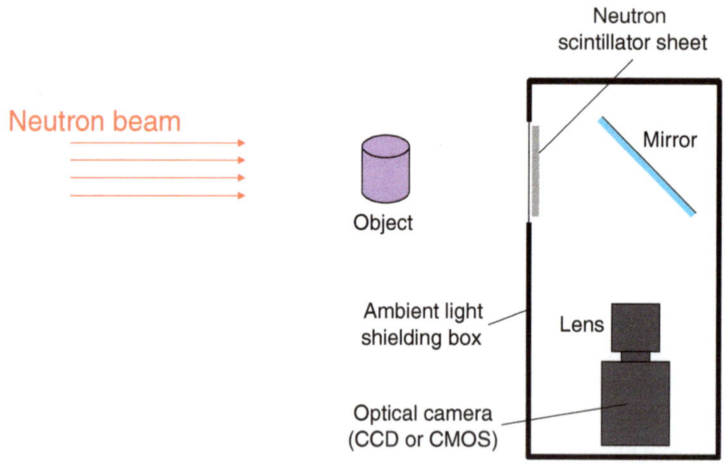

Figure 12.23 *General configuration of the neutron camera system using the neutron scintillator sheet.*

12.5.2 Fast Neutron Detectors

For thermal neutron detection, we use the exothermic nuclear reactions emitting energetic charged particle, such as a $^6Li(n,t)$ reaction. However, for fast neutron detection, the elastic scattering can also be applied. When a neutron has sufficient energy, a recoil nucleus can also have enough energy to ionize or excite the surrounding atoms. In other words, the recoil nucleus can be a directly ionizing radiation particle. The kinetic energy of the recoil nucleus E_{recoil} is given by

$$E_{recoil} = \frac{4m_n M}{(m_n + M)^2} \cos^2\theta \, E_n, \tag{12.15}$$

where m_n and M are the masses of an incident neutron and recoil nucleus, θ is the angle between directions of the incident neutron and recoil nucleus, and E_n is the incident neutron energy. The recoil nucleus has the maximum kinetic energy in a head-on collision, in which $\theta = 0$. In addition, when the mass of the nucleus is the same as the neutron, the recoil nucleus energy would be the maximum. Therefore, the most important nucleus for fast neutron detection is ^1H or a proton, which has almost the same mass as a neutron, because protons can most efficiently receive the incident neutron energy. In the case of a head-on collision, the entire neutron energy can be transferred to the recoil proton. Therefore, organic scintillators, such as plastic or liquid organic scintillators, are frequently applied for fast neutron detection.

Fast neutron detectors using proton recoil are used in nuclear fusion experiments. Typical nuclear fusion reactions are DD and DT reactions, which are written by

$$\begin{aligned} d + d &\rightarrow p + t + 4.03\,MeV\,(p: 3.02\,MeV, t: 1.01\,MeV) \\ &\rightarrow {}^3He + n + 3.27\,MeV\,({}^3He: 0.82MeV, n: 2.45\,MeV), \end{aligned}$$

$$d + t \rightarrow {}^4He + n + 17.6\,MeV\,({}^4He: 3.5\,MeV, n: 14.1\,MeV).$$

In the DT and DD fusion reactions, neutrons are generated as their reaction products. Since neutrons are direct evidence of fusion reactions, neutron detectors can be used to evaluate the number of nuclear fusion reactions. Since emitted neutrons originally have mono energy, direct neutrons, which enter the detector without any interaction before being detected, are discriminated from scattered neutrons by using neutron energy information. Figure 12.24a shows a schematic drawing of the proton recoil neutron detector.

The energy of the recoil proton E_{recoil} is written as

$$E_{recoil} = E_n \cos^2\theta, \tag{12.16}$$

where E_n is the incident neutron energy, and θ is the recoil angle. As shown in the above equation, the recoil proton energy depends on the recoil angle. Since the neutron scattering cross-section of ^1H is approximately isotropic in the center of the mass system, the recoil proton energy distribution shows a rectangular distribution, as shown in Figure 12.24b, when

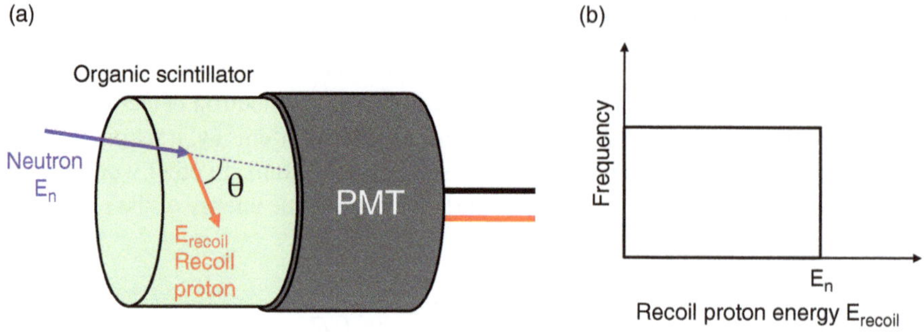

Figure 12.24 *(a) Schematic drawing of the proton recoil neutron detector. (b) Recoil proton energy distribution when irradiating with mono-energy neutrons.*

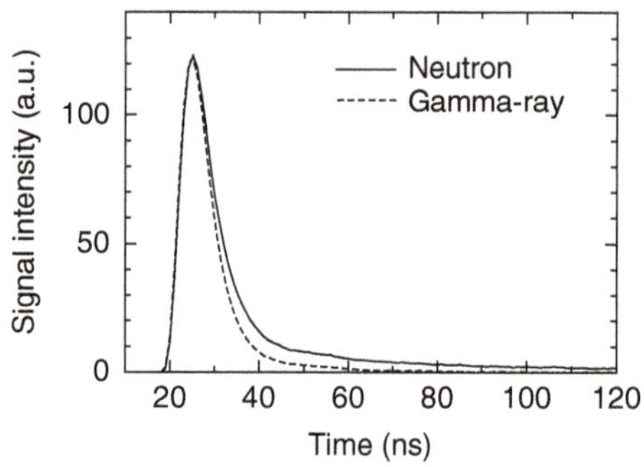

Figure 12.25 *Signal pulse shapes obtained from the stilbene scintillator.*

irradiating with mono-energy neutrons [1]. Although the proton recoil neutron detector cannot show the signal pulse height directly corresponding to the neutron energy itself, the neutron energy information can be estimated from the pulse height distribution. The fusion neutron profile monitor, which combines the proton recoil neutron detectors and neutron collimators, are applied to estimate the spatial distribution of the fusion reactions in the fusion reactor.

Since the organic scintillators have significant gamma-ray sensitivity, the proton recoil neutron detectors also suffer from gamma-ray interference. To reject gamma-ray events in the organic scintillators, the pulse shape discrimination (PSD) technique is frequently applied. In some organic scintillators, the scintillation decay time depends on the LET of directly ionizing radiation particles, such as protons and electrons. Figure 12.25 shows the signal pulse shapes obtained from the stilbene scintillator, which is one of the famous organic crystalline scintillators [30]. The pulse shape of neutron events has a longer decay component compared with that of gamma-ray events. The gamma-ray events can be discriminated from the neutron events by using the difference in the pulse shape.

Although there are various techniques for the PSD, one of the methods used in the PSD is the charge integration method [30]. In this method, signal pulse shapes are digitized and then the digital data are processed. In the pulse shapes, as shown in Figure 12.25, the fast and slow components are integrated. Figure 12.26 shows the 2-D plot of the fast and slow

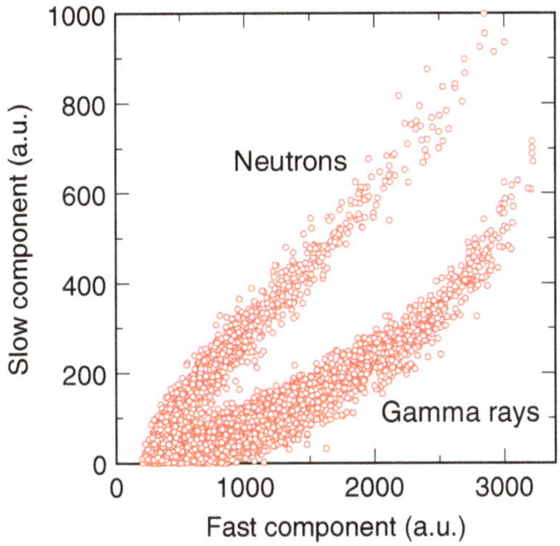

Figure 12.26 Two-dimensional plot of the fast and slow components obtained from the stilbene scintillator when irradiating with neutrons emitted from a ^{252}Cf source surrounded by polyethylene moderator. The time windows of the fast and slow components were set to 0–20 ns and 20–130 ns, respectively.

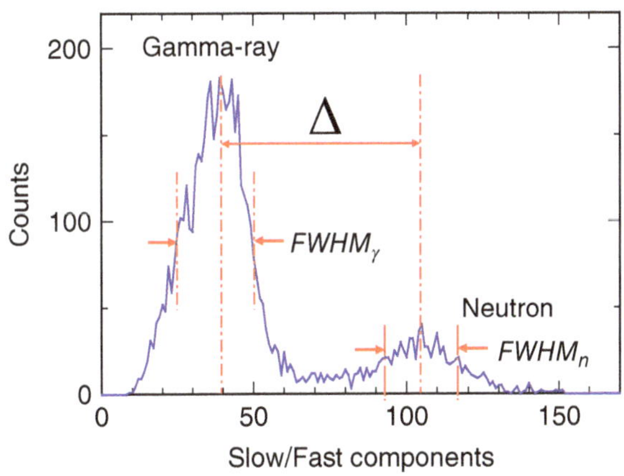

Figure 12.27 PSD spectrum obtained from the stilbene scintillator. The ratio of the slow to fast components was used as the PSD index.

components. The time windows of the fast and slow components were set to 0–20 and 20–130 ns, respectively. The neutron and gamma-ray components are clearly separated. To evaluate the PSD ability, the PSD spectrum, as shown in Figure 12.27, is often used. In this case, the ratio of the slow to fast components was used as the PSD index. As the performance index of the PSD, the figure of merit (FOM) is often used. The FOM is defined as

$$FOM = \frac{\Delta}{FWHM_{\gamma} + FWHM_{n}}, \tag{12.17}$$

where Δ is the interval between the peak positions of gamma-ray and neutron components, and $FWHM\gamma$, $FWHM_n$ are the FWMH of gamma-ray and neutron peaks.

12.6 Personal Dosimeters

Among the radiation induced luminescence materials, the delayed luminescence materials are generally used for personal dosimeters. The delayed luminescence materials, such as thermally stimulated luminescence (TL), OSL, and RPL materials, can accumulate deposition energy or dose into the material. The accumulated information can be read out at an arbitrary timing. Therefore, this kind of passive detector is suitable for long-term personal dosimetry. Although there are various definitions of the radiation dose, the dose measured by the personal dosimeter is, of course, the personal dose equivalent, which is one of the operational quantities in radiation protection. The dosimeters should ideally have the same energy response as the conversion coefficient to the personal dose equivalent. For this purpose, generally, composition of the dosimeter material is desired to be tissue equivalent. However, bright or sensitive dosimeter materials are generally not tissue equivalent. Unlike the scintillator-based active radiation monitors, the dosimeters using the delayed luminescence materials cannot gain any information on individual radiation quanta, such as pulse height. Therefore, an active correction method like the G-function method, which is used for the scintillation survey meter, cannot be applied to delayed luminescence materials. To correct the energy response, a number of dosimeter elements with different energy responses are combined. Some metal filters are used to change the energy response.

The detailed properties of each luminescence material were described in previous chapters. In this section, we focus on the personal dosimetry system using the delayed luminescence materials.

12.6.1 TL-Based Dosimetry System

In thermally stimulated luminescence (TL) materials, the information of accumulated radiation dose can be read out by thermally stimulating or heating the material. Dosimeter elements are used in the form of powder, solid pellets, and chips. The TL reader consists of a TL material heating, light collection and detection, and data recording and control units [31]. Figure 12.28 shows the basic configuration of the TL reader system.

The heating system is important to obtain a reproducible TL readout. Various heating methods can be applied. One of the simplest methods is the resistive heating method. The TL element is placed on the metal holder. The temperature of the holder can be controlled by adjusting the heater current. The spatial distribution of the heater temperature should be uniform. The thermal contact between the holder and TL element should be taken into account. Another heating method is the hot gas stream method. The heating medium is hot nitrogen gas. This method can achieve the almost perfect thermal contact between the TL element and the heating medium, which helps to provide a reproducible TL readout. An optical heating using an infrared pulsed laser is also developed. Generally, the temperature of the TL element gradually increases with a constant ramp-up rate to obtain the glow curve.

To convert TL photons into an electrical signal, a photodetector, such as a PMT, is applied. The sensitivity and/or gain of the PMT depends on the temperature. Normally, thermal separation between the heating unit and the light detection system is essential. An optical system

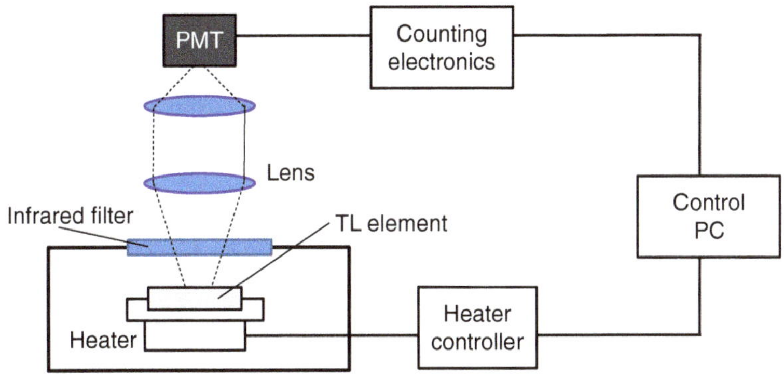

Figure 12.28 *Basic configuration of the TL reader system.*

is therefore designed preferably with heat or infrared photon absorbing filter and focusing lenses. Cooling the PMT is sometimes required to reduce the dark current and to maintain a stable amplification. Frequently, the PMT is used in the photon counting mode. The glow curve is created by recording the time trend of the counting rate of TL photons during heating.

As a method to minimize the fading effect of TL elements, the pre-heating technique is often used. The pre-heating is performed after radiation irradiation and before the TL read-out process. By annealing the TL element at a relatively low temperature, for example at 100 °C, electrons trapped in shallow traps, which can be naturally released at room temperature and cause the fading effect, can be released.

To erase the information accumulated in the TL element, we need to perform a thermal treatment or annealing process. The annealing process consists of heating up the TL elements to a given temperature, keeping them for a given time, and cooling down them to room temperature. By performing the annealing process, thc TL elements can be re-used repeatedly.

12.6.2 OSL-Based Dosimetry System

In OSL materials, the information of accumulated radiation dose can be read out by optically stimulating the OSL material or irradiating the material with stimulation light. The OSL reader system basically consists of a stimulation light source, light transmission optics, stimulation light rejection optics, photodetector, and associated electronics [32]. Figure 12.29 shows the basic configuration of the OSL reader system.

The OSL reader system has a similar configuration to the TL reader system. In the OSL system, the scattered stimulation light could be a disturbing factor for luminescence detection. Instead of the thermal separation in the TL reader, the elimination mechanism of the scattered stimulation light is essential in the OSL reader system. Generally, since the stimulation light intensity is many orders of magnitude higher than the OSL signal, it is important to carefully choose the filter used in front of the PMT and the wavelength spectrum of the stimulation light. The long-pass filter is also used to cut the obstructive wavelength component in the stimulation light, if necessary. Another elimination method of the scattered stimulation light is temporal separation between irradiating the stimulation light and detecting the luminescence [33]. This is also known as the pulsed optically stimulated luminescence

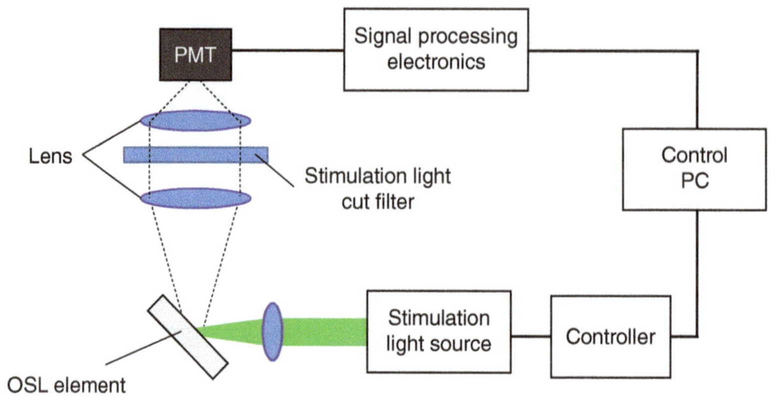

Figure 12.29 *Basic configuration of the OSL reader system.*

(POSL). As a stimulation light source, the pulsed laser is applied. The luminescence measurements are conducted during the intervals of laser pulses.

During irradiation of the constant stimulation light, the OSL intensity exponentially decreases because the trapped electrons are released at a constant rate. Since the release rate of the trapped electrons is proportional to the stimulation light intensity, this decay time is inversely proportional to the stimulation light intensity. If a significant signal level can be secured, the stimulation light intensity is desired to be as small as possible. In some cases, the reduction rate per single readout trial can be at a negligible level, which is less than 0.1%. This means that the OSL signal reading can be repeated several times. This feature plays an important role in improving the reliability of the OSL measurements.

To erase the information in the OSL element, there are two choices: annealing and optical illumination. In some OSL materials, a powder material is formed into a certain shape, such as a disk or plate, by mixing with some binder materials. In many cases, since these binder materials have less resistance to high temperatures, the optical illumination is preferably used instead of annealing. In some OSL materials, such as $C:Al_2O_3$, ultraviolet light might induce additional trapped electrons, which is a negative process in the erasing process. The optical spectrum of the illumination in the OSL erasing process should be taken into account.

12.6.3 RPL-Based Dosimetry System

In RPL materials, the information on the accumulated radiation dose can be read out by irradiating the material with excitation light. The basic configuration of the RPL reader system is almost the same as the OSL reader. The RPL reader system consists of the excitation light source, light transmission optics, excitation light rejection optics, photodetector, and associated electronics. The typical RPL material for personal dosimetry is Ag doped phosphate glass. Since this element is transparent unlike other dosimeter elements, the transmission layout can also be adopted. The temporal separation technique, in which luminescence detection is conducted during the intervals between the excitation light pulses, can also be applied to the RPL reader system. Since the RPL is one of the photoluminescence phenomena, the RPL signal can be read out repeatedly. This is one of the most important advantages of the RPL. By repeatedly reading the RPL signals, the precision and reliability of the measurement can be improved.

To erase the information of the RPL element, the annealing process is necessary because the accumulated information does not disappear by illuminating the element with light.

12.7 OSL-Based Imaging System

The OSL materials are also used in radiography applications. The plate type OSL elements are available as storage phosphor plates or imaging plates, instead of the traditional X-ray films. In this application, the OSL is often called photostimulated luminescence (PSL). The OSL or PSL phosphor powder is formed into a plate shape by mixing with binder materials. A typical material of the OSL imaging plate is Eu:BaFBr. Since this material includes a relatively heavy element, such as barium, it is more sensitive to low energy X-ray compared with the traditional X-ray films. After irradiation of ionizing radiation, the trapped electrons are accumulated in the OSL materials as a latent image. They are read out by irradiating with stimulation light. The OSL imaging plate can be used repeatedly, because the latent image can be erased by illuminating the plate with light.

Figure 12.30 shows a schematic drawing of the OSL imaging plate reader system [34, 35]. As a stimulation light source, a He-Ne laser or a laser diode is applied. Generally, the laser beam is scanned in 1-D, for example, by using the high-speed Galvano mirror system. The imaging plate itself is slowly moved in a perpendicular direction. The OSL photons are collected by a light guide, such as bundled optical fibers, and then detected by the PMT. In front of the PMT, the optical filter is placed to eliminate the scattered stimulation light. The PMT signal is fed into the signal processing circuits and finally digitized. The luminescence intensity at each pixel is recorded one by one. Therefore, the recorded information of the system is a digital image. Since the digital image data can be processed with various methods in the computer, this technique is often called the computed radiography (CR). This technique can achieve high spatial resolution down to a few tens micrometers.

The OSL imaging plate is usually stored into the cassette during exposure to prevent undesired light illumination. In the commercial reader system, the OSL imaging plate is automatically removed from the cassette before the reading process and re-inserted after reading. The fading effect should also be taken into account when quantitative analysis is necessary.

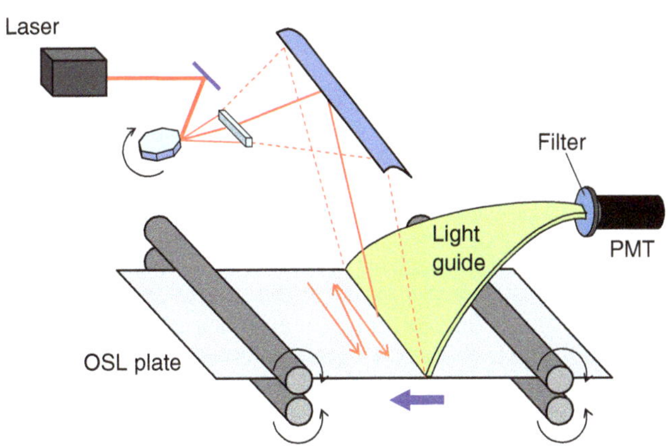

Figure 12.30 *Schematic drawing the OSL imaging plate reader system.*

References

1. Glenn, G. (2010). *Radiation Detection and Measurement*, 4e. Wiley.
2. Dorenbos, P., de Haas, J.T.D., and van Eijk, C.W.V. (1995). Non-proportionality in the scintillation response and the energy resolution obtainable with scintillation crystals. *IEEE Trans. Nucl. Sci.* 42: 2190–2202. https://doi.org/10.1109/23.489415.
3. Mengesha, W. (1998). Light yield nonproportionality of CsI(Tl), CsI(Na), and YAP. *IEEE Trans. Nucl. Sci.* 45: 456–461. https://doi.org/10.1109/23.682426.
4. Beck, P.R., Payne, S.A., Hunter, S. et al. (2015). Nonproportionality of scintillator detectors. V: Comparing the gamma and electron response. *IEEE Trans. Nucl. Sci.* 62: 1429–1436. https://doi.org/10.1109/TNS.2015.2414357.
5. Hofstadter, R. and Mcintyre, J.A. (1950). Measurement of gamma-ray energies with single crystals of NaI(T1). *sssPhysiol. Rev.* 80: 631–637. https://doi.org/10.1103/PhysRev.80.631.
6. Holl, I., Loren, E., and Mageras, G. (1988). A measurement of the light yield of common inorganic scintillators. *IEEE Trans. Nucl. Sci.* 35: 105–109. https://doi.org/10.1109/23.12684.
7. Sakai, E. (1987). Recent measurements on scintillator-photodetector systems. *IEEE Trans. Nucl. Sci.* 34: 418–422. https://doi.org/10.1109/TNS.1987.4337375.
8. Van Loef, E.V.D., Dorenbos, P., Van Eijk, C.W.E. et al. (2002). Scintillation properties of $LaBr_3$:Ce^{3+} crystals: fast, efficient and high-energy-resolution scintillators. *Nucl. Instruments Methods Phys. Res. Sect. A: Accel. Spectrometers, Detect. Assoc. Equip.* 486: 254–258. https://doi.org/10.1016/S0168-9002(02)00712-X.
9. van Loef, E.V., Wilson, C.M., Cherepy, N.J. et al. (2009). Crystal growth and scintillation properties of strontium iodide scintillators. *IEEE Trans. Nucl. Sci.* 56: 869–872. https://doi.org/10.1109/TNS.2009.2013947.
10. ICRP International Commission on Radiological Protection (2003) Annals of the ICRP Annals of the ICRP Annals of the ICRP.
11. ICRU (1985) Report 39, J. Int. Comm. Radiat. Units Meas. os20, NP-NP. https://doi.org/10.1093/jicru/os20.2.Report39.
12. Kim, E., Endo, A., Yamaguchi, Y. et al. (2002). Measurement of neutron dose with an organic liquid scintillator coupled with a Spectrum weight function. *Radiat. Prot. Dosim.* 102: 31–40. https://doi.org/10.1093/oxfordjournals.rpd.a006071.
13. Nunomiya, T., Nakamura, T., Yamamura, S. et al. (2017). Development of lightweight portable neutron survey meter. *J. Nucl. Sci. Technol.* 54: 1215–1222. https://doi.org/10.1080/00223131.2017.1359115.
14. Yamamura, S., Aoyama, K., Masui, K. et al. (2007) Development of wide-energy range X/ray survey meter, Proc. 21st Work. Radiat. Detect. Their Uses. 69–78.
15. Nagarkar, V.V., Gupta, T.K., Miller, S.R. et al. (1998). Structured CsI(Tl) scintillators for X-ray imaging applications. *IEEE Trans. Nucl. Sci.* 45: 492–496. https://doi.org/10.1109/23.682433.
16. Humm, J.L., Rosenfeld, A., and Del Guerra, A. (2003). From PET detectors to PET scanners. *Eur. J. Nucl. Med. Mol. Imaging* 30: 1574–1597. https://doi.org/10.1007/s00259-003-1266-2.
17. Murayama, H., Ishibashi, H., Uchida, H. et al. (1998). Depth encoding multicrystal detectors for pet. *IEEE Trans. Nucl. Sci.* 45: 1152–1157. https://doi.org/10.1109/23.681994.
18. Vandenberghe, S., Mikhaylova, E., D'Hoe, E. et al. (2016). Recent developments in time-of-flight PET. *EJNMMI Phys.* 3 https://doi.org/10.1186/s40658-016-0138-3.
19. IAEA (1971). *Nuclear Well Logging in Hydrology*. Vienna: IAEA https://www.iaea.org/publications/1227/nuclear-well-logging-in-hydrology.
20. Al Hamrashdi, H., Monk, S.D., and Cheneler, D. (2019). Passive gamma-ray and neutron imaging systems for national security and nuclear non-proliferation in controlled and uncontrolled detection areas: review of past and current status. *Sensors (Switzerland)* 19 https://doi.org/10.3390/s19112638.
21. Martin, C.J. and Sutton, D.G. (2015). *Practical Radiation Protection in Healthcare*. Oxford University Press https://doi.org/10.1093/med/9780199655212.001.0001.
22. Shibata, K., Iwamoto, O., Nakagawa, T. et al. (2011). JENDL-4.0: a new library for nuclear science and engineering. *J. Nucl. Sci. Technol.* 48: 1–30. https://doi.org/10.1080/18811248.2011.9711675.
23. Sugimoto, D., Watanabe, K., Hirota, K. et al. (2014). Neutron TOF experiments using transparent rubber sheet type neutron detector with dispersed small pieces of $LiCaAlF_6$ scintillator. *Physics Procedia* 60: 349–355. https://doi.org/10.1016/j.phpro.2014.11.047.
24. Yoshikawa, A., Yanagida, T., Yokota, Y. et al. (2009). Single crystal growth, optical properties and neutron response of Ce^{3+}doped LiCaAlFg. *IEEE Trans. Nucl. Sci.* 56 https://doi.org/10.1109/TNS.2009.2033115.

25. Yanagida, T., Yamaji, A., Kawaguchi, N. et al. (2011). Europium and sodium codoped $LiCaAlF_6$ scintillator for neutron detection. *Appl. Phys Express* 4: 106401. https://doi.org/10.1143/APEX.4.106401.
26. Watanabe, K., Kawabata, Y., Yamazaki, A. et al. (2015). Development of an optical fiber type detector using a Eu:$LiCaAlF_6$ scintillator for neutron monitoring in boron neutron capture therapy. *Nucl. Instruments Methods Phys. Res. Sect. A: Accel. Spectrometers, Detect. Assoc. Equip.* 802: 1–4. https://doi.org/10.1016/j.nima.2015.08.056.
27. Ishikawa, A., Yamazaki, A., Watanabe, K. et al. (2020). Sensitivity and linearity of optical fiber-based neutron detectors using small ^{6}Li-based scintillators. *Nucl. Instruments Methods Phys. Res. Sect. A: Accel. Spectrometers, Detect. Assoc. Equip.* 954: 161661. https://doi.org/10.1016/j.nima.2018.11.120.
28. Watanabe, K., Yamazaki, T., Sugimoto, D. et al. (2015). Wavelength-shifting fiber signal readout from transparent RUbber SheeT (TRUST) type LiCaAlF6 neutron scintillator. *Nucl. Instruments Methods Phys. Res. Sect. A: Accel. Spectrometers, Detect. Assoc. Equip.* 784: 260–263. https://doi.org/10.1016/j.nima.2014.11.109.
29. Rhodes, N.J. (2006). Scientific reviews: status and future development of neutron scintillation detectors. *Neutron News* 17: 16–18. https://doi.org/10.1080/10448630500517597.
30. Yanagida, T., Watanabe, K., and Fujimoto, Y. (2015). Comparative study of neutron and gamma-ray pulse shape discrimination of anthracene, stilbene, and p-terphenyl. *Nucl. Instruments Methods Phys. Res. Sect. A: Accel. Spectrometers, Detect. Assoc. Equip.* 784: 111–114. https://doi.org/10.1016/j.nima.2014.12.031.
31. Humphreys, J.C. and. Chappell, S.E. (1979) Standard Procedure for Use of Thermoluminescence Dosimetry in Radiation-Hardness Testing of Electronic Devices
32. Yukihara, E.G. and McKeever, S.W. (2008). Optically stimulated luminescence (OSL) dosimetry in medicine. *Phys. Med. Biol.* 53: 351–379. https://doi.org/10.1088/0031-9155/53/20/R01.
33. Akselrod, M.S. and McKeever, S.W.S. (1999). A radiation dosimetry method using pulsed optically stimulated luminescence. *Radiat. Prot. Dosim.* 81: 167–175. https://doi.org/10.1093/oxfordjournals.rpd.a032583.
34. Sonoda, M., Takano, M., Miyahara, J. et al. (1983). Computed radiography utilizing scanning laser stimulated luminescence. *Radiol.* 148: 833–838. https://doi.org/10.1148/radiology.148.3.6878707.
35. Leblans, P., Vandenbroucke, D., and Willems, P. (2011). Storage phosphors for medical imaging. *Materials (Basel)* 4: 1034–1086. https://doi.org/10.3390/ma4061034.

Index

Phosphors for Radiation Detectors, First Edition. Edited by Takayuki Yanagida and Masanori Koshimizu.

www.ingramcontent.com/pod-product-compliance
Ingram Content Group UK Ltd.
Pitfield, Milton Keynes, MK11 3LW, UK
UKHW050842101026
15624UKWH00018B/433

9 781119 583325